Insulin Resistance as a Risk Factor in Visceral and Neurological Disorders

Insulin Resistance as a Risk Factor in Visceral and Neurological Disorders

AKHLAQ A. FAROOQUI
Department of Molecular and Cellular Biochemistry,
The Ohio State University, Columbus, OH, United States

ELSEVIER

ACADEMIC PRESS
An imprint of Elsevier

Academic Press is an imprint of Elsevier
125 London Wall, London EC2Y 5AS, United Kingdom
525 B Street, Suite 1650, San Diego, CA 92101, United States
50 Hampshire Street, 5th Floor, Cambridge, MA 02139, United States
The Boulevard, Langford Lane, Kidlington, Oxford OX5 1GB, United Kingdom

British Library Cataloguing-in-Publication Data
A catalogue record for this book is available from the British Library

Library of Congress Cataloging-in-Publication Data
A catalog record for this book is available from the Library of Congress

ISBN: 978-0-12-819603-8

For Information on all Academic Press publications
visit our website at https://www.elsevier.com/books-and-journals

Publisher: Nikki Levy
Acquisitions Editor: Melanie Tucker
Editorial Project Manager: Kristi Anderson
Production Project Manager: Sujatha Thirugnana Sambandam
Cover Designer: Miles Hitchen

Typeset by MPS Limited, Chennai, India

Dedication

This monograph is dedicated to my beloved father "late Sharafyab Ahmed Sahab" whose guidance and influence continue to inspire and support me.

Akhlaq A. Farooqui

Contents

About the author *xi*
Preface *xiii*
Acknowledgments *xvii*
List of abbreviations *xix*

1. Insulin resistance and obesity **1**

Introduction 1
Insulin signaling in the brain 3
Insulin resistance 6
Insulin resistance as a protective mechanism 21
Effects of diet on microbiota population 30
Microbiota and short chain fatty acids 34
Effects of gut microbiota and obesity on the brain 38
Link among insulin signaling, obesity, and insulin resistance 42
Contribution of leptin in the development of obesity 43
Contribution of adiponectin in obesity 45
Insulin signaling obesity and neurological disorders 47
Conclusion 49
References 50
Further reading 69

2. Insulin resistance, diabetes, and metabolic syndrome **71**

Introduction 71
Biomarkers for type 2 diabetes 77
miRNAs, diabetes, and vascular complications 78
Comlications caused by type 2 diabetes 79
Molecular mechanisms contributing to type 2 diabetes complications 83
Risk factors contributing to metabolic syndrome 88
Pathogenesis of metabolic syndrome 89
Link between type 2 diabetes and metabolic syndrome 92
Type 2 diabetes and metabolic syndrome as risk factors for Alzheimer's disease 94
Conclusion 100
References 101
Further reading 111

3. Insulin resistance and heart disease **113**

Introduction 113
Insulin signaling in vasculature 121
Molecular mechanism of atherosclerosis 133
Carbohydrate metabolism, insulin resistance, and heart disease 136
Biochemical links between insulin resistance and heart disease 140
Contribution of lipid mediators in heart disease 140
Conclusion 143
References 144
Further reading 155

4. Insulin resistance and sleep apnea **157**

Introduction 157
Obstructive sleep apnea 160
Biochemical changes in obstructive sleep apnea 162
Oxidative stress, insulin resistance, and obstructive sleep apnea 162
Obstructive sleep apnea—mediated changes in lipid metabolism 167
Inflammation, insulin resistance, and obstructive sleep apnea 168
Biomarkers for obstructive sleep apnea 170
Obstructive sleep apnea and heart disease 172
Obstructive sleep apnea and hypertension 175
Obstructive sleep apnea and metabolic syndrome 176
Obstructive sleep apnea and stroke 179
Obstructive sleep apnea and its relationship with various diseases 182
Obstructive sleep apnea and Alzheimer's disease 184
Obstructive sleep apnea and depression 189
Conclusion 191
References 191
Further reading 205

5. Insulin resistance and stroke **207**

Introduction 207
Hypertension and pathogenesis of stroke 212
Contribution of diet, microbiota, and insulin resistance in the pathogenesis of stroke 215
Stroke-mediated changes in the brain 218
Molecular link among advanced glycated end products, insulin resistance, and stroke 230
Adipokines, insulin resistance, and stroke 233

Molecular link between hypertension and brain damage 234
Effect of microbiota composition on stroke outcome 236
Conclusion 238
References 239
Further reading 248

6. Insulin resistance and Alzheimer's disease 249

Introduction 249
Insulin signaling in the brain 250
Pathogenesis of Alzheimer's disease 255
Insulin receptor, insulin signaling, and insulin resistance in the brain 264
Neurochemical links among insulin resistance, type 2 diabetes, and Alzheimer's disease 265
Cerebral blood flow in type 2 diabetes and Alzheimer's disease 271
Ceramode-mediated insulin resistance in type 2 diabetes and Alzheimer's disease 273
Conclusion 277
References 278
Further reading 291

7. Insulin resistance and Parkinson's disease 293

Introduction 293
Familial Parkinson's disease 298
Sporadic Parkinson's disease 299
Gut microbiota, insulin resistance, and Parkinson's disease 303
Insulin resistance, type 2 diabetes, and Parkinson's disease 308
Overlap between Parkinson's disease and Krabbe's disease 316
Overlap between Parkinson's disease and Gaucher's disease 318
Biomarkers for Parkinson's disease 319
Oxidative stress in Parkinson's disease 320
Neuroinflammation in Parkinson's disease 322
Depression and Parkinson's disease 324
Conclusion 326
References 327
Further reading 342

8. Insulin resistance, dementia, and depression 349

Introduction 349
Normal aging and cognitive function 352

Neurochemical aspects of dementia 354
Classification of dementias 355
Alzheimer's type of dementia 356
Neurochemical aspects of vascular dementia 359
Neurochemical aspects of Lewy body dementia 361
Neurochemical aspects of frontotemporal dementia 365
Insulin resistance, stress, and depression 366
Effects of diet and exercise on dementia and depression 373
Conclusion 374
References 374
Further reading 384

9. Use of phytochemicals for the treatment of insulin resistance–linked visceral and neurological disorders 385

Introduction 385
Effects of various types of diets on insulin resistance 389
Effects of curcumin on insulin resistance and insulin resistance–linked diseases 400
Effect of green tea on insulin resistance–linked diseases 404
Effect of resveratrol on insulin resistance and insulin resistance–linked diseases 407
Effect of n-3 fatty acids in insulin resistance and insulin resistance–linked diseases 413
Effects of cinnamon on insulin resistance and insulin resistance–linked diseases 415
Effects of garlic on insulin resistance and insulin resistance–linked diseases 419
Conclusion 424
References 425
Further reading 437

10. Summery and perspective for future research on insulin resistance and insulin resistance–linked visceral and neurological disorders 439

Introduction 439
Chronic insulin resistance: a common link between visceral and neurological disorders 443
Direction for future research 453
Conclusion 455
References 456

Index *463*

About the author

Dr. Akhlaq A. Farooqui is a leader in the field of signal transduction, brain phospholipases A_2, bioactive ether lipid metabolism, polyunsaturated fatty acid metabolism, glycerophospholipid-, sphingolipid-, and cholesterol-derived lipid mediators, glutamate-induced neurotoxicity, and modulation of signal transduction by phytochemicals. He has discovered the stimulation of plasmalogen-selective phospholipase A_2 (PlsEtn-PLA_2) and diacyl- and monoacylglycerol lipases in brains from patients with Alzheimer's disease. Stimulation of PlsEtn-PLA_2 produces plasmalogen deficiency and increases levels of eicosanoids that may be related to the loss of synapses in brains of patients with Alzheimer's disease. He has published cutting edge research on the generation and identification of glycerophospholipid-, sphingolipid-, and cholesterol-derived lipid mediators in kainic acid–mediated neurotoxicity by lipidomics. He has authored 12 monographs: *Glycerophospholipids in Brain: Phospholipase A_2 in Neurological Disorders* (2007); *Neurochemical Aspects of Excitotoxicity* (2008); *Metabolism and Functions of Bioactive Ether Lipids in Brain* (2008); *Hot Topics in Neural Membrane Lipidology* (2009); *Beneficial Effects of Fish Oil in Human Brain* (2009); *Neurochemical Aspects of Neurotraumatic and Neurodegenerative Diseases* (2010); *Lipid Mediators and their Metabolism in the Brain* (2011); *Phytochemicals, Signal Transduction, and Neurological Disorders* (2012); *Metabolic Syndrome: An Important Risk Factor for Stroke, Alzheimer's Disease, and Depression* (2013); *Inflammation and Oxidative Stress in Neurological Disorders* (2014); *High Calorie Diet and the Human Brain* (2015); and *Therapeutic Potentials of Curcumin For Alzheimer's Disease* (2016). All above monographs have been published by Springer, New York and Springer International Publishing Switzerland. Monographs on *Neurochemical Aspects of Alzheimer's Disease* (2017), *Ischemic, Traumatic Brain, and Spinal Cord Injuries: Mechanisms and Potential Therapies* (2018), and *Molecular Mechanisms of Dementia: Biomarkers, Neurochemistry, and Therapy* have been published by Academic Press: An imprint of Elsevier, San Diego, CA.

In addition, he has edited 12 books: *Biogenic Amines: Pharmacological, Neurochemical and Molecular Aspects in the CNS* (2010), Nova Science Publisher, Hauppauge, New York; *Molecular Aspects of Neurodegeneration and Neuroprotection* (2011), Bentham Science Publishers Ltd; *Phytochemicals and Human Health: Molecular and pharmacological Aspects* (2011), Nova

Science Publisher, Hauppauge, New York; *Molecular Aspects of Oxidative Stress on Cell Signaling in Vertebrates and Invertebrates* (2012), Wiley Blackwell Publishing Company, New York; *Beneficial Effects of Propolis on Human Health in Chronic Diseases* (2012) Vol. 1, Nova Science Publishers, Hauppauge, New York; *Beneficial Effects of Propolis on Human Health in Chronic Diseases* (2012) Vol 2, Nova Science Publishers, Hauppauge, New York; *Metabolic Syndrome and Neurological Disorders* (2013), Wiley Blackwell Publishing Company, New York; *Diet and Exercise in Cognitive Function and Neurological Diseases* (2015), Wiley Blackwell Publishing Company, New York; *Trace Amines and Neurological Disorders: Potential Mechanisms and Risk Factors* (2016), Elsevier, New York; *Neuroprotective Effects of Phytochemicals in Neurological Disorders* (2017), Wiley Blackwell, John Wiley and Sons, Inc., Hoboken, NJ; *Role of the Mediterranean Diet in the Brain and Neurodegenerative Diseases* (2018); and *Curcumin for Neurological and Psychiatric Disorders: Neurochemical and Pharmacological Properties* (2019), Elsevier, New York.

Preface

At the end of 2011, the United Nations declared for the first time in the history of humanity that noncommunicable diseases have outpaced infectious diseases as the main threat to human health globally. Among noncommunicable diseases, insulin-linked visceral diseases [type 2 diabetes, metabolic syndrome (MetS), sleep apnea, cardiovascular diseases (CVD), and chronic kidney diseases] and neurological disorders [stroke, Alzheimer's disease (AD), Parkinson's disease (PD), and various types of dementias] are of paramount importance. Type 2 diabetes, MetS, sleep apnea, and CVD are characterized by clustering of insulin resistance (insulin desensitization), hyperinsulinemia, obesity, hypertension, dyslipidemia, impaired glucose metabolism, and induction of low-grade inflammation. Similarly, insulin resistance–linked neurological disorders are accompanied by impaired insulin signaling in brain, accumulation of mis-folded proteins (β-amyloid and α-synclein), induction of neuroinflammation, and loss of glucose homeostasis. Changes in human dietary habits in recent years have led to the consumption of Western diet, which is enriched in saturated fats, high in simple sugars (sucrose, glucose, and fructose), high in salt, and low in fiber. The insulin resistance–linked visceral and neurological disorders are highly prevalent pathological conditions that affect a considerate number of adult humans. Approximately one-fourth of European, American, and Canadian adults suffer from insulin resistance–linked visceral and neurological disorders. Insulin resistance, hyperinsulinemia, hypertension, dyslipidemia, impaired glucose homeostasis, and abdominal obesity reflect overnutrition, physical inactivity, and resultant excess adiposity. At the molecular level, insulin-linked visceral and neurological disorders are accompanied not only by dysregulation in the expression of adipocytokines and chemokines, but also by increase in levels of lipids and lipid mediators (free fatty acids, diacylglycerol, triacylglycerols, and ceramide). These changes not only modulate immune function and responses, but also inflammation that lead to alterations in the hypothalamic function "bodyweight/appetite/satiety set point" resulting in the initiation and development of insulin-linked visceral and neurological disorders. The mechanisms underlying the relationship between insulin resistance–linked visceral and neurological disorders are not fully understood. However, many studies have indicated that impairment of insulin signaling, glucose toxicity, elevated levels of lipid mediators, generation of advanced glycation endproducts, impairment of endothelial cell function, and vascular low-grade inflammation may be major molecular mechanistic bridges between insulin resistance and insulin resistance–linked visceral diseases and neurological disorders. Thus having type 2 diabetes and MetS triple the risk of having a stroke, AD, PD, and various types of dementia compared with age-matched healthy humans.

Information on molecular links between insulin resistance–linked visceral diseases and neurological disorders is scattered throughout the literature mainly in the form of original papers and some reviews. As Baby Boomer generation grows older, enormous impact of insulin resistance–linked visceral diseases and neurological disorders will be felt by the American society. At present more than 795,000 people suffer from stroke every year (approximately one person every 45 seconds), approximately 5.6 million people over the age of 65 in the United States suffer from AD, and about 50 million US adults have experienced dementia and depression at some point during their lifetime. The projected cost to Medicare for treating these chronic diseases is estimated to be about 5–8 trillion dollars by 2050. This number does not include other visceral and other neurological diseases, which do not involve insulin resistance. Such an amount will not only burst National Institute of Health budget, but will seriously affect the US economy. Available drugs may not reverse the insulin resistance–linked visceral and neurological disorders. However, healthy diet and regular exercise may produce beneficial effects not only on motor and cognitive functions, but also on memory deficits that occur to some extent during normal aging and to a large extent in stroke, AD, PD, and dementia. This monograph provides readers with a comprehensive and cutting-edge description of links among insulin resistance–linked visceral and neurological disorders in a manner that is useful not only to students and teachers but also to researchers, dietitians, nutritionists, and physicians.

This monograph has 10 chapters. The first chapter describes the information on biochemical aspects of insulin resistance and obesity. Chapter 2, Insulin resistance, diabetes, and metabolic syndrome, describes the contribution of insulin resistance in the pathogenesis of diabetes, and MetS. Chapter 3, Insulin resistance and heart disease, narrates the contribution of insulin resistance in heart disease. Chapter 4, Insulin resistance and sleep apnea, deals with the contribution of insulin resistance in the pathogenesis of sleep apnea. Chapter 5, Insulin resistance and stroke, describes cutting-edge information on the involvement of insulin resistance in the pathogenesis of stroke. Chapter 6, Insulin resistance and Alzheimer's disease, focuses on the contribution of insulin resistance in the pathogenesis of AD. Chapter 7, Insulin resistance and Parkinson's disease, deals with information on the involvement of insulin resistance in the pathogenesis of PD. Chapter 8, Insulin resistance, dementia, and depression, reports on the contribution of insulin resistance in the pathogenesis of various types of dementias. Chapter 9, Use of phytochemicals for the treatment of insulin resistance–linked visceral and neurological disorders, describes the effect of dietary phytochemicals on insulin-linked visceral and neurological disorders. Chapter 10, Summary and perspective for future research on insulin resistance and insulin resistance–linked visceral and neurological disorders, provides readers with a perspective that will be important for future research work on the relationship between insulin resistance and insulin-linked visceral and neurological disorders. My presentation and demonstrated ability to present complicated information on signal transduction processes on insulin resistance and visceral and neurological disorders will make this book particularly accessible to neuroscience graduate students, teachers, and fellow researchers. It can be used as a supplemental text for a range of neuroscience and biochemistry courses. Clinicians, neuroscientists, neurologists, and pharmacologists will find this book useful for understanding the contribution of insulin resistance in the pathogenesis of visceral and neurological disorders. To the best of my knowledge, this monograph will be the first to provide a comprehensive description of signal transduction processes associated with the relationship between insulin resistance and insulin resistance–linked visceral and neurological disorders.

The choices of topics presented in this monograph are personal. They are not only based on my interest in the biochemistry of insulin resistance and insulin resistance–linked visceral and neurological disorders, but also in areas where major progress has been made. I have tried to ensure uniformity and mode of presentation as well as a logical progression of subjects from one topic to another and have provided the extensive bibliography. For the sake of simplicity and uniformity, a large number of figures with chemical structures of drugs used for the treatment of insulin resistance and insulin resistance–linked visceral and neurological disorders along with line diagrams of colored signal transduction pathways are also included. I hope that my attempt to integrate and consolidate the knowledge on insulin resistance and insulin resistance–linked visceral and neurological disorders will initiate more studies on this field. This knowledge may be useful in developing novel therapeutic interventions for insulin resistance and insulin resistance–linked visceral and neurological disorders.

Akhlaq A. Farooqui
Columbus, OH, United States

Acknowledgments

I thank my wife, Tahira, for critical reading of this monograph, offering valuable advice, useful discussion, and evaluation of subject matter. Without her help and participation, this monograph neither could nor would have been completed. I would also like to express my gratitude to Melanie Tucker and Kristi Anderson of Elsevier/Academic Press for their quick responses to my queries and professional manuscript handling. Finally, I would like to thank Production Project Manager Sujatha Thirugnana Sambandam for her help and patience and cooperation in production of this monograph and Miles Hitchen for designing the cover page.

Akhlaq A. Farooqui

List of abbreviations

Aβ	β-amyloid
AD	Alzheimer's disease
ADDLs	Aβ-derived diffusible ligands
AGEs	advanced glycation endproducts
APP	amyloid precursor protein
ARA	arachidonic acid
BBB	blood—brain barrier
CAT	catalase
COX	cyclooxygenase
CSF	cerebrospinal fluid
DHA	docosahexaenoic acid
EPA	eicosapentaenoic acid
GPx	glutathione peroxidase
GSK-3	glycogen synthase-3
IGF	insulin growth factor
IL	interleukin
Ins-1,4,5-P_3	inositol-1,4,5-trisphosphate
LOX	lipoxygenase
MetS	metabolic syndrome
NFTs	neurofibrillary tangles
NO	nitric oxide
PD	Parkinson's disease
PET	positron emission tomography
PKC	protein kinase C
PLA_2	phospholipase A_2
PtdCho	phosphatidylcholine
PtdEtn	phosphatidylethanolamine
PtdIns	phosphatidylinositol
PtdIns(4,5)P_2	phosphatidylinositol 4,5-bisphosphate
PtdIns4P	phoshatidylinositol 4-phosphate
RAGEs	receptor for advanced glycation endproducts
RNS	reactive nitrogen species
ROS	reactive oxygen species
α-Syn	α-synuclein
SOD	superoxide dismutase
SPECT	single-photon emission computed tomography
TNF-α	tumor necrosis factor-alpha

CHAPTER 1

Insulin resistance and obesity

Introduction

Insulin is an anabolic hormone (51 amino acid containing peptide with mol mass of 5800 kDa), which is synthesized and secreted by β-cells of the islets of Langerhans located in the pancreas and its serum concentration increases in a direct proportion to the glucose concentration. Insulin is coded on the short arm of chromosome 11 and synthesized in the β-cells of the pancreatic islets of Langerhans as its precursor, proinsulin (Wilcox, 2005). Removal of the signal peptide forms proinsulin results in the synthesis of insulin. Insulin plays an important role in carbohydrate and lipid metabolism in the body by modulating the uptake of glucose and its storage as glycogen in the liver, muscles, and fat cells (Duckworth et al., 1997). Insulin also plays an important role in maintenance of mitochondrial function, promotes a microcirculation, and induces the cell proliferation (Ye, 2007). In the brain, neural cells produce small amount of insulin (Banks, 2004). However, most insulin's action in the brain is probably due to the circulating peripheral insulin, which crosses the blood–brain barrier (BBB) and produces its neurochemical actions (Banks, 2004).

BBB is not simply a physical barrier but a regulatory interface between the brain and immune system. The BBB both affects and is affected by the immune system and connects at many levels with the brain. The permeability of peripheral insulin to the brain across BBB vary considerably among different regions of brain. It is shown that insulin crosses the BBB two to eight times faster in the olfactory bulb than other brain regions (Banks et al., 2012). Insulin not only regulates glucose and lipid metabolism in the brain, but also modulates neurotransmission and synaptic activities (Zhao and Alkon, 2001; Zhao et al., 2004) such as long-term potentiation (LTP) (Nisticò et al., 2011), as well as promoting long-term depression (LTD) (Labouèbe et al., 2013). In addition, brain insulin regulates dendritic sprouting, neuronal stem cell activation, cell growth, repair, synaptic maintenance, and neuroprotection (Fig. 1.1) (Craft and Watson, 2004; Van Dam and Aleman, 2004; Kleinridders et al., 2014). Insulin enhances

Insulin Resistance as a Risk Factor in Visceral and Neurological Disorders
DOI: https://doi.org/10.1016/B978-0-12-819603-8.00001-8

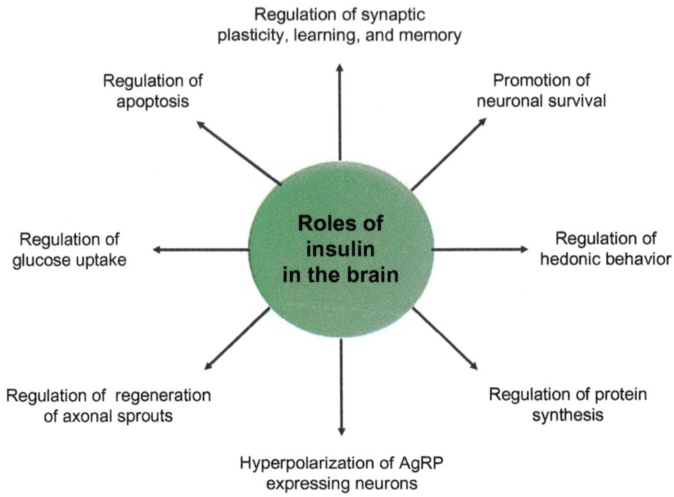

Figure 1.1 Roles of insulin in the brain.

cognitive abilities via activation of insulin receptors (IRs) in the hippocampal region of brain. Insulin also stimulates translocation of glucose transporter type 4 (GLUT4) to hippocampal plasma membranes thereby enhancing the glucose uptake in the time-dependent manner (Ren et al., 2014). Insulin is stored in synaptic vesicles at nerve endings in rat brain and is released on depolarization conditions (Blázquez et al., 2014). Insulin also potentiates the brain transport of molecules such as leptin (Kastin and Akerstrom, 2001) and amino acids (Tagliamonte et al., 1976). Insulin signaling in the brain is associated with neuronal survival, neurotransmission, and modulation of synaptic activities (Zhao and Alkon, 2001). In addition, insulin is also involved in the regulation of synaptic plasticity and modulation of LTP (Nisticò et al., 2011), as well as promoting LTD (Labouèbe et al., 2013). These processes are involved in learning and memory. Insulin also potentiates the brain transport of molecules such as leptin (Kastin and Akerstrom, 2001) and amino acids (Tagliamonte et al., 1976). In streptozotocin-treated mice, insulin increases cerebral microvessels expression of occludin, claudin-5, and ZO-1 (Sun et al., 2015). In specific types of hypothalamic neurons, insulin decreases the expression of orexigenic neuropeptides such as neuropeptide Y (NYP) or agouti-related peptide (AgRP) thereby promoting the decrease in food intake (Fick and Belsham, 2010; Kleinridders et al., 2014; Posey et al., 2009). Insulin also inhibits food intake by promoting expression of anorexigenic neuropeptides such as pro-opiomelanocorticotropin (POMC). Insulin is also involved in regulation of hedonic behavior and nonhomeostatic control of intake of food and other substances via reward processing. Insulin also supports neuronal protein synthesis and cytoskeletal protein expression (Schuling Kemp et al., 2000), neurite outgrowth (Dickson, 2003; Song et al., 2003), migration, and differentiation in the absence of other growth factors, and nascent synapse formation (Schuling Kemp et al., 2000). Besides inhibiting AgRP synthesis, insulin induces the hyperpolarization of the AgRP-expressing arcuate neurons reduces the firing rate of these neurons. Finally, insulin also modulates receptor for advanced glycation end products (RAGE) expression. Levels of soluble RAGE are inversely correlated with plasma insulin concentration during an oral glucose tolerance test in healthy human subjects (Forbes et al., 2014). In isolated brain microvessels from streptozotocin-injected mice, insulin reduces the concentration of RAGE compared to diabetic mice (Sun et al., 2015) supporting the view that insulin modulates RAGE.

Insulin signaling in the brain

Insulin produces its effects by interacting with the IR, a transmembrane receptor with tyrosine kinase activity. It is made up of two α-subunits and two β-subunits (Fig. 1.2). The α-subunits (120–135 kDa) contain the insulin-binding sites. The α-subunit of IR is predominantly hydrophilic in nature, lacks membrane anchor regions, and contains 15 potential *N*-glycosylation sites and 37 cysteine residues. The β-subunits (95 kDa) form transmembrane and intracellular parts of the receptor (White, 2003). The intracellular part of the β-subunits contains ATP-binding motifs, autophosphorylation sites, and tyrosine-specific protein kinase activity, which facilitate rapid autophosphorylation upon ligand-binding. This results in the recruitment and tyrosine phosphorylation of adaptor proteins, including insulin receptor substrates (IRSs) such as IRS-1 and IRS-2 (Long et al., 2011). In the brain, IRS-2 signaling plays an important role in brain growth, nutrient sensing, and life span regulation, whereas IRS-1 may be less important for these functions (Taguchi and White, 2008). The metabolic action of insulin is linked with IRS through phosphatidyl-inositol 3 kinase (PtdIns 3K) and Akt pathway (Fig. 1.2). Upon activation, the IR phosphorylates IRS proteins. These proteins represent a critical node of activation in insulin and IGF-1 signaling cascades (Shaw, 2011). In addition to their activation of the Ras–mitogen-activated protein kinase (MAPK) pathway, activated IRS proteins

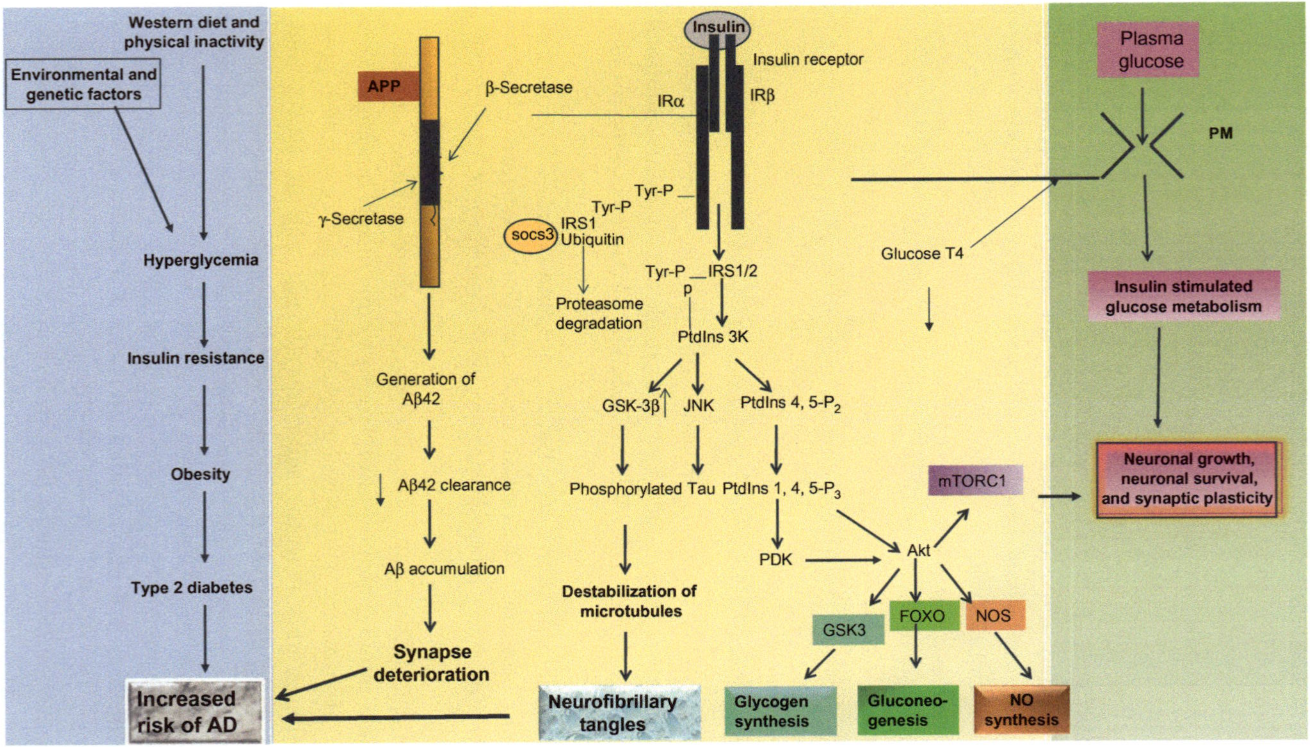

Figure 1.2 Hypothetical diagram showing insulin receptor signaling and its role in glucose uptake and in pathogenesis of type 2 diabetes and Alzheimer's disease.

Aβ, beta amyloid; *AD*, Alzheimer's disease; *Akt*, serine/threonine protein kinase; *APP*, amyloid precursor protein; *GSK3*, glycogen synthase kinase 3; *IRS*, insulin receptor substrate; *JNK*, c-Jun NH(2)-terminal kinase; *PtdIns 3K*, phosphatidyl-inositol 3 kinase; *upward arrow* indicates increase and *downward arrow* indicates decrease.

serve as docking sites for the assembly and activation of, among others, PtdIns 3K, which generates the lipid second messenger phosphatidylinositol 1,4,5-trisphosphate (PtdIns 1,4,5-P_3). PtdIns 3K represents another critical node of cross-talk with other signaling pathways, including the c-Jun-N-terminal kinase (JNK) stress signaling pathway (Fig. 1.2). Elevated levels of PIP_3 activate phosphoinositide-dependent protein kinase-1 (PDK1) and Akt. Akt represents yet another critical node of interaction with the mammalian target of rapamycin (mTOR) nutrient signaling pathway. Under physiological conditions, the binding of Akt targets rapamycin (mTOR), and extracellular signal-regulated kinases (ERK). This activation of kinases eventually results in phosphorylation of the IRS leading to inhibition of insulin signaling in a negative feedback regulation (Talbot et al., 2012; Biessels and Reagan, 2015; Pearson-Leary and McNay, 2012; Di Domenico et al., 2017). In neurons, the PtdIns 3K, Akt, glycogen synthase kinase 3β, BCL-2 agonist of cell death, transcription factor fork-head box class O (FOXO), mTOR, and the MAPK pathways are critical for cell survival signaling and are regulated by the activity of the IR (Craft, 2005; McCrimmon et al., 2012). Among these pathways, mTORC1 pathway is a major pathway contributing to cellular energy sensing. This pathway serves as a key energy sensor that controls several important cellular processes such as mitochondrial biogenesis and function, cellular proliferation, and autophagy. The mTORC1 consists of a highly conserved catalytic subunit (mTOR), the unique regulatory subunit Raptor (regulatory associated protein of mTOR), and several accessory proteins that integrate a variety of energy signals such as growth factors, hormones, amino acids, and glucose/insulin signaling. The mTORC1 integrates these signals through modulation of cellular transcription and translation processes including the p70 ribosomal S6 kinase (S6K) and its downstream target the S6 ribosomal protein conferring gene and protein programming within cells. Converging evidence suggests that the mTOR signaling network regulates critical cellular and developmental processes such as cell growth, differentiation, cell survival, and metabolism.

Insulin performs a wide range of functions in the brain and peripheral tissues. Thus in peripheral tissues, insulin is important for cell growth and survival. The mechanisms involved in impairing the ability of insulin to lower blood glucose levels not only include the activation of the transcription factor FOXO1 in the liver (Klotz et al., 2015), but also the disruption of GLUT4 glucose transporter translocation to the surface of plasma membrane in skeletal muscle (Ryder et al., 2001; Pendergrass et al., 2007). FOXO1 is a transcription factor that not only a central regulator of cellular homeostasis, but is also associated with increases in the expression of key enzymes involved in gluconeogenesis, hence its upregulation in the liver is associated with the increased conversion of incoming substrates to glucose. A decrease in GLUT4 levels at the surface plasma membrane in muscle contributes to reduction in glucose uptake from the circulation. In the liver, insulin signaling normally results in phosphorylation, which suppresses the function of FOXO1 through the action of the protein kinase (Akt). This process promotes the presence of FOXO1 in the cytoplasm where it is inactive (Nakae et al., 2000; Gross et al., 2008). However, in obesity, FOXO1 expression is upregulated and the protein apparently modified to become insensitive to insulin regulation (Titchenell et al., 2016). Under peripheral insulin-resistant conditions, the overproduction of free fatty acids (FFAs) results in the release of proinflammatory cytokines into the blood circulation. These cytokines activate several types of serine kinase such as IκB kinase (IKK) and JNK (Sethi and Vidal-Puig, 2007; Osborn and Olefsky, 2012). In addition, these cytokines negatively regulate IRS proteins, particularly in suppressor of cytokine signaling (SOCS) protein. This protein binds with the phosphorylated IR, consequently blocking an activation of the IRS proteins (de Luca and Olefsky, 2008; Johnson and Olefsky, 2013). In addition, the SOCS proteins promote the ubiquitination of IRS proteins resulting in IRS degradation through the proteasomal complex. Those findings suggest that inflammation following insulin resistance can lead to impaired insulin signaling in the target organs.

Insulin resistance

Insulin resistance is a pathological condition involving a failed response to normal levels of insulin in target tissues. It is defined by reduction of its capacity of insulin to stimulate glucose utilization, either due to insulin deficiency or by impairment in its secretion and/or utilization. The peripheral insulin resistance causes pancreatic β cells to secrete more insulin, in a process known as compensatory hyperinsulinemia. However, during insulin resistance often there is β cell depletion, which results in sustained hyperglycemia and type 2 diabetes (Shulman, 2000).

The exact molecular mechanisms leading to insulin resistance have not been elucidated so far. However, it is known that the amount of IR expression on target tissues is diminished due to insulin's cellular internalization and reduced tyrosine kinase activity. Furthermore, postreceptor alterations in insulin receptor substrate-1 (IRS-1), which regulate phosphorylation and dephosphorylation, may also play a predominant role in the development of insulin resistance. Converging evidence suggests that an imbalance between IRS-1 tyrosine and serine phosphorylation (Petersen and Shulman, 2018) may be closely associated with the pathogenesis of insulin resistance. Diminished IRS-1 tyrosine phosphorylation is associated with reduction in translocation of GLUT4 to the plasma membrane, which enables glucose influx into the cells. Simultaneously, enhanced IRS-1 serine phosphorylation facilitates the activation of mitogen-activated proteins, whose action is not involved in metabolic but in mitotic insulin activity and proinflammatory pathways activation. This process not only results in intramitochondrial stress, but also in the enhancement of insulin resistance. This process may also contribute to the onset of diabetes-related micro- and macrovascular complications. In summary, insulin resistance consists of two tightly coupled mechanisms: lack of suppression of glucose production and lack of glucose uptake by peripheral tissues, primarily muscles (Petersen and Shulman, 2018). Development of insulin resistance contributes to the induction of other conditions such as dyslipidemia, hypertension, and atherosclerosis. This pathological state is called as metabolic syndrome (MetS). Little is known about biomarkers of insulin resistance. These biomarkers (adiponectin, RBP4, chemerin, A-FABP, FGF21, fetuin A, myostatin, IL-6, and irisin) play significant roles in determining insulin sensitivity (Park et al., 2015). In the insulin sensitivity assays, insulin resistance shows following characteristics: hyperinsulinemia and hyperglycemia in fasting condition, increased glycosylated hemoglobin (HbA1C), postprandial hyperglycemia, hyperlipidemia, impaired glucose tolerance, impaired insulin tolerance, decreased glucose infusion rate, increased hepatic glucose production, loss of first phase secretion of insulin, hypo-ponectinemia, and increased inflammatory markers in plasma. HbA1C is a marker of longer term glycemic control in diabetes patients. A1C formation occurs when an excess of circulating glucose covalently binds to hemoglobin of erythrocytes. Indeed, an elevation in A1C imparts a 50% increased risk of retinopathy and is a globally recommended diagnostic test for type 2 diabetes (Florkowski, 2013). In prediabetic subjects, hyperglycemia, insulin resistance, inflammation, and metabolic derangements are associated with endothelial vasodilator and fibrinolytic dysfunction. This leads to increase

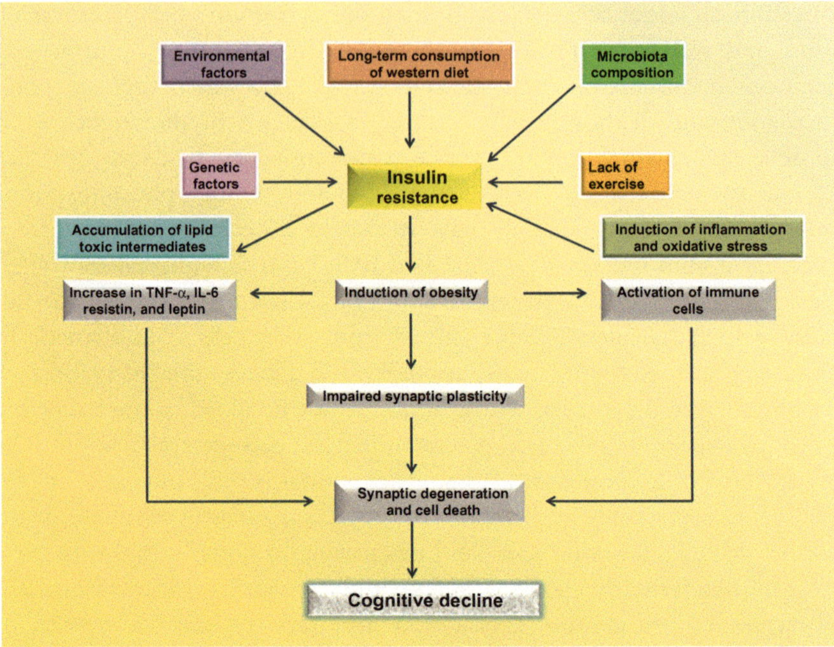

Figure 1.3 Factors modulating insulin resistance and obesity and their effect on cognition.

in risk of cardiovascular and renal disease. Importantly, the microvasculature affects insulin sensitivity by affecting the delivery of insulin and glucose to skeletal muscle; thus endothelial dysfunction and extracellular matrix remodeling promote the progression from prediabetes to type 2 diabetes. Weight loss is the major way to treat prediabetes patients, but type 2 diabetes therapies, which improve endothelial function and vasodilation, may also prevent cardiovascular disease and slow progression of this condition.

Insulin resistance is modulated by both genetic and acquired factors (Figs. 1.2 and 1.3). Although very little is known about the genetic causes or predispositions of insulin resistance in prediabetic populations, but it is proposed that defects in oxidative metabolism and inherited defects in the basic insulin signaling cascade are closely associated with genetic causes of insulin resistance (Morino et al., 2006, 2008; Thaler and Schwartz, 2010; Thaler et al., 2012; Brown and Walker, 2016). Many candidate genes are closely associated with insulin resistance (Table 1.1). It is also proposed that genetic component may interact with environmental factors to

Table 1.1 Candidate genes contributing to the pathogenesis of insulin resistance.

SNP	Nearby gene	Chromosome	References
rs13081389	PPARG	3	Deeb et al. (1998)
rs972283	KLF14	7	Voight et al. (2010)
rs2943641	IRS-1	2	Rung et al. (2009)
rs780094	GCKR	2	Dupuis et al. (2010)
rs8050136	FTO	16	Do et al. (2008)
rs7903146	TCF7L2	10	Elbein et al. (2007)
rs1208	NAT2	8	Knowles et al. (2015)
rs6723108	TMEM163	2	Dupuis et al. (2010)
rs35767	IGF-1	12	Tabassum et al. (2013)
rs12970134	MC4R	18	Chambers et al. (2008)
Rs17046216	SC4MOL	4	Chen et al. (2012)
Rs17077836	TCERGIL	10	Chen et al. (2012)
Rs702634	ARL15	12	Mahajan et al. (2014)

promote a pronounced pathophysiologically abnormality during insulin resistance (Thaler and Schwartz, 2010; Thaler et al., 2012). To this end, it is known that insulin sensitivity can be enhanced with drugs that primarily act through transcription factor targets, such as thiazolidinediones, an activator of peroxisome proliferator-activated receptor gamma (PPARγ; Ahmadian et al., 2013; Soccio et al., 2014) and glucocorticoids, which activate the glucocorticoid receptor. Second, drugs, which contribute to chromatin remodeling, such as certain HDAC inhibitors modulate insulin sensitivity in cells, animal models, and human subjects (Masuccio et al., 2010). Third, mice with genetic alterations in chromatin modifying enzymes, such as Jhdm2a and Ehmt1, develop obesity and insulin resistance (Inagaki et al., 2009; Ohno et al., 2013). Finally, many studies have indicated that the development of insulin resistance in later life is strongly affected by nutritional conditions experienced in utero. For example, pregnant rodents that undergo caloric restriction give birth to offspring that have a significantly greater chance of developing insulin resistance as adults (Rando and Simmons, 2015). The same phenomenon has been reported in human populations, as with offspring of Dutch women who were pregnant during the "hunger winter" of 1944–45 (Kyle and Pichard, 2006). Such examples of "metabolic memory" have been proposed to be associated with epigenetic factors (Raychaudhuri et al., 2008).

Mammals have two types of adipose tissue, white adipose tissue (WAT) and brown adipose tissue (BAT) (Gil et al., 2011). WAT is most common and found as subcutaneous tissue, around the abdomen, thighs, and waist. In contrast, BAT is found particularly as perivascular, epicardial, supra-adrenal, and suprascapular tissue. The major difference between WAT and BAT is that BAT has higher number of mitochondria and small fat droplets whereas WAT has a big fat droplet and less mitochondria. High accumulation of WAT leads to obesity. The most common acquired factors that induce insulin resistance are obesity, sedentary lifestyle, and aging. All these factors are interrelated (Mokdad et al., 2003; Hamilton et al., 2007; Thaler and Schwartz, 2010). The molecular mechanisms contributing to insulin resistance are not fully understood. However, it is becoming increasingly evident that the accumulation of lipids [saturated FFAs, diacylglycerol (DAG), and triacylglycerol (TAG)] and lipid mediators such as long-chain acyl-CoAs, acylcarnitines, uric acid (2,6,8-trioxypurine), isoprostane (IsoP), and ceramide (Fig. 1.4) may contribute to the molecular mechanisms of insulin resistance (Aguer et al., 2015; Ikonen and Vainio, 2005; Inokuchi, 2006; Itani et al., 2002; Adams et al., 2004, 2009; Holland et al., 2007a,b; Farooqui, 2013) (Table 1.2). Among abovementioned metabolites, there is a strong correlation between

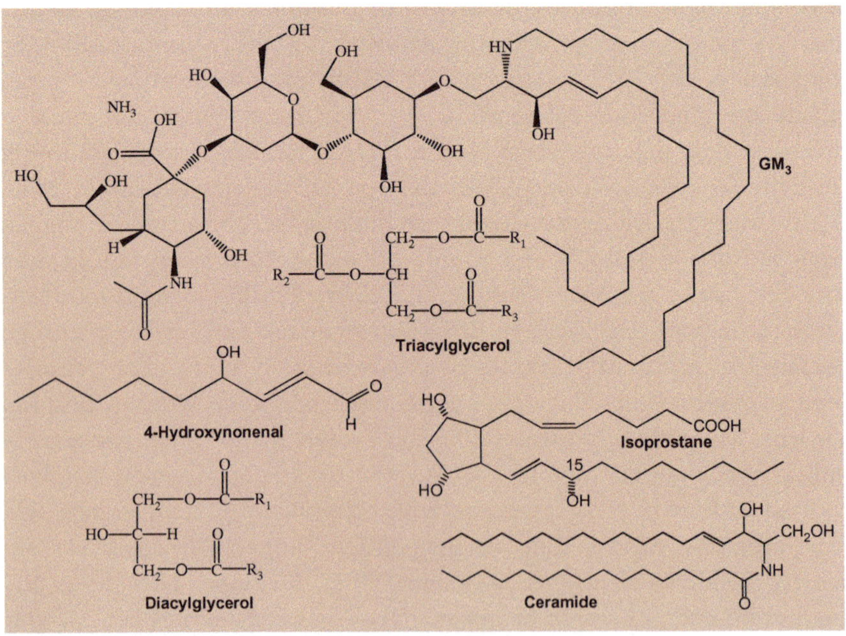

Figure 1.4 Chemical structures of lipid and lipid mediators contributing to insulin resistance.

Table 1.2 Lipid mediators associated with the pathogenesis of insulin resistance.

Lipid mediator	Levels	References
Free fatty acids	Increased	Kraegen et al. (2001) and Delarue and Magnan (2007)
Diacylglycerol	Increased	Erion and Shulman (2010)
Triacylglycerol	Increased	Kraegen et al. (2001)
Acylcarnitine	Increased	Aguer et al. (2015)
Uric acid	Increased	Farooqui (2013)
Isoprostane	Increased	Farooqui (2013)
Ceramide	Increased	Holland et al. (2007a,b)

intramyocellular TAG concentrations and the severity of insulin resistance. The molecular mechanism of the induction of insulin resistance in various body tissues by TAG and DAG is still unclear. However, it is proposed that increased levels of DAG are associated with protein kinase C activation (PKCθ and PKCε), which can regulate insulin-mediated signal transduction via serine phosphorylation of IRS-1 (Yu et al., 2002). Furthermore, DAG is an intermediate in the synthesis of TAG from fatty acids and glycerol, its level can be lowered by either improving the oxidation of cellular fatty acids or by accelerating the incorporation of fatty acids into TAG (Timmers et al., 2008; Muoio, 2010). Other mechanisms of TAG-mediated insulin resistance may involve: (1) activation of DNA-dependent protein kinase to stimulate upstream stimulatory factor (USF) 1/USF2 heterodimers, enhancing the lipogenic transcription factor sterol regulatory element binding protein 1c (SREBP1c); (2) stimulation of fatty acid synthase through adenosine monophosphate (AMP) kinase modulation; (3) mobilization of lipid droplet proteins to promote retention of TAG; and (4) upregulation of a novel carbohydrate response element binding protein β isoform that potently stimulates transcription of lipogenic enzymes. Additionally, insulin signaling through mTOR to activate transcription and processing of SREBP1c described in liver may apply to adipose tissue (Czech et al., 2013). In contrast, uric acid is an end product of purine metabolism. It is major natural antioxidant in human plasma. Elevated uric acid levels have been hypothesized to protect against oxidative damage in several neurodegenerative disorders (Irizarry et al., 2009). On the other hand, for each molecule of uric acid produced, the enzymatic degradation of xanthine simultaneously produces superoxide anions, which are among the most powerful pro-oxidants (Glantzounis et al.,

2005). Presence of uric acid (6.8–7 mg/mL) in serum not only contributes to increased risk of gout (Dalbeth et al., 2016), but also promotes chronic inflammation leading to pain and swelling.

Carnitine (Fig. 1.4) is mainly absorbed from the diet, but can be formed through biosynthesis (Vaz and Wanders, 2002). FFAs are activated by esterification with CoA. Long-chain acyl-CoAs (such as palmitic acid) are transported into mitochondria, where long-chain acyl-CoAs are converted into acylcarnitines by carnitine palmitoyltransferase 1 (CPT1), which is located at the outer mitochondrial membrane (Vaz and Wanders, 2002). In the brain, CPT1a is expressed in the endoplasmic reticulum (ER) (and not the mitochondria) of neurons in the brain. Insulin resistance is linked with incomplete fatty acid β-oxidation and the subsequent increase in acylcarnitines. The physiological role of acylcarnitine efflux to the plasma compartment is not known. However, there are several scenarios. Acylcarnitine formation prevents CoA trapping, allowing continuation of CoA-dependent metabolic processes (Ramsay et al., 2001). It is also proposed that plasma acylcarnitines may serve as a means of transportation between cells or organs or sink for cellular/tissue acylcarnitine sequestration. It is generally accepted that long-term consumption of western diet increases cytosolic lipid content of insulin-responsive tissues (such as liver, skeletal muscle, and brain). This negatively affects the insulin sensitivity of these tissues by inhibiting insulin signaling via lipids and lipid-derived intermediates such as ceramide, DAG, gangliosides, 4-hydroxynonenal (4-HNE), and possible other long-chain FA-derived metabolites (Figs. 1.4 and 1.5) (Holland et al., 2007a,b; Timmers et al., 2008; Samuel and Shulman, 2012; Farooqui, 2015). High levels of lipids and lipid mediators impair insulin signaling and promote insulin resistance (Farooqui, 2013). Chronic insulin resistance is also related to diet-induced inflammation. The molecular mechanisms of insulin resistance are quite complex. They are based on the ability of increased cellular inflammation to interrupt insulin's action by disrupting signaling mechanisms within the cell in particular by the enhancing the phosphorylation of IRS (Farooqui, 2013). Inflammation is a physiological process, which is characterized by increase in number of white blood cells (WBCs) or increase in levels of proinflammatory cytokines in the circulation or tissue (Ye and McGuinness, 2013; Farooqui, 2013). In general, inflammation contributes to organ remodeling, tissue repairing, wound healing, and immunity against infections. Inflammation is a protective mechanism, which controls the harmful insults and initiates the healing process. Uncontrolled

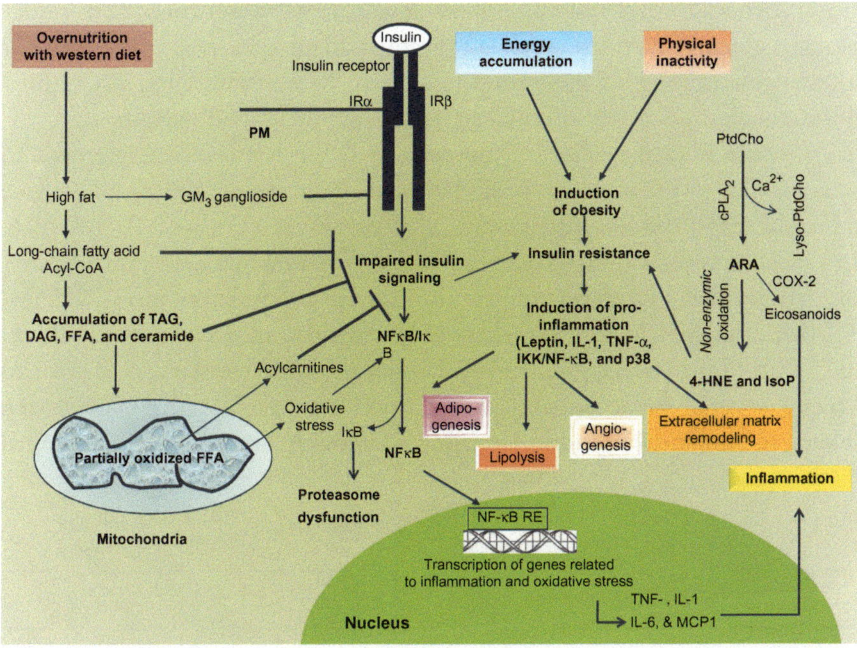

Figure 1.5 Effects of western diet consumption on insulin resistance.
DAG, diacylglycerol; *FFAs*, free fatty acids; *I-κB*, inhibitory subunit of NF-κB; *IL-1β*, interleukin-1β; *IL-6*, interleukin-6; *LCFA*, long-chain fatty acid; *NF-κB*, nuclear factor-κB; *NF-κB-RE*, nuclear factor-κB-response element; *PM*, plasma membrane; *TAG*, triacylglycerol; *TNF-α*, tumor necrosis factor-α.

inflammatory response usually leads to multiple side effects such as tissue injury and organ dysfunction. Obesity-mediated inflammation starts in adipose tissue and liver with elevated macrophage infiltration and expression of proinflammatory cytokines. The proinflammatory cytokines enter the blood stream to cause systemic inflammation. In obesity, inflammation has both beneficial and detrimental effects (Ye and McGuinness, 2013; Farooqui, 2013).

Reactive oxygen species (ROS) is a collective term that refers to oxygen radicals such as superoxide and hydroxyl radical, •OH and to nonradical derivatives of O_2, including H_2O_2 and ozone (O_3) in cells and tissue. It is determined not only by cellular production but also by the levels of antioxidant defenses. Indeed, activities of antioxidant enzymes such as superoxide dismutase (SOD), catalase (CAT), glutathione peroxidase (GPx), thioredoxin, peroxiredoxins, and heme oxygenase-1 regulate and often reduce the level of ROS in biological systems. Generation of ROS also influences insulin resistance. Low levels of ROS contribute to fundamental cellular functions, such as growth, adaptation responses, and for optimal functioning of the immune system. In contrast, high ROS promotes the translocation of nuclear factor kappa B (NF-κB) from cytoplasm to the nucleus, where it interacts with NF-κB response element to facilitate the expression of proinflammatory enzymes [$sPLA_2$, COX-2, inducible nitric oxide synthase (iNOS)], cytokines (TNF-α, IL-1β, IL-6, IL-12), chemokines (MIP-1α, MCPP1), growth factors, cell cycle regulatory molecules, adhesion molecule leading to inflammation (ICAM, VCAM, and E-selectin), and antiinflammatory molecules (Cai et al., 2005; Farooqui, 2012) (Fig. 1.6). In addition, oxidative stress is known to activate several serine kinases and transcription factors that have been linked to impaired insulin signaling, including c-jun amino-terminal

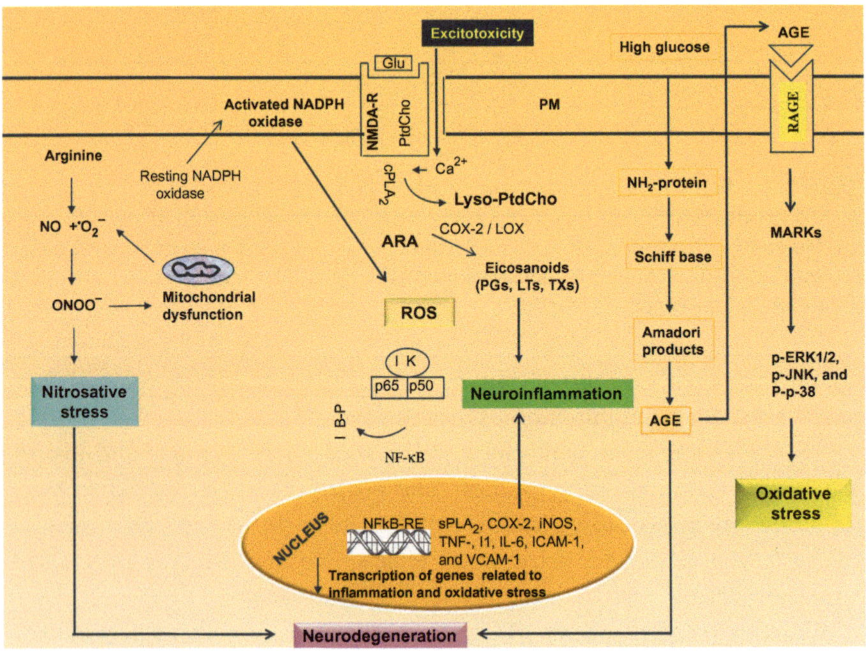

Figure 1.6 Signal transduction diagram showing the effect of oxidative stress on neurodegeneration.

AGE, advanced glycated end product; *ARA*, arachidonic acid; *COX-2*, cyclooxygenase-2; *cPLA₂*, cytosolic phospholipase A₂; *eNOS*, endothelial nitric oxide synthase; *ERK1/2*, extracellular signal-regulated kinase-1/2; *Glu*, glutamate; *MAPK*, mitogen-activated protein kinase; *NMDA-R*, NMDA receptor; *NO*, nitric oxide; *p-38*, signaling pathway; *PtdCho*, phosphatidylcholine; *RAGE*, receptor for advanced glycation end products; *ROS*, reactive oxygen species.

kinases (JNKs), IkB kinase catalytic subunit β (IKK-β), NF-κB, and protein kinase C (Bloch-Damti and Bashan, 2005). Furthermore, mitochondrial dysfunction plays an important role in the induction of oxidative stress and insulin resistance. These processes have a common connection to redox imbalance. Furthermore, ROS signaling and redox sensing rely heavily on the interdependent glutathione and thioredoxin reducing systems (Muoio and Neufer, 2012). Both use the reducing power of nicotinamide adenine dinucleotide phosphate (NADPH) to mitigate oxidative stress and to modulate reversible oxidation/reduction of protein thiols/disulfides. They are named as "sulfur switches." They play important regulatory roles in cell signaling, mitochondrial function, and metabolic control (Brandes et al., 2009; Jones, 2008). These redox circuits intersect and regulate insulin signaling molecules such as PTEN, SHIP2, and PTP1B (Brandes et al., 2009; Jones, 2008).

Ceramides is a sphingolipid that plays an active role in glucose homeostasis and insulin signaling (Holland and Summers, 2008; Farooqui, 2011; Górski, 2012). Two primary pathways contribute to the synthesis of ceramides. The first pathway involves the condensation of palmitate and serine (called de novo synthesis) and the second pathway synthesizes ceramide either through the reacylation of sphingosine (salvage pathway) or through the N-acylation of a sphingoid base, a reaction, which is catalyzed by ceramide synthase (CERS) (Fig. 1.7). The primary mechanism through which ceramide contributes to insulin resistance involves the inhibition of Akt/PKB activity. This enzyme is an essential facilitator of glucose transport into the cell. Ceramide inhibits Akt/PKB activity by two independent mechanisms: (1) first mechanism involves stimulation of Akt dephosphorylation via protein phosphatase 2A (PP2A) and (2) the second mechanism the blocked of Akt translocation via involvement of PKCζ (Stratford et al., 2004). Ceramide also activates PP2A, which inhibits the action of Akt/PKB by impairing Akt serine phosphorylation. The process decreases the translocation of GLUT4 to the plasma membrane and hence decreased uptake of glucose. In addition, ceramide initiates inflammatory signaling pathways, leading to the activation of both c-jun NH_2-terminal kinase (JNK) and NF-κB/inducer of κ kinase (Ruvolo, 2003). All these processes are closely associated with the development of insulin resistance (Cai et al., 2005; Chung et al., 2008; Ikonen and Vainio, 2005). Ceramide can enter brain by crossing the BBB. Once in the brain, ceramide induce oxidative stress, inflammation, and insulin resistance. Induction of these processes leads to neurodegeneration

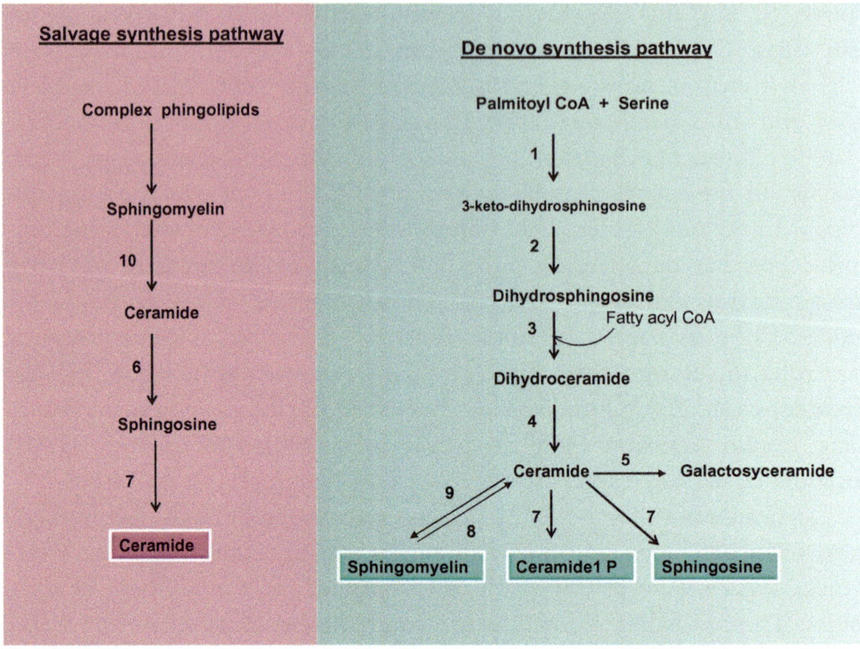

Figure 1.7 De novo synthesis and salvage pathways of ceramide production. 1, serine palmitoyltransferase; 2, ketodihydrosphingosine reductase; 3, ceramide synthase; 4, dihydroceramide desaturase; 5, UDP-Gal transferase; 6, ceramidase; 7, ceramide kinase; 8, sphingomyelinase; 9, sphingomyelin synthase; 10, ceramide synthase.

(Tong and de la Monte, 2009; Stumvoll et al., 2005; Thaler and Schwartz, 2010).

Like ceramide, GM_3 ganglioside (Fig. 1.4) also contributes to the development of insulin resistance. This sphingolipid not only inhibits tyrosine phosphorylation of IRs and IRS-1, but also retards the activity of phospholipase A_2 (Kabayama et al., 2005). It is proposed that GM_3 induces TNF-α-mediated insulin resistance causes enhancement of tyrosine phosphorylation of IRs (Tagami et al., 2002; Kabayama et al., 2005; Yamashita et al., 2003). Improved phosphorylation of IRs is also observed in GM_3 synthase knockout mice (Zhao et al., 2007). These studies indicate that increased levels of complex sphingolipids (ceramide and GM_3 ganglioside) play an important role in insulin resistance.

4-HNE is a nine carbon α,β-unsaturated aldehyde, which is derived from peroxidation of arachidonic acid (Fig. 1.4). This aldehyde contains three functional groups, which often act in concert and help to explain its high reactivity (Poli and Schaur, 2000). 4-HNE contains a C=C double bond and a C=O carbonyl group which provide a partial positive charge to carbon 3 due to the presence of mobile pi-electrons. This positive charge is further enhanced by the inductive effect of the hydroxy group at carbon 4. Therefore 4-HNE is considered to be soft electrophiles and is prone to be attacked by nucleophiles, such as thiol or amino groups. This reaction occurs primarily at carbon 3 and secondarily at the carbonyl carbon 1 (Esterbauer et al., 1991). 4-HNE contributes to insulin resistance not only by damaging pancreatic β-cells, but impairs the ability of muscle and liver cells to respond to insulin. These processes may contribute to insulin resistance (Mattson, 2009; Pillon et al., 2012). 4-HNE not only promotes oxidative stress, impairs adipogenesis, alter the expression of adipokines, but also upregulates genes for expression of lipolytic enzymes resulting in increase in FFA release (Dasuri et al., 2013). Increase in circulating levels of lipids and lipid mediators produce lipotoxicity. This process plays important roles not only in insulin resistance, but also in pancreatic β-cell dysfunction. Different signaling pathways, such as novel protein kinase c pathway and the JNK-1 pathway, are involved as the mechanisms of how lipotoxicity leads to insulin resistance in liver, muscles, and brain.

Branched chained amino acids (BCAAs: valine, leucine, and isoleucine) are essential components of the human diet and important nutrient signals, which regain particular interest in recent years with the avenue of metabolomics studies. Reducing dietary BCAAs leads to improvements in western diet–induced obesity and glucose intolerance in mice (Cummings et al., 2018). Additionally, the supplementation of BCAAs abolishes the effect of protein restriction on glucose metabolism and induces inflammation in visceral adipose tissue in mice (Cummings et al., 2018). It is suggested that BCAAs not only play an important role in progression of type 2 diabetes and whole-body insulin resistance (Newgard et al., 2009; Sun et al., 2016a,b). The molecular mechanism associated with BCAAs-mediated insulin resistance is not fully understood. However, recent studies have indicated that BCAAs induce whole-body insulin resistance by increasing muscle BCAA oxidation, and decreasing glucose and fatty acid oxidation (Newgard et al., 2009). In addition, BCAAs-mediated increase in obesity may also due to the accumulation of toxic BCAA metabolites that, in turn, triggers mitochondrial dysfunction, contributing to insulin resistance, type 2 diabetes, and cardiovascular disease (Lynch and Adams, 2014; Karwi et al., 2019; Wang et al., 2016). Detailed mechanistic investigations have indicated that the accumulation of BCAAs act by activating mTOR (Lynch and Adams, 2014) and impair insulin signaling, due to its ability to phosphorylate IRS-1 via directly activating p70S6K through mTOR (Newgard et al., 2009; Dodd and Tee, 2012; Han et al., 2012; Wolfson et al., 2015; Fillmore et al., 2018). Thus a decrease in cardiac BCAA oxidation has the potential to increase cardiac BCAA levels, which may activate mTOR signaling and reduce cardiac insulin sensitivity.

Among various subcellular organelles mitochondrial dysfunction and ER stress also contribute to the development of insulin resistance in overnourished or obese rodents (Sripetchwandee et al., 2018). Elevation in mitochondrial oxidative phosphorylation due to increase in influx of nutrients stimulates the production of mitochondrial superoxide ($O_2^{\cdot -}$) contributing to the pathogenesis of insulin resistance in animal and cellular models (Hoehn et al., 2009). On the basis of many studies, it is suggested that mitochondrial $O_2^{\cdot -}$ production may represent a link between mitochondrial function and insulin resistance (Hoehn et al., 2009). It is also reported that treatment with the mitochondrial SOD mimetics (Hoehn et al., 2009) and mitochondria-specific free radical scavengers (Anderson et al., 2009) protect rodents from developing insulin resistance following high-fat overfeeding.

Collective evidence suggests that induction of insulin resistance is the primary cause of type 2 diabetes. Several processes are associated with insulin resistance. They include: (1) obesity, (2) inflammation, (3) mitochondrial dysfunction, (4) hyperinsulinemia, (5) lipotoxicity/hyperlipidemia, (6) genetic background, (7) ER stress, (8) aging, (9) oxidative stress, (10) fatty liver, (11) hypoxia, (12) lipodystrophy, and (13) pregnancy. Many of these factors are associated with obesity and aging, which are the major risk factors for insulin resistance in the general population. Induction of insulin resistance also contributes to hypertension, and heart disease (Chang et al., 2018), stroke, Alzheimer's disease (AD), Parkinson's disease (PD), and vascular dementia (Vanorsdall et al., 2008). All these conditions are accompanied by insulin resistance and significant decrease in cognitive function. Thus accumulation of lipids and lipid mediators act by engaging stress-responsive serine kinases that impede insulin activation of cell surface receptor, as well as downstream signaling molecules such as IRS-1 and protein kinase B/Akt leading to disruption of insulin signaling cascade and inducing insulin resistance (Muoio, 2010). In addition, ER stress also plays an important role in the onset of insulin resistance. ER is a well-organized protein-folding machine composed of protein chaperones, proteins that catalyze protein folding, and sensors that detect the presence of misfolded or unfolded proteins (Malhotra and Kaufman, 2007). Furthermore, ER also contains a sensitive surveillance mechanism that prevents misfolded proteins from transiting the secretory pathway. The efficiency of protein-folding reactions not only depends on appropriate environmental and genetic factors, but also on metabolic conditions. Conditions that disrupt protein folding threaten cells with decrease in viability and longevity. Accumulation of unfolded proteins in ER lumen initiates activation of an adaptive signaling cascade known as the unfolded protein response (UPR) (Malhotra and Kaufman, 2007). Appropriate adaptation to misfolded protein accumulation in the ER lumen requires regulation at all levels of gene expression including transcription, translation, translocation into the ER lumen, and ER-associated degradation (ERAD) leading to the activation of protective, apoptotic, and inflammatory responses (Malhotra and Kaufman, 2007). Several major transducers of the UPR have been identified. These include PKR-like ER kinase, inositol-requiring enzyme 1 (IRE1), and activating transcription factor 6. The activation of these factors transmits signals from the ER to the cytoplasm or nucleus, and activate three pathways: (1) suppression of protein translation to avoid the generation of more unfolded proteins (Harding et al., 2000); (2) induction of genes encoding ER molecular chaperones to facilitate protein folding (Li et al., 2000); and (3) activation of ERAD to reduce unfolded protein accumulation in the ER (Ng et al., 2000). If these strategies fail, the cells are unable to maintain ER homeostasis and undergo apoptosis due to increase in ER stress (Urano et al., 2000), which activates metabolic pathways that trigger insulin resistance, release of macrophage chemoattractant proteins, and initiate chronic inflammation. The infiltrated macrophages in turn release inflammatory proteins causing further recruitment of macrophages to adipose tissue and the release of inflammatory cytokines. IRE1 also induces an inflammatory signaling cascade by activating IKK, the MAPKs p38 and JNK, and finally the major inflammatory transcription factor NF-κB. Consequently, obesity-induced ER stress leads to IRSs serine phosphorylation and inhibits insulin signaling (Ozcan et al., 2004; Ozcan et al., 2006; Xu et al., 2003; Lumeng et al., 2007; Thaler and Schwartz, 2010). Collective evidence suggests that mitochondrial oxidative stress, ER stress, intracellular ceramide accumulation, and the induction of JNK, IKK, or PKCθ may contribute to the development of insulin resistance in overnourished or obese rodents (Savage et al., 2007). It should also be noted that inflammatory response to dietary fat is mediated and supported by ER stress and toll-like receptor (TLR) signaling, which results in the activation of NF-κB and production of inflammatory cytokines, such as IL-1β, IL-6, and TNF-α (Akira et al., 2006; Hayden and Ghosh, 2008). In addition, many other signaling pathways support and contribute to the development of insulin resistance. These include (1) signaling pathway associated with TLR4 through the involvement of inhibitor of NF-κB and kappa B kinase- (IKK-) signaling (Kim and Sears, 2010; Onyango, 2017), (2) advanced glycation end products (AGEs) or uric acid–induced RAGE signaling via NF-κB (Sakaguchi et al., 2011; Cai et al., 2017), (3) upregulation of NADPH oxidase (Nox) expression and activity (Pereira et al., 2014; Sukumar et al., 2013), (4) increased mitochondrial ROS generation (Anderson et al., 2009), (5) upregulation of iNOS (Cha et al., 2011), (6) ER stress and the UPR (Zhang et al., 2013; Diaz et al., 2015), (7) dysregulation of the heat shock response (Chung et al., 2008), (8) autophagy dysregulation (Yang et al., 2010), (9) activation of p53 (Derdak et al., 2011), and (10) inflammasome activation (Nov et al., 2010; Stienstra et al., 2012). These signaling pathways contribute to insulin resistance through defects in IR function, abnormalities in insulin signaling, alterations in glucose metabolism, induction of hyperinsulinemia, hyperglycemia, and inflammation, but also increase blood pressure (Wang and Jin, 2009). These processes not only play an important role in the pathogenesis of cardiovascular diseases, type 2 diabetes, MetS, hypertension, dyslipidemia, myocardial infarction, but also contribute to certain cancers, sleep apnea, osteoarthritis, and neurological disorders (stroke and AD) is increasing with an alarming rate (Raffaitin et al., 2011; Flegal et al., 2007; Farooqui, 2012). Insulin resistance also associated with impaired fibrinolysis, and hypercoagulability. Insulin resistance may result from abnormalities in key molecules of the insulin signaling pathways, including overexpression of phosphatases and downregulation and/or activation of protein kinase cascades (Avramoglu et al., 2006), leading to abnormalities in the expression and action of various cytokines, growth factors, and peptides, and overproduction of very low-density lipoprotein (Fonseca et al., 2004). Insulin resistance can be improved by increased levels of irisin, a myokine, which is associated with increased energy expenditure through the stimulation of "browning" of WAT (Gizaw et al., 2017). Irisin is produced through the breakdown of fibronectin type 3 domain. Increase in levels of circulating irisin contributes to improvement in glucose homeostasis through reduction in insulin resistance (Gizaw et al., 2017). Several studies have indicated that irisin improves insulin resistance and type 2 diabetes by increasing sensitization of the IR in skeletal muscle and heart by improving hepatic glucose and lipid metabolism, promoting pancreatic β cell functions, and transforming WAT to BAT.

Insulin resistance contributes to several related diseases such as glucose intolerance, type 2 diabetes, obesity, dyslipidemia, hypertension, and MetS (Bruce and Hanson, 2010; Petersen and Shulman, 2006). In addition, type 2 diabetes is a major risk factors for the development of heart disease, stroke, and AD (Taubes, 2009; Farooqui, 2013) supporting the view that insulin resistance is a core feature of type 2 diabetes mellitus, cardio- and cerebrovascular diseases, AD, depression, and anxiety (Farooqui, 2017). In diabetes, insulin resistance is a result of accumulation of intracellular lipid metabolites (e.g., fatty acyl CoAs, DAG, TAG, ceramide) not only in visceral tissues (skeletal muscle and hepatocytes), but also in neurological disorders (stroke and AD). In neurological disorders, brain insulin resistance is defined as the failure of brain cells to respond to insulin as they normally would, resulting in impairments in synaptic, metabolic, and immune response functions. Brain insulin resistance is defined as the failure of brain cells to respond to insulin as they normally do, resulting in impairments in synaptic, metabolic, and immune response functions (Farooqui, 2017). In contrast, the increase in hepatic insulin sensitivity observed in patients with type 2 diabetes following weight loss is accompanied by a significant reduction in intrahepatic fat without any changes in circulating adipocytokines (interleukin-6, resistin, and leptin). The molecular mechanism underlying defective insulin-stimulated glucose transport activity is not fully understood. However, it is proposed that molecular mechanism may be attributed to increase in intramyocellular lipid metabolites such as fatty acyl CoAs and DAG, which in turn activate a serine/threonine kinase cascade, resulting in defects in insulin signaling through Ser/Thr phosphorylation of IRS-1.

Insulin resistance as a protective mechanism

Several in vivo and in vitro studies in rats have indicated long-term exposure cells to high levels of glucose result in insulin resistance through the impairment of GLUT4 action (Rossetti et al., 1990; Baron et al., 1995). This proposal is supported not only by impairment GLUT4 translocation and induction of insulin resistance by glucosamine (Baron et al., 1995), but also the inhibition of glucose uptake by prolonged exposure of cells to FFAs (Roden et al., 1996; Hawkins et al., 1997). On the basis of these studies, it is proposed that the insulin resistance is a protective response of cells to hyperglycemia and lipotoxicity-mediated stress, and damage. It results in the exclusion of glucose from cells, which are heavily loaded with lipid mediators. It not only reduces the lipotoxicity-mediated damage, but also slows the onset of oxidative stress (Unger, 2003). Alternatively, insulin resistance may be induced partly in response to oxidative stress, a process caused by increased ROS production or reduction in ROS clearance (Azzi, 2007). Levels of ROS are increased in response to increased consumption of macronutrient (Codoñer-Franch et al., 2011). Excessive nutrient intake results in increased superoxide ($O_2^{\cdot-}$) generation by mitochondria, which are capable of chemically altering all major classes of biomolecules by modifying their structure and function (Tiganis, 2011). In addition, mitochondria produce most of the cell's energy, and in this capacity take up the majority of intracellular oxygen (Finkel and Holbrook, 2000). Several signaling pathways are activated in response to increasing oxidative stress; these pathways include MAPK, NF-κB, and the PtdIns 3K–Akt pathway (Huang et al., 2018). It is also suggested that production of mitochondrial $O_2^{\cdot-}$ may act as a nutrient sensor, which regulates nutrient intake under conditions of overnutrition (Hoehn et al., 2009).

Molecular aspects of obesity

Obesity is a chronic and multifactorial condition, which results in the pathological growth of adipocytes causing excess body weight. According to WHO, over 600 million people worldwide are obese, and obesity-attributable healthcare expenditures in the US approach $147 billion annually (Trogdon et al., 2012). Obesity is caused by an imbalance of food intake and energy expenditure, but what causes this imbalance is unclear as the regulation of these processes is complex. At the least, these processes involve communication between the peripheral tissues that acquire, sense, or store nutrients with specialized nuclei in the brain that regulate feeding and metabolism (Schwartz et al., 2000; Sikaris, 2004). Leptin and insulin play central roles in the regulation of energy balance in the body. Insulin and leptin receptors (LepRs) are expressed by brain neurons involved in energy intake, and administration of either peptide, directly into the brain reduces food intake, whereas deficiency of either hormone results in the converse. Although humans with severe obesity, due to mutation of leptin or its receptor, have been described, most forms of human obesity are characterized by normal or increased circulating levels of insulin and leptin, and are devoid of any known defect in receptors for these two adiposity signals. Thus common forms of obesity in both humans and animal models are hypothesized to involve insulin resistance and leptin resistance downstream of their neuronal receptors, affecting the neural pathways that mediate their effects on energy balance, but the means by which they become resistant to the inputs from insulin and leptin signals are not completely known. Leptin is known to induce satiety and augments energy expenditure to limit weight gain. In opposition, insulin induces energy accumulation and promotes weight gain. Yet, despite the marked increase in leptin in obesity, animals gain weight, presumably from central leptin resistance that is responsible for impaired control of food intake and energy expenditure. The energy surplus or weight gain is a major risk factor for insulin resistance, which triggers hyperinsulinemia. Adipose tissue inflammation is generally believed to contribute to the pathogenesis of insulin resistance through proinflammatory cytokines (Olefsky and Glass, 2010; Shoelson et al., 2006). However, the same cytokines stimulate energy expenditure and induce satiety and thus limit adiposity and improve glucose homeostasis. The functional resistance to leptin and insulin in the hypothalamus is a consequence of diet-induced activation of inflammatory signaling, specifically in this site of the brain, which leads to the molecular impairment of leptin and insulin signal transduction by at least four distinct mechanisms; induction of SOCS3 expression (Howard et al., 2004), activation of JNK and IKK (De Souza et al., 2005), and induction of protein tyrosine phosphatase 1B (PTP1B) (Bence et al., 2006).

Induction of oxidative stress is the critical factor linking obesity with its complications. In obesity, the induction of systemic oxidative stress supported by several biochemical mechanisms including increase in Ca^{2+} level, superoxide generation from NADPH oxidases, oxidative phosphorylation, glyceraldehyde auto-oxidation, protein kinase C activation, and polyol and hexosamine pathways (Farooqui, 2013). Other factors that also contribute to oxidative stress in obesity include hyperleptinemia, low antioxidant defense, chronic inflammation, and postprandial ROS generation. The development of obesity also involves the interactions of gene and environment with food intake and physical activity. These interactions regulate energy balance (Pigeyre et al., 2016). Both central and peripheral signals are involved in the regulation of short- and long-term energy balance. Central signals start from brain regions inside and outside the hypothalamus presiding over cognitive processes, hedonic effects of food consumption, memory, and attention. Peripheral signals arise from each district of the body including the microbiome and cells within adipose tissue, stomach, pancreas, and other organs (Van der Klaauw and Farooqi, 2015). The disruption of these interactions counteracts variations of food intake and/or physical activity in order to keep body weight constant leads to the constitutional activation of orexigenic signals with a chronic positive energy balance leading to the deposition of energy oversupply in TAG form. During the development of obesity, the lipid deposition in adipose tissue can exceed the storage capacity of adipocytes, resulting in elevated circulating concentrations and inappropriate accumulation in multiple tissues, most notably liver and skeletal muscle. As stated above, insulin resistance and obesity are accompanied with the ectopic accumulation of reactive lipid species such as DAG, FFAs, free cholesterol, acylcarnitine, and ceramides. These metabolites have been reported to impair metabolism not only through the induction of local tissue inflammation, but also via the induction of ER stress (Virtue and Vidal-Puig, 2008; Symons and Abel, 2013; Contreras et al., 2014). In addition, genetic predisposition and several environmental factors (consumption of large portions of energy dense foods such as refined carbohydrates and high fat), intake of fructose-sweetened beverages, sedentary lifestyle, and induction of endocrine dysfunction such as hypothyroidism also contribute to the development of obesity (Farooqui, 2013). These factors not only induce increase in appetite, but also promote increase in body weight (Taheri et al., 2004; Thomas et al., 2007). Approximately 35% of adults (Flegal et al., 2012) and 17% of adolescents and children in the United States are obese (Ogden et al., 2012).

At the molecular level, obesity is accompanied by a diverse range of stress-responsive and counter-regulatory signaling pathways, including activation of JNK, IKKβ, IRE1, mTOR, ERKs, PKCθ, SOCS proteins, and PKR (Qatanani and Lazar, 2007; Samuel and Shulman, 2012; Hotamisligil, 2010; Nakamura et al., 2010). These pathways collaborate to produce two metabolically important effects: (1) Each pathway facilitates the inhibition of insulin signaling pathways, primarily through serine phosphorylation of IRS proteins, which not only blunts insulin action in stressed target tissues, but also stems the influx of nutrients into already overwhelmed cells. Moreover, the induction of two main inflammatory signaling pathways (JNK and IKKβ) initiates, supports, and augments an inflammatory response within metabolic tissues. (2) Furthermore, overnutrition in obese subjects can bypass cellular stress responses entirely and trigger inflammatory activation through a variety of mechanisms, including triggering of innate immune receptors (e.g., saturated fatty acids—fetuin A ligation of TLR4) (Pal et al., 2012), increased gut-derived lipopolysaccharide translocation, and intestinal dysbiosis (see below) (Tremaroli and Bäckhed, 2012).

As stated above, obesity is accompanied by marked increase in FFAs and induction of chronic inflammatory pathways. These inflammatory signaling pathways in obesity are causally linked to insulin resistance (Wellen and Hotamisligil, 2005). However, the fundamental mechanisms responsible for activating inflammatory pathways in obesity are poorly understood. It has been hypothesized that insulin resistance in obesity is caused by elevated levels of FFAs, which originate from adipose tissue. Indeed, FFA infusion in vivo has been shown to impair the ability of insulin to suppress hepatic glucose production and to stimulate glucose uptake into skeletal muscle (Boden et al., 2005), which in turn may lead to insulin resistance. Potential intracellular mechanisms by which FFAs contribute to insulin resistance may involve inflammatory signaling networks. For examples, intracellular kinases, which are linked to inflammatory signaling include PKCθ, IKKα, and JNK may play important roles in fatty acid—induced insulin resistance. This suggestion is based on the finding that *PKCθ*-, *IKKα*-, and *JNK*-knockout mice are substantially protected from FFA-induced insulin resistance (Hirosumi et al., 2002).

Obesity is also accompanied by high prevalence of comorbidities. Common comorbidities not only include cardiometabolic disorders but also mood and cognitive disorders. Obese subjects often show deficits in memory, learning, and executive functions compared to normal weight subjects (Agustí et al., 2018). Common comorbidities of obesity have different etiologies, but share common pathological mechanisms. Gut microbiota is a mediating factor between the environmental pressures (e.g., diet, lifestyle) and host physiology, and its alteration can partly explain the cross-link between those pathologies (Agustí et al., 2018). Thus the consumption of western diet and lack of exercise are major causes of the obesity epidemic. Experimental studies in animal models and some human studies suggest that microbiota may produce endocrine, neurochemical, and inflammatory alterations during development of obesity and its comorbidities. These alterations include dysregulation of the HPA-axis with overproduction of glucocorticoids, alterations in levels of neuroactive metabolites [e.g., neurotransmitters, short chain fatty acids (SCFAs)] and activation of a proinflammatory milieu that can cause neuroinflammation. In addition, neural networks, such as the enteric nervous system and vagus nerve also support the function of gut–brain axis by convey in information to the host. Converging evidence suggests that obesity epidemic is fueling a rise not only in type 2 diabetes, MetS, and heart disease (Eeg-Olofsson et al., 2009), but also in neurological disorders such as stroke, AD, and depression (Farooqui, 2013). It is proposed that lifestyle changes or use of some drugs can prevent or limit obesity would significantly impact the health of the nation as well as reduce the socioeconomic burden on society.

Obesity is measured by body mass index (BMI), a simple index of weight-for-height that is commonly used to classify underweight, overweight, and obesity in adult humans (ages of 18 and 65 years). BMI of 18.5–24.9 is considered as normal. Individuals below BMI of 18.5 are considered as underweight whereas individuals above BMI of 25 are considered as overweight (Fig. 1.8). Obesity ($BMI > 30$) not only increases the risk of many visceral and neurological diseases, but also enhances

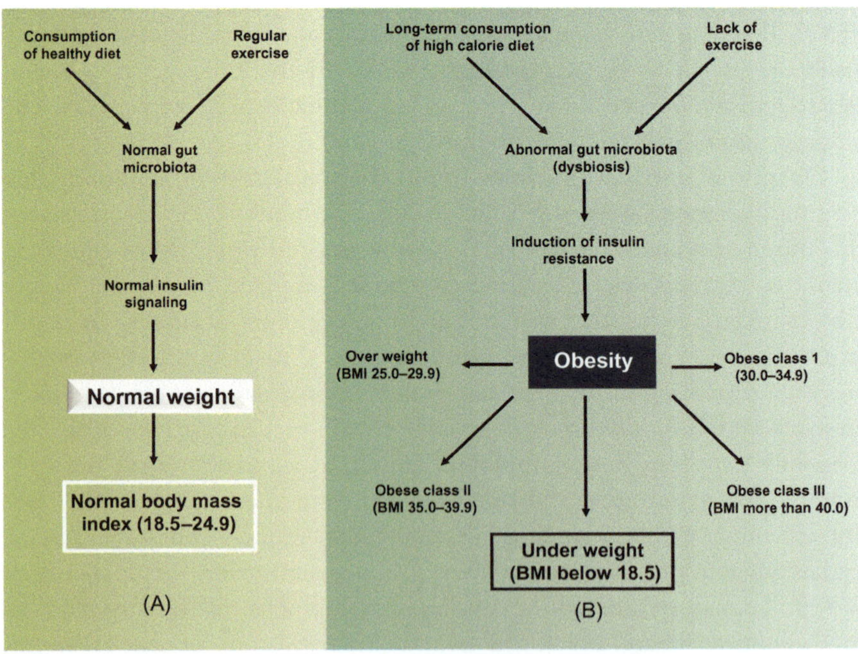

Figure 1.8 Relationship between body mass index and weight.

all-causes of mortality and reduces life expectancy (Fontaine et al., 2003). Caucasian individuals, who have a BMI > 40 between the ages of 20 and 29 years, are expected to decrease their life expectancy by approximately 6 and 12 years, respectively (Fontaine et al., 2003; Huffman and Barzilai, 2009). Obesity is closely linked with the etiology of type 2 diabetes mellitus. Earlier studies have shown that the relative risk for overweight ($25.0 \leq \text{BMI} \leq 29.9\ \text{kg/m}^2$) in adult humans contribute to the development type 2 diabetes mellitus. There is a 4.6-fold in women and 3.5-fold in men compared with their normal weight ($18.5 \leq \text{BMI} \leq 24.9\ \text{kg/m}^2$) the same sex peers (Field et al., 2001). A Japanese study indicates that an increase in BMI of $1\ \text{kg/m}^2$ (corresponding to a body weight gain of 2.4–2.9 kg) may increase the risk of diabetes by 25% (Nagaya et al., 2005). In addition, weight reduction achieved through a hypocaloric diet or bariatric surgery increases the probability of remission from type 2 diabetes (Lean et al., 2018). Furthermore, a lifestyle intervention has been shown to prevent the development of type 2 diabetes in overweight individuals at high risk for this outcome, with weight loss being the dominant predictor of reduced diabetes risk (Hamman et al., 2006). BMI is a commonly used indicator of obesity. It is associated with an unfavorable lipid profile consisting of elevated triglycerides, total cholesterol, and low-density lipoprotein cholesterol and low high-density lipoprotein cholesterol in men and women as young as 20 years of age (Denke et al., 1994). As for the association between BMI and longevity concerned, there is predominant evidence from longitudinal observational studies have indicated that the overweight or obese individuals with diabetes have lower mortality compared with normal weight individuals (Carnethon et al., 2014). In addition, plasma total cysteine (tCys) levels are consistently correlated with BMI, fat mass and odds of obesity in large human studies (Elshorbagy et al., 2009), and higher consumption of the cysteine precursor methionine is associated with increased BMI and prevalence of diabetes, MetS, and cardiovascular disease (El-Khairy et al., 2001; Giral et al., 2008). It is also indicated that high dietary cystine inhibits metabolic rate, lowers insulin sensitivity, and increases visceral fat deposition, in conjunction with changes in the expression of several genes involved in lipid and glucose metabolism. Cystine supplementation is known to enhance total fat mass, lean mass, and bone mineral content, with no effect on bone mineral density. Fasting plasma tCys and total glutathione are not increased, but levels of plasma taurine are elevated in the cystine-supplemented animals. Although the mechanism involved in sulfur amino acid–mediated adiposity in human is not fully understood. In vitro studies have indicated that in rat adipocytes cysteine not only inhibits lipolysis in a concentration-dependent manner, but high cystine intake lowers energy expenditure and decreases glucose tolerance, and upregulates lipogenic and diabetogenic enzymes (Elshorbagy et al., 2009).

Obesity is accompanied by inflammatory response, which not only involves systemic enhancement in circulating inflammatory cytokines, adipokines, and acute phase proteins (C-reactive protein, haptoglobin, and Aβ) and recruitment of WBCs to inflamed tissues, but also activation of tissue WBC and induction of reparative tissue responses along with accumulation of macrophages in adipose tissue. However, the nature of obesity-mediated inflammation is unique in comparison to other inflammatory paradigms of infections and autoimmune diseases (Alam et al., 2012). Several mechanisms may be associated with obesity-mediated inflammation in peripheral tissues and hypothalamus. These mechanisms include the activation of TLR4, induction of ER stress, and activation of serine/threonine kinases, such as IKKβ (Thaler and Schwartz, 2010; Thaler et al., 2012). Although the relative contribution made by these mechanisms remains unknown, but early onset of inflammation in hypothalamus relative to that in peripheral tissues suggests that different processes may cause inflammation in peripheral tissues and hypothalamus. Moreover, the nature of the hypothalamic inflammation occurring during the first days of high-fat diet (HFD) feeding may differ fundamentally from that involved with chronic HFD exposure. Based on several studies, it is proposed that the rapid onset of mediobasal hypothalamus inflammation is a manifestation of neuron injury and associated neuroprotective responses is consistent with previous evidence of apoptosis and glial ensheathment of hypothalamic arcuate nucleus neurons in animals exposed to rendered obese b chronic HFD feeding (Horvath et al., 2010; Moraes et al., 2009).

Obesity is linked with decrease in life expectancy due to the prevalence of cardiovascular diseases, diabetes, colon cancer, and other chronic neurological diseases. In addition, obesity also has negative impact on cognitive function due to vascular defects, impaired insulin metabolism and defect in glucose transport mechanisms in brain. Molecular mechanisms underlying cognitive dysfunction and early death in obesity are not fully understood. However, it is proposed that exaggerated metabolic demands trigger stress responses in adipocytes that may contribute to oxidative stress, chronic inflammation, cell injury, and death (Ozcan et al., 2004).

It is also indicated that the disequilibrium between oxygen demand and supply in expanding adipose tissues in obesity can lead to local hypoxia, which may cause adipocyte dysfunction by inducing alterations in adipokine expression, and promoting chronic inflammation in adipose tissue (Hosogai et al., 2007). Secretion of adipokines by adipose tissues promotes the production of ROS.

Thus many factors (genetic, environmental, socioeconomical, behavioral, lack of physical activity, and psychological factors) modulate the onset of obesity (Fig. 1.9), but the main cause of the development of obesity is a positive energy balance, which consists in imbalance between energy intake and expenditure, lasting for several years (Businaro et al., 2012). Such a balance is regulated by a complex network of signal transduction processes that connect the endocrine system with the brain signaling (Straub et al., 2011). As stated above, the diet composition regulates the population of microbiota in the intestine. The animal-based diet increases the abundance of bile-tolerant microorganisms and decreases the levels of Firmicutes. In contrast, plant-based diet increases that population of *Roseburia*, *Eubacterium rectale*, and *Ruminococcus bromii* suggesting that

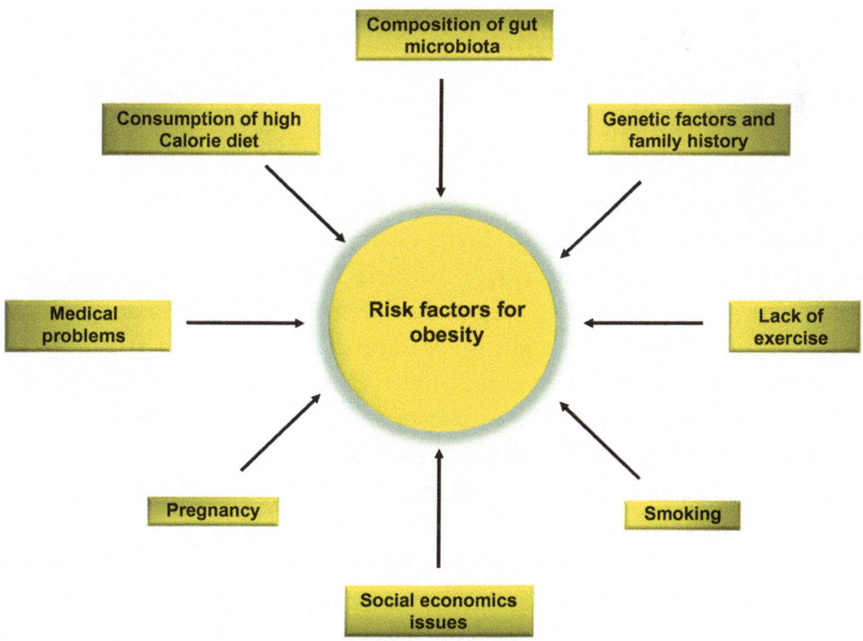

Figure 1.9 Factors effecting obesity.

herbivorous and carnivorous mammals can be distinguished on the basis of diet consumption.

Weight loss is recommended for overweight and obese individuals with type 2 diabetes. This recommendation is based on short-term studies showing numerous benefits of weight loss, including improvements in glycemic control, cardiovascular disease risk factors, quality of life, and other obesity-related comorbidities. However, it is unknown whether weight loss reduces the risk of cardiovascular morbidity and mortality in individuals with type 2 diabetes. Epidemiological studies in individuals with diabetes provide conflicting results, perhaps due to confounding from unintentional weight loss. A recent meta-analysis (Williamson, 1998) of cohort studies concluded that moderate intentional weight loss was associated with reduced mortality in "unhealthy" individuals, which included individuals with diabetes. The Swedish Obesity Subjects (SOS) study (ClinicalTrials.gov number NCT01479452) reported lower cardiovascular event rates over a mean follow-up of 13.3 years in patients with type 2 diabetes who underwent bariatric surgery (Romeo et al., 2012); however, this was a nonrandomized study, and the results achieved through surgery cannot be generalized to other approaches to weight loss.

Effects of diet on microbiota population

The human gastrointestinal tract harbors a complex and dynamic population of microorganisms called the gut microbiota. Adult human microbiota population consist of hundred trillion bacteria in the body, about 4 million distinct bacterial genes, with more than 95% of them located in the large intestine (Qin et al., 2010; Galland, 2014). It is suggested that about 10^{14} different populations of bacteria in the human intestinal tract, which are at least 100 times larger than the number of human genes in the body, and their total weight is approximately 2 kg (Picca et al., 2018). Since most of these genes encode for enzymes and structural proteins that influence the functioning of mammalian cells, the gut microbiome can be viewed as an anaerobic bioreactor programmed to synthesize molecules which direct the mammalian immune system (Hooper, 2004), modify the mammalian epigenome (Li et al., 2008), and regulate host metabolism (Jacobsen et al., 2013; Picca et al., 2018). Microbiota exert a marked influence on the host during homeostasis and disease. Multiple factors contribute to the establishment of the human gut microbiota during infancy. Diet is considered as one of the main drivers in shaping the gut microbiota across the lifetime. Intestinal bacteria play a crucial role in maintaining immune and metabolic homeostasis and protecting against pathogens. The gut microbiota offers many benefits to the host, through a range of physiological functions such as strengthening gut integrity or shaping the intestinal epithelium (Natividad and Verdu, 2013), harvesting energy (den Besten et al., 2013), protecting against pathogens (Bäumler and Sperandio, 2016), and regulating host immunity (Gensollen et al., 2016). The composition of the gut microbiota is also influenced by the environment, diet, medications, age, geographic factors, surgical interventions, and host genetics, particularly genes related to the immune system and metabolism (Yatsunenko et al., 2012; Dabrowska and Witkiewicz, 2016; Goodrich et al., 2016). Among above factors, diet plays a predominant role in modulating the composition of gut microbiota and promoting obesity-associated dysbiosis, parallel initiatives are required to elucidate dietary patterns and diet components (e.g., prebiotics, probiotics) that promote healthy gut microbiota (Stecher et al., 2013; Chassaing et al., 2017). The composition of diet regulates the population of microbiota in the intestine. Thus western diet, rich in calories and refined sugar, is associated with lower richness in microbial communities at individual level (alpha diversity) and higher variation among individuals (beta diversity) when compared with diets high in fiber and relatively low in calories (Martínez et al., 2015). Individuals who consume a western type diet with high-energy and high-fat intake present changes in metabolic and immune biomarkers, such as a higher BMI and higher levels of inflammatory markers than those who follow a high-fiber, low-calorie diet (Cani et al., 2009). Furthermore, the consumption of animal-based diet (western or HFD) also increases the abundance of bile-tolerant microorganisms (*Alistipes*, *Bilophila*, and *Bacteroides*) and decreases the levels of Firmicutes that metabolize dietary plant polysaccharides (*Roseburia*, *E. rectale*, and *R. bromii*) supporting the view that microbial composition mirrors differences between herbivorous and carnivorous mammals. Converging evidence suggests that the consumption of "Western" diet increases the level of unique class of bacteria called bacteroides. The molecular mechanisms associated with western diet–mediated detrimental effects on human metabolism are not fully understood. However, it is suggested that the ingestion of western diet results in development of inflammation not only in the hypothalamus, but also in the peripheral tissues including the liver, adipose tissue, skeletal muscle, and intestine (Guillemot-Legris et al., 2016). Concerning the development of chronic systemic inflammation, it is suggested that consumption of western diet results in alterations in the gut microbiota due to the direct effects of FFAs on intestinal cells, which may be the first step in the development of inflammation (Sanmiguel et al., 2015). This proposal is supported by the hypothesis that germ-free mice exhibit neither obesity nor upregulation of intestinal TNF-α level compared to conventionalized mice when fed a western diet, whereas reconstitution of the gut microbiota from obese mice in the germ-free mice produce an increase in body fat (Turnbaugh et al., 2009). In contrast, long-term consumption of diet enriched in plant-based food increases the level of *Prevotella* (Wu et al., 2011; Devkota et al., 2012; David et al., 2014; Hamaker and Tuncil, 2014; El Kaoutari et al., 2013). Indeed, regular ingestion of plant polysaccharides is integral for maintaining a healthy balance of microbes in our lower gastrointestinal tract (De Filippo et al., 2010; Sonnenburg and Sonnenburg, 2014). As stated above, members of the bacteroidetes, a dominant phylum in the human gut, possess an arsenal of Polysaccharide Utilization Loci (PUL) to target a wide range of complex glycans (El Kaoutari et al., 2013). Alterations in gut bacterial composition (dysbiosis) are associated with the pathogenesis of many inflammatory diseases and infections. Thus it is becoming increasingly evident that the composition of gut microbiota plays an important role in the development of obesity, obesity-associated inflammation, induction of insulin resistance, and onset of diabetes. Emerging evidence suggests the human gut microbiota may regulate weight gain through several interdependent pathways including energy harvesting, SCFAs signaling, behavior modifications, controlling satiety, and modulating inflammatory responses within the host supporting the view that gut microbiota may not only regulate obesity by promoting or regulating energy expenditure, but also by controlling many pathological conditions such as type 2 diabetes, obesity, heart disease, cancer, kidney disease, and inflammatory bowel disease (Matsuzawa-Nagata et al., 2008; Zhang and Zhang, 2013). Recent studies have also indicated that the gut microbiome interacts with the brain and regulates many brain functions through the maintenance of "gut—brain axis" homeostasis (Hamilton and Raybould, 2016), modulation of bidirectional neurohumoral communication, production of neuroactive molecules, and regulation of the circulating levels of some cytokines. It has been demonstrated that some microbiota and their metabolites may target the brain directly, via vagal stimulation, or indirectly through immune-neuroendocrine mechanisms, triggering the release of proinflammatory cytokines, which promote neuroinflammation (Agustí et al., 2018).

The other mechanism of interaction between microbiome and brain is through the BBB. BBB is a component of the neurovascular unit (NVU). The BBB is not just a physical barrier (due to the presence of specialized tight junctions and other changes that prevent unregulated leakage) but also acts more selectively as a transport interface (with specific transporters present on luminal and abluminal membranes), a secretory body, and a metabolic barrier (containing and releasing certain enzymes locally) (Abbott et al., 2006). BBB acts on the blood–brain interface, mediating communication between the brain and the periphery. The BBB separates the circulation from the brain, allowing for protection from and transport regulation of serum factors and neurotoxins. Microbiota and their release factors may enter into the systemic circulation from the gut. Once in the blood, the microbiota-derived factors can alter peripheral immune cells. This process may facilitate interactions with the BBB and ultimately with other elements of the NVU. On the basis of these studies, it is suggested that gut microbiota-derived factors can cross the BBB not only by altering BBB integrity, changing BBB transport rates, but also by inducing the release of neuroimmune substances from the BBB cells (Logsdon et al., 2018). Metabolic products produced by the microbiota, such as SCFAs, can cross the BBB to affect brain function (see below) (Logsdon et al., 2018). Converging evidence suggests that gut microbiota contribute to many host functions that impact the development and maintenance of the obese state not only by regulating host food intake behavior, energy harvest, energy expenditure, and fat storage (David et al., 2014). Diverse signaling pathways, which regulate gut permeability and microbiota's activities contribute to systemic inflammation ("metabolic endotoxemia") that is a hallmark of obesity and its complications (Chassaing et al., 2015; Turnbaugh et al., 2008; Stecher et al., 2013). There occurs biochemical cross-talk between host and microbiota. This cross-talk not only promotes the metabolic health, but also contributes to superorganism and whose dysregulation is a hallmark of the obese state. There are differences in microbiota community composition, their functional genes, and metabolic activities of the gut microbiota, which can distinguish lean humans from obese individuals, suggesting that gut "dysbiosis" contributes to the development of obesity and/or its complications. To this end, the major problem is to determine the relative importance of obesity-associated compositional and functional changes in the composition of microbiota and to identify the relevant taxa and functional gene, which modulate that promote leanness and metabolic health. Gut microbiota derives their energy by breaking down complex dietary macromolecules, synthesizing micronutrients, fermenting indigestible food substances (fiber), assisting in absorption of electrolytes, and growth and differentiation of the intestinal and colonic epithelium through regulating the diverse aspects of cellular differentiation and gene expression (Hooper et al., 2002; Gill et al., 2006).

Microbiota and short chain fatty acids

Gut microbiota produces a wide range of metabolites, including SCFAs. These SCFAs are absorbed in the large bowel and are defined as 1–6 carbon volatile fatty acids (acetate, propionate, butyrate, and other SCFAs) (Table 1.3). Their production is influenced by the pattern of food intake and diet-mediated changes in the gut microbiota. SCFAs have distinct physiological roles in host metabolism: they contribute to shaping the gut environment, influence the physiology of the colon, they can be used as energy sources by host cells and the intestinal microbiota and they also participate in different host-signaling mechanisms and are also used to

Table 1.3 Names of microbiota releasing short chain fatty acids.

Name	SCFA released	References
Eubacterium rectate	Butyric acid	den Besten et al. (2013) and Miyamoto et al. (2016)
Roseburia spp.		den Besten et al. (2013) and Miyamoto et al. (2016)
Faecalibacterium prausnitzil	Acetic acid	den Besten et al. (2013) and Miyamoto et al. (2016)
Ruminococcus bromii		den Besten et al. (2013) and Miyamoto et al. (2016)
Prevotella spp.	Propionic acid	den Besten et al. (2013) and Miyamoto et al. (2016)
Bifidobacterium spp.		den Besten et al. (2013) and Miyamoto et al. (2016)
Dialister spp.		den Besten et al. (2013) and Miyamoto et al. (2016)
Veillonella spp.		den Besten et al. (2013) and Miyamoto et al. (2016)

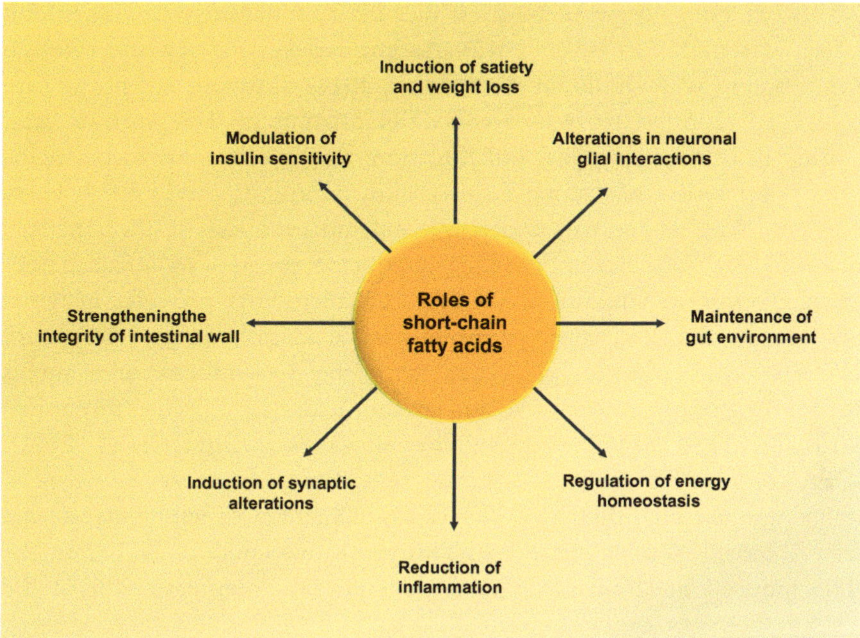

Figure 1.10 Roles of short chain fatty acids in peripheral tissues and the brain.

produce energy (Macfarlane and Macfarlane, 2011). In addition to above functions, SCFAs have multiple regulatory roles in energy homeostasis, insulin sensitivity, and glucose and lipid metabolism (Fig. 1.10) (Canfora et al., 2015).

Among SCFA, butyrate and acetate protect against diet-induced obesity without causing hypophagia, while propionate was shown to reduce food intake. The underlying mechanisms for these effects are unclear. However, it is suggested that SCFAs produce their effects on signal transduction processes not only by interacting with G-protein coupled receptors (FFAR2, FFAR3, OLFR78, GPR109A), but also by acting as epigenetic regulators of gene expression. This process involves the inhibition of histone deacetylase (HDAC) (Kasubuchi et al., 2015). Butyrate and propionate not only inhibit HDAC, but also alter the expression of specific genes via conformational changes in the active site of HDAC leading to its inactivation (Aoyama et al., 2010; Dashwood et al., 2006). In vitro studies have also shown that butyrate amplifies the antioxidant properties of glutathione-*S*-transferase (Ebert et al., 2003). Studies on administration SCFA in mice have indicated that butyrate, propionate, and acetate not only protect against diet-induced obesity, but also inhibit insulin resistance (Lin et al., 2012). Acetate reduces the appetite through its interaction with the brain (Frost et al., 2014). Butyrate and propionate have been reported to induce the differentiation of T-regulatory cells, assisting to control intestinal inflammation; this effect seems to be mediated via inhibition of histone deacetylation (Donohoe et al., 2014; Louis et al., 2014). This control of intestinal inflammation may result beneficial in terms of gut barrier maintenance, reducing the risk of inflammatory bowel disease. In addition to SCFA, gut microbiota provides enzymes involved in the utilization of nondigestible carbohydrates, cholesterol reduction, and enzymes associated with the biosynthesis of vitamins (K and B group), isoprenoids, and amino acids (e.g., lysine, threonine) (Fig. 1.11) (Hooper et al., 2002; Gill et al., 2006; De Preter et al., 2011). SCFA contribute to reduction in obesity and insulin resistance in experimental animals on HFD after dietary supplementation with butyrate has been observed (Gao et al., 2009). This protective effect of SCFA on the HFD-induced metabolic alterations depends on downregulation of the

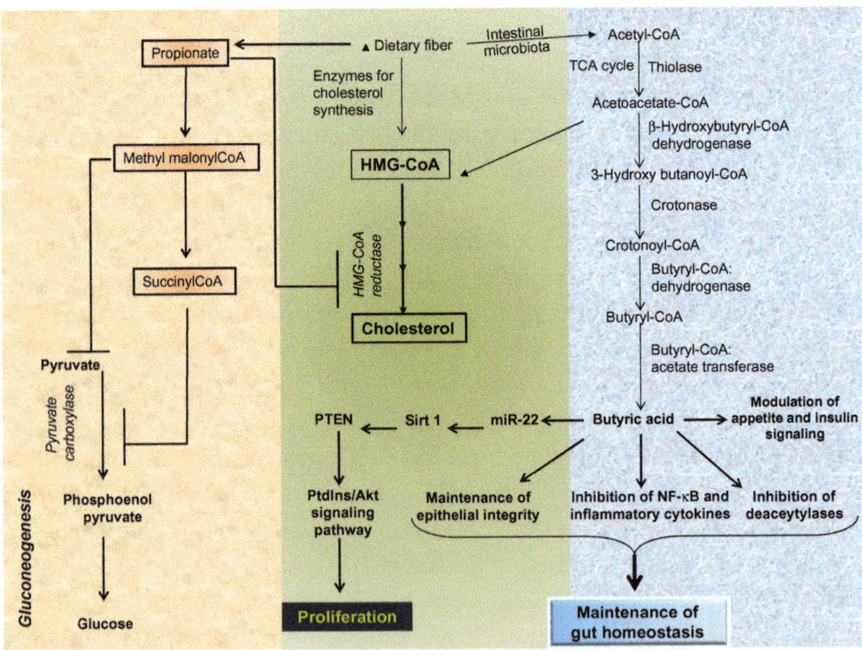

Figure 1.11 Generation of acetate, propionate, butyrate, and cholesterol from dietary fiber.

PPAR-γ, therefore promoting a change from lipid synthesis to lipids oxidation (den Besten et al., 2015). Several other mechanisms have also been proposed to explain the effects of SCFA on host tissues. These mechanisms may involve activation of protein kinases, such as AMP-activated protein kinase (AMPK; Gao et al., 2009; Peng et al., 2009; den Besten et al., 2015) or MAPK (Jung et al., 2015). Collective evidence suggests that SCFAs are important metabolites associated with production of energy by gut mucosal cells or transferred to the circulation to generate an important source of calorie and energy for the organism and to act as signaling molecules. Upon synthesis by the gut microbiota, both propionate and butyrate have local effects as the primary energy source in by gut mucosal cells (butyrate) and by activating intestinal gluconeogenesis (propionate) through distinct mechanisms (De Vadder et al., 2014; Donohoe et al., 2011). Distal effects of SCFAs are illustrated by propionate-mediated stimulation of liver gluconeogenesis, de novo lipid synthesis, and protein synthesis, whereas acetate is a precursor for cholesterol synthesis.

Gut microbiota are also known to cleave some dietary trimethylamine (TMA) containing compounds to produce TMA, which can be further oxidized as trimethylamine N oxide (TMAO) in the host liver by flavin monooxygenases (Wang et al., 2011; Koeth et al., 2013). In addition, gut microbiota also produces choline, phosphatidylcholine, carnitine, γ-butyrobetaine, betaine, crotonobetaine, and glycerophosphocholine (Koeth et al., 2013; Wang et al., 2016). Many studies have indicated that TMAO show the proatherogenic properties. Circulating TMAO level is associated with prevalence of cardiovascular disease and can independently predict incident risk for major adverse cardiac events, including myocardial infarction, stroke or death after adjustment for traditional cardiac risk factors and renal function (Wang et al., 2011; Tang et al., 2013). Molecular mechanisms by which how TMAOs promote atherosclerosis and thrombosis have been studied at the molecular level. TMAO activates vascular smooth muscle cell and endothelial cell MAPK, NF-κB signaling, leading to inflammatory gene expression and endothelial cell adhesion of leukocytes (Seldin et al., 2016). Meanwhile, TMAO can also activate the NLRP3 inflammasome (Sun et al., 2016a,b; Boini et al., 2017; Chen et al., 2017). In vivo, TMAOs have been reported to increase scavenger receptor, CD36 and SR-A1 expression, leading to more uptake of modified LDL for macrophage to form foam cell (Wang et al., 2011). On the other hand, TMAO decreases expression of two key enzymes, CYP7A1 and CYP27A1, essential for bile acid biosynthesis and multiple bile acid transporters (OATP1, OATP4, MRP2, and NTCP) in the liver, which decreases bile acid pool, resulting in decreased reverse cholesterol efflux (Koeth et al., 2013). Moreover, TMAO increases endoplasmic recticulum calcium release in platelet cell, consequently leading to platelet aggregation and thrombosis (Zhu et al., 2016). Decreasing the consumption of TMA precursors in diet and inhibition of TMAO production has been used to protect from harmful effects of microbiota.

Effects of gut microbiota and obesity on the brain

The hypothalamus, a key processing region in the brain, regulates food intake through the production of regulatory proteins such as leptin, cholecystokinin (CCK), ghrelin, orexin (ORX), insulin, and NPY (Dietrich and Horvath, 2009; Blouet and Schwartz, 2010). Peripheral signals from adipose stores, the gastrointestinal tract, and endocrine system communicate and modulate neurons within the arcuate nucleus of the hypothalamus. When fat stores are reduced and energy levels are low, hunger signals are induced through elevation in the ghrelin, and decrease in levels of insulin, glucose, leptin, and CCK. These changes increase the activity of NPY and AgRP neurons, which in turn produce reduction in the melanocortin system, producing an increase in melanin-concentrating hormone and ORX signaling causing a marked orexigenic effect. High levels of glucose, insulin, CCK, and reduction in levels of ghrelin result in an increase in POMC. POMC increases α-melanocyte-stimulating hormone (α-MSH) leading to termination of feeding signals (Coll et al., 2007; Valassi et al., 2008; Abizaid and Horvath, 2008; Schwartz et al., 2000). Thus insulin and leptin control food intake via regulating POMC and AgRP expression. FOXO1 is a downstream effecter of insulin signaling and Sirt1 is an NAD^+-dependent deacetylase, both are closely associated with the regulation of metabolism not only in liver, pancreas, muscle, and adipose tissue, but also in the hypothalamus (Sasaki and Kitamura, 2010). Both FOXO1 and SIRT1 are expressed in AgRP and POMC neurons (Fig. 1.12). In fasted rats, FOXO1 is localized in the nucleus, whereas in refed rats, it is localized in the cytoplasm. Unlike FOXO1, hypothalamic SIRT1 protein is decreased during fasting due to increase in ubiquitination of SIRT1. In rodents, overexpression of FOXO1 in the hypothalamus through the microinjection of adenovirus produces hyperphagia and body weight gain, and simultaneous overexpression of SIRT1 suppresses

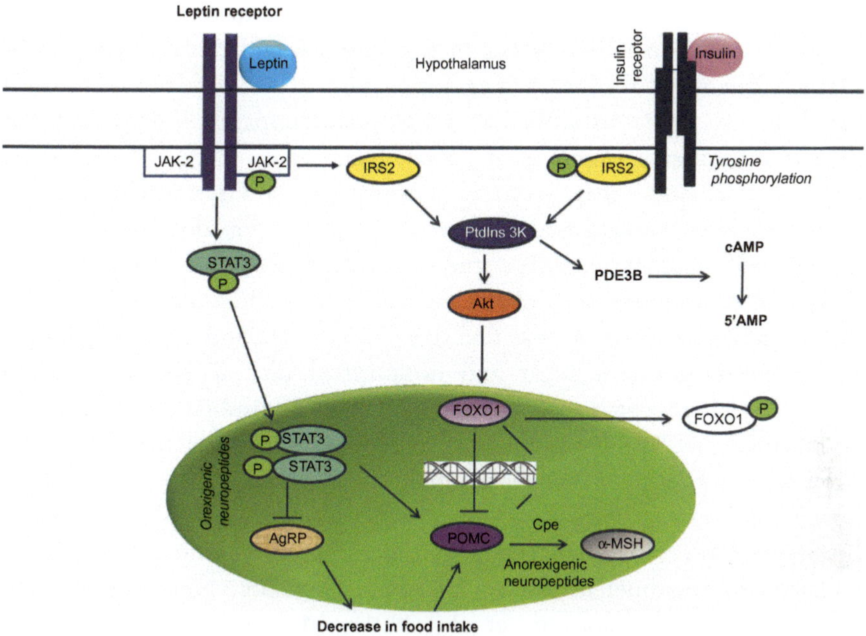

Figure 1.12 Modulation of appetite by insulin and leptin signaling in the hypothalamus. *Cpe*, carboxypeptidase; *IRS-2*, insulin receptor substrate 2; *JAK2*, Janus kinase 2; *α-MSH*, α-melanocyte-stimulating hormone; *POMC*, pro-opiomelanocorticotropin; *PtdIns 3K*, phosphatidyl-inositol 3 kinase; *STAT3*, signal transducer and activator of transcription 3.

these processes (Sasaki and Kitamura, 2010). Collective evidence suggests that FOXO1 and the transcription factor STAT3 exert opposing effects on the expression of AgRP and POMC through transcriptional squelching, and SIRT1 suppresses AgRP expression (Sasaki and Kitamura, 2010). These studies support the view that food intake is regulated by a complex signaling network associated with hypothalamic structures in the brain (Williams et al., 2011).

The macronutrient components of HFD, which promote obesity, may also impact neurotransmitter signaling in the brain. Thus high-fat obesogenic diet reduces dopamine levels in the nucleus accumbens, as well as shift the reactivity of the meso-corticolimbic system in such a manner that more palatable diet is required to achieve similar food-induced increases in extracellular dopamine as seen in chow-fed controls (Archer and Mercer, 2007; Geiger et al., 2008; Volkow et al., 2011). Certain foods, particularly those rich in sugars and fat, are potent reward inducers that promote eating (even in the absence of an energetic requirement) and trigger learned associations between the stimulus and the reward (conditioning) (Lenoir et al., 2007). The stimulation of orosensory system with either sucrose or fat is sufficient to stimulate dopamine release in the nucleus accumbens (Hajnal et al., 2004; Liang et al., 2006). It should be noted that fat and sugars seem to modulate reward systems differently. Sugar produces addictive-like behaviors whereas fat does not (Avena et al., 2009). Autoradiographic studies reveal an increase in D_1 receptor binding in the nucleus accumbens and decrease in D_2 receptor binding in the striatum relative to nonpurified diet-fed rats (Colantuoni et al., 2001). Rats with intermittent sugar and nonpurified diet access also show a decrease in D_2 receptor mRNA in the nucleus accumbens, and increased D_3 receptor mRNA in the nucleus accumbens and dorsal striatum compared with nonpurified diet-fed controls (Spangler et al., 2004). Similarly, sugar-bingeing rats show a significant decrease in enkephalin mRNA (Spangler et al., 2004), whereas μ-opioid receptor binding is significantly enhanced in the nucleus accumbens shell, cingulate, hippocampus, and locus coeruleus (Colantuoni et al., 2001). Maintaining a healthy gut microbiota state is necessary to support the metabolic activities of the brain (Goyal et al., 2015; Subramanian et al., 2015). Recent research has also linked microbial dysbiosis to neurological disorders, such as PD, and AD, multiple sclerosis (MS), and autism. The CNS connects with the gut via sympathetic and parasympathetic nerves. However, the importance of these connections has not been studied in depth.

Adipose tissues are considered independent factors for the generation of systemic oxidative stress. There are several mechanisms by which obesity produces oxidative stress. The first the oxidation of fatty acids by mitochondria and peroxisomes can produce ROS. Another mechanism for ROS generation involves the overconsumption of oxygen, which generates free radicals in the mitochondrial respiratory chain that is coupled with oxidative phosphorylation in mitochondria (Esposito et al., 2006). Lipid and carbohydrate-rich diets may also contribute to the generation of ROS because they can alter oxygen metabolism.

Finally, increase of adipose tissue is accompanied by diminished activities of antioxidant enzymes such as SOD, CAT, and GPx (Chrysohoou et al., 2007). In obesity, the immune system produces both superoxide and nitric oxide, which may react together to produce significant amounts of peroxynitrite anion ($ONOO^-$). Peroxynitrite is a potent oxidizing agent that can cause DNA fragmentation and lipid peroxidation (Pacher et al., 2007).

Many studies have demonstrated the formation of peroxynitrite in diabetic vasculature, both in experimental models and in humans (Pacher and Szabo, 2008; Molnar et al., 2006). Peroxynitrite induces diabetic complications and vascular alterations through activation of the nuclear enzyme poly(ADP-ribose) polymerases (PARP enzymes). Activated PARP-1 cleaves NAD^+ into nicotinamide and ADP-ribose and polymerizes the latter on nuclear acceptor proteins. Peroxynitrite-mediated overactivation of PARP utilizes NAD^+ and consequently ATP culminating in cell dysfunction, apoptosis, or necrosis (Virag and Szabo, 2002). It is also suggested that peroxynitrite produces it toxic effect through the cytotoxic effects of high glucose (Du et al., 2001).

Studies on oxidative markers in obesity subjects indicate that oxidative damage is associated with increased BMI and percentage of body fat (Vincent and Taylor, 2006). Conversely, parameters of antioxidant capacity are inversely related to the amount of body fat and central obesity (Hartwich et al., 2007). Thus mounting evidence suggests that possible mechanisms of obesity-related oxidative stress include increase in oxygen consumption and subsequent production of ROS-derived from the increase in mitochondrial respiration, diminished antioxidant capacity, fatty acid oxidation, lipid peroxidation, and cell injury producing increase in rates of ROS production (Vincent and Taylor, 2006; Brown et al., 2009). The increase in obesity-mediated ROS production is due to the induction of proinflammatory cytokines (TNF-α, IL-1, and IL-6), which can lead to further increase in ROS production. In addition, obesity also stimulates NADPH oxidase activity, which contributes to ROS production (Bastard et al., 2006). High concentrations of ROS modulate platelet function through different mechanisms, including reduction in NO bioavailability, calcium mobilization abnormalities and overexpression of membrane glycoproteins (Krötz et al., 2004). Based on these observations, it is proposed that obesity can lead to stressful proinflammatory state that increases the risk of diabetes along with cardiovascular and neurological diseases decreasing life expectancy (Altintas et al., 2011; Farooqui, 2012). These suggestions are also supported by the presence of shorter telomeres (240 bp) in obese women compared to lean women of similar age (Valdes et al., 2005) supporting the view that shorter telomeres can result in shorter age.

Adipose tissue is an active endocrine organ, which secretes numerous bioactive molecules, collectively known as adipocytokines (leptin, adiponectin, and resistin), proinflammatory cytokines (TNF-α, IL-1β, and

IL-6), and chemokines (MCP-1). These molecules not only regulate carbohydrate and lipid metabolism, but also modulate immune function, blood coagulability, and serve as blood biomarkers of cardiometabolic risk. Adepocytokines also regulate metabolic and inflammatory processes in an autocrine and paracrine manner (Ahima et al., 2000). Adipocytokines, cytokines, and chemokines contribute to low-grade inflammation and initial adipose macrophage infiltration, which promote impairment in adipocyte insulin signaling and induction of insulin resistance through the interference of insulin/insulin-like growth factor I receptor (IGF-IR) signaling pathways (Jiao et al., 2009). Normally, insulin/IGF-IR signaling is accompanied by Tyr phosphorylation of IRS followed by activation of two major downstream pathways: the PtdIns 3K–Akt pathway largely responsible for insulin-mediated glucose uptake and suppression of gluconeogenesis. The other major pathway involves MAPK signaling. This pathway regulates gene expression and is closely associated with cell growth and differentiation.

Link among insulin signaling, obesity, and insulin resistance

A metabolic consequence of obesity (especially central adiposity) produces insulin resistance, which is mediated by an increase in the secretion of insulin from the pancreas. Increased insulin production leads to compensatory hyperinsulinemia, which may contribute to insulin resistance (Johansen et al., 2010). The development of insulin resistance is one of the earliest negative effects of obesity, and is associated with the early alterations in glucose metabolism, chronic inflammation, oxidative stress and decreased levels of adipose hormone adiponectin and PPAR-γ, key regulators for adipogenesis (Jazet et al., 2003). Accumulating evidence supports the view that a major cause of obesity is insulin resistance. As stated above, the release of FFAs and accumulation of lipids are critical mechanistic links between obesity and insulin resistance (Cornier et al., 2008). Under normal metabolism, insulin-mediated inhibition of adipocyte hormone-sensitive lipase activity reduces the release of FFAs from adipose tissue. This process is disrupted in obese and insulin-resistant individuals, leading to persistent FFA elevations (Craft, 2009). Normalizing FFA levels cause 50% increase in insulin sensitivity in obese adults (Kivipelto and Solomon, 2008). FFAs also produce defect in insulin-stimulated glucose transport and/or phosphorylation that is caused by a defect in insulin signaling (Boden and Chen, 1995). Plasma FFAs can easily enter cells, where they are oxidized to generate energy in the form of ATP or are re-esterified for storage as triglycerides. Furthermore, FFAs may also interfere with insulin-mediated stimulation of glucose transport by modulating glucose transporter gene transcription and mRNA stability (Armoni et al., 2005). The metabolic relationship between FFA elevations and development of type 2 diabetes is supported by the observation that normoglycemic individuals with a family history of diabetes show high fasting FFA levels (Whitmer et al., 2005).

Obesity is also accompanied by increase in systemic and local levels of proinflammatory cytokines. These cytokines are released from cells of innate immune system (macrophages, mast cells, and neutrophils). Cytokines not only accumulate in the adipose tissue, but also stimulate the release of fatty acids through the stimulation of lipases and phospholipases. In addition, cells of adaptive immunity (regulatory T cells, $CD8^+$ T cells, and natural killer T cells) also contribute to adipose tissue inflammation in obesity. Studies on the role of adiponectin and leptin provide an important link between obesity and insulin resistance (Tilg and Moschen, 2006). The onset of the insulin resistant is not only accompanied by weight gain, but also alterations in levels of adiponectin, leptin, and other proinflammatory cytokines (Muoio and Newguard, 2005).

Contribution of leptin in the development of obesity

Leptin is a prototypical adipokine, which has 167-amino acid with a four-helix bundle motif (Brennan and Mantzoros, 2006; Dardeno et al., 2010). Leptin is primarily synthesized and secreted by WATs. It acts in several peripheral tissues as well as the brain. Leptin plays important roles in the regulation of food intake, energy expenditure, metabolism, improves insulin sensitivity, neuroendocrine axis, and immune function (Dardeno et al., 2010). The binding of leptin to its specific receptor in the brain leads to activation of multiple signal transduction pathways. A number of intracellular signaling pathways are involved in modulating leptin's central and peripheral effects. Levels of leptin in circulation correlate with both the BMI and total amount of body fat (Frederich et al., 1995). Levels of leptin fluctuate according to changes in calorie intake with a marked decrease during starvation and an increase in overfed and obese states (Dalamaga et al., 2013). Leptin levels are increased by insulin, glucocorticoids, and proinflammatory cytokines and decreased by catecholamines (Ahima and

Osei, 2004). Leptin produces its effects by interacting with LepRs, which are expressed in the hippocampus, neocortex, hypothalamic and brainstem autonomic nuclei, choroid plexus as well as peripheral tissues. In choroid plexus, LepRs play a role in leptin uptake or efflux from the cerebrospinal fluid and in receptor-mediated transport of leptin across the BBB into the brain (Oral et al., 2002). Various alternatively spliced isoforms of LepR have been described, but the long isoform of leptin receptor (LepRb) is primarily responsible for leptin signaling. The binding of leptin to LepRb activates a number of signaling pathways, JAK2/STAT 3 and STAT5, IRS/PtdIns 3K, SHP2/MAPK, and AMPK/acetylCoA carboxylase (ACC) (Fig. 1.12). The leptin signaling cascade is terminated by the induction of a SOCS3. SOCS3 inhibits JAK2/STAT3 signaling, providing a negative feedback mechanism. In addition, PTP1B is implicated in the negative regulation of leptin signaling LepRb is strongly expressed in the hypothalamus and other areas of the brain, where it regulates energy homeostasis, hedonic regulation of feeding, neuroendocrine function, and memory and learning (Fig. 1.13) (Tanaka et al., 2010, 2011). Leptin-deficient ob/ob mice and LepRb-deficient db/db mice develop hyperphagia, morbid obesity, infertility, and reduced linear growth (Kelesidis

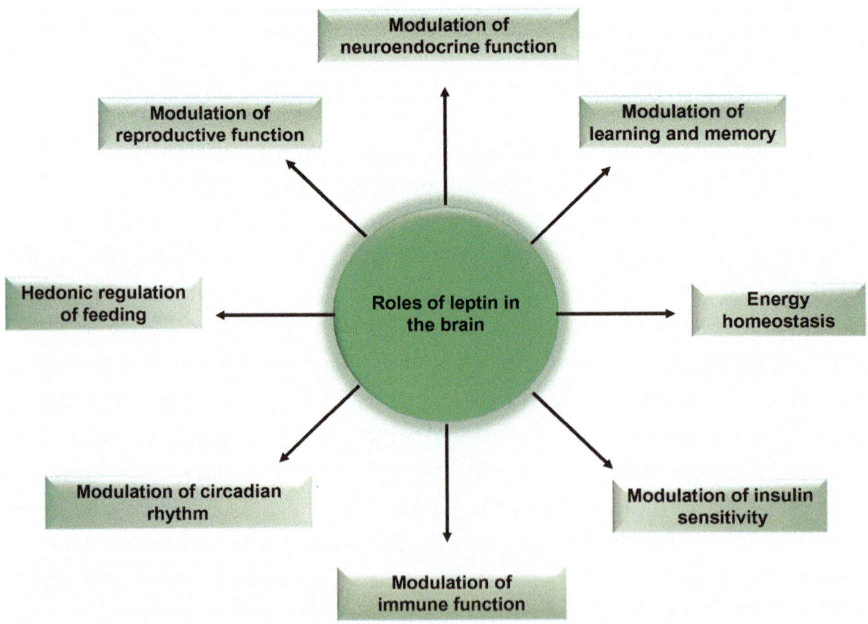

Figure 1.13 Roles of leptin in the brain.

et al., 2010). Congenital leptin deficiency or loss-of-function mutations of the LepR in humans also result in hyperphagia and morbid obesity. Collective evidence suggests that leptin suppresses appetite and increases the metabolic rate, mainly through hypothalamic orexigenic and anorexigenic factors, and it also prevents excessive weight gain and the accumulation of fat (Dietrich and Horvath, 2013; Stem et al., 2016). Leptin also plays an important role in maintaining circadian rhythm. The binding of leptin to its receptors in the brain stimulates the production of POMC. The two products of POMC are α-MSH and adrenocorticotropin. α-MSH binds to melanocortin-4 receptors in the hypothalamic paraventricular nucleus that cause a decrease in food intake.

Contribution of adiponectin in obesity

Adiponectin, an adipocyte-derived 30 kDa secretory protein, which consists of a signal sequence followed by a nonconserved N-terminal domain, 22 collagen repeats, and a C-terminal globular domain (gAd) which is structurally related to TNF-α (Hug and Lodish, 2005). Adiponectin is the most abundant plasma adipokine, which plays an important role not only in the regulation of glucose and lipid metabolism but also has ability to produce insulin-sensitizing, antiinflammatory, angiogenic, and vasodilatory effects in visceral tissues. Levels of adiponectin are decreased in obesity. Furthermore, levels of adiponectin are explicitly correlated with fat cell size and are found to be negatively related to BMI (Johnson et al., 2001). The decrease in adiponectin levels is modulated by interactions between genetic factors and environmental factors causing obesity and leading to the development of insulin resistance, type 2 diabetes, MetS, and atherosclerosis (Kadowaki et al., 2008). Adiponectin has ability to induce insulin-sensitizing, antiinflammatory, angiogenic, and vasodilatory properties in visceral organs and the brain. Adiponectin produces these effects by binding with its receptors (AdipoR1 and AdipoR2) (Bloemer et al., 2018). The interactions between adiponectin and its receptors influence energy homeostasis.

The binding of adiponectin with AdipoR1 and R2 initiates a series of tissue-dependent signal transduction processes, including phosphorylation of AMPK and p38 MAPK, and increase in PPARα ligand activity. These signal transduction processes are modulated by adaptor protein containing a pleckstrin homology domain, phosphotyrosine binding domain, and leucine zipper motif (APPL1). The adaptor binds directly to the intracellular regions of AdipoR1 and R2. It is interesting to note that AdipoR1 and R2 also contain inherent ceramidase activity, which modulate levels of intracellular ceramide, a sphingolipid that has been implicated in insulin resistance, cell death, inflammation, and atherosclerosis (Mao et al., 2006; Cheng et al., 2007). Adiponectin regulates lipid and glucose metabolism through AMPK phosphorylation (Kadowaki and Yamauchi, 2005). Activated AMPK stimulates phosphorylation of ACC (key enzyme in fatty acids de novo synthesis) as well as fatty acid oxidation and glucose uptake in muscle. Moreover, AMPK inhibits enzymes involved in gluconeogenesis in liver, which leads to a reduction of plasma glucose concentration (Kadowaki and Yamauchi, 2005). In addition, adiponectin stimulates fatty acid oxidation in skeletal muscle and inhibits glucose production in the liver, resulting in an improvement in whole-body energy homeostasis. Insulin resistance and obesity are accompanied by the induction of moderate inflammation, in which adipocytes and immune cells residing in adipose tissue contribute by increasing the level of circulating proinflammatory cytokines. One such cytokine, which contribute to inflammatory reaction, is TNF-α (Hotamisligil et al., 1995). TNF-α has been shown to inhibit insulin-stimulated tyrosine kinase activity of the IR as well as IRS-1, by inducing a serine phosphorylation of IRS-1 (Hotamisligil et al., 1994). Adiponectin also acts as classic antiinflammatory agent, reducing inflammation in various cell types through AdipoR1 and R2 signaling mechanisms. Adiponectin's antiinflammatory and antiapoptotic properties results in protection of the vasculature, heart, lung, and colon (Arita et al., 2002; Ouchi et al., 1999, 2001; Furukawa et al., 2004).

Regular intake of docosahexaenoic acid (DHA) increases the expression of adiponectin in adipose tissues contributing to the insulin-sensitizing and antisteatotic effects of DHA. The increase in adiponectin is also accompanied by DHA-mediated phosphorylation of AMPK, a fuel-sensing enzyme downstream the adiponectin receptor that acts as a gatekeeper of the systemic energy balance by modulating glucose and lipid homeostasis in adipose, liver, and muscle tissues (Long and Zierath, 2006). In addition, adiponectin contributes to insulin sensitivity through the involvement of AMPK-dependent PPARγ activation (Nawrocki et al., 2006). Adiponectin also increases fat oxidation resulting in reduced circulating fatty acid levels and reduction in intramyocellular or liver TAG content. In addition to its insulin-sensitizing actions, adiponectin has central actions in the regulation of energy homeostasis. In addition, adiponectin inhibits endothelial cell proliferation and migration and possesses antiangiogenesis and antitumor activity (Bråkenhielm et al., 2004). Adiponectin also decreases gluconeogenesis, increases glucose uptake, and stimulates β-oxidation and TAG clearance (Yamauchi et al., 2003). Emerging evidence suggests that adiponectin is protective against vascular dysfunction caused by obesity, through its multiple favorable effects on glucose and lipid metabolism as well as on vascular function. Adiponectin not only improves insulin sensitivity but also induces metabolic profiles. These processes decrease risk of heart disease. In addition, adiponectin protects the vasculature through its pleiotropic actions on endothelial cells, endothelial progenitor cells, smooth muscle cells, and macrophages (Li et al., 2011). It is proposed that adiponectin is an important component of the adipo-vascular axis that mediates the cross-talk between adipose tissue and vasculature. Human cross-sectional studies have indicated that plasma adiponectin levels are negatively correlated with obesity (Hotta et al., 2000), waist to hip ratio (Nakamura et al., 2004), insulin resistance (Hotta et al., 2000), dyslipidemia (Rothenbacher et al., 2005), diabetes (Xydakis et al., 2004), and cardiovascular disease (Bjorbaek and Kahn, 2004). A low plasma adiponectin level is observed in the MetS (Considine et al., 1996) and is considered as an independent risk factor for future development of type 2 diabetes (Sader et al., 2003).

Insulin signaling obesity and neurological disorders

Very little information is available on the direct link among insulin resistance, obesity, and neurological disorders. However, it is becoming increasingly evident that obesity is accompanied by insulin resistance, dyslipidemia, metabolic dysfunction, oxidative stress, and neuroinflammation (Fig. 1.14) (Farooqui and Farooqui, 2018). These processes are associated with the development of variety of alterations in brain activity and function. Recent evidence suggests that insulin resistance, dyslipidemia, metabolic dysfunction, oxidative stress, and neuroinflammation are not only risk factors for obesity, but also contribute to the pathogenesis of stroke, AD, and PD (Farooqui, 2013; Chen et al., 2014; Cai et al., 2015). Induction of above processes and related metabolic changes can alter the synaptic plasticity, which may lead to neurodegeneration, either by apoptosis or cell necrosis and induction of cognitive dysfunction (Fig. 1.14) (Bhat et al., 2017).

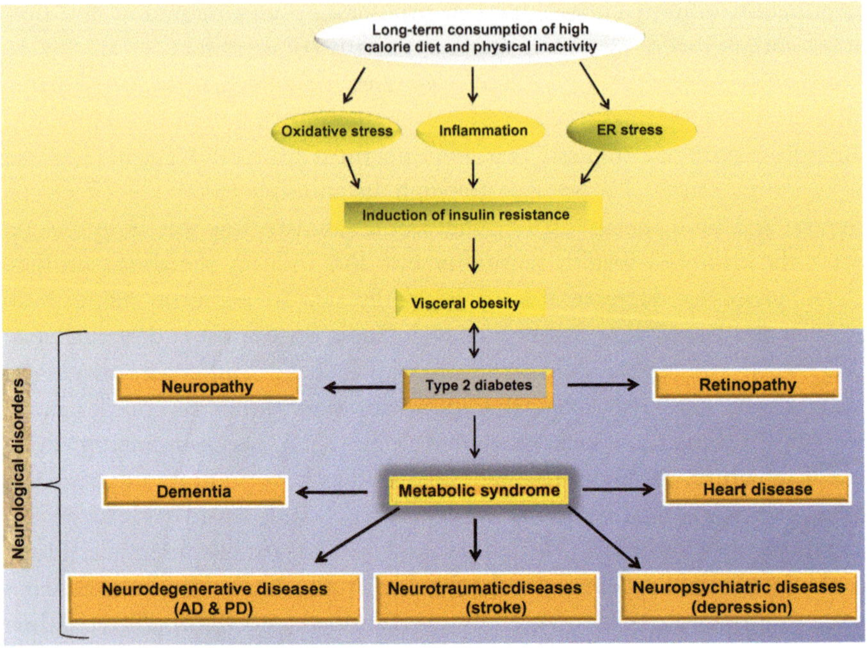

Figure 1.14 Diagram showing relationship among high calorie diet consumption, insulin resistance, and neurological disorders.

At present, it is difficult to relate obesity with impairment in cognitive function at the molecular level; however, collective evidence suggests that obesity is closely associated not only with a variety of cardiovascular and cerebrovascular risk factors but also with long-term cognitive performance. In addition, lower cognitive abilities are a risk factor for obesity. Converging evidence suggests that cognitive performance may influence the pathogenesis of obesity and being overweight may induce the development of cognitive impairment. Obesity and diabetes increase the risk of dementia, a condition of acquired cognitive defects sufficient to interfere with social or occupational functioning. It affects memory, judgment, speech, comprehension, execution, orientation, and learning (Sosa-Ortiz et al., 2012; Farooqui, 2019). Dementia is a major cause of disability, which is characterized by impairment in memory and activities of daily living, altered behavior, personality, and other cognitive dysfunctions (Sosa-Ortiz et al., 2012; Farooqui, 2019). Dementia mainly affects older people: only 2% of cases start before the age of 65 years. After this the prevalence doubles with every 5-year increment in age. Dementia is one of the major causes of disability in later life. Several types of dementia have been reported to occur in human population including Alzheimer type of dementia (30%), vascular dementia (26%), mixed dementia (21%), Lewy body dementia (11%), frontotemporal dementia/degeneration (7%), and infective dementia (5%). Secondary causes of dementia include vascular, CNS infections, trauma, metabolic derangements, and other reversible/treatable causes such as type 2 diabetes, stroke, AIDS, or MS (Kabasakalian and Finney, 2009; Farooqui, 2019).

Conclusion

Obesity is a multifactorial condition which is a result of the interaction of host and environmental factors and its prevalence in high income and upper middle-income countries is more than double that of low and lower middle-income countries. Obesity is controlled by several factors including the dysfunction of feeding center in the hypothalamus, imbalance in energy intake and expenditure, and genetic variations. Obesity is accompanied by chronic low-grade inflammation and insulin resistance, which is defined as a decrease in tissue response to insulin stimulation thus insulin resistance. At the molecular level insulin resistance is characterized by defects in uptake and oxidation of glucose, a decrease in glycogen synthesis, to a lesser extent, the ability to suppress lipid oxidation, and the activation of both JNK and IKK pathways leading to disruption of serine phosphorylation of IRS-1, a protein that connects the IR to the PtdIns 3K signaling cascade. In parallel to the activation of these kinases and their downstream signaling cascades is accompanied by alterations in expression and levels of adipokines (TNF-α, IL-6, and MCP-1) and inflammatory cytokines, and decrease in antiinflammatory factors.

Obesity and insulin resistance also involve a variety of adipocytokines (leptin, adiponectin, and expression of different adipocytokines) in peripheral tissues and the brain. Among adipocytokines, leptin not only acts on key neuronal circuits within the brain to reduce food intake and limit body adiposity, but also enhances peripheral glucose metabolism. Leptin increases thermogenesis, hepatic lipid oxidation, and lipolysis in adipocytes and skeletal muscle. Obese individuals typically have high levels of leptin; however, they seem to be resistant to this hormone's action. The absence of leptin leads to obesity, indicating that leptin plays an essential role in the maintenance of energy homeostasis. In contrast, adiponectin is an antiatherosclerotic and insulin-sensitizing protein. It inhibits hepatic glucose production, increases the uptake of glucose by muscle and the oxidation of fatty acids by muscle and liver cells, and increases energy expenditure. Adiponectin also seems to have a protective effect against insulin resistance, atherogenesis, and proinflammatory status. Adiponectin not only produces antiinflammatory effects but also reduces obesity by regulating food intake and therefore exerting a direct effect on energy balance and weight control. Other cytokines and chemokines contribute to low-grade inflammation and initial adipose macrophage infiltration, which promote impairment in adipocyte insulin signaling and induction of insulin resistance through the interference of insulin/IGF-IR signaling pathways. Collective evidence suggests that obesity is associated with an enormous medical, social, and economic burden. Obesity contributes to the metabolic dysfunction, dyslipidemia, and inflammation caused by insulin resistance, a process, which contributes to the development of a wide variety of visceral and neurological disorders. In the CNS, mild cognitive impairment can be attributed to obesity-induced alterations in hippocampal structure and function in some patients. Likewise, compromised hypothalamic function and subsequent defects in maintaining whole-body energy balance might be early events that contribute to weight gain and obesity development. In the peripheral nervous system, obesity-driven alterations in the autonomic nervous system prompt imbalances in sympathetic—parasympathetic activity, while alterations in the sensory-somatic nervous system underlie peripheral polyneuropathy, a common complication of diabetes. Pharmacotherapy and bariatric surgery are promising interventions for people with obesity that can improve neurological function. However, lifestyle interventions via dietary changes and exercise are the preferred approach to combat obesity and reduce its associated health risks: https://www.nutritionaction.com/daily/diet-and-weight-loss/extra-pounds-means-extra-cancer-risk/.

References

Abbott, N.J., Ronnback, L., Hansson, E., 2006. Astrocyte-endothelial interactions at the blood-brain barrier. Nat. Rev. Neurosci. 7, 41—53.

Abizaid, A., Horvath, T.L., 2008. Brain circuits regulating energy homeostasis. Regul. Pept. 149, 3—10.

Adams II, J.M., Pratipanawatr, T., Berria, R., Wang, E., DeFronzo, R.A., Sullards, M.C., et al., 2004. Ceramide content is increased in skeletal muscle from obese insulin-resistant humans. Diabetes 53, 25—31.

Adams, S.H., Hoppel, C.L., Lok, K.H., Zhao, L., Wong, S.W., Minkler, P.E., et al., 2009. Plasma acylcarnitine profiles suggest incomplete long-chain fatty acid beta-oxidation and altered tricarboxylic acid cycle activity in type 2 diabetic African-American women. J. Nutr. 139, 1073–1081.

Aguer, C., McCoin, C.S., Knotts, T.A., Thrush, A.B., Ono-Moore, K., McPherson, R., et al., 2015. Acylcarnitines: potential implications for skeletal muscle insulin resistance. FASEB J. 29, 336–345.

Agustí, A., García-Pardo, M.P., López-Almela, I., Campillo, I., Maes, M., Romaní-Pérez, M., et al., 2018. Interplay between the gut-brain axis, obesity and cognitive function. Front. Neurosci. 12, 155.

Ahima, R.S., Osei, S.Y., 2004. Leptin signaling. Physiol. Behav. 81, 223–241.

Ahima, R.S., Prabakaran, D., Flier, J.S., 2000. Leptin. Annu. Rev. Physiol. 62, 413–437.

Ahmadian, M., Suh, J.M., Hah, N., Liddle, C., Atkins, A.R., Downes, M., et al., 2013. PPARgamma signaling and metabolism: the good, the bad and the future. Nat. Med. 19, 557–566.

Akira, S., Uematsu, S., Takeuchi, O., 2006. Pathogen recognition and innate immunity. Cell 124, 783–801.

Alam, I., Ng, T.P., Larbi, N.A., 2012. Does inflammation determine whether obesity is metabolically healthy or unhealthy? The aging perspective. Mediators Inflamm. 456456.

Altintas, M.M., Rossetti, M.A., Nayer, B., Puig, A., Zagallo, P., Ortega, L.M., et al., 2011. Apoptosis, mastocytosis, and diminished adipocytokine gene expression accompany reduced epididymal fat mass in long-standing diet-induced obese mice. Lipids Health Dis. 10, 198.

Anderson, E.J., Lustig, M.E., Boyle, K.E., et al., 2009. Mitochondrial H_2O_2 emission and cellular redox state link excess fat intake to insulin resistance in both rodents and humans. J. Clin. Invest. 119, 573–581.

Aoyama, M., Kotani, J., Usami, M., 2010. Butyrate and propionate induced activated or non-activated neutrophil apoptosis via HDAC inhibitor activity but without activating GPR-41/GPR-43 pathways. Nutrition 26, 653–661.

Archer, Z.A., Mercer, J.G., 2007. Brain responses to obesogenic diets and diet-induced obesity. Proc. Nutr. Soc. 66, 124–130.

Arita, Y., Kihara, S., Ouchi, N., Maeda, K., Kuriyama, H., Okamoto, Y., et al., 2002. Adipocyte-derived plasma protein adiponectin acts as a platelet-derived growth factor-BB-binding protein and regulates growth factor-induced common postreceptor signal in vascular smooth muscle cell. Circulation 105, 2893–2898.

Armoni, M., Harel, C., Bar-Yoseph, F., Milo, S., Karnieli, E., 2005. Free fatty acids repress the Glut4 gene expression in cardiac muscle via novel response elements. J. Biol. Chem. 280, 34786–34795.

Avena, N.M., Rada, P., Hoebel, B.G., 2009. Sugar and fat bingeing have notable differences in addictive-like behavior. J. Nutr. 139, 623–628.

Avramoglu, R.K., Basciano, H., Adeli, K., 2006. Lipid and lipoprotein dysregulation in insulin resistant states. Clin. Chim. Acta. 368, 1–19.

Azzi, A., 2007. Oxidative stress: a dead end or a laboratory hypothesis? Biochem. Biophys. Res. Commun. 362, 230–232.

Banks, W.A., 2004. The source of cerebral insulin. Eur. J. Pharmacol. 490, 5–12.

Banks, W.A., Owen, J.B., Erickson, M.A., 2012. Insulin in the brain: there and back again. Pharmacol. Ther. 136, 82–93.

Baron, A.D., Zhu, J.S.H., Weldon, H., Maianu, L., Garvey, W.T., 1995. Glucosamine induces insulin resistance in vivo by affecting GLUT 4 translocation in skeletal muscle. Implications for glucose toxicity. J. Clin. Invest. 96, 2792–2801.

Bastard, J.P., Maachi, M., Lagathu, C., Kim, M.J., Caron, M., Vidal, H., et al., 2006. Recent advances in the relationship between obesity, inflammation, and insulin resistance. Eur. Cytokine Network 17, 4–12.

Bäumler, A.J., Sperandio, V., 2016. Interactions between the microbiota and pathogenic bacteria in the gut. Nature 535, 85–93.

Bence, K.K., Delibegovic, M., Xue, B., Gorgun, C.Z., Hotamisligil, G.S., et al., 2006. Neuronal PTP1B regulates body weight, adiposity and leptin action. Nat. Med. 12, 917–924.

Bhat, Z.F., Morton, J.D., Mason, S., Bekhit, A.E.A., Bhat, H.F., 2017. Obesity and neurological disorders: dietary perspective of a global menace. Crit. Rev. Food Sci. Nutr. 19, 1–17.

Biessels, G.J., Reagan, L.P., 2015. Hippocampal insulin resistance and cognitive dysfunction. Nat. Rev. Neurosci. 16, 660–671.

Bjorbaek, C., Kahn, B.B., 2004. Leptin signaling in the central nervous system and the periphery. Recent Prog. Hormone Res. 59, 305–331.

Blázquez, E., Velázquez, E., Hurtado-Carneiro, V., Ruiz-Albusac, J.M., 2014. Insulin in the brain: its pathophysiological implications for states related with central insulin resistance, type 2 diabetes and Alzheimer's disease. Front. Endocrinol. 5, 161.

Bloch-Damti, A., Bashan, N., 2005. Proposed mechanisms for the induction of insulin resistance by oxidative stress. Antioxid. Redox. Signal. 7, 1553.

Bloemer, J., Pinky, P.D., Govindarajulu, M., Hong, H., Judd, R., Amin, R.H., et al., 2018. Role of adiponectin in central nervous system disorders. Neural Plast. 2018, 4593530.

Blouet, C., Schwartz, G.J., 2010. Hypothalamic nutrient sensing in the control of energy homeostasis. Behav. Brain Res. 209, 1–12.

Boden, G., Chen, X., 1995. Effects of fat on glucose uptake and utilization in patients with noninsulin-dependent diabetes. J. Clin. Invest. 96, 1261–1268.

Boden, G., She, P., Mozzoli, M., Cheung, P., Gumireddy, K., Reddy, P., et al., 2005. Free fatty acids produce insulin resistance and activate the proinflammatory nuclear factor-κB pathway in rat liver. Diabetes 54, 3458–3465.

Boini, K.M., Hussain, T., Li, P.L., Koka, S., 2017. Trimethylamine-N-oxide instigates NLRP3 inflammasome activation and endothelial dysfunction. Cell. Physiol. Biochem. 44, 152–162.

Bråkenhielm, E., Veitonmäki, N., Cao, R., Kihara, S., Matsuzawa, Y., Zhivotovsky, B., et al., 2004. Adiponectin-induced antiangiogenesis and antitumor activity involve caspase-mediated endothelial cell apoptosis. Proc. Natl. Acad. Sci. U.S.A. 101, 2476–2481.

Brandes, N., Schmitt, S., Jakob, U., 2009. Thiol-based redox switches in eukaryotic proteins. Antioxid. Redox. Signal. 11, 997.

Brennan, A.M., Mantzoros, C.S., 2006. Drug Insight: the role of leptin in human physiology and pathophysiology—emerging clinical applications. Nat. Clin. Pract. Endocrinol. Metab. 2, 318–327.

Brown, A.E., Walker, M., 2016. Genetics of insulin resistance and the metabolic syndrome. Curr. Cardiol. Rep. 18, 75.

Brown, L.A., Kerr, C.J., Whiting, P., 2009. Oxidant stress in healthy normal-weight, overweight, and obese individuals. Obesity (Silver Spring) 17, 460–466.

Bruce, K.D., Hanson, M.A., 2010. The developmental origins, mechanisms, and implications of metabolic syndrome. J. Nutr. 140, 648–652.

Businaro, R., Ippoliti, F., Ricci, S., Canitano, N., Fuso, A., 2012. Alzheimer's disease promotion by obesity: induced mechanisms-molecular links and perspectives. Curr. Gerontol. Geriatr. Res. 2012, 986823.

Cai, D., Yuan, M., Frantz, D.F., Melendez, P.A., Hansen, L., Lee, J., et al., 2005. Local and systemic insulin resistance resulting from hepatic activation of IKK-beta and NF-kappaB. Nat. Med. 11, 183–190.

Cai, Z., Xiao, M., Chang, L., Yan, L.J., 2015. Role of insulin resistance in Alzheimer's disease. Metab. Brain Dis. 30, 839–851.

Cai, W., Duan, X.-M., Liu, Y., et al., 2017. Uric acid induces endothelial dysfunction by activating the HMGB1/RAGE signaling pathway. BioMed Res. Int. 2017, 11.

Cha, H.-N., Song, S.E., Kim, Y.-W., Kim, J.-Y., Won, K.-C., Park, S.-Y., 2011. Lack of inducible nitric oxide synthase prevents lipid-induced skeletal muscle insulin resistance without attenuating cytokine level. J. Pharmacol. Sci. 117, 77–86.

Chambers, J.C., Elliott, P., Zabaneh, D., et al., 2008. Common genetic variation near MC4R is associated with waist circumference and insulin resistance. Nat. Genet. 40, 716–718.

Canfora, E.E., Jocken, J.W., Blaak, E.E., 2015. Short-chain fatty acids in control of body weight and insulin sensitivity. Nat. Rev. Endocrinol. 11, 577–591.

Cani, P.D., Possemiers, S., Van de Wiele, T., Guiot, Y., Everard, A., Rottier, O., et al., 2009. Changes in gut microbiota control inflammation in obese mice through a mechanism involving GLP-2-driven improvement of gut permeability. Gut 58, 1091–1103.

Carnethon, M.R., Rasmussen-Torvik, L.J., Palaniappan, L., 2014. The obesity paradox in diabetes. Curr. Cardiol. Rep. 16, 446.

Chang, C.C., Wu, C.H., Liu, L.K., Chou, R.H., Kuo, C.S., Huang, P.H., et al., 2018. Association between serum uric acid and cardiovascular risk in non-hypertensive and nondiabetic individuals: the Taiwan I-Lan Longitudinal Aging Study. Sci. Rep. 8, 5234.

Chassaing, B., Koren, O., Goodrich, J.K., Poole, A.C., Srinivasan, S., Ley, R.E., et al., 2015. Dietary emulsifiers impact the mouse gut microbiota promoting colitis and metabolic syndrome. Nature 519, 92–96.

Chassaing, B., Van de Wiele, T., Gewirtz, A., 2017. O-013 dietary emulsifiers directly impact the human gut microbiota increasing its pro-inflammatory potential and ability to induce intestinal inflammation. Inflamm. Bowel Dis. 23 (Suppl. 1), S5.

Chen, G., Bentley, A., Adeyemo, A., et al., 2012. Genome-wide association study identifies novel loci association with fasting insulin and insulin resistance in African Americans. Hum. Mol. Genet. 21, 4530–4536.

Chen, Y., Deng, Y., Zhang, B., Gong, C.X., 2014. Deregulation of brain insulin signaling in Alzheimer's disease. Neurosci. Bull. 30, 282–294.

Chen, M.L., Zhu, X.H., Ran, L., Lang, H.D., Yi, L., Mi, M.T., 2017. Trimethylamine-N-oxide induces vascular inflammation by activating the NLRP3 inflammasome through the SIRT3-SOD2-mtROS signaling pathway. J. Am. Heart Assoc. 6, pii: e006347.

Cheng, K.K., Lam, K.S., Wang, Y., Huang, Y., Carling, D., Wu, D., et al., 2007. Adiponectin-induced endothelial nitric oxide synthase activation and nitric oxide production are mediated by APPL1 in endothelial cells. Diabetes 56, 1387–1394.

Chrysohoou, C., Panagiotakos, D.B., Pitsavos, C., Skoumas, I., Papademetriou, L., Economou, M., et al., 2007. The implication of obesity on total antioxidant capacity apparently healthy men and women: the ATTICA study. Nutr. Metab. Cardiovasc. Dis. 17, 590–597.

Chung, J., Nguyen, A.K., Henstridge, D.C., Holmes, A.G., Chan, M.H., Mesa, J.L., et al., 2008. HSP72 protects against obesity-induced insulin resistance. Proc. Natl. Acad. Sci. U.S.A. 105, 1739–1744.

Codoñer-Franch, P., Valls-Bellés, V., Arilla-Codoñer, A., Alonso-Iglesias, E., 2011. Oxidant mechanisms in childhood obesity: the link between inflammation and oxidative stress. Transl. Res. 158, 369–384.

Colantuoni, C., Schwenker, J., McCarthy, J., Rada, P., Ladenheim, B., Cadet, J.L., et al., 2001. Excessive sugar intake alters binding to dopamine and mu-opioid receptors in the brain. Neuroreport 12, 3549–3552.

Coll, A.P., Farooqi, I.S., O'Rahilly, S., 2007. The hormonal control of food intake. Cell. 129, 251–262.

Considine, R.V., Sinha, M.K., Heiman, M.L., Kriauciunas, A., Stephens, T.W., Nyce, M. R., et al., 1996. Serum immunoreactive-leptin concentrations in normal-weight and obese humans. N. Engl. J. Med. 334, 292–295.

Contreras, C., González-García, I., Martínez-Sánchez, N., Seoane-Collazo, P., Jacas, J., Morgan, D.A., et al., 2014. Central ceramide-induced hypothalamic lipotoxicity and ER stress regulate energy balance. Cell Rep. 9, 366–377.

Cornier, M.A., Dabelea, D., Hernandez, T.L., Lindstrom, R.C., Steig, A.J., Stob, N.R., et al., 2008. The metabolic syndrome. Endocr. Rev. 29, 777–822.

Craft, S., Watson, G.S., 2004. Insulin and neurodegenerative disease: shared and specific mechanisms. Lancet Neurol. 3, 169–178.

Craft, S., 2005. Insulin resistance syndrome and Alzheimer's disease: age- and obesity-related effects on memory, amyloid, and inflammation. Neurobiol. Aging 26, 65–69.

Craft, S., 2009. The role of metabolic disorders in Alzheimer disease and vascular dementia: two roads converged. Arch. Neurol. 66, 300–305.

Cummings, N.E., Williams, E.M., Kasza, I., Konon, E.N., Schaid, M.D., Schmidt, B.A., 2018. Restoration of metabolic health by decreased consumption of branched-chain amino acids. J. Physiol. 596, 623–645.

Czech, M.P., Tencerova, M., Pedersen, D.J., Aouadi, M., 2013. Insulin signalling mechanisms for triacylglycerol storage. Diabetologia 56, 949–964.

Dabrowska, K., Witkiewicz, W., 2016. Correlations of host genetics and gut microbiome composition. Front. Microbiol. 2016, 1357.

Dalamaga, M., Chou, S.H., Shields, K., Papageorgiou, P., Polyzos, S.A., Mantzoros, C.S., 2013. Leptin at the intersection of neuroendocrinology and metabolism: current evidence and therapeutic perspectives. Cell. Metab. 18, 29–42.

Dalbeth, N., Merriman, T.R., Stamp, L.K., 2016. Gout. Lancet. 388, 2039–2052.

Dardeno, T.A., Chou, S.H., Moon, H., Chamberland, J.P., Fiorenza, C.G., Mantzoros, C.S., 2010. Leptin in human physiology and therapeutics. Front. Neuroendocrinol. 31, 377–393.

Dashwood, R.H., Myzak, M.C., Ho, E., 2006. Dietary HDAC inhibitors: time to rethink weak ligands in cancer chemoprevention? Carcinogenesis 27, 344–349.

Dasuri, K., Ebenezer, P., Fernandez-Kim, S.O., Zhang, L., Gao, Z., Bruce-Keller, A.J., et al., 2013. Role of physiological levels of 4-hydroxynonenal on adipocyte biology: implications for obesity and metabolic syndrome. Free Radic. Res. 47, 8–19.

David, L.A., Maurice, C.F., Carmody, R.N., Gootenberg, D.B., Button, J.E., Wolfe, B. E., et al., 2014. Diet rapidly and reproducibly alters the human gut microbiome. Nature 505, 559–563.

Deeb, S.S., Fajas, L., Nemoto, M., et al., 1998. A Pro12Ala substitution in PPARgamma2 associated with decreased receptor activity, lower body mass index and improved insulin sensitivity. Nat. Genet. 20, 284–287.

De Filippo, C., Cavalieri, D., Di Paola, M., Ramazzotti, M., Poullet, J.B., Massart, S., et al., 2010. Impact of diet in shaping gut microbiota revealed by a comparative study in children from Europe and rural Africa. Proc. Natl. Acad. Sci. U.S.A. 107, 14691–14696.

de Luca, C., Olefsky, J.M., 2008. Inflammation and insulin resistance. FEBS Lett. 582, 97–105.

De Preter, V., Geboes, K.P., Bulteel, V., Vandermeulen, G., Suenaert, P., Rutgeerts, P., et al., 2011. Kinetics of butyrate metabolism in the normal colon and in ulcerative colitis: the effects of substrate concentration and carnitine on the beta-oxidation pathway. Aliment Pharmacol. Ther. 34, 526–532.

den Besten, G., van Eunen, K., Groen, A.K., Venema, K., Reijngoud, D.J., Bakker, B. M., 2013. The role of short-chain fatty acids in the interplay between diet, gut microbiota, and host energy metabolism. J. Lipid Res. 54, 2325–2340.

den Besten, G., Bleeker, A., Gerding, A., Van Eunen, K., Havinga, R., Van Dijk, T.H., et al., 2015. Short-chain fatty acids protect against high-fat diet-induced obesity via a PPARgamma-dependent switch from lipogenesis to fat oxidation. Diabetes 64, 2398–2408.

Delarue, J., Magnan, C., 2007. Free fatty acids and insulin resistance. Curr. Opin. Clin. Nutr. Metab. Care. 10, 142–148.

Denke, M.A., Sempos, C.T., Grundy, S.M., 1994. Excess body weight. An under-recognized contributor to dyslipidemia in white American women. Arch. Intern. Med. 154, 401–410.

Derdak, Z., Lang, C.H., Villegas, K.A., et al., 2011. Activation of p53 enhances apoptosis and insulin resistance in a rat model of alcoholic liver disease. J. Hepatol. 54, 164–172.

De Souza, C.T., Araujo, E.P., Bordin, S., Ashimine, R., Zollner, R.L., et al., 2005. Consumption of a fat-rich diet activates a proinflammatory response and induces insulin resistance in the hypothalamus. Endocrinology 146, 4192–4199.

De Vadder, F., Kovatcheva-Datchary, P., Goncalves, D., Vinera, J., Zitoun, C., Duchampt, A., et al., 2014. Microbiota-generated metabolites promote metabolic benefits via gut–brain neural circuits. Cell 156, 84–96.

Devkota, S., Devkota, S., Wang, Y., Musch, M.W., Leone, V., Fehlner-Peach, H., et al., 2012. Dietary-fat-induced taurocholic acid promotes pathobiont expansion and colitis in Il10 − / − mice. Nature 487, 104–108.

Diaz, B., Fuentes-Mera, L., Tovar, A., et al., 2015. Saturated lipids decrease mitofusin 2 leading to endoplasmic reticulum stress activation and insulin resistance in hypothalamic cells. Brain Res. 1627, 80–89.

Dickson, B.J., 2003. Development. Wiring the brain with insulin. Science 300, 440–441.

Di Domenico, F., Barone, E., Perluigi, M., Butterfield, D.A., 2017. The triangle of death in Alzheimer's disease brain: the aberrant cross-talk among energy metabolism, mammalian target of rapamycin signaling, and protein homeostasis revealed by redox proteomics. Antioxid. Redox. Signal. 26, 364–387.

Dietrich, M., Horvath, T., 2009. Feeding signals and brain circuitry. Eur. J. Neurosci. 30, 1688–1696.

Dietrich, M.O., Horvath, T., 2013. Hypothalamic control of energy balance: insights into the role of synaptic plasticity. Trends Neurosci. 36, 65–73.

Do, R., Bailey, S.D., Desbiens, K., et al., 2008. Genetic variants of FTO influence adiposity, insulin sensitivity, leptin levels, and resting metabolic rate in the Quebec family study. Diabetes 57, 1147–1150.

Dodd, K.M., Tee, A.R., 2012. Leucine and mTORC1: a complex relationship. Am. J. Physiol. Endocrinol. Metab. 302, E1329–E1342.

Donohoe, D.R., Garge, N., Zhang, X., Sun, W., O'Connell, T.M., Bunger, M.K., et al., 2011. The microbiome and butyrate regulate energy metabolism and autophagy in the mammalian colon. Cell. Metab. 13, 517–526.

Donohoe, D.R., Holley, D., Collins, L.B., Montgomery, S.A., Whitmore, A.C., Hillhouse, A., et al., 2014. A gnotobiotic mouse model demonstrates that dietary fiber protects against colorectal tumorigenesis in a microbiota- and butyrate-dependent manner. Cancer Discov. 4, 1387–1397.

Du, X.L., Edelstein, D., Dimmeler, S., Ju, Q., Sui, C., Brownlee, M., 2001. Hyperglycemia inhibits endothelial nitric oxide synthase activity by posttranslational modification at the Akt site. J. Clin. Invest. 108, 1341–1348.

Duckworth, W.C., Bennett, R.G., Hamel, F.G., 1997. The significance of intracellular insulin to insulin action. J. Investig. Med. 45, 20–27.

Dupuis, J., Langenberg, C., Prokopenko, I., et al., 2010. New genetic loci implicated in fasting glucose homeostasis and their impact on type 2 diabetes risk. Nat. Genet. 42, 105–116.

Ebert, M.N., Klinder, A., Peters, W.H., Schaferhenrich, A., Sendt, W., Scheele, J., et al., 2003. Expression of glutathione S-transferases (GSTs) in human colon cells and inducibility of GSTM2 by butyrate. Carcinogenesis 24, 1637–1644.
Eeg-Olofsson, K., Cederholm, J., Nilsson, P., Zethelius, B., Nunez, L., et al., 2009. Risk of cardiovascular disease and mortality in overweight and obese patients with type 2 diabetes: an observational study in 13,087 patients. Diabetologia 52, 65–73.
Elbein, S.C., Chu, W.S., Das, S.K., et al., 2007. Transcription factor 7-like 2 polymorphisms and type 2 diabetes, glucose homeostasis traits and gene expression in US participants of European and African descent. Diabetologia 50, 1621–1630.
El Kaoutari, A., Armougom, F., Gordon, J.I., Raoult, D., Henrissat, B., 2013. The abundance and variety of carbohydrate-active enzymes in the human gut microbiota. Nat. Rev. Microbiol. 11, 497–504.
El-Khairy, L., Ueland, P.M., Refsum, H., Graham, I.M., Vollset, S.E., 2001. Plasma total cysteine as a risk factor for vascular disease: the European Concerted Action Project. Circulation 103, 2544–2549.
Elshorbagy, A.K., Refsum, H., Smith, A.D., Graham, I.M., 2009. The association of plasma cysteine and gamma-glutamyltransferase with BMI and obesity. Obesity (Silver Spring) 17, 1435–1440.
Erion, D.M., Shulman, G.I., 2010. Diacylglycerol-mediated insulin resistance. Nat. Med. 16, 400–402.
Esposito, K., Ciotola, M., Giugliano, D., 2006. Oxidative stress in the metabolic syndrome. J. Endocrinol. Invest. 29, 791–795.
Esterbauer, H., Schaur, R.J., Zollner, H., 1991. Chemistry and biochemistry of 4-hydroxynonenal, malonaldehyde and related aldehydes. Free Radic. Biol. Med. 11, 81–128.
Farooqui, A.A., 2011. Lipid Mediators and Their Metabolism in the Brain. Springer Science + Business Media, LLC, New York.
Farooqui, A.A., 2012. Generation of reactive oxygen species in the brain: signaling for neural cell survival or suicide. In: Farooqui, T., Farooqui, A.A. (Eds.), Oxidative Stress in Vertebrate and Invertebrate. Wiley-Blackwell, Hoboken, NJ.
Farooqui, A.A., 2013. Metabolic Syndrome. Springer Science + Business, New York.
Farooqui, A.A., 2015. High Calorie Diet and the Human Brain. Springer International Publishing, Switzerland.
Farooqui, A.A., 2017. Neurochemical Aspects of Alzheimer's Disease. Academic Press, an imprint of Elsevier, San Diego, CA.
Farooqui, A.A., 2019. Molecular Mechanisms of Dementia: Mechanisms of Dementia, Biomarkers, Neurochemistry, and Therapy. Academic Press, an imprint of Elsevier, San Diego, CA.
Farooqui, T., Farooqui, A.A., 2018. Role of the Mediterranean Diet in the Brain and Neurodegenerative Disease. Academic Press, an imprint of Elsevier, San Diego, CA.
Fick, L.J., Belsham, D.D., 2010. Nutrient sensing and insulin signaling in neuropeptide-expressing immortalized, hypothalamic neurons: a cellular model of insulin resistance. Cell Cycle 9, 3186–3193.
Field, A.E., Coakley, E.H., Must, A., Spadano, J.L., Laird, N., Dietz, W., et al., 2001. Impact of overweight on the risk of developing common chronic diseases during a 10-year period. Arch. Intern. Med. 161, 1581–1586.
Fillmore, N., Wagg, C.S., Zhang, L., Fukushima, A., Lopaschuk, G.D., 2018. Cardiac branched-chain amino acid oxidation is reduced during insulin resistance in the heart. Am. J. Physiol. Endocrinol. Metab. 315, E1046–E1052.
Finkel, T., Holbrook, N.J., 2000. Oxidants, oxidative stress and the biology of ageing. Nature 408, 239–247.

Flegal, K.M., Graubard, B.I., Williamson, D.F., Gail, M.H., 2007. Cause-specific excess deaths associated with underweight, overweight, and obesity. JAMA 298, 2028–2037.
Flegal, K.M., Carroll, M.D., Kit, B.K., Ogden, C.L., 2012. Prevalence of obesity and trends in the distribution of body mass index among US adults, 1999–2010. JAMA 307, 491–497.
Florkowski, C., 2013. HbA(1c) as a diagnostic test for diabetes mellitus—reviewing the evidence. Clin. Biochem. Rev. 34, 75–83.
Fonseca, V., Desouza, C., Asnani, S., Jialal, I., 2004. Nontraditional risk factors for cardiovascular disease in diabetes. Endocr. Rev. 25, 153–175.
Fontaine, K.R., Redden, D.T., Wang, C., Westfall, A.O., Allison, D.B., 2003. Years of life lost due to obesity. J. Am. Med. Assoc. 289, 187–193.
Forbes, J.M., Sourris, K.C., de Courten, M.P., Dougherty, S.L., Chand, V., Lyons, J.G., et al., 2014. Advanced glycation end products (AGEs) are cross-sectionally associated with insulin secretion in healthy subjects. Amino Acids 46, 321–326.
Frederich, R.C., Hamann, A., Anderson, S., Lollmann, B., Lowell, B.B., Flier, J.S., 1995. Leptin levels reflect body lipid content in mice: evidence for diet-induced resistance to leptin action. Nat. Med. 1, 1311–1314.
Frost, G., Sleeth, M.L., Sahuri-Arisoylu, M., Lizarbe, B., Cerdan, S., Brody, L., et al., 2014. The short-chain fatty acid acetate reduces appetite via a central homeostatic mechanism. Nat. Commun. 5, 3611.
Furukawa, K., Hori, M., Ouchi, N., Kihara, S., Funahashi, T., Matsuzawa, Y., et al., 2004. Adiponectin down-regulates acyl-coenzyme A:cholesterol acyltransferase-1 in cultured human monocyte-derived macrophages. Biochem. Biophys. Res. Commun. 317, 831–836.
Galland, L., 2014. The gut microbiome and the brain. J. Med. Food 17, 1261–1272.
Gao, Z., Yin, J., Zhang, J., Ward, R.E., Martin, R.J., Lefevre, M., et al., 2009. Butyrate improves insulin sensitivity and increases energy expenditure in mice. Diabetes 58, 1509–1517.
Geiger, B.M., Behr, G.G., Frank, L.E., Caldera-Siu, A.D., Beinfeld, M.C., Kokkotou, E. G., et al., 2008. Evidence for defective mesolimbic dopamine exocytosis in obesity-prone rats. FASEB J. 22, 2740–2746.
Gensollen, T., Iyer, S.S., Kasper, D.L., Blumberg, R.S., 2016. How colonization by microbiota in early life shapes the immune system. Science 352, 539–544.
Gil, A., Olza, J., Gil-Campos, M., Gomez-Llorente, C., Aguilera, C.M., 2011. Is adipose tissue metabolically different at different sites? Int. J. Pediatr. Obes. 6, 13–20.
Gill, S.R., Pop, M., Deboy, R.T., et al., 2006. Metagenomic analysis of the human distal gut microbiome. Science 312, 1355–1359.
Giral, P., Jacob, N., Dourmap, C., Hansel, B., Carrie, A., Bruckert, E., 2008. Elevated gamma-glutamyltransferase activity and perturbed thiol profile are associated with features of metabolic syndrome. Arterioscler. Thromb. Vasc. Biol. 28, 587–593.
Gizaw, M., Anandakumar, P., Debela, T., 2017. A review on the role of irisin in insulin resistance and type 2 diabetes mellitus. J. Pharmacopuncture 20, 235–242.
Glantzounis, G.K., Tsimoyiannis, E.C., Kappas, A.M., et al., 2005. Uric acid and oxidative stress. Curr. Pharm. Des. 11, 4145–4151.
Goodrich, J.K., Davenport, E.R., Waters, J.L., Clark, A.G., Ley, R.E., 2016. Cross-species comparisons of host genetic associations with the microbiome. Science 352, 532–535.
Górski, J., 2012. Ceramide and insulin resistance: how should the issue be approached? Diabetes 61, 3081–3308.
Goyal, M.S., Venkatesh, S., Milbrandt, J., et al., 2015. Feeding the brain and nurturing the mind: linking nutrition and the gut microbiota to brain development. Proc. Natl. Acad. Sci. U.S.A. 112, 14105–14112.

Gross, D.N., van den Heuvel, A.P., Birnbaum, M.J., 2008. The role of FoxO in the regulation of metabolism. Oncogene 27, 2320–2336.

Guillemot-Legris, O., Masquelier, J., Everard, A., Cani, P.D., Alhouayek, M., Muccioli, G.G., 2016. High-fat diet feeding differentially affects the development of inflammation in the central nervous system. J. Neuroinflamm. 13, 206.

Hajnal, A., Smith, G.P., Norgren, R., 2004. Oral sucrose stimulation increases accumbens dopamine in the rat. Am. J. Physiol. Regul. Integr. Comp. Physiol. 286, R31–R37.

Hamaker, B.R., Tuncil, Y.E., 2014. A perspective on the complexity of dietary fiber structures and their potential effect on the gut microbiota. J. Mol. Biol. 426, 3838–3850.

Hamilton, M.K., Raybould, H.E., 2016. Bugs, guts and brains, and the regulation of food intake and body weight. Int. J. Obes. Suppl. 6, S8–S14.

Hamilton, M.T., Hamilton, D.G., Zderic, T.W., 2007. The role of low energy expenditure and sitting on obesity, metabolic syndrome, Type 2 diabetes, and cardiovascular disease. Diabetes 56, 2655–2667.

Hamman, R.F., Wing, R.R., Edelstein, S.L., Lachin, J.M., Bray, G.A., Delahanty, L., et al., 2006. Effect of weight loss with lifestyle intervention on risk of diabetes. Diabetes Care 29, 2102–2107.

Han, J.M., Jeong, S.J., Park, M.C., Kim, G., Kwon, N.H., Kim, H.K., et al., 2012. Leucyl-tRNA synthetase is an intracellular leucine sensor for the mTORC1-signaling pathway. Cell. 149, 410–424.

Harding, H.P., Novoa, I., Zhang, H., Wek, R., Schapira, M., Ron, D., 2000. Regulated translation initiation controls stress-induced gene expression in mammalian cells. Mol. Cell 2000 (6), 1099–1108.

Hartwich, J., Góralska, J., Siedlecka, D., Gruca, A., Trzos, M., Dembinska-Kiec, A., 2007. Effect of supplementation with vitamin E and C on plasma hsCRP level and cobalt-albumin binding score as markers of plasma oxidative stress in obesity. Genes Nutr. 2, 151–154.

Hawkins, M., Barzilai, N., Liu, R., Hu, M., Chen, W., Rossetti, L., 1997. Role of the glucosamine pathway in fat-induced insulin resistance. J. Clin. Invest. 99, 2173–2182.

Hayden, M.S., Ghosh, S., 2008. Shared principles in NF-κB signaling. Cell 132, 344–362.

Hirosumi, J., Tuncman, G., Chang, L., Görgün, C.Z., Uysal, K.T., Maeda, K., et al., 2002. A central role for JNK in obesity and insulin resistance. Nature 420, 333–336.

Hoehn, K.L., Salmon, A.B., Hohnen-Behrens, C., Turner, N., Hoy, A.J., Maghzal, G.J., et al., 2009. Insulin resistance is a cellular antioxidant defense mechanism. Proc. Natl. Acad. Sci. U.S.A. 106, 17787–17792.

Holland, W.L., Summers, S.A., 2008. Sphingolipids, insulin resistance, and metabolic disease: new insights from in vivo manipulation of sphingolipid metabolism. Endocr. Rev. 29, 381–402.

Holland, W.L., Brozinick, J.T., Wang, L.P., Hawkins, E.D., Sargent, K.M., Liu, Y., et al., 2007a. Inhibition of ceramide synthesis ameliorates glucocorticoid-, saturated-fat- and obesity-induced insulin resistance. Cell. Metab. 5, 167–179.

Holland, W.L., Knotts, T.A., Chavez, J.A., Wang, L.P., Hoehn, K.L., Summers, S.A., 2007b. Lipid mediators of insulin resistance. Nutr. Rev. 65, S39–S46.

Hooper, L.V., 2004. Bacterial contributions to mammalian gut development. Trends Microbiol. 12, 129–134.

Hooper, L.V., Midtvedt, T., Gordon, J.I., 2002. How host-microbial interactions shape the nutrient environment of the mammalian intestine. Annu. Rev. Nutr. 22, 283–307.

Horvath, T.L., Sarman, B., García-Cáceres, C., Enriori, P.J., Sotonyi, P., Shanabrough, M., et al., 2010. Synaptic input organization of the melanocortin system predicts diet-induced hypothalamic reactive gliosis and obesity. Proc. Natl. Acad. Sci. U.S.A. 107, 14875–14880.

Hosogai, N., Fukuhara, A., Oshima, K., Miyata, Y., Tanaka, S., Segawa, K., et al., 2007. Adipose tissue hypoxia in obesity and its impact on adipocytokine dysregulation. Diabetes 56, 901–911.

Hotamisligil, G.S., 2010. Endoplasmic reticulum stress and the inflammatory basis of metabolic disease. Cell. 140, 900–917.

Hotamisligil, G.S., Murray, D.L., Choy, L.N., Spiegelman, B.M., 1994. Tumor necrosis factor alpha inhibits signaling from the insulin receptor. Proc. Natl. Acad. Sci. U.S.A. 91, 4854–4858.

Hotamisligil, G.S., Arner, P., Caro, J.F., Atkinson, R.L., Spiegelman, B.M., 1995. Increased adipose tissue expression of tumor necrosis factor-alpha in human obesity and insulin resistance. J. Clin. Investig. 95, 2409–2415.

Hotta, K., Funahashi, T., Arita, Y., Takahashi, M., Matsuda, M., Okamoto, Y., et al., 2000. Plasma concentrations of a novel adipose-specific protein, adiponectin, in type 2 diabetic patients. Arterioscler. Thromb. Vasc. Biol. 20, 1595–1599.

Howard, J.K., Cave, B.J., Oksanen, L.J., Tzameli, I., Bjorbaek, C., et al., 2004. Enhanced leptin sensitivity and attenuation of diet-induced obesity in mice with haploinsufficiency of Socs3. Nat. Med. 10, 734–738.

Huang, A., Liu, G., Guo, J., Su, Z., 2018. The PI3K/Akt pathway in obesity and type 2 diabetes. Int. J. Biol. Sci. 14, 1483–1496.

Huffman, D.M., Barzilai, N., 2009. Role of visceral adipose tissue in aging. Biochim. Biophys. Acta 1790, 1117–1123.

Hug, C., Lodish, H.F., 2005. The role of the adipocyte hormone adiponectin in cardiovascular disease. Curr. Opin. Pharmacol. 5, 129–134.

Ikonen, E., Vainio, S., 2005. Lipid microdomains and insulin resistance: is there a connection? Sci. STKE pe3.

Inagaki, T., Tachibana, M., Magoori, K., Kudo, H., Tanaka, T., Okamura, M., et al., 2009. Obesity and metabolic syndrome in histone demethylase JHDM2a-deficient mice. Genes Cells 14, 991–1001.

Inokuchi, J., 2006. Insulin resistance as a membrane microdomain disorder. Biol. Pharm. Bull. 29, 1532–1537.

Irizarry, M.C., Raman, R., Schwarzschild, M.A., et al., 2009. Plasma urate and progression of mild cognitive impairment. Neurodegener. Dis. 6, 23–28.

Itani, S.I., Ruderman, N.B., Schmieder, F., Boden, G., 2002. Lipid-induced insulin resistance in human muscle is associated with changes in diacylglycerol, protein kinase C, and IkappaB-alpha. Diabetes 51, 2005–2011.

Jacobsen, U.P., Nielsen, H.B., Hildebrand, F., Raes, J., Sicheritz-Ponten, T., Kouskoumvekaki, I., et al., 2013. The chemical interactome space between the human host and the genetically defined gut metabotypes. ISME J. 7, 730–742.

Jazet, I.M., Pijl, H., Meinders, A.E., 2003. Adipose tissue as an endocrine organ: impact on insulin resistance. Neth. J. Med. 61, 194–212.

Jiao, P., Chen, Q., Shah, S., Tao, B., Tzameli, I., Yan, W., 2009. Obesity-related upregulation of monocyte chemotactic factors in adipocytes: involvement of nuclear factor-κB and c-Jun NH2-terminal kinase pathways. Diabetes 58, 104–115.

Johansen, D., Stocks, T., Jonsson, H., Lindkvist, B., Bjorge, T., Concin, H., et al., 2010. Metabolic factors and the risk of pancreatic cancer: a prospective analysis of almost 580,000 men and women in the metabolic syndrome and cancer project. Cancer Epidemiol. Biomark. Prev. 19, 2307–2317.

Johnson, A.M., Olefsky, J.M., 2013. The origins and drivers of insulin resistance. Cell. 152, 673–684.

Johnson, J.A., Ynada, Y., Flancbaum, L., Albu, J., 2001. Cytokine release in deep subcutaneous, abdominal and visceral tissue depots. Diabetes 50, A88.
Jones, D.P., 2008. Radical-free biology of oxidative stress. Am. J. Physiol. Cell. Physiol. 295, C849–C868.
Jung, T.H., Park, J.H., Jeon, W.M., Han, K.S., 2015. Butyrate modulates bacterial adherence on LS174T human colorectal cells by stimulating mucin secretion and MAPK signaling pathway. Nutr. Res. Pract. 9, 343–349.
Kabasakalian, A., Finney, G.R., 2009. Reversible dementias. Int. Rev. Neurobiol. 84, 283–302.
Kabayama, K., Sato, T., Kitamura, F., Uemura, S., Kang, B.W., Igarashi, Y., et al., 2005. TNFalpha-induced insulin resistance in adipocytes as a membrane microdomain disorder: involvement of ganglioside GM3. Glycobiology 15, 21–29.
Kadowaki, T., Yamauchi, T., 2005. Adiponectin and adiponectin receptors. Endocr. Rev. 26, 439–451.
Kadowaki, T., Yamauchi, T., Kubota, N., 2008. The physiological and pathophysiological role of adiponectin and adiponectin receptors in the peripheral tissues and CNS. FEBS Lett. 582, 74–80.
Karwi, Q.G., Zhang, L., Wagg, C.S., Wang, W., Ghandi, M., Thai, D., et al., 2019. Targeting the glucagon receptor improves cardiac function and enhances insulin sensitivity following a myocardial infarction. Cardiovasc. Diabetol. 18, 1.
Kastin, A.J., Akerstrom, V., 2001. Glucose and insulin increase the transport of leptin through the blood-brain barrier in normal mice but not in streptozotocin-diabetic mice. Neuroendocrinology 73, 237–242.
Kasubuchi, M., Hasegawa, S., Hiramatsu, T., Ichimura, A., Kimura, I., 2015. Dietary gut microbial metabolites, short-chain fatty acids, and host metabolic regulation. Nutrients 7, 2839–2849.
Kelesidis, T., Kelesidis, I., Chou, S., Mantzoros, C.S., 2010. Narrative review: the role of leptin in human physiology: emerging clinical applications. Ann. Intern. Med. 152, 93–100.
Kim, J.J., Sears, D.D., 2010. TLR4 and insulin resistance. Gastroenterol. Res. Pract. 2010, 11.
Kivipelto, M., Solomon, A., 2008. Alzheimer's disease—the ways of prevention. J. Nutr. Health Aging 12, 89S–94S.
Kleinridders, A., Ferris, H.A., Cai, W., Kahn, C.R., 2014. Insulin action in brain regulates systemic metabolism and brain function. Diabetes 63, 2232–2243.
Klotz, L.O., Sánchez-Ramos, C., Prieto-Arroyo, I., Urbánek, P., Steinbrenner, H., Monsalve, M., 2015. Redox regulation of FoxO transcription factors. Redox Biol. 6, 51–72.
Knowles, J.W., Xie, W., Zhang, Z., et al., 2015. Identification and validation of *N*-acetyltransferase 2 as an insulin sensitivity gene. J. Clin. Invest. 125, 1739–1751.
Koeth, R.A., Wang, Z., Levison, B.S., Buffa, J.A., Org, E., Sheehy, B.T., et al., 2013. Intestinal microbiota metabolism of l-carnitine, a nutrient in red meat, promotes atherosclerosis. Nat. Med. 19, 576–585.
Kraegen, E.W., Cooney, G.J., Ye, J.M., Thompson, A.L., Furler, S.M., 2001. The role of lipids in the pathogenesis of muscle insulin resistance and beta cell failure in type II diabetes and obesity. Exp. Clin. Endocrinol. Diabetes 109 (Suppl 2), S189–S201.
Krötz, F., Sohn, H.-Y., Pohl, U., 2004. Reactive oxygen species: players in the platelet game. Arterioscler. Thromb. Vasc. Biol. 24, 1988–1996.
Kyle, U.G., Pichard, C., 2006. The Dutch Famine of 1944–1945: a pathophysiological model of long-term consequences of wasting disease. Curr. Opin. Clin. Nutr. Metab. Care. 9, 388–394.

Labouèbe, G., Liu, S., Dias, C., Zou, H., Wong, J.C., et al., 2013. Insulin induces long-term depression of ventral tegmental area dopamine neurons via endocannabinoids. Nat. Neurosci. 16, 300–308.

Lean, M.E., Leslie, W.S., Barnes, A.C., Brosnahan, N., Thom, G., McCombie, L., et al., 2018. Primary care-led weight management for remission of type 2 diabetes (DiRECT): an open-label, cluster-randomised trial. Lancet 391, 541–551.

Lenoir, M., Serre, F., Cantin, L., Ahmed, S.H., 2007. Intense sweetness surpasses cocaine reward. PLoS One. 2, e698.

Li, M., Baumeister, P., Roy, B., Phan, T., Foti, D., Luo, S., et al., 2000. ATF6 as a transcription activator of the endoplasmic reticulum stress element. Thapsigargin stress-induced changes and synergistic interactions with NF-Y and YY1. Mol. Cell. Biol. 20, 5096–5106.

Li, M., Wang, B., Zhang, M., Rantalainen, M., Wang, S., Zhou, H., et al., 2008. Symbiotic gut microbes modulate human metabolic phenotypes. Proc. Natl. Acad. Sci. U.S.A. 105, 2117–2122.

Li, F.Y., Cheng, K.K., Lam, K.S., Vanhoutte, P.M., Xu, A., 2011. Cross-talk between adipose tissue and vasculature: role of adiponectin. Acta Physiol. Oxf. 203, 167–180.

Liang, N.C., Hajnal, A., Norgren, R., 2006. Sham feeding corn oil increases accumbens dopamine in the rat. Am. J. Physiol. Regul. Integr. Comp. Physiol. 291, R1236–R9.

Lin, H.V., Frassetto, A., Kowalik Jr, E.J., Nawrocki, A.R., Lu, M.M., et al., 2012. Butyrate and propionate protect against diet-induced obesity and regulate gut hormones via free fatty acid receptor 3-independent mechanisms. PLoS One. 7 (4), e35240.

Logsdon, A.F., Erickson, M.A., Rhea, E.M., Salameh, T.S., Banks, W.A., 2018. Gut reactions: how the blood-brain barrier connects the microbiome and the brain. Exp. Biol. Med. (Maywood). 243, 159–165.

Long, Y.C., Zierath, J.R., 2006. AMP-activated protein kinase signaling in metabolic regulation. J. Clin. Invest. 116, 1776–1783.

Long, Y.C., Cheng, Z., Copps, K.D., White, M.F., 2011. Insulin receptor substrates Irs1 and Irs2 coordinate skeletal muscle growth and metabolism via the Akt and AMPK pathways. Mol. Cell. Biol. 31, 430–441.

Louis, P., Hold, G.L., Flint, H.J., 2014. The gut microbiota, bacterial metabolites and colorectal cancer. Nat. Rev. Microbiol. 12, 661–672.

Lumeng, C.N., Deyoung, S.M., Bodzin, J.L., Saltiel, A.R., 2007. Increased inflammatory properties of adipose tissue macrophages recruited during diet-induced obesity. Diabetes 56, 16–23.

Lynch, C.J., Adams, S.H., 2014. Branched-chain amino acids in metabolic signalling and insulin resistance. Nat. Rev. Endocrinol. 10, 723.

Macfarlane, G.T., Macfarlane, S., 2011. Fermentation in the human large intestine: its physiologic consequences and the potential contribution of prebiotics. J. Clin. Gastroenterol. 45 (Suppl), S120–S127.

Mahajan, A., Go, M.J., Zhang, W., et al., 2014. Genome-wide trans-ancestry meta-analysis provides insight into the genetic architecture of type 2 diabetes susceptibility. Nat. Genet. 46, 234–244.

Malhotra, J.D., Kaufman, R.J., 2007. The endoplasmic reticulum and the unfolded protein response. Semin. Cell. Dev. Biol. 18, 716–731.

Mao, X., Kikani, C.K., Riojas, R.A., Langlais, P., Wang, L., Ramos, F.J., et al., 2006. APPL1 binds to adiponectin receptors and mediates adiponectin signalling and function. Nat. Cell Biol. 8, 516–523.

Martínez, I., Stegen, J.C., Maldonado-Gómez, M.X., Eren, A.M., Siba, P.M., Greenhill, A.R., et al., 2015. The gut microbiota of rural Papua New Guineans: composition, diversity patterns, and ecological processes. Cell Rep. 11, 527–538.

Masuccio, F., Verrotti, A., Chiavaroli, V., de Giorgis, T., Giannini, C., Chiarelli, F., et al., 2010. Weight gain and insulin resistance in children treated with valproate: the influence of time. J. Child Neurol. 25, 941–947.
Mattson, M.P., 2009. Roles of the lipid peroxidation product 4-hydroxynonenal in obesity, the metabolic syndrome, and associated vascular and neurodegenerative disorders. Exp. Gerontol. 44, 625–633.
Matsuzawa-Nagata, N., Takamura, T., Ando, H., Nakamura, S., Kurita, S., et al., 2008. Increased oxidative stress precedes the onset of high-fat diet–induced insulin resistance and obesity. Metabolism 57, 1071–1077.
McCrimmon, R.J., Ryan, C.M., Frier, B.M., 2012. Diabetes and cognitive dysfunction. Lancet 379, 2291–2299.
Miyamoto, J., Hasegawa, S., Kasubuchi, M., Ichimura, A., Nakajima, A., Kimura, I., 2016. Nutritional signaling via free fatty acid receptors. Int. J. Mol. Sci. 17, 450.
Mokdad, A.H., Ford, E.S., Bowman, L.A., Dietz, W.H., Vinicor, F., Bales, V.S., et al., 2003. Prevalence of obesity, diabetes, and obesity-related health risk factors, 2001. JAMA 289, 76–79.
Molnar, A., Toth, A., Bagi, Z., Papp, Z., Edes, I., Vaszily, M., et al., 2006. Activation of the poly(ADP-ribose) polymerase pathway in human heart failure. Mol. Med. 12, 143–152.
Moraes, J.C., Coope, A., Morari, J., Cintra, D.E., Roman, E.A., et al., 2009. High-fat diet induces apoptosis of hypothalamic neurons. PLoS One. 4, e5045.
Morino, K., Petersen, K.F., Shulman, G.I., 2006. Molecular mechanisms of insulin resistance in humans and their potential links with mitochondrial dysfunction. Diabetes 55 (Suppl. 2), S9–S15.
Morino, K., Neschen, S., Bilz, S., Sono, S., Tsirigotis, D., Reznick, R.M., et al., 2008. Muscle-specific IRS-1 Ser→Ala transgenic mice are protected from fat-induced insulin resistance in skeletal muscle. Diabetes 57, 2644–2651.
Muoio, D.M., 2010. Intramuscular triacylglycerol and insulin resistance: guilty as charged or wrongly accused? Biochim. Biophys. Acta 1801, 281–288.
Muoio, D.M., Neufer, P.D., 2012. Lipid-induced mitochondrial stress and insulin action in muscle. Cell. Metab. 15, 595–605.
Muoio, D.M., Newguard, C.B., 2005. Metabolism: A is for adipokine. Nature 436, 337–338.
Nagaya, T., Yoshida, H., Takahashi, H., Kawai, M., 2005. Increases in body mass index, even within non-obese levels, raise the risk for type 2 diabetes mellitus: a follow-up study in a Japanese population. Diabet. Med. 22, 1107–1111.
Nakae, J., Barr, V., Accili, D., 2000. Differential regulation of gene expression by insulin and IGF-1 receptors correlates with phosphorylation of a single amino acid residue in the forkhead transcription factor FKHR. EMBO J. 19, 989–996.
Nakamura, Y., Shimada, K., Fukuda, D., Shimada, Y., Ehara, S., Hirose, M., et al., 2004. Implications of plasma concentrations of adiponectin in patients with coronary artery disease. Heart 90, 528–533.
Nakamura, T., Furuhashi, M., Li, P., Cao, H., Tuncman, G., Sonenberg, N., et al., 2010. Double-stranded RNA-dependent protein kinase links pathogen sensing with stress and metabolic homeostasis. Cell. 140, 338–348.
Natividad, J.M.M., Verdu, E.F., 2013. Modulation of intestinal barrier by intestinal microbiota: pathological and therapeutic implications. Pharmacol. Res. 69, 42–51.
Nawrocki, A.R., Rajala, M.W., Tomas, E., Pajvani, U.B., Saha, A.K., Trumbauer, M.E., et al., 2006. Mice lacking adiponectin show decreased hepatic insulin sensitivity and reduced responsiveness to peroxisome proliferator-activated receptor gamma agonists. J. Biol. Chem. 281, 2654–2660.

Newgard, C.B., An, J., Bain, J.R., Muehlbauer, M.J., Stevens, R.D., Lien, L.F., et al., 2009. A branched-chain amino acid-related metabolic signature that differentiates obese and lean humans and contributes to insulin resistance. Cell. Metab. 9, 311–326.

Ng, D.T., Spear, E.D., Walter, P., 2000. The unfolded protein response regulates multiple aspects of secretory and membrane protein biogenesis and endoplasmic reticulum quality control. J. Cell. Biol. 150, 77–88.

Nisticò, R., Dargan, S.L., Amici, M., Collingridge, G.L., Bortolotto, Z.A., 2011. Synergistic interactions between kainate and mGlu receptors regulate bouton Ca signalling and mossy fibre LTP. Sci. Rep. 1, 103.

Nov, O., Kohl, A., Lewis, E.C., et al., 2010. Interleukin-1β may mediate insulin resistance in liver-derived cells in response to adipocyte inflammation. Endocrinology. 151, 4247–4256.

Ogden, C.L., Carroll, M.D., Kit, B.K., Flegal, K.M., 2012. Prevalence of obesity and trends in body mass index among US children and adolescents, 1999–2010. JAMA 307, 483–490.

Ohno, H., Shinoda, K., Ohyama, K., Sharp, L.Z., Kajimura, S., 2013. EHMT1 controls brown adipose cell fate and thermogenesis through the PRDM16 complex. Nature 504, 163–167.

Olefsky, J.M., Glass, C.K., 2010. Macrophages, inflammation, and insulin resistance. Annu. Rev. Physiol. 72, 219–246.

Onyango, A.N., 2017. The contribution of singlet oxygen to insulin resistance. Oxid. Med. Cell Longev. 2017, 14.

Oral, E.A., Simha, V., Ruiz, E., Andewelt, A., Premkumar, A., Snell, P., et al., 2002. Leptin—replacement therapy for lipodystrophy. N. Engl. J. Med. 346, 570–578.

Osborn, O., Olefsky, J.M., 2012. The cellular and signaling networks linking the immune system and metabolism in disease. Nat. Med. 18, 363–374.

Ouchi, N., Kihara, S., Arita, Y., Maeda, K., Kuriyama, H., Okamoto, Y., et al., 1999. Novel modulator for endothelial adhesion molecules: adipocyte-derived plasma protein adiponectin. Circulation 100, 2473–2476.

Ouchi, N., Kihara, S., Arita, Y., Nishida, M., Matsuyama, A., Okamoto, Y., et al., 2001. Adipocyte-derived plasma protein, adiponectin, suppresses lipid accumulation and class A scavenger receptor expression in human monocyte-derived macrophages. Circulation 103, 1057–1063.

Ozcan, U., Cao, Q., Yilmaz, E., Lee, A.H., Iwakoshi, N.N., Ozdelen, E., et al., 2004. Endoplasmic reticulum stress links obesity, insulin action, and type 2 diabetes. Science 306, 457–461.

Ozcan, U., Yilmaz, E., Ozcan, L., Furuhashi, M., Vaillancourt, E., Smith, R.O., et al., 2006. Chemical chaperones reduce ER stress and restore glucose homeostasis in a mouse model of type 2 diabetes. Science 313, 1137–1140.

Pacher, P., Szabo, C., 2008. Role of the peroxynitrite-poly(ADP-ribose) polymerase pathway in human disease. Am. J. Pathol. 173, 2–13.

Pacher, P., Beckman, J.S., Liaudet, L., 2007. Nitric oxide and peroxynitrite in health and disease. Physiol. Rev. 87, 315–424.

Pal, D., Dasgupta, S., Kundu, R., Maitra, S., Das, G., Mukhopadhyay, S., et al., 2012. Fetuin-A acts as an endogenous ligand of TLR4 to promote lipid-induced insulin resistance. Nat. Med. 18, 1279–1285.

Park, S.E., Park, C.Y., Sweeney, G., 2015. Biomarkers of insulin sensitivity and insulin resistance: past, present and future. Crit. Rev. Clin. Lab. Sci. 52, 180–190.

Pearson-Leary, J., McNay, E.C., 2012. Intrahippocampal administration of amyloid-β(1-42) oligomers acutely impairs spatial working memory, insulin signaling, and hippocampal metabolism. J. Alzheimer's Dis 30, 413–422.

Pendergrass, M., Bertoldo, A., Bonadonna, R., Nucci, G., Mandarino, L., Cobelli, C., et al., 2007. Muscle glucose transport and phosphorylation in type 2 diabetic, obese nondiabetic, and genetically predisposed individuals. Am. J. Physiol. Endocrinol. Metab. 292, E92–100.

Peng, L., Li, Z.R., Green, R.S., Holzman, I.R., Lin, J., 2009. Butyrate enhances the intestinal barrier by facilitating tight junction assembly via activation of AMP-activated protein kinase in Caco-2 cell monolayers. J. Nutr. 139, 1619–1625.

Pereira, S., Park, E., Mori, Y., et al., 2014. FFA-induced hepatic insulin resistance in vivo is mediated by PKCδ, NADPH oxidase, and oxidative stress. Am. J. Physiol. Endocrinol. Metab. 307, E34–E46.

Petersen, M.C., Shulman, G.I., 2018. Mechanisms of insulin action and insulin resistance. Physiol. Rev. 98, 2133–2223.

Petersen, K.F., Shulman, G.I., 2006. Etiology of insulin resistance. Am. J. Med. 119, S10–S16.

Picca, A., Fanelli, F., Calvani, R., et al., 2018. Gut dysbiosis and muscle aging: searching for novel targets against sarcopenia. Mediators Inflamm. 2018, 15.

Pigeyre, M., Yazdi, F.T., Kaur, Y., Meyre, D., 2016. Recent progress in genetics, epigenetics and metagenomics unveils the pathophysiology of human obesity. Clin. Sci. 130, 943–986.

Pillon, N.J., Croze, M.L., Vella, R.E., Soulère, L., Lagarde, M., Soulage, C.O., 2012. The lipid peroxidation by-product 4-hydroxy-2-nonenal (4-HNE) induces insulin resistance in skeletal muscle through both carbonyl and oxidative stress. Endocrinology 153, 2099–2111.

Poli, G., Schaur, R.J., 2000. 4-Hydroxynonenal in the pathomechanisms of oxidative stress. IUBMB Life 50, 315–321.

Posey, K.A., Clegg, D.J., Printz, R.L., Byun, J., Morton, G.J., Vivekanandan-Giri, A., et al., 2009. Hypothalamic proinflammatory lipid accumulation, inflammation, and insulin resistance in rats fed a high-fat diet. Am. J. Physiol. Endocrinol. Metabolism. 296, E1003–E1012.

Qatanani, M., Lazar, M.A., 2007. Mechanisms of obesity-associated insulin resistance: many choices on the menu. Genes Dev. 21, 1443–1455.

Qin, J., Li, R., Raes, J., Arumugam, M., Burgdorf, K.S., Manichanh, C., et al., 2010. A human gut microbial gene catalogue established by metagenomic sequencing. Nature 464, 59–65.

Raffaitin, C., Feart, C., Le Goff, M., Amieva, H., Helmer, C., Akbaraly, T.N., et al., 2011. Metabolic syndrome and cognitive decline in French elders: the Three-City Study. Neurology 76, 518–525.

Ramsay, R.R., Gandour, R.D., van der Leij, F.R., 2001. Molecular enzymology of carnitine transfer and transport. Biochim. Biophys. Acta 1546, 21–43.

Rando, O.J., Simmons, R.A., 2015. I'm eating for two: parental dietary effects on offspring metabolism. Cell. 161, 93–105.

Raychaudhuri, N., Raychaudhuri, S., Thamotharan, M., Devaskar, S.U., 2008. Histone code modifications repress glucose transporter 4 expression in the intrauterine growth-restricted offspring. J. Biol. Chem. 283, 13611–13626.

Ren, H., Yan, S., Zhang, B., Lu, T.Y., Arancio, O., et al., 2014. Glut4 expression defines an insulin-sensitive hypothalamic neuronal population. Mol. Metab. 3, 452–459.

Roden, M., Price, T.B., Perseghin, G., Petersen, K.F., Rothman, D.L., Cline, G.W., et al., 1996. Mechanism of free fatty acid-induced insulin resistance in humans. J. Clin. Invest. 97, 2859–2865.

Romeo, S., Maglio, C., Burza, M.A., et al., 2012. Cardiovascular events after bariatric surgery in obese subjects with type 2 diabetes. Diabetes Care 35, 2613–2617.

Rossetti, L., Giaccari, A., DeFronzo, R.A., 1990. Glucose toxicity. Diabetes Care 13, 610–630.

Rothenbacher, D., Brenner, H., Marz, W., Koenig, W., 2005. Adiponectin, risk of coronary heart disease and correlations with cardiovascular risk markers. Euro. Heart J. 26, 1640–1646.

Rung, J., Cauchi, S., Albrechtsen, A., et al., 2009. Genetic variant near IRS1 is associated with type 2 diabetes, insulin resistance and hyperinsulinemia. Nat. Genet. 41, 1110–1115.

Ruvolo, P.P., 2003. Intracellular signal transduction pathways activated by ceramide and its metabolites. Pharmacol. Res. 47, 383–392.

Ryder, J.W., Gilbert, M., Zierath, J.R., 2001. Skeletal muscle and insulin sensitivity: pathophysiological alterations. Front. Biosci. 6, D154–163.

Sader, S., Nian, M., Liu, P., 2003. Leptin: a novel link between obesity, diabetes, cardiovascular risk, and ventricular hypertrophy. Circulation 108, 644–646.

Sakaguchi, M., Murata, H., Yamamoto, K., et al., 2011. TIRAP, an adaptor protein for TLR2/4, transduces a signal from RAGE phosphorylated upon ligand binding. PLoS One 6, e23132.

Samuel, V.T., Shulman, G.I., 2012. Mechanisms for insulin resistance: common threads and missing links. Cell. 148, 852–871.

Sanmiguel, C., Gupta, A., Mayer, E.A., 2015. Gut microbiome and obesity: a plausible explanation for obesity. Curr. Obes. Rep. 4, 250–261.

Sasaki, T., Kitamura, T., 2010. Roles of FoxO1 and Sirt1 in the central regulation of food intake. Endocrinol. J. 57, 939–946.

Savage, D.B., Petersen, K.F., Shulman, G.I., 2007. Disordered lipid metabolism and the pathogenesis of insulin resistance. Physiol. Rev. 87, 507–520.

Schuling Kemp, R.J., Pagano, T.C., Hung, D., Raffa, R.B., 2000. Insulin receptors and insulin action in the brain: review and clinical implications. Neurosci. Biobehav. Rev. 24, 855–872.

Schwartz, M.W., Woods, S.C., Porte, D., Seeley, R.J., Baskin, D.G., 2000. Central nervous system control of food intake. Nature 404, 661–671.

Seldin, M.M., Meng, Y., Qi, H., Zhu, W., Wang, Z., Hazen, S.L., et al., 2016. Trimethylamine N-oxide promotes vascular inflammation through signaling of mitogen-activated protein kinase and nuclear factor-kappaB. J. Am. Heart Assoc. . Available from: https://doi.org/10.1161/JAHA.115.002767.

Sethi, J.K., Vidal-Puig, A.J., 2007. Thematic review series: adipocyte biology. Adipose tissue function and plasticity orchestrate nutritional adaptation. J. Lipid Res. 48, 1253–1262.

Shaw, L.M., 2011. The insulin receptor substrate (IRS) proteins: at the intersection of metabolism and cancer. Cell Cycle 10, 1750–1756.

Shoelson, S.E., Lee, J., Goldfine, A.B., 2006. Inflammation and insulin resistance. J. Clin. Invest. 116, 1793–1801.

Shulman, G.I., 2000. Cellular mechanisms of insulin resistance. J. Clin. Invest. 106, 171–176.

Sikaris, K.A., 2004. The clinical biochemistry of obesity. Clin. Biochem. Rev. 25, 165–181.

Soccio, R.E., Chen, E.R., Lazar, M.A., 2014. Thiazolidinediones and the promise of insulin sensitization in type 2 diabetes. Cell. Metab. 20, 573–591.

Sonnenburg, E.D., Sonnenburg, J.L., 2014. Starving our microbial self: the deleterious consequences of a diet deficient in microbiota-accessible carbohydrates. Cell. Metab. 20, 779–786.

Song, J., Wu, L., Chen, Z., Kohanski, R.A., Pick, L., 2003. Axons guided by insulin receptor in Drosophila visual system. Science 300, 502–505.

Sosa-Ortiz, A.L., Acosta-Castillo, I., Prince, M.J., 2012. Epidemiology of dementias and Alzheimer's disease. Arch. Med. Res. 43, 600–608.

Spangler, R., Wittkowski, K.M., Goddard, N.L., Avena, N.M., Hoebel, B.G., Leibowitz, S.F., 2004. Opiate-like effects of sugar on gene expression in reward areas of the rat brain. Brain. Res. Mol. Brain. Res. 124, 134–142.

Sripetchwandee, J., Chattipakorn, N., Chattipakorn, S.C., 2018. Links between obesity-induced brain insulin resistance, brain mitochondrial dysfunction, and dementia. Front. Endocrinol. (Lausanne). 9, 496.

Stecher, B., Maier, L., Hardt, W.-D., 2013. "Blooming" in the gut: how dysbiosis might contribute to pathogen evolution. Nat. Rev. Microbiol. 11, 277–284.

Stem, J.H., Rutkowski, J.M., Scherer, P.E., 2016. Adiponectin, leptin, and fatty acids in the maintenance of metabolic homeostasis through adipose tissue crosstalk. Cell. Metab. 23, 770–784.

Stienstra, R., Tack, C.J., Kanneganti, T.-D., Joosten, L.A.B., Netea, M.G., 2012. The inflammasome puts obesity in the danger zone. Cell. Metab. 15, 10–18.

Stratford, B.A.P.K., Hoehn, S., Liu, K.L.F., Summers, S.A., 2004. Regulation of insulin action by ceramide. J. Biol. Chem. 279, 36608–36615.

Straub, R.H., Buttgereit, F., Cutolo, M., 2011. Alterations of the hypothalamic-pituitary-adrenal axis in systemic immune diseases—a role for misguided energy regulation. Clin. Exp. Rheumatol. 29 (5 Suppl. 68), S23–S31.

Stumvoll, M., Goldstein, B.J., van Haeften, T.W., 2005. Type 2 diabetes: principles of pathogenesis and therapy. Lancet 365, 1333–1346.

Subramanian, S., Blanton, L.V., Frese, S.A., et al., 2015. Cultivating healthy growth and nutrition through the gut microbiota. Cell. 161, 36–48.

Sukumar, P., Viswambharan, H., Imrie, H., et al., 2013. Nox2 NADPH oxidase has a critical role in insulin resistance–related endothelial cell dysfunction. Diabetes 62, 2130–2134.

Sun, Y.N., Liu, L.B., Xue, Y.X., Wang, P., 2015. Effects of insulin combined with idebenone on blood-brain barrier permeability in diabetic rats. J. Neurosci. Res. 93, 666–677.

Sun, H., Olson, K.C., Gao, C., Prosdocimo, D.A., Zhou, M., Wang, Z., et al., 2016a. Catabolic defect of branched-chain amino acids promotes heart failure. Circulation 133, 2038–2049.

Sun, X., Jiao, X., Ma, Y., Liu, Y., Zhang, L., He, Y., et al., 2016b. Trimethylamine N-oxide induces inflammation and endothelial dysfunction in human umbilical vein endothelial cells via activating ROS-TXNIP-NLRP3 inflammasome. Biochem. Biophys. Res. Commun. 481, 63–70.

Symons, J.D., Abel, E.D., 2013. Lipotoxicity contributes to endothelial dysfunction: a focus on the contribution from ceramide. Rev. Endocr. Metab. Disord. 14, 59–68.

Tabassum, R., Chauhan, G., Dwivedi, O.P., et al., 2013. Genome-wide association study for type 2 diabetes in Indians identifies a new susceptibility locus at 2q21. Diabetes 62, 977–986.

Tagami, S., Inokuchi, Ji. J., Kabayama, K., Yoshimura, H., Kitamura, F., Uemura, S., et al., 2002. Ganglioside GM3 participates in the pathological conditions of insulin resistance. J. Biol. Chem. 277, 3085–3092.

Tagliamonte, A., DeMontis, M.G., Olianas, M., Onali, P.L., Gessa, G.L., 1976. Possible role of insulin in the transport of tyrosine and tryptophan from blood to brain. Adv. Exp. Med. Biol. 69, 89–94.

Taguchi, A., White, M.F., 2008. Insulin-like signaling, nutrient homeostasis, and life span. Annu. Rev. Physiol. 70, 191–212.

Taheri, S., Lin, L., Austin, D., Young, T., Mignot, E., 2004. Short sleep duration is associated with reduced leptin, elevated ghrelin, and increased body mass index. PLoS Med. 1, e62.

Talbot, K., Wang, H., Kazi, H., Han, L., Bakshi, K.P., Stucky, A., et al., 2012. Demonstrated brain insulin resistance in Alzheimer's disease patients is associated with IGF1 resistance, IRS1 dysregulation, and cognitive decline. J. Clin. Investig. 122, 1316–1338.

Tanaka, M., Suganami, T., Sugita, S., Shimoda, Y., Kasahara, M., Aoee, S.S., et al., 2010. Role of central leptin signaling in renal macrophage infiltration. Endocr. J. 57, 61–72.

Tanaka, M., Suganami, T., Kim-Saijo, M., Toda, C., Tsuiji, M., Ochi, K., et al., 2011. Role of central leptin signaling in the starvation-induced alteration of B-cell development. J. Neurosci. 31, 8373–8380.

Tang, W.H., Wang, Z., Levison, B.S., Koeth, R.A., Britt, E.B., Fu, X., et al., 2013. Intestinal microbial metabolism of phosphatidylcholine and cardiovascular risk. N. Engl. J. Med. 368, 1575–1584.

Taubes, G., 2009. Insulin resistance. Prosperity's plague. Science 325, 256–260.

Thaler, J.P., Schwartz, M.W., 2010. Minireview: inflammation and obesity pathogenesis: the hypothalamus heats up. Endocrinology 151, 4109–4115.

Thaler, J.P., Yi, C.-X., Schur, E.A., Guyenet, S.J., Hwang, B.H., Dietrich, M.O., et al., 2012. Obesity is associated with hypothalamic injury in rodents and humans. J. Clin. Invest. 122, 153–162.

Thomas, L.A., De Bellis, M.D., Graham, R., LaBar, K.S., 2007. Development of emotional facial recognition in late childhood and adolescence. Dev. Sci. 10, 547–558.

Tiganis, T., 2011. Reactive oxygen species and insulin resistance: the good, the bad and the ugly. Trends Pharmacol. Sci. 32, 82–89.

Timmers, S., Schrauwen, P., de Vogel, J., 2008. Muscular diacylglycerol metabolism and insulin resistance. Physiol. Behav. 94, 242–251.

Tilg, H., Moschen, A.R., 2006. Adipocytokines: mediators linking adipose tissue, inflammation and immunity. Nat. Rev. Immunol. 6, 772–783.

Titchenell, P.M., Quinn, W.J., Lu, M., Chu, Q., Lu, W., Li, C., et al., 2016. Direct hepatocyte insulin signaling is required for lipogenesis but is dispensable for the suppression of glucose production. Cell. Metab. 23, 1154–1166.

Tong, M., de la Monte, S.M., 2009. Mechanisms of ceramide-mediated neurodegeneration. J. Alzheimers. Dis. 16, 705–714.

Tremaroli, V., Bäckhed, F., 2012. Functional interactions between the gut microbiota and host metabolism. Nature 489, 242–249.

Trogdon, J.G., Finkelstein, E.A., Feagan, C.W., Cohen, J.W., 2012. State and payer-specific estimates of annual medical expenditures attributable to obesity. Obesity. 20, 214–220.

Turnbaugh, P.J., Bäckhed, F., Fulton, L., Gordon, J.I., 2008. Diet-induced obesity is linked to marked but reversible alterations in the mouse distal gut microbiome. Cell. Host. Microbe. 3, 213–223.

Turnbaugh, P.J., Hamady, M., Yatsunenko, T., Cantarel, B.L., Duncan, A., Ley, R.E., et al., 2009. A core gut microbiome in obese and lean twins. Nature 457, 480–484.

Unger, R., 2003. Lipid overload and overflow: metabolic trauma and the metabolic syndrome. Trends Endocrinol. Metab. 14, 398–403.

Urano, F., Wang, X., Bertolotti, A., Zhang, Y., Chung, P., Harding, H.P., et al., 2000. Coupling of stress in the ER to activation of JNK protein kinases by transmembrane protein kinase IRE1. Science 287, 664–666.

Valassi, E., Scacchi, M., Cavagnini, F., 2008. Neuroendocrine control of food intake. Nutr. Metab. Cardiovasc. Dis. 18, 158–168.

Valdes, A.M., Andrew, T., Gardner, J.P., Kimura, M., Oelsner, E., Cherkas, L.F., et al., 2005. Obesity, cigarette smoking, and telomere length in women. Lancet 366, 662–664.

Van Dam, P.S., Aleman, A., 2004. Insulin-like growth factor-I, cognition and brain aging. Eur. J. Pharmacol. 490, 87–95.

Van der Klaauw, A.A., Farooqi, I.S., 2015. The hunger genes: pathways to obesity. Cell. 161, 119–132.

Vanorsdall, T.D., Jinnah, H.A., Gordon, B., Kraut, M., Schrelten, D., 2008. Cerebral ischemia mediates the effect of serum uric acid on cognitive function. Stroke 39, 3418–3420.

Vaz, F.M., Wanders, R.J., 2002. Carnitine biosynthesis in mammals. Biochem. J. 361, 417–429.

Vincent, H.K., Taylor, A.G., 2006. Biomarkers and potential mechanisms of obesity-induced oxidant stress in humans. Int. J. Obes. 30, 400–418.

Virag, L., Szabo, C., 2002. The therapeutic potential of poly(ADP-ribose) polymerase inhibitors. Pharmacol. Rev. 54, 375–429.

Virtue, S., Vidal-Puig, A., 2008. It's not how fat you are, it's what you do with it that counts. PLoS Biol. 6, e237.

Voight, B.F., Scott, L.J., Steinthorsdottir, V., et al., 2010. Twelve type 2 diabetes susceptibility loci identified through large-scale association analysis. Nat. Genet. 42, 579–589.

Volkow, N.D., Wang, G.J., Baler, R.D., 2011. Reward, dopamine and the control of food intake: implications for obesity. Trends Cognit. Sci. 15, 37–46.

Wang, Q., Jin, T., 2009. The role of insulin signaling in the development of beta-cell dysfunction and diabetes. Islets. 1, 95–101.

Wang, Z., Klipfell, E., Bennett, B.J., Koeth, R., Levison, B.S., Dugar, B., et al., 2011. Gut flora metabolism of phosphatidylcholine promotes cardiovascular disease. Nature 472, 57–63.

Wang, W., Zhang, F., Xia, Y., Zhao, S., Yan, W., Wang, H., et al., 2016. Defective branched chain amino acid catabolism contributes to cardiac dysfunction and remodeling following myocardial infarction. Am. J. Physiol. Heart Circ. Physiol. 311, H1160–H1169.

Wellen, K.E., Hotamisligil, G.S., 2005. Inflammation, stress, and diabetes. J. Clin. Invest. 115, 1111–1119.

White, M.F., 2003. Insulin signaling in health and disease. Science 302, 1710–1711.

Wilcox, G., 2005. Insulin and insulin resistance. Clin. Biochem. Rev. 26, 19–39.

Williams, K.W., Scott, M.M., Elmquist, J.K., 2011. Modulation of the central melanocortin system by leptin, insulin, and serotonin: co-ordinated actions in a dispersed neuronal network. Eur. J. Pharmacol. 660, 2–12.

Williamson, D.F., 1998. Weight loss and mortality in persons with type-2 diabetes mellitus: a review of the epidemiological evidence. Exp. Clin. Endocrinol. Diabetes. 106 (Suppl. 2), 14–21.

Whitmer, R.A., Sidney, S., Selby, J., Johnston, S.C., Yaffe, K., 2005. Midlife cardiovascular risk factors and risk of dementia in late life. Neurology 64, 277–281.

Wolfson, R.L., Chantranupong, L., Saxton, R.A., Shen, K., Scaria, S.M., Cantor, J.R., et al., 2015. Sestrin2 is a leucine sensor for the mTORC1 pathway. Science 351, 43–48.

Wu, G.D., Chen, J., Hoffmann, C., Bittinger, K., Chen, Y.Y., Keilbaugh, S.A., et al., 2011. Linking long-term dietary patterns with gut microbial enterotypes. Science 334, 105–108.

Xu, H., Barnes, G.T., Yang, Q., Tan, G., Yang, D., Chou, C.J., et al., 2003. Chronic inflammation in fat plays a crucial role in the development of obesity-related insulin resistance. J. Clin. Invest. 112, 1821–1830.

Xydakis, A.M., Case, C.C., Jones, P.H., Hoogeveen, R.C., Liu, M.Y., Smith, E.O., et al., 2004. Adiponectin, inflammation and the expression of the metabolic syndrome in

obese individuals: the impact of rapid weight loss through caloric restriction. J. Clin. Endocrinol. Metab. 89, 2697–2703.
Yamashita, T., Hashiramoto, A., Haluzik, M., Mizukami, H., Beck, S., Norton, A., et al., 2003. Enhanced insulin sensitivity in mice lacking ganglioside GM3. Proc. Natl. Acad. Sci. U.S.A. 100, 3445–3449.
Yamauchi, T., Kamon, J., Waki, H., Imai, Y., Shimozawa, N., Hioki, K., 2003. Globular adiponectin protected ob/ob mice from diabetes and ApoE-deficient mice from atherosclerosis. J. Biol. Chem. 278, 2461–2468.
Yang, L., Li, P., Fu, S., Calay, E.S., Hotamisligil, G.S., 2010. Defective hepatic autophagy in obesity promotes ER stress and causes insulin resistance. Cell Metab. 11, 467–478.
Yatsunenko, T., Rey, F.E., Manary, M.J., Trehan, I., Dominguez-Bello, M.G., Contreras, M., et al., 2012. Human gut microbiome viewed across age and geography. Nature 486, 222–227.
Ye, J., 2007. Role of insulin in the pathogenesis of free fatty acid-induced insulin resistance in skeletal muscle. Endocr. Metab. Immune. Disord. Drug Targets. 7, 65–74.
Ye, J., McGuinness, O.P., 2013. Inflammation during obesity is not all bad: evidence from animal and human studies. Am. J. Physiol. Endocrinol. Metab. 304, E466–E477.
Yu, C., Chen, Y., Cline, G.W., Zhang, D., Zong, H., Wang, Y., et al., 2002. Mechanism by which fatty acids inhibit insulin activation of insulin receptor substrate-1 (IRS-1)-associated phosphatidylinositol 3-kinase activity in muscle. J. Biol. Chem. 277, 50230–50236.
Zhang, Y., Zhang, H., 2013. Microbiota associated with type 2 diabetes and its related complications. Food Sci. Hum. Wellness. 2, 167–172.
Zhang, W., Hietakangas, V., Wee, S., Lim, S.C., Gunaratne, J., Cohen, S.M., 2013. ER stress potentiates insulin resistance through PERK-mediated FOXO phosphorylation. Genes Dev. 27, 441–449.
Zhao, W.Q., Alkon, D.L., 2001. Role of insulin and insulin receptor in learning and memory. Mol. Cell. Endocrinol. 177, 125–134.
Zhao, W.Q., Chen, H., Quon, M.J., Alkon, D.L., 2004. Insulin and the insulin receptor in experimental models of learning and memory. Eur. J. Pharmacol. 490, 71–81.
Zhao, H., Przybylska, M., Wu, I.H., Zhang, J., Siegel, C., Komarnitsky, S., et al., 2007. Inhibiting glycosphingolipid synthesis improves glycemic control and insulin sensitivity in animal models of type 2 diabetes. Diabetes 56, 1210–1218.
Zhu, W., Gregory, J.C., Org, E., Buffa, J.A., Gupta, N., Wang, Z., et al., 2016. Gut microbial metabolite TMAO enhances platelet hyperreactivity and thrombosis risk. Cell. 165, 111–124.

Further reading

Abraham, M.A., Filippi, B.M., Kang, G.M., Kim, M.S., Lam, T.K., 2014. Insulin action in the hypothalamus and dorsal vagal complex. Exp. Physiol. 99, 1104–1109.
Harrington, M., Gibson, S., Cottrell, R.C., 2009. A review and meta-analysis of the effect of weight loss on all-cause mortality risk. Nurt. Res. Rev. 22, 93–108.
Mazon, J.H., de Mello, A.D., Ferreira, G.K., Rezin, G.T., 2017. The impact of obesity on neurodegenerative diseases. Life Sci. 182, 22–28.
Summers, S.A., Garza, L., Zhou, H., Birnbaum, M.J., 1998. Regulation of insulin-stimulated glucose transporter GLUT4 translocation and Akt kinase activity by ceramide. Mol. Cell. Biol. 18, 5457–5464.

Uysal, K.T., Wiesbrock, S.M., Marino, M.W., Hotamisligil, G.S., 1997. Protection from obesity-induced insulin resistance in mice lacking TNF-alpha function. Nature 389, 610–614.

Van Houten, M., Posner, B.I., Kopriwa, B.M., Brawer, J.R., 1979. Insulin binding sites in the rat brain: in vivo localisation to the circumventricular organs by quantitative autoradiography. Endocrinology. 105, 666–673.

CHAPTER 2

Insulin resistance, diabetes, and metabolic syndrome

Introduction

Glucose is a major circulating sugar and primary fuel for the animal tissues for energy production. Glucose enters in cells by the membrane transporter glucose transporter type 1 (GLUT1) and GLUT4. In the brain, glucose is metabolized in a similar manner to liver. Hypothalamus plays an important role in control of food intake and energy expenditure (Cha et al., 2008). Insulin receptor–expressing neurons in the rat are localized in important hypothalamic and hindbrain areas that modulate glucose homeostasis, energy intake and expenditure, and neuroendocrine and autonomic functions (Unger et al., 1991). Insulin signaling produces different effects in the brain and peripheral organs. In brain, insulin receptor signaling leads to increase in serum glucose levels, decreases in food intake and body weight, and produces decrease in reproductive function (Bruning et al., 2000). Activation of insulin receptors in hypothalamus involves the activation of phosphoinositide 3-kinase (PtdIns 3K) and mitogen-activated protein kinase (MAPK) signaling pathways (Unger et al., 1991). This activation produces both short- and long-term changes in neuronal activity. The prolonged changes in neuronal activity include increase in gene transcription, changes in neuroplasticity, and modulation of cognitive function (Spanswick et al., 2000; Levin et al., 2006).

Diabetes is a chronic metabolic condition, which is characterized by chronic hyperglycemia, with alterations in carbohydrate, fat, and protein metabolism due to defects in secretion and/or insulin action (American Diabetes Association, 2014a,b). Insulin resistance is defined as reduction or lack of insulin sensitivity to the target tissues, such as adipose tissue, muscles, liver, and the brain (Wilcox, 2005). Diabetes has become an epidemic and remains a major public health issue. Two types of diabetes, namely type 1 and type 2 diabetes, are known to occur in human population. The molecular mechanisms associated with insulin resistance in type 1 and type 2 diabetes are not clearly understood. However, it is becoming

Insulin Resistance as a Risk Factor in Visceral and Neurological Disorders
DOI: https://doi.org/10.1016/B978-0-12-819603-8.00002-X

increasingly evident that type 1 diabetes is an autoimmune disease, which is characterized by the selective destruction of the insulin-producing islet β-cells in the pancreas leading to the deficiency of insulin and insulin resistance (Kawasaki et al., 1999; Stumvoll et al., 2005). In type 1 diabetes, in spite of excess of glucose in the blood, cells become starved of energy. This is then followed by life-threatening conditions of hypoglycemia, low blood sugar, and hyperglycemia, high blood sugar. When hypoglycemia develops, cells do not get enough glucose and patients suffer of confusion, loss of consciousness, and coma. Even death can result when the brain is deprived of glucose for too long. The hallmark of autoimmune type 1 diabetes is T cell–mediated destruction of β-cells, which results from an imbalance in the activity between disease promoting autoreactive effector T cells and protective elements, which is now known as regulatory T cells (Kawasaki et al., 1999). Furthermore, insulitis, the other pathologic hallmark of type 1 diabetes, is an inflammatory lesion consisting of immune cell infiltrates around and within the islets (Krishna and Srikanta, 2015; Pugliese, 2016). Type 1 diabetes accounts for only 5%–10% cases of diabetes. Markers of the immune destruction of the β-cell in type 1 diabetes include islet cell autoantibodies, autoantibodies to insulin, autoantibodies to glutamic acid decarboxylase (GAD_{65}), and autoantibodies to the tyrosine phosphatases IA-2 and IA-2β. One and usually more of these autoantibodies are present in 85%–90% of individuals when fasting hyperglycemia is initially detected. Also, the disease has strong HLA associations, with linkage to the *DQA* and *DQB* genes, and it is influenced by the *DRB* genes. These *HLA-DR/DQ* alleles can be either predisposing or protective (Krishna and Srikanta, 2015; Paschou et al., 2018).

In contrast, type 2 diabetes (diabetes mellitus) is a complex and disorder of chronic lifestyle, which is accompanied by persistent hyperglycemia, elevations in free fatty acids (FFAs), and increases in circulating cytokines (Stumvoll et al., 2005). Chronic exposure of β cell to these conditions induces excessive production of reactive oxygen species (ROS) and activation of caspases, which inhibit insulin secretion and promote apoptosis of pancreatic β cells (Andersson et al., 2001). Type 2 diabetes affects brain not only by altering glycolysis (Zheng et al., 2017), decreasing neuronal mitochondrial activity, attenuating brain-derived neurotrophic factor levels, desensitizing insulin resistance possibly as a consequence of depressed adiponectin levels (Ng et al., 2016), but also high levels of proinflammatory cytokines (Geng et al., 2018), exaggerated β-site APP-cleaving enzyme 1-mediated amyloid biogenesis (Lee et al., 2016), and tau phosphorylation (Platt et al., 2016) along with induction of hyperglycemia and increase in blood–brain barrier permeability (Zlokovic, 2011). Induction of these process result in cerebral extravasation of plasma molecules, neuroinflammatory and oxidative sequalae, and thereafter, progressive loss of neuronal function and neuronal apoptosis (Ramirez et al., 2009).

Type 2 diabetes affects 8.3% of the adult population of the world. According to the International Diabetes Federation in 2015, an estimated 415 million people globally were suffering from this condition (International Diabetes Federation, 2015). The clinical manifestation of a complex metabolic disease like type 2 diabetes is delayed by years, thereby restricting its timely diagnosis. Complications of type 2 diabetes account for increased morbidity, disability, and mortality and represent a threat for the economies of all countries, especially the developing ones (Kautzky-Willer et al., 2016; Papatheodorou et al., 2016). The etiology of type 2 diabetes is linked with a number of unhealthy lifestyle factors such as excess weight (obesity), genes, inactivity, and smoking (Pan et al., 2015). Among them excess weight is the strongest risk factor according to a meta-analysis based on 18 prospective cohort studies (Abdullah et al., 2010). Low birth weight, proposedly reflecting fetal malnutrition, is also associated with an increased risk of type 2 diabetes (Whincup et al., 2008). The number of people with this disorder is increasing at an alarming rate (Cho et al., 2018; Nolan et al., 2011). The fact that the number of subjects with type 2 diabetes has doubled over the past three decades has made this disease a global challenge (Shaw et al., 2010). The number of diabetes mellitus patients is projected to increase from 382 million in 2013 to 592 million by 2035, denoting a net increase of 55% (Cho et al., 2018).

At the molecular level, type 2 diabetes is characterized by impaired insulin secretion, β-cell dysfunction, aberrant insulin signaling, and insulin resistance (Stumvoll et al., 2005). Type 2 diabetes often occurs in adulthood when the body becomes resistant to insulin or fails to make sufficient amounts of insulin. Many factors contribute to type 2 diabetes including insulin resistance, hyperglycemia, abdominal obesity, genetic and epigenetic factors, and environmental factors (Fig. 2.1). Excess consumption of nutrients promotes the pathogenesis and progression of type 2 diabetes and chronic inflammation not only via the activation of toll-like receptors (TLRs) and increased expression of adipocytokines, proinflammatory cytokines, and chemokines but also through the induction of

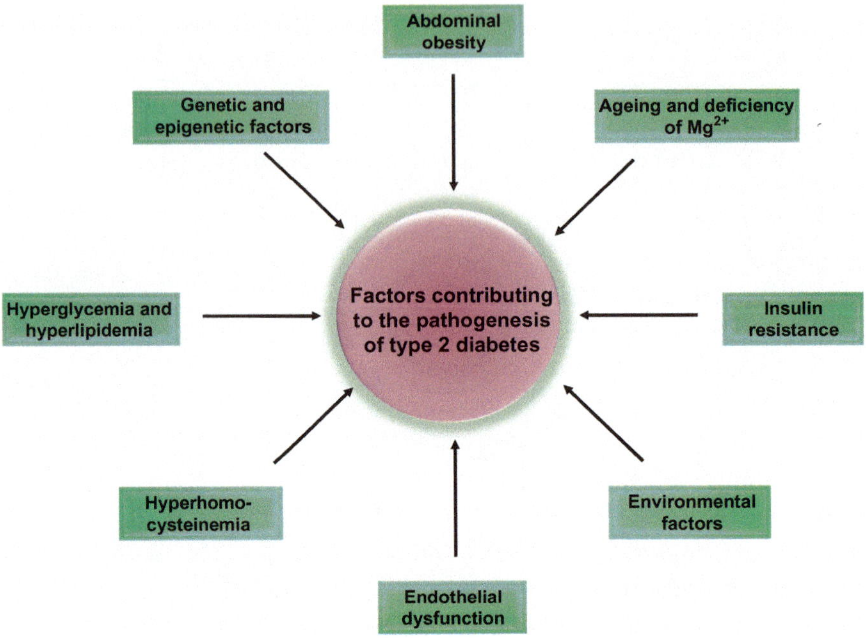

Figure 2.1 Factors contributing to the pathogenesis of type 2 diabetes.

endoplasmic reticulum stress. The involvement of inflammation in the pathogenesis of insulin resistance is traced by the integration of metabolism and innate immunity via nutrient-sensing pathways mutual to pathogen-sensing pathways. Various components of nutrition (FFAs, glucose, and amino acids) are metabolized to proinflammatory molecules such as cytokines.

The pathogenesis of type 2 diabetes starts with the induction of insulin resistance, which is promoted with the accumulation of high levels of long-chain FFAs, diacylglycerol, ceramide, GM_3 ganglioside, and adipocytokines in cells (Fig. 2.2). These mediators block glucose uptake leading to hyperglycemia (Farooqui and Farooqui, 2013). Insulin resistance can also be induced by different environmental factors, including vitamin D deficiency (i.e., hypovitaminosis D) (Leung, 2016). The consumption of energy-dense/high-fat and high-carbohydrate diet also promotes obesity and insulin resistance, which negatively effects insulin sensitivity, particularly when the excess of body fat is located in abdominal region. However, the link between consumption of high-fat diet and obesity is not limited to the high-energy content of fatty foods. In some individuals, the ability to oxidize and metabolize dietary fat under hyperglycemic

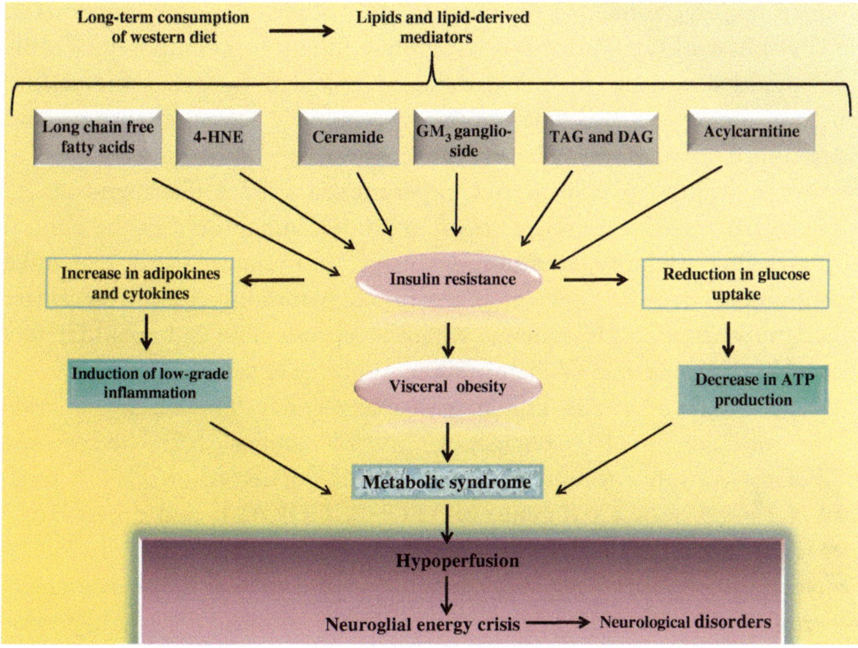

Figure 2.2 Contribution of lipid mediators in the development of insulin resistance and metabolic syndrome.

conditions is impaired genetically predisposing subjects to obesity and insulin resistance (Farooqui, 2013).

As stated in Chapter 1, Insulin resistance and obesity, the molecular mechanism, which is associated with the development of insulin resistance, involves impairment in the insulin signaling pathway in insulin-responsive cells (adipocytes, myocytes, hepatocytes, β-cells, and neural cells). Normally, when insulin binds to the insulin receptor on these cells, the insulin receptor is autophosphorylated at its Tyr residues and its tyrosine kinase is activated (Pilch and Lee, 2004; Huang et al., 2018). The insulin receptor then phosphorylates tyrosine residues on the insulin receptor substrates (IRSs), which then serve as docking proteins for SH2-containing enzymes such as p85 subunit of PtdIns 3K or protein tyrosine phosphatase 2. This leads to linear signaling cascades that result in Akt activation (Huang et al., 2018). The activation of Akt induces the translocation of GLUT4 and glycogen synthesis and thus plays an important role in metabolic signaling. Hence, disruption of this insulin signaling cascade can induce insulin resistance and is associated with the development of obesity and type 2 diabetes. These alterations in insulin signaling also produce negative influence on memory and other cognitive functions in the later phase of type 2 diabetes. Cognitive function is composed of multiple cognitive domains including memory, calculation, spatial orientation, structure ability and executive ability, language comprehension, and expression (Dolan, 2002). The influence of diabetes on cognitive domains varies significantly in clinical and experimental studies (Sadanand et al., 2016). A battery of screening tests for cognitive impairment in the clinical setting include memory, attention, language, and visuospatial or executive functioning. The Morris water maze test is commonly used to evaluate the cognitive function in rodents, which is reproducible and quantitatively countable (Vorhees and Williams, 2006; Webster et al., 2014).

Type 2 diabetes is also linked with many genes. These genes include genes for calpain 10, potassium inward-rectifier 6.2, peroxisome proliferator-activated receptor gamma (PPAR-γ), IRS-1, *KCNJ11* gene, *TCF7L2* gene, *ABCCC8* gene, NOTCH2, CDKAL1, and others along with genes for adiponectin (Stumvoll et al., 2005; Kampoli et al., 2011). Converging evidence suggests that hyperglycemia, insulin resistance, hyperinsulinemia, hyperlipidemia (in particular elevated FFAs), hyperhomocysteinemia, and deficiency of magnesium (Mg^{2+}) are important pathophysiological components of type 2 diabetes, cardiovascular, and kidney disease. In addition, long-term consumption of western diet, genetic and epigenetic factors, ageing, and physical inactivity may also lead to not only obesity and insulin resistance, but also ultimately type 2 diabetes. These parameters also trigger systemic inflammation and impair nitric oxide (NO) bioavailability, with consequent impair endothelial function (Kampoli et al., 2011; Tang et al., 2014). In type 2 diabetes, impairment in NO production, increase oxidative stress, and changes in function of endothelial progenitor cells contribute to the accelerated atherosclerotic process. Abnormalities in endothelial function worsen with increased weight burden owing to several mechanisms associated with excess fat mass including impaired glucose tolerance, insulin resistance, metabolic dysregulation, adipocytokine release, and systemic inflammation that play a key role in the evolution and clinical expression of cardiovascular disease (Arkin et al., 2008). Although endothelial dysfunction is a strong predictor of cardiovascular events, restoring arterial homeostasis reduces vascular risk (Modena et al., 2002). Recent studies have indicated that short-term weight loss improves endothelial function within weeks via mechanisms that relate more closely to metabolic changes than degree or mode of weight intervention (Gokce et al., 2005; Sciacqua et al., 2003).

Biomarkers for type 2 diabetes

As mentioned earlier, lifestyle, environment, and socioeconomic factors have different impacts on human males and females. It is well known that sex hormones have different impacts on energy metabolism, body composition, vascular function, and inflammatory responses in human males and females. Thus endocrine imbalances relate to unfavorable cardiometabolic traits, observed in women with androgen excess or men with hypogonadism. Both biological and psychosocial factors are responsible for sex and gender differences in diabetes risk and outcome (Kautzky-Willer et al., 2016). It is stated that overall, psychosocial stress has greater impact on women rather than on men. In addition, women have greater increases of cardiovascular risk, myocardial infarction, and stroke mortality than men, compared with nondiabetic subjects (Kautzky-Willer et al., 2016). The prevalence, progression, and pathophysiology of both microvascular (nephropathy, neuropathy, and retinopathy) and macrovascular (coronary heart disease, myocardial infarction, peripheral arterial disease, and stroke) disease are different in man and women. In general, men appear to be at a higher risk for diabetic microvascular complications, while the consequences of macrovascular complications may be greater in women. Interestingly, in the absence of diabetes, women have a far lower risk of either micro- or macro-vascular disease compared with men for much of their lifespan. Currently, glycated hemoglobin (HbA1c), fructosamine, glycated albumin, and blood glucose levels are used as biomarkers. These classical biomarkers have their own limitations in terms of moderate sensitivity, specificity, and inaccuracies in certain clinical conditions. HbA1c is formed when glucose attaches to the amino-terminal group of the β subunit of hemoglobin (American Diabetes Association, 2014a). HbA1c reflects chronic glycemia rather than glucose levels at a single time point. Currently, the adenosine deaminase criteria for diabetes are HbA1c $\geq$6.5% (48 mmol/mol) and 5.7%–6.4% (39–46 mmol/mol) for prediabetes (Bookchin and Gallop, 1968). Increased HbA1c levels are associated with increased morbidity and mortality. The monitoring of HbA1c represents the exposure of hemoglobin to circulating glycemia in the previous 3 months, glycated albumin and fructosamine represent exposure for a shorter period, which may be beneficial to monitor rapid metabolic alterations or changes in diabetes treatment. Based on this information, it is proposed that prognosis of disease at early stages and prediction of population at a higher risk require identification of specific markers that are sensitive enough to be detected at early stages of type 2 diabetes (Vaishya et al., 2018). Recent studies have indicated that in diabetic women hepatokine fetuin A is an important biomarker for type 2 diabetes only in women (Laughlin et al., 2013). Other biomarkers in women are copeptin (the C-terminal portion of the precursor of vasopressin) (Abbasi et al., 2012), proneurotensin (precursor molecule of neurotensin) (Daniels and Maisel, 2015), netrin (Yimer et al., 2018), and vitamin D3 deficiency (Stadlmayr et al., 2015). Among these biomarkers, netrin, a family of extracellular, laminin-related proteins, has attracted considerable attention (Ranganathan et al., 2014). Netrin has four members namely Netrin-1, Netrin-3, Netrin-4, and dual glycosylphosphatidylinositol-attached membrane peptides (Netrin G1 and G2) (Dun and Parkinson, 2017). The biological roles of netrins are not fully understood. However, it is proposed that netrin regulates cell migration, cell—cell interactions, and cell—extracellular matrix adhesion during the embryonic development of multiple tissues, including the nervous system, vasculature, lung, pancreas, muscle, and mammary gland. It can be used as a novel biomarker and therapeutic modality for early identification of type 2 diabetes and related complications (Yimer et al., 2018). Netrin mediates it action by interacting with two well-recognized receptor families, namely, the deleted receptors in colorectal cancer and the uncoordinated 5 receptors (Gao et al., 2016; Övünç Hacıhamdioğlu et al., 2016). In addition, netrin also interacts with CD46 receptors such as CD146 (also known as melanoma cell adhesion molecule and Down syndrome cell adhesion molecule) (Sun et al., 2011; Tu et al., 2015).

miRNAs, diabetes, and vascular complications

miRNAs are a group of small (20—25 nt long) noncoding RNAs, which normally binds to the 3′ end of its target mRNAs to inhibit its translation, eventually leading to a reduced gene expression (Ebert and Sharp, 2012). Due to their stability and presence in various body fluids, miRNAs emerged as potential biomarkers for type 2 diabetes and related complications. Besides, the differential expression of miRNAs in various tissues has been reported in type 2 diabetes and related complications. A number of pancreatic B-cell—specific miRNAs have been identified, including miR-375, miR-124a, miR-96, miR-7a, miR7a2, miR-30d, miR-9, miR-200, miR-184, and let-7 (Banerjee et al., 2017). These miRNAs are associated with pancreatic function, insulin secretion, and glucose tolerance.

Differential miRNA signatures have been identified among prediabetic individuals, patients with diabetes, and patients with diabetes and vascular complications, supporting the view that miRNAs may be novel biomarkers. Diabetic cardiovascular complications are associated with increased levels of miR-223, miR-320, miR-501, miR504, and miR1 and decreased levels of miR-16, miR-133, miR-492, and miR-373 (Zhang et al., 2017). Whether these changes in miRNA are simply biomarkers of disease or whether they are directly related to the vasculopathy of diabetes is not known, and more studies are needed in young and aged type 2 diabetes patients.

Comlications caused by type 2 diabetes

Studies on epidemiology of type 2 diabetes have indicated that gender, age, and ethnic background are important factors when considering the development of type 2 diabetes and its complications (Li and Aronow, 2011). As mentioned earlier, genes and epigenetic mechanisms, nutritional factors, and sedentary lifestyle modulate the risk of type 2 diabetes. During the pathogenesis of type 2 diabetes, the onset of chronic hyperglycemia is accompanied by high susceptibility of patients to various forms of both short- and long-term complications. Among the short-range diabetes complications, ketoacidosis, hyperosmolar hyperglycemic state, and comma are the commonly encountered problems. Long-term complications of type 2 diabetes are cardiovascular diseases, cerebrovascular diseases, renal disorders, inflammation and immunity, and obesity (Uppu and Parinandi, 2011). These complications affect many organs such as liver, kidney, and brain. They can be categorized into vascular or nonvascular complications. Vascular complications can be microvascular (microangiopathy), such as retinopathy or nephropathy, or macrovascular (macroangiopathy), including cerebrovascular disease, coronary artery disease, and peripheral vascular disease (Papatheodorou et al., 2016). The vascular complications of diabetes are the most serious manifestations of the disease. Atherosclerosis is the main reason for impaired life expectancy in patients with diabetes, whereas diabetic nephropathy and retinopathy are the largest contributors to end-stage renal disease, sexual dysfunction, and blindness (Fig. 2.3). Nonvascular complications include forms of chronic neuropathy. In addition to hyperglycemia, other factors, including hypertension, dyslipidemia, genetic factors, glucose oscillations, hypoglycemia, obesity, coagulopathy and smoking, contribute to the development of

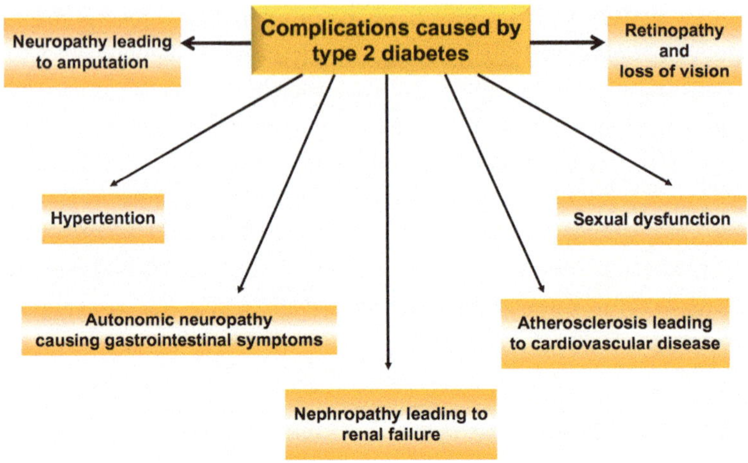

Figure 2.3 Complications caused by type 2 diabetes.

chronic complications of diabetes (Haffner and Cassells, 2003; Gan, 2003; Ceriello, 1993). The leading cause of death in people with type 2 diabetes is heart and blood vessel disease (70%–80%). The risk of heart and blood vessel disease is eight times higher in patients with diabetes. Patients with type 2 diabetes without previous heart attacks are at higher risk of a heart attack as compared with nondiabetic patients who have previously had a heart attack. Diabetes is the most common risk factor for stroke, especially in women (5.4 times higher risk). Diabetic retinopathy is a significant cause of blindness, nephropathy is the most important cause of kidney failure, and diabetic foot is the primary cause of lower extremity amputations and the most important cause of disability in patients (Haffner and Cassells, 2003; Brownlee, 2001; Wild et al., 2004; Howard et al., 2002). Type 2 diabetes complications also include sexual dysfunction (Fig. 2.4) (American Diabetes Association, 2009). Additional complications type 2 diabetics are higher body mass index (BMI), metabolic syndrome (MetS), atherosclerosis, heart disease, and increase in bone fracture (Moseley, 2012). Collective evidence suggests that complications of type 2 diabetes account for increased morbidity, disability, and mortality (Kautzky-Willer et al., 2016).

In type 2 diabetes patients, excess of nutrients activates, macrophages, and adipocytes through common receptors, such as TLRs sense broad classes of molecular structures common to pathogen groups and are central to innate immunity and inflammation (Fig. 2.5). TLR signaling pathways involve not only receptor dimerization but also recruitment of

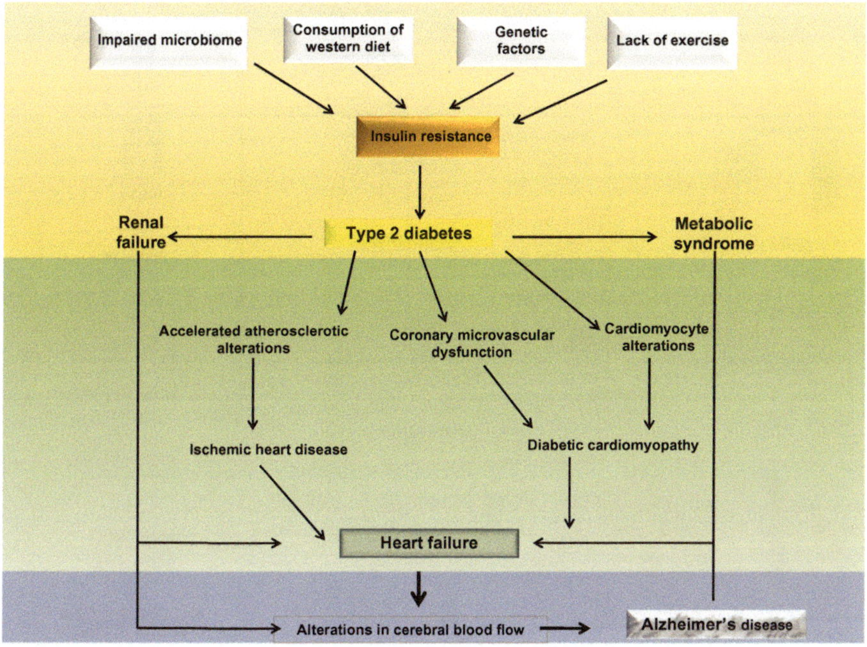

Figure 2.4 Progression of type 2 diabetes–mediated changes in the development of heart failure.

adapter proteins that mediate other protein–protein interactions, such as myeloid differentiation primary-response protein 88 (MyD88) and Toll/interleukin 1 receptor domain-containing adapter interferon-β (TRIF; Takeda et al., 2003; Yamamoto et al., 2003). All TLR signaling pathways require MyD88, except TLR3, which initiates signaling via the TRIF adapter. Furthermore, the activation of TLR4 requires both MyD88 and TRIF-related adapter molecule (Fitzgerald et al., 2003; Ruckdeschel et al., 2004). The MyD88-dependent pathway TLR signaling initiates with the recruitment of tumor necrosis factor receptor–associated factor 6 (TRAF6) and members of the IL-1R–associated kinases family. The activation of TRAF6 allows the translocation of nuclear factor kappaB (NF-κB) to the nucleus, where it interacts with NF-κB response element and promotes the expression of different proinflammatory cytokines, chemokines, and proinflammatory enzymes such as cyclooxygenase-2 and inducible nitric oxide synthase (iNOS) (Takeda and Akira, 2004; Broad et al., 2007; Kawai and Akira, 2010; http://www.nf-kb.org/target/index.html). NF-κB also drives expression of target genes that mediate cell proliferation and release of antimicrobial molecules to activate the immune

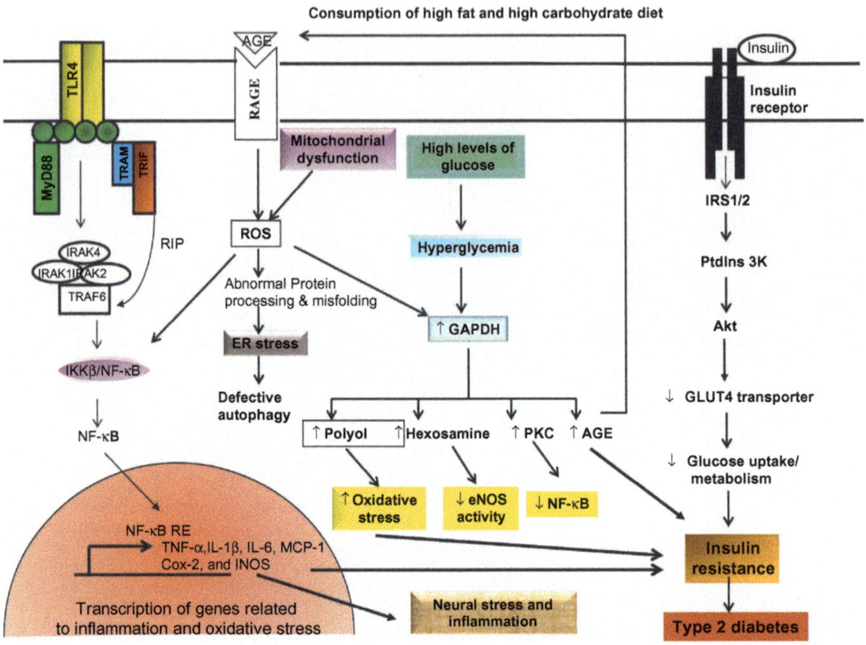

Figure 2.5 Contribution of toll-like receptor, advanced glycated end products receptors, and insulin receptors in the pathogenesis of type 2 diabetes.

response (Hayden and Ghosh, 2008). Induction of inflammation is also modulated by the fatty acid composition of the diet. An increase in FFA levels, and more importantly, the relative amounts of saturated and unsaturated fatty acids contribute to the development of inflammation and insulin resistance, which occurs in several tissues, including the liver, muscle, adipose tissue, and the brain. Other inflammatory pathways are induced by the excess of western diet nutrients include c-Jun N-terminal kinase (JNK; also known as MAPK8) and IκB kinase-β. These pathways lead to the serine phosphorylation of IRS1, the disruption of the insulin signaling pathway, and altered metabolic responses (Hotamisligil, 2006). The disruption of the insulin signaling pathway in this manner can also be mediated by extracellular signal–regulated kinase (ERK), ribosomal protein S6 kinase (S6K), mammalian target of rapamycin (mTOR), protein kinase C (PKC), and glycogen synthase kinase 3β, all of which can be activated by immune signaling pathways, although the exact nature of the modifications and the metabolic outcomes in each scenario are still not completely understood (Zick, 2005). It is probable that many other immune signaling pathways and proteins will be linked to altered metabolic responses.

Molecular mechanisms contributing to type 2 diabetes complications

Increased production of superoxide during the progression of type 2 diabetes results in the stimulation of four mechanisms: (1) increased production of advanced glycation end product (AGE); (2) increased activity of polyol pathway; (3) activation of PKC isoforms; and (4) increased hexosamine pathway flux (Brownlee, 2001, 2005). Increase in synthesis of superoxide in type 2 diabetes also directly inactivates two critical antiatherosclerotic enzymes, endothelial NOS and prostacyclin synthase (Giacco and Brownlee, 2010; Folli et al., 2011). Increased production of intracellular ROS also promotes defective angiogenesis in response to ischemia. ROS also activate a number of proinflammatory pathways and cause long-lasting epigenetic changes that result in persistent expression of proinflammatory genes after glycemia is normalized. Atherosclerosis and cardiomyopathy in type 2 diabetes are caused in part by pathway-selective insulin resistance, which increases mitochondrial ROS production from FFAs and by inactivation of antiatherosclerosis enzymes by ROS (Giacco and Brownlee, 2010). It should be noted that overexpression of superoxide dismutase in transgenic diabetic mice prevents diabetic retinopathy, nephropathy, and cardiomyopathy. Overproduction of ROS alters insulin sensitivity by several mechanisms. Oxidative stress activates stress signaling kinases, among which JNK1 plays an important role in the etiology of insulin resistance: by phosphorylating IRS-1 at inhibitory sites, JNK1 prevents recruitment of this protein to the activated insulin receptor and disrupts downstream events of the insulin signaling pathway. Alternatively, insulin resistance may induce chronic inflammation by promoting oxidative stress. The proinflammatory cytokines tumor necrosis factor-alpha (TNF-α) and interlukin-6 (IL-6) have been shown to trigger phosphorylation of IRS-1, thus exerting an inhibitory action on insulin signaling; accordingly, genetic ablation either of TNF-α or of its receptor can improve insulin resistance caused by obesity in rodent models (Styskal et al., 2012). Finally, oxidative damage to critical proteins in insulin-sensitive tissues can potentially affect their function and therefore the propagation of insulin-stimulated signals for instance, in vitro studies have shown that oxidative stress reduces the ability of insulin receptor to correctly bind insulin, and insulin resistance developed in older humans is often accompanied by reduced function of proteins involved in insulin signaling (Bryan et al., 2013). Generation and accumulation of ROS may also contribute to obesity-induced hypertension. Deregulation of insulin signaling may also alter the activity of endothelial PtdIns 3K/protein kinase B (Akt) pathway, which lessens endothelial cells lessen NO synthesis leading to decrease in vasodilatation and increase in blood pressure (Whaley-Connell and Sowers, 2012).

The polyol pathway promotes the reduction of glucose into sorbitol via aldose reductase in a nicotinamide adenine dinucleotide phosphate (NADPH)-dependent manner. Sorbitol is then oxidized to fructose by the enzyme sorbitol dehydrogenase, with NAD^+ reduced to NADH. The main function of aldose reductase is to reduce toxic aldehydes formed by ROS or other substrates to inactive alcohols. Under normal conditions, aldose reductase has a low affinity for glucose, with a very small percentage of total glucose converted to sorbitol via this pathway. Under hyperglycemic conditions, there is an increase in the enzymatic activity and production of sorbitol, resulting in an overall decrease in NADPH. NADPH is an essential cofactor for the production of glutathione (GSH), a critical intracellular antioxidant (Brownlee, 2001; Ahmad et al., 2005; Giacco and Brownlee, 2010). It has also been proposed that the increase in sorbitol and its conversion to fructose increases the $NADH:NAD^+$ ratio, which can lead to PKC activation and inhibition of the enzyme glyceraldehyde-3-phosphate dehydrogenase (GADPH) (Brownlee, 2001; Giacco and Brownlee, 2010). Increased glucose flux through the polyol pathway does not produce ROS directly, but contributes greatly to an overall redox imbalance in the cell that leads to oxidative stress. Under normal conditions, PARP resides in the nucleus in an inactive form. In type 2 diabetes increased production of ROS causes DNA damage through the activation of PARP in the nucleus. PARP promotes the hydrolysis of NAD^+ into nicotinic acid and adenosine diphosphate (ADP)-ribose. PARP then proceeds to make polymers of ADP-ribose, which accumulate on GAPDH and other nuclear proteins. GAPDH is commonly thought to reside exclusively in the cytosol. However, it normally shuttles in and out of the nucleus, where it plays a critical role in DNA repair. In type 2 diabetes, hyperglycemia and insulin resistance promote the conversion of glucose into fructose, which is then metabolized into AGEs. In AGE production pathway, elevated blood glucose levels contribute to the glycation of proteins and lipids, resulting in the formation of AGEs (Fig. 2.6) (Basta et al., 2004; Brownlee, 2001; Giacco and Brownlee, 2010; Farooqui et al., 2012; Farooqui, 2013). AGEs bind with two main types of cell surface receptors: (1) scavenger receptors, which

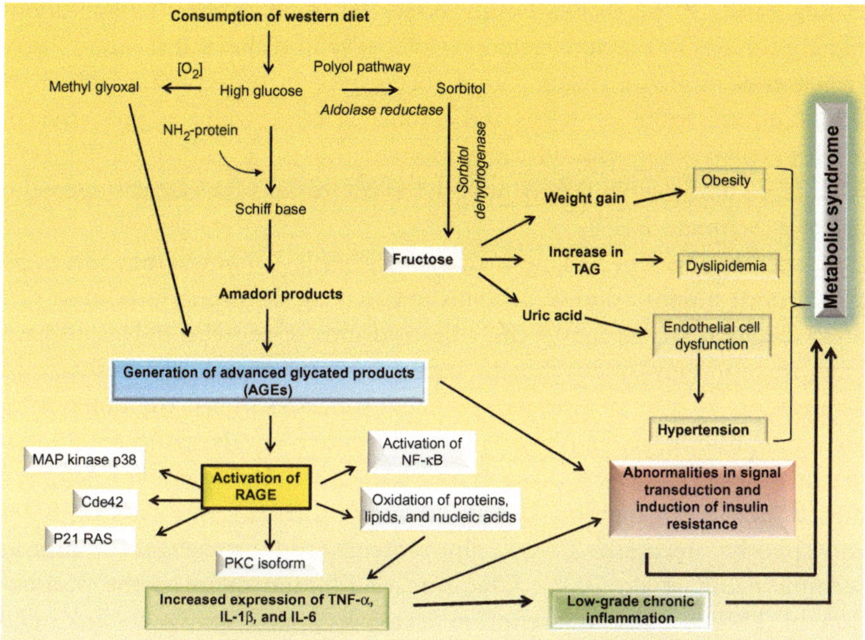

Figure 2.6 Consumption of western diet and its role in generation of advanced glycated products.
MAP kinase, mitogen-activated protein kinase; *NF-κB*, nuclear factor κappaB; *PKC*, protein kinase C; *TAG*, triacylglycerol.

remove and degrade AGEs and (2) receptors for AGEs (RAGEs), which trigger specific cellular signaling responses on AGE binding. RAGE is a member of the immunoglobulin family and interacts many ligands including AGEs, high mobility group protein B1, S100 calcium binding proteins (including calgranulin), amyloid-*b*-protein, and amphotericin. RAGEs are expressed in many different tissues and cell types, including endothelial cells, vascular smooth muscle cells, neurons, glial cells, and macrophages (Wendt et al., 2002). AGE–RAGE signaling involves NF-κB, MAPKs (ERK1/2 and p38MAPK), and NADPH oxidases. AGE–RAGE signaling also induces expression of vascular adhesion molecule 1, eselectin, vascular endothelial growth factor, and proinflammatory cytokines [interleukin-1 beta (IL-1β), IL-6, and TNF-α] (Manigrasso et al., 2014). In diabetes, activation of these signaling pathways is increased in vascular smooth muscle cells, leading to vascular fibrosis, calcification, inflammation, prothrombotic effects, and vascular damage, processes underlying diabetic nephropathy, retinopathy, neuropathy, and atherosclerotic cardiovascular disease (CVD)

(Koulis et al., 2015). Coexisting hypertension amplifies these complications and contributes to the accelerated vasculopathy in diabetes (Farooqui, 2013; Frimat et al., 2017). Patients with diabetes have increased tissue and circulating concentrations of AGEs and soluble RAGE, which is predictive of cardiovascular events and all-cause mortality. As such, urinary and plasma AGE levels and soluble RAGE may act as biomarkers for vascular disease in diabetes (Schmidt, 2017).

Collective evidence suggests that glycation is a nonenzymic and post-translational modification of proteins initiated by the primary addition of a sugar aldehyde or ketone to the amino groups of proteins. Glycated proteins are not only susceptible to oxidative damage, but also resistant to degradation by lysosomal enzymes. These processes impair the function of glycated proteins. Example of glycated protein is glycated-hemoglobin that is used as marker for diabetes and MetS. Another example is glycated protein that is glycated low-density lipoprotein (LDL). In this form, LDLs are poorly recognized by lipoprotein and scavenger receptors (Zimmermann et al., 2001). Glycation of albumin increases the production of TNF-α. This transcription factor has been shown to link with insulin resistance via induction of proinflammatory mechanisms that suppress insulin signal transduction (Naitoh et al., 2001; Miele et al., 2003). Other targets of glycation are nucleotides and lipids. Modifications of these targets result in DNA mutation and decrease in cell membrane integrity. In mitochondria, the consequences of glycation can alter bioenergy production. Under physiological conditions, antiglycation defenses are sufficient, with proteasomes preventing accumulation of glycated proteins, while lipid turnover clears glycated products and nucleotide excision repair removes glycated nucleotides (Fournet et al., 2018). If this does not occur, glycation damage accumulates, and pathologies may develop. Over time, AGE products and their debris accumulate in the blood serum and along arterial walls. AGEs play a critical role in aging, diabetes, atherosclerosis and cardiovascular diseases, and in neurodegenerative diseases. Hyperglycemia is closely linked with accelerated AGEs formation (Méndez et al., 2010). The major AGEs in vivo are formed from highly reactive intermediate carbonyl groups, known as α-dicarbonyls or oxoaldehydes, including 3-deoxyglucosone, glyoxal, and methylglyoxal (Brownlee, 2001; Kim et al., 2005). Within tissues, glycation results in protein aggregates due to bonds created by three distinct mechanisms: (1) the formation of covalent bonds between glycation end products, (2) the oxidation of sulfur groups (sulfhydryl groups) into disulfide bridges, and (3) the formation of new reactive groups within a protein. Chemical bridges formed by AGEs also result in the reticulation of proteins and their cross-linking (assembly), a phenomenon that occurs within the extracellular matrix and significantly increases the structural rigidity. Glycation can also inhibit the biological effects of some proteins including the effects of hormones (e.g., insulin), growth factors, and antibacterial peptides such as lysozyme and lactoferrin (Song and Schmidt, 2012; Alavi et al., 2013).

Major complication of hyperglycemia is atherosclerosis, a condition, which hardens and narrows the blood vessels. Improperly managed type 2 diabetes is a cause of premature death to a number of health problems, including heart disease, stroke, kidney disease, retinopathy, and neuropathy. Hyperglycemia can also cause nerve damage (Boulton et al., 2005), which may lead to the need of limb amputation (Brownlee, 2005). Such ailments reduce the patients' quality of life, and potentially relationship with others around them.

The polyol pathway transforms excess of glucose into sorbitol, which in turn produces a variety of intracellular changes in vascular tissues. This reaction is promoted by sorbitol dehydrogenase. In polyol pathway, aldose reductase has a low affinity for glucose. At the normal glucose concentration (nondiabetic subjects), this reaction is very slow and a very small percentage of glucose is utilized through this pathway. However, in diabetic patients, high levels of glucose induce an overproduction of ROS via the mitochondrial electron transport chain, leading to the activation of aldose reductase and the subsequent elevation in conversion of glucose to sorbitol. This reaction is catalyzed by aldose reductase (Madonna and De Caterina, 2011). High glucose (hyperglycemia) promotes oxidative stress not only through the auto-oxidation of glucose, but also through the generation of nonenzymic glycation products (Fig. 2.6) (Kikuchi et al., 2003).

The liver helps to maintain fasting glucose levels through gluconeogenesis and glycogenolysis. However, when the liver is insulin resistant, the suppression of hepatic glucose production is impaired, and thus gluconeogenesis and glycogenolysis continue at inappropriately high levels despite normal or high circulating glucose levels. Adipose tissue and muscle are similarly affected by insulin resistance, although the problem here relates more to the impaired ability of insulin to promote glucose disposal. To compensate for the insulin resistance in these tissues, pancreatic β-cells produce more insulin. However, there is a limit to how much can be produced, and when this has been reached, the β-cells fail. Type 2 diabetes occurs when an inappropriately low level of insulin is produced in response to a given concentration of glucose. At the molecular level, type 2 diabetes is accompanied by not only with increased sorbitol synthesis, oxidative and nitrosative stress, and endogenous antioxidant depletion, but also with enhanced lipid peroxidation and alterations in hormonal responses (Brownlee, 2005). Induction of above processes is supported by increase in intracellular glucose level and its downstream metabolic metabolism. Excessive levels of glucose not only disrupt the electron transport chain in the mitochondria, but also promote the production of superoxide anions (Nishikawa et al., 2000).

Risk factors contributing to metabolic syndrome

The MetS is a common and complex disorder combining abdominal obesity, dyslipidemia, endothelial dysfunction, hypertension, insulin resistance, genetic susceptibility, elevated blood pressure, hypercoagulable state, and chronic stress (Bruce and Byrne, 2009). MetS is also accompanied by hyperglycemia, elevation in triglycerides (TAG), dyslipidemia, and low high-density lipoprotein cholesterol (HDL-C). These factors predispose the individual to increased risk of developing not only type 2 diabetes, MetS, CVD, and fatty liver disease, but also neurological disorders such as stroke, Alzheimer's disease (AD), vascular dementia, and depression (Lakka et al., 2002). Pathophysiological abnormalities that contribute to the development of the MetS include impaired mitochondrial oxidative phosphorylation and mitochondrial biogenesis, dampened insulin metabolic signaling, endothelial dysfunction, and associated myocardial functional abnormalities (Ren et al., 2010). Subjects with MetS also have autonomic nervous system dysfunction characterized by predominance of the sympathetic nervous system in many organs, including heart, kidneys, vasculature, adipose tissue, and muscles.

Worldwide prevalence of MetS ranges from $<10\%$ to as much as 84%, depending on the region, urban or rural environment, composition (sex, age, race, and ethnicity) of the population studied, and the definition of the syndrome used (Desroches and Lamarche, 2007; Kolovou et al., 2007). The International Diabetes Federation estimates that one-quarter of the world's adult population has the MetS (Cho et al., 2018). Higher socioeconomic status, sedentary lifestyle, and high BMI are significantly associated with the onset of MetS. It is concluded that the differences in genetic background, diet, levels of physical activity, smoking, family history of diabetes, and education all influence the prevalence of the MetS and its components (Cameron et al., 2004). The prevalence of MetS varies from 8% to 43% in men and from 7% to 56% in women around the world (Cameron et al., 2004).

Pathogenesis of metabolic syndrome

Molecular mechanisms contributing to the pathogenesis of MetS are not fully understood. However, it is well known that long-term consumption of western diet (high fat) with processed carbohydrate, and abundant protein, along with corn sirup (fructose) containing soft drinks, lack of exercise, and family history may contribute to the pathogenesis of MetS (Fig. 2.7) (Grundy et al., 2005). Clinically, long-term consumption of western diet not only contributes to dyslipidemia and low HDLs, and increase in arterial blood pressure, but also dysregulation of glucose homeostasis, increase in abdominal obesity, and induction of insulin

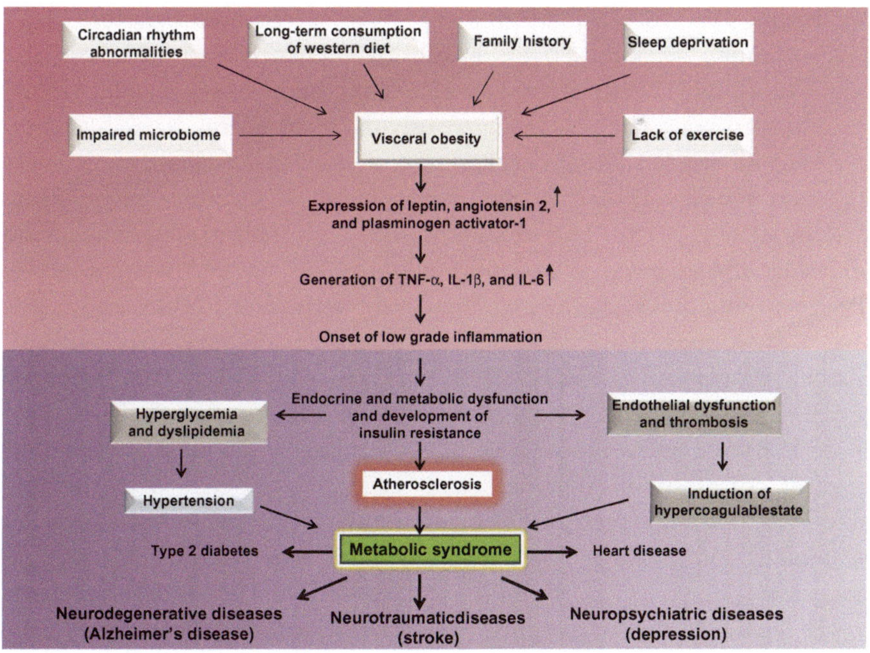

Figure 2.7 Contribution of western diet, family history, age, and sedentary lifestyle (lack of exercise) in the development of metabolic syndrome.
Upward arrow indicates increase.

resistance. These clinical changes not only produce chronic proinflammatory and prothrombotic states and nonalcoholic fatty liver disease, but also promote sleep apnea suggesting that molecular mechanisms of MetS are quite complex. It is well known that the entry of glucose entry into cells is controlled by insulin. The excess of glucose promotes hyperinsulinemia and insulin resistance, which is characterized by the failure of peripheral tissues to appropriately regulate glucose homeostasis in response to insulin (Parillo and Riccardi, 2004; Hwu et al., 2009). The available systems of energy disposal are unable to cope with the excess of substrates (glucose), since they are geared for saving not for spendthrift. This process promotes and facilitates the conversion and storage of glucose into fat in adipose tissues. Thus adipose tissues are an important energy sink, which store the energy that cannot be used otherwise. The growth of adipose tissue has limits, and the excess of energy induces inflammation and induce the ineffective intervention of the immune system. However, even under this acute situation, the presence of excess glucose remains in favor of its conversion to fat. In addition to energy storage, adipose tissue plays an active role in many homeostatic processes including energy expenditure, appetite regulation, and glucose regulation. Fat tissue is critical for thyroid function, immune response, bone health maintenance, reproduction, and blood clotting. The maintenance of energy homeostasis is achieved by sophisticated brain circuits that rigorously maintain energy levels by affecting food intake and energy expenditure. Adipose tissue has a central role in leptin production and in the management of systemic energy stores (Caron et al., 2018). The adipose tissue is an active endocrine organ secreting FFAs, leptin, adiponectin, adipsin, complement factor 3, IL-6, TNF-α, angiotensinogen, plasminogen activation inhibitor-1, etc. (Fig. 2.8). Among above secretory products, increased levels of proinflammatory adipokines, such as IL-1β, IL-6, TNF-α, and leptin, and decreased levels of antiinflammatory adipokines, such as adiponectin, in obesity promote a chronic state of low-grade inflammation, which induces the development of insulin resistance and type 2 diabetes, hypertension, atherosclerosis and other cardiovascular diseases, and some types of cancer (Odegaard and Chawla, 2013; Hotamisligil, 2017). Moreover, since adiponectin also acts as an insulin-sensitizing hormone in muscle and liver, lower levels of adiponectin further contribute to peripheral insulin resistance in obesity (Liu et al., 2015; Saltiel and Olefsky, 2017). Finally, increased circulating levels of leptin in obesity lead to hypothalamic leptin resistance, turning down anorexigenic and energy expenditure signals and

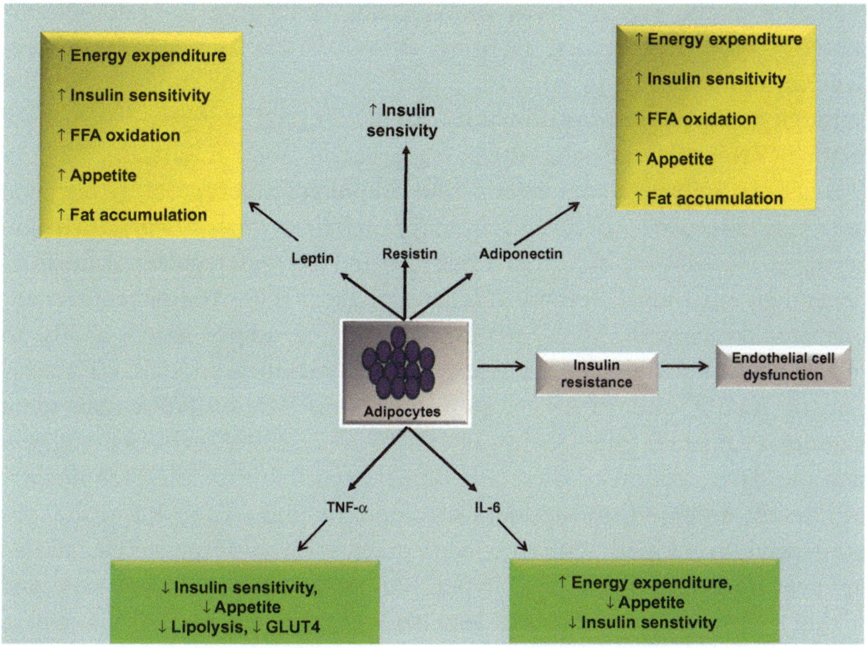

Figure 2.8 Release of adipokines from adipose tissue and their role in metabolic processes.
Upward and downward arrows indicate increase and decrease, respectively.

further contributing to aggravate obesity (Waterson and Horvath, 2015). Among above adipokines, the leptin/adiponectin ratio plays an important role in the development of MetS. It has been demonstrated that abdominal obesity was uniquely related to decrease in plasma concentrations of adiponectin and increase in leptin levels. Leptin/adiponectin imbalance contributes not only to increase in waist circumference and decrease in vascular response to acetylcholine, but also to increase in vasoconstriction due to the involvement of angiotensin II (Finucane et al., 2009; López-Jaramillo et al., 2014). Another mechanism, which contributes to obesity involves IL-6 and leptin-mediated stimulation of dopamine uptake. Induction of this mechanism leads to feeling of fullness via the involvement limbic system. These adipokines also induce the production of ROS and induction of oxidative stress. The increase in adipose tissue is also accompanied by decrease in the activity of antioxidant enzymes such as superoxide dismutase, catalase, and GSH peroxidase. Finally, high ROS production and the decrease in antioxidant capacity lead to various abnormalities including endothelial dysfunction, which is characterized by a reduction in the bioavailability of NO, and an increase in endothelium-derived contractile factors, favoring atherosclerotic disease (Fernández-Sánchez et al., 2011). In contrast, adiponectin hormone promotes insulin sensitivity, displays antiinflammatory properties, and is a pro-cell survival factor (Yamauchi et al., 2003; Yamauchi and Kadowaki, 2013). Interactions between adiponectin and adiponectin receptor1 (AdipoR1) not only stimulate adenosine monophosphate-activated kinase (Iwabu et al., 2010; Mao et al., 2006), but also induce and regulate AdipoR1-dependent ceramidase activity (Holland et al., 2011). Adiponectin receptors are involved in MetS, mainly due to the adiponectin's ability to restore insulin sensitivity in obese, diabetic preclinical models via activation of AMPK and PPAR-α pathways in an AdipoR1/R2-dependent manner (Yamauchi and Kadowaki, 2013). Collective evidence suggests that the dysregulation of these adipokines contributes to the pathogenesis of obesity. Adipose tissue-resident macrophages and adipocytes in the adipose tissue combined with the consequences of hyperglycemia, altered lipoprotein metabolism, and hyperinsulinemia in the vasculature and within organ microcirculation lead to dysfunctional endothelia and a proinflammatory state. Thus MetS represents a combination of synergistic vascular pathologies that lead to an accelerated atherogenic state that compromises the ability of the patient to satisfactorily respond to humoral, cellular, and mechanical stresses (Ayyobi and Brunzell, 2003; Hwu et al., 2009). In addition, there is compelling evidence that small size at birth in full-term pregnancies is linked with the subsequent development of the major features of the MetS, namely glucose intolerance, insulin resistance, type 2 diabetes, hypertension, dyslipidemia, and increased mortality from cardiovascular disease (Levitt et al., 2000). Levels of adiponectin and the ratio between leptin and adiponectin have also been suggested as surrogate markers of insulin resistance (Finucane et al., 2009). It is proposed that the leptin:adiponectin ratio may be a stable indicator in nonfasted subjects since the fluctuations in their levels are modest (Finucane et al., 2009).

Link between type 2 diabetes and metabolic syndrome

Type 2 diabetes is a component of MetS. As mentioned earlier, type 2 diabetes involves multiple organ system dysfunction, including impaired insulin action in muscle and adipose tissues, defective control of hepatic glucose production, and insulin deficiency produced by loss of β-cell mass and function (Muoio and Newgard, 2008). Type 2 diabetes and MetS patients also develop CNS pathology suggesting the ability of adipose tissue to communicate with the brain and impact brain function. It is not clear yet how this cross-talk between adipose tissue and brain occurs, but it is becoming increasingly evident that adipokines play an important role in this process. Adipokines may impact brain physiology through different mechanisms. Some adipokines such as leptin and TNF-α can cross the blood–brain barrier (BBB) and act directly in the brain, while other adipokines may produce their effects not only by interacting with endothelial brain cells, and regulating BBB permeability, but also through accessing other circulating mediators into the brain. Importantly, in pathological states such as neuroinflammation, the BBB integrity is compromised allowing the penetration of adipokines and other substances to which the brain is normally inaccessible. It is reported that the induction of insulin resistance in brains of type 2 diabetes and MetS patients produces atrophy and electro-physiological changes causing deficits in learning and memory, attention, executive function, and psychomotor efficiency (Wrighten et al., 2009). These changes in the brain may, over time, lead to an acceleration in brain ageing and increase the risk of age-related neurodegenerative and neurovascular disorders (such as AD) (Biessels et al., 2002; Farooqui, 2013; Kiliaan et al., 2014).

Neurodegenerative diseases are caused by complex interactions between genetic and environmental factors, which promote the pathogenesis of type 2 diabetes, MetS, and neurodegenerative diseases. Recent studies have also indicated that human type 2 diabetes and MetS are constellation of disorders associated with polymorphisms in a wide array of genes, with each individual gene accounting for $<1\%$ of disease risk (Ridderstråle and Groop, 2009). Factors such as obesity, hyperglycemia, insulin resistance, adiposity, hypertension, and inflammation are related to the onset of AD (Haan, 2006). Mice fed with a high-fat diet for a long time develop obesity, insulin resistance, inflammation, and elevated levels of beta amyloid (Aβ) peptides in the brain (Busquets et al., 2017; Nuzzo et al., 2015). Consumption of high calorie diet by rats produces MetS and a deterioration of the cognitive process, which is linked to the loss of synaptic connections (Treviño et al., 2017) supporting the view that onset of MetS may contribute to the pathogenesis of AD. Collective evidence suggests that neuropathologies triggered by MetS often result from increased permeability of the BBB. The BBB, a system designed to restrict entry of toxins, immune cells, and pathogens to the brain, is vital for proper neuronal function. Local and systemic inflammation induced by obesity or type 2 diabetes may cause disruption of BBB, decrease in the removal of waste products, and increase infiltration of immune cells. This process may not only contribute to the disruption of glial and neuronal cells interaction, but also in hormonal dysregulation and increase in immune sensitivity along with impairment in cognitive function depending on the affected brain region.

Type 2 diabetes and metabolic syndrome as risk factors for Alzheimer's disease

In addition to abnormal amyloid precursor protein (APP) processing, accumulation of Aβ enriched senile plaques and neurofibrillary tangles (enriched in hyperphosphorylated tau protein), AD is characterized by abnormal insulin receptor signaling and impairment in brain glucose utilization (Tu et al., 2014; Viola and Klein, 2015; Selkoe, 2008; Bloom, 2014; Iqbal et al., 2009). All these processes contribute to the pathogenesis of AD not only by inducing inhibition of synaptic plasticity, synaptic loss and progressive cognitive decline, but also by generating oxidative stress and inducing neuroinflammation due to induction of astrogliosis and microglial cell activation along with increased expression of proinflammatory cytokines and chemokines (Fang et al., 2010; Borger et al., 2013; Tu et al., 2014; Viola and Klein, 2015; Selkoe, 2008; Bloom, 2014) (Fig. 2.9). Little is known about the consequences of cytokine and chemokine-mediated changes on brain function and neurodegeneration relevant to AD. However, a growing number of studies in mice have indicated that these immune proteins not only induce, but have potent effects on amyloidosis, neurodegeneration, and cognition (Wyss-Coray, 2006). Among cytokines, TNF-α has received much attention because of its ability to promote Parkinson's disease (PD) progression (McCoy et al., 2006), whereas TNF-α receptor 1 knockout protects against AD- and PD-like disease in mice (Sriram et al., 2002; He et al., 2007). Similarly, potent effects are exerted by transforming growth factor (TGF)-β1, which is increased in human AD brains at the transcript and protein level, whereas TGF receptor expression is decreased (Tesseur et al., 2006). TGF-β1 promotes cerebrovascular amyloidosis but delays parenchymal Aβ accumulation in APP mice that also overproduce this cytokine from astrocytes (Wyss-Coray et al., 2001). As mentioned earlier, formation of neurofibrillary tangles in an important neuropathological feature of AD. Similar to AD, type 2 diabetes is also considered as tauopathy-associated

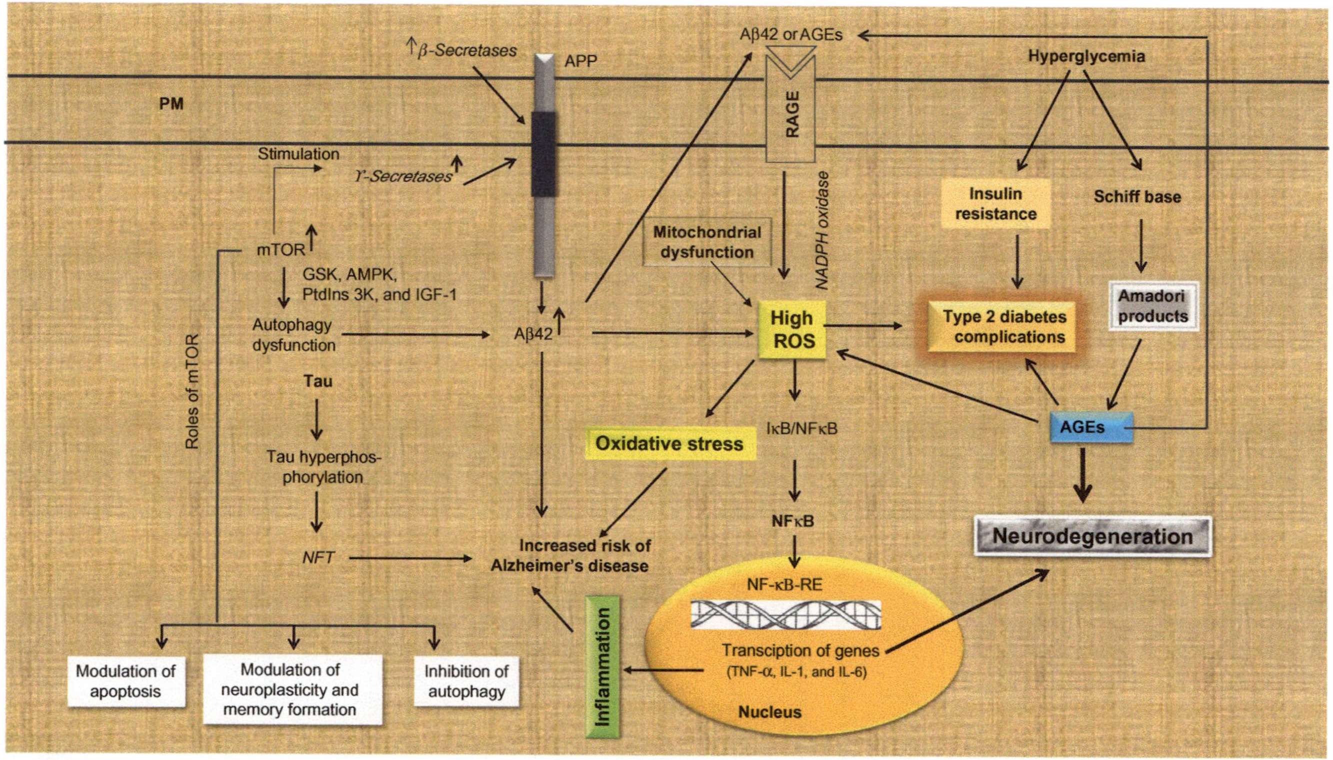

Figure 2.9 Synthesis of Aβ from APP, stimulation of RAGE receptor by Aβ and its roles of mTOR in the brain.
AGEs, advanced glycation products; *AMPK*, AMP-activated protein kinase; *APP*, amyloid precursor protein; *Aβ*, beta amyloid; *GSK*, glycogen synthase kinase; *IκB*, inhibitory subunit of NF-κB; *IGF-1*, insulin-like growth factor 1; *IL-1β*, interleukin-1 beta; *IL-6*, interleukin-6; *mTOR*, mammalian target of rapamycin; *NF-κB-RE*, nuclear factor kappaB response element; *PtdIns 3K*, phosphoinositide 3-kinase; *PM*, plasma membrane; *ROS*, reactive oxygen species; *RAGE*, receptor for advanced glycation end products; *TNF-α*, tumor necrosis factor-α.

disease based on the accumulation of hyperphosphorylated tau in brains of type 2 diabetes patients (Planel et al., 2007; Platt et al., 2016; Papon et al., 2013). Inhibition of the formation of tau hyperphosphorylation is known to reverse the development of cognitive dysfunction in type 2 diabetes (Qu et al., 2014).

At the cellular level, AD is also accompanied by a progressive loss of pyramidal cells in the entorhinal cortex and CA1 region of the hippocampus that are responsible for maintenance of higher cognitive functions and gradual loss of memory due to disruption disrupts connectivity between neural circuits (Serrano-Pozo et al., 2011). Early symptoms of AD are also marked by synaptic dysfunction that disrupts connectivity between neural circuits, thereby initiating the gradual loss of memory.

Long-term consumption of western diet and lack of exercise are linked to an increased risk of insulin resistance, altered insulin signaling, obesity, MetS, and AD. MetS has been further linked mechanistically to AD pathogenesis based on abnormal metabolism of the obesity-related protein leptin (Fewlass et al., 2004). Leptin has been shown to attenuate β-secretase–mediated processing of APP in neuronal cells, possibly through mechanisms involving altered lipid composition of membrane lipid rafts. Most strikingly, chronic administration of leptin to AD-transgenic animals and in animal models of MetS reduce the brain Aβ load, supporting the view that leptin can be used as potential therapeutic agent for MetS and AD (Fewlass et al., 2004). To this end, it has shown that circulating leptin is transported into the brain by binding to the lipoprotein receptor megalin at the choroid plexus epithelium (Dietrich et al., 2007). Attenuation of megalin expression in physiological and pathological conditions, such as during aging or in AD dementia, correlates with poor entry of leptin into the brain (Dietrich et al., 2007). In addition, the pathogenesis of MetS is linked with a pattern of multimorbidity that increases risk for subsequent stroke and cardiovascular disease and may be predictive of both AD and vascular dementia. MetS is also a powerful predictor of cerebrovascular morbidity (Eckel et al., 2005) including decrease in cerebral blood flow, hypertension, insulin resistance, infarcts, and white matter hyperintensities. These changes have been linked to increased risk for development of MetS and AD (Portet et al., 2012) and AD-related pathology (Grimmer et al., 2012). Thus significant reduction in components of the insulin signaling pathway, including IRS-1, PtdIns 3K, and phospho-Akt, is observed in the frontal cortex and hippocampus of autopsied MetS and AD brains, but not in control brains (Steen et al., 2005). These changes are considered as a core syndrome that increases the risk of type 2 diabetes, MetS, AD, and vascular dementia. This suggestion is supported by positron emission tomography and magnetic resonance imaging studies, which show marked impairment of glucose and energy metabolism in type 2 diabetes, MetS, and AD (Umegaki, 2014). In addition, amyloidogenesis remains a salient feature of all three conditions. As mentioned earlier, extracellular β-amyloid plaques form one of the characteristic features of AD. Likewise, deposits of amyloidogenic peptide (IAPP) are detected in the pancreatic islets of Langerhans of type 2 diabetes and MetS patients (Haataja et al., 2008). Interestingly, diabetic mice overexpressing IAPP develop oligomers and fibrils with more severe diabetic and MetS trait similar to AD mouse models that overexpress APP (Marzban et al., 2003). Furthermore, as mentioned earlier, RAGEs accumulate in the sites of diabetic and MetS complications such as kidney, retina, and atherosclerotic plaques under conditions of ER and oxidative stress (Nowotny et al., 2015). Similarly, glycated products of Aβ and tau form in transgenic AD models as well as in postmortem brains of AD patients under similar stress conditions and form an important component of neurofibrillary tangles (Schedin-Weiss et al., 2014). Moreover, additional traits of synaptic dysfunction, activation of the inflammatory response pathways, and impairment of autophagy are pathological features common to AD, type 2 diabetes, and MetS (De Felice and Ferreira, 2014; Carvalho et al., 2015).

Recent studies have also indicated that the involvement of mTOR signaling also contributes to the pathogenesis of type 2 diabetes, MetS, AD, and PD. Two forms of mTOR (mTORC1 and mTORC2) are found in various mammalian tissues. mTORC1 is a critical regulator of translation initiation and ribosome biogenesis and plays an evolutionarily conserved role in cell growth control and cell proliferation through several downstream effectors, including 4E-BP1 and S6K1 (Erol, 2007). Nutrients and cellular metabolism regulate these mTORC1 effectors (Hsu et al., 2011). In contrast, mTORC2 is associated with downstream effector Akt in the insulin signaling pathway. Interestingly, loss of S473 phosphorylation of Akt after rictor knockdown reduces the phosphorylation of some, but not all, Akt substrates. Only fully active, doubly phosphorylated Akt inhibits FoxO1 and FoxO3a transcription factors, thus promoting cell survival upon growth factor stimulation (Kumar et al., 2008). The activation of mTOR enhances Aβ generation and deposition through the activation of regulating β- and γ-secretase (Spilman et al., 2010;

Son et al., 2012; Chen et al., 2009). In addition, mTOR also interacts with several key signaling pathways and regulate Aβ generation or Aβ clearance, including PtdIns 3K/Akt (Damjanac et al., 2008; Bhaskar et al., 2009), GSK-3 (Maiese et al., 2012), AMPK (Cai et al., 2012), and insulin/IGF-1 (Pei and Hugon, 2008). These studies indicate that mTOR not only contributes to synaptic plasticity, but also facilitates memory formation in the hippocampus and the inhibition of mTOR impairs memory consolidation (Slipczuk et al., 2009). An increase in phosphorylation of mTOR by GSK-3β and tau protein by p70 ribosome S6 kinase (p70S6K) has been reported to occur in AD brain, suggesting that phosphorylation of mTOR and hyperphosphorylation of tau may promote AD progression (An et al., 2003; Griffin et al., 2005). Furthermore, inhibition of mTOR has been reported to increase autophagy in murine models of AD. This process not only improves memory, but also contribute to the decrease in Aβ levels (Spilman et al., 2010). It is interesting to note that loss of mTOR signaling impairs long-term potentiation and decreases synaptic plasticity in models of AD (d'Abramo et al., 2006). In addition, Aβ, which is toxic to neuronal cells, inhibits the activation of mTOR and p70S6K in neuroblastoma cells and in lymphocytes of patients with AD (Dal Col and Dolcetti, 2008) supporting the view that the activation of mTOR and p70S6K may prevent neurodegeneration following Aβ exposure in microglia (Zemke et al., 2007). Other studies have indicated that blockade of mTOR activity may lead to neuronal atrophy and apoptosis in AD (Prada et al., 2007). A growing body of literature suggests that the vascular risk factors comprising MetS are associated with Aβ accumulation. Specifically, amyloid deposition has been associated with obesity (Godzik et al., 2016), diabetes (Yang and Song, 2013), cholesterol levels (Reed et al., 2014; Hughes et al., 2014), and hypertension (Toledo et al., 2012; Langbaum et al., 2012).

mTOR also downregulates autophagy, a lysosome dependent, homeostatic process, in which organelles and proteins are degraded and recycled into energy. Autophagy can be classified into three categories based on the mechanism by which intracellular constituents are supplied into lysosome for degradation: microautophagy, chaperone-mediated autophagy, and macroautophagy. In microautophagy, the cytoplasmic material is absorbed into lysosome by direct invagination of the lysosomal membrane (Marzella et al., 1981). The chaperone-mediated autophagy facilitates the degradation of cytosolic proteins by directly targeting them to lysosomes and into the lysosomal lumen (Kaushik and Cuervo, 2012). In macroautophagy, degradable contents of cytoplasm are encapsulated in subcellular double-membrane structures named "autophagosomes." Autophagosomes transport the cell "waste" to the lysosomes for degradation (Settembre et al., 2013). Autophagy decreases the Aβ clearance by autophagy/lysosome system. This system is responsible for the clearance of abnormal proteins (Pei and Hugon, 2008). It is proposed that chronic deterioration of the autophagy/lysosome pathway is an important factor in the failure of Aβ clearance from the AD brain (Spilman et al., 2010; Bhaskar et al., 2009). Furthermore, inhibition of mTOR, activity by mTOR inhibitor, retards autophagy. This process is closely associated with high levels of Aβ (Spilman et al., 2010). In contrast, upregulation of mTOR signaling by lifestyle choices may at least partially block the pathogenesis of AD in patients and animal models (Caccamo et al., 2010; Oddo, 2012) suggesting MetS increases the risk of AD by an mTOR-dependent mechanism (Ma et al., 2013; Orr et al., 2014).

Genes controlling the neurochemical aspects of type 2 diabetes and MetS may also contribute to the pathogenesis of AD (Hao et al., 2015). This suggestion is supported by neuroimaging studies. Using MRI, it is shown that older patients with type 2 diabetes and MetS have a moderately increased risk for developing hippocampal atrophy and that the severity of lesions parallels the progression of AD and mild cognitive impairment (MCI; Ferreira et al., 2010; Janson et al., 2004). Conversely, AD patients exhibit significantly increased prevalence of type 2 diabetes and MetS. Patients with type 2 diabetes, MetS, and AD show cognitive dysfunction, which is defined as the loss of intellectual functions such as thinking, remembering, and reasoning that interfere with daily activities. Patients with cognitive dysfunction also loss ability to learn, recall, concentrate, and problem solving. Cognitive function is regulated not only by neurochemical and intricate synaptic changes, but also by neuronal and glial interactions (Morrison and Baxter, 2012). The intensity of cognitive decline is markedly increased not only in patients with type 2 diabetes, MetS, and AD, but also in patients of neuropsychiatric diseases (Schuh et al., 2011; Farooqui, 2013). At the molecular level cognitive decline is controlled by several factors such as genes for oxidative stress, neuroinflammation, immune response, mitochondrial functions, growth factors, neuronal survival, and calcium homeostasis (Lu et al., 2004; Loerch et al., 2008). In general, genes that are stress responsive and related to inflammation and DNA repair are upregulated, while genes associated with neuronal growth and survival and mitochondrial functions are downregulated with advancing age in several organisms (Yankner et al., 2008).

Furthermore, there is growing epidemiological evidence indicating that MetS and its components (impaired glucose tolerance, abdominal or central obesity, hypertension, hypertriglyceridemia, and reduced HDL-C) may contribute to the development of age-related cognitive decline (ARCD), MCI, vascular dementia, and AD (Frisardi et al., 2010). The signal transduction pathways associated with ARCD, MCI in MetS, and AD remain unknown. However, cognitive performance may be influenced by the pathogenesis of insulin resistance, obesity and being overweight may induce the development of cognitive impairment. Insulin resistance, obesity, and MetS increase the risk of dementia, a condition of acquired cognitive defects sufficient to interfere with social or occupational functioning. As stated in Chapter 1, Insulin resistance and obesity, it affects memory, judgment, speech, comprehension, execution, orientation, and learning (Xu et al., 2004; Qaseem et al., 2008). A major risk factor for dementia is advancing age. After the age of 65, the prevalence and onset of dementia double every 5 years (Alzheimers Disease International, 2010). Onset of insulin resistance and obesity is known to accelerate its rate.

Conclusion

Type 2 diabetes mellitus is a complex syndrome that consists of high serum glucose, insulin resistance, impaired function of vascular endothelial cells, atherosclerosis, and obesity. Among the various features of type 2 diabetes mellitus, hyperglycemia is of course a central feature. The pathogenesis of type 2 diabetes involves insulin resistance, which is linked to factors such as genes, obesity, age, and lifestyle. Classical complications of diabetes elicited by microangiopathy (nephropathy, retinopathy, and neuropathy) are closely associated with hyperglycemia, and glucose-lowering therapies have been shown to prevent the development of these complications. Similarly, MetS is a constellation of cardiac, kidney, and metabolic disorders including insulin resistance, obesity, metabolic dyslipidemia, high blood pressure, and evidence of early cardiac and kidney diseases. The consequences of MetS may result in intracellular ATP depletion, increased uric acid production, oxidative stress, inflammation, and increased lipogenesis, which are associated with endothelial dysfunction. Endothelial dysfunction is an early manifestation of vascular disease and a driver for the development of MetS.

Onset of type 2 diabetes and MetS predisposes the individual to increased risk of CVD, and fatty liver disease, stroke, AD, vascular dementia, and depression. Pathophysiological abnormalities that contribute to the development of above neurological disorders include induction of oxidative stress, impaired mitochondrial biogenesis, dampened insulin metabolic signaling, endothelial dysfunction, and induction of low-grade inflammation.

References

Abbasi, A., Corpeleijn, E., Meijer, E., et al., 2012. Sex differences in the association between plasma copeptin and incident type 2 diabetes: the Prevention of Renal and Vascular Endstage Disease (PREVEND) study. Diabetologia 55, 1963–1970.

Abdullah, A., Peeters, A., de Courten, M., Stoelwinder, J., 2010. The magnitude of association between overweight and obesity and the risk of diabetes: a meta-analysis of prospective cohort studies. Diabetes Res. Clin. Pract. 89, 309–319.

Ahmad, F.K., Zhiheng, H., King, G.L., 2005. Molecular targets of diabetic cardiovascular complications. Curr. Drug Targets 6, 487–494.

Alavi, P., Yousefi, R., Amirghofran, S., Karbalaei-Heidari, H.R., Moosavi-Movahedi, A. A., 2013. Structural analysis and aggregation propensity of reduced and nonreduced glycated insulin adducts. Appl. Biochem. Biotechnol. 170, 623–638.

Alzheimer's Disease International, 2010. World Alzheimer Report 2010. ADI, Illinois.

American Diabetes Association, 2009. Diagnosis and classification of diabetes mellitus. Diabetes Care 32 (Suppl. 1), S62–S67.

American Diabetes Association, 2014a. Standards of medical care in diabetes—2014. Diabetes Care 37 (Suppl. 1), S14–S80.

American Diabetes Association, 2014b. Diagnosis and classification of diabetes mellitus. Diabetes Care 37 (Suppl. 1), S81–S90.

An, W.L., Cowburn, R.F., Li, L., Braak, H., Alafuzoff, I., Iqbal, K., et al., 2003. Up-regulation of phosphorylated/activated p70 S6 kinase and its relationship to neurofibrillary pathology in Alzheimer's disease. Am. J. Pathol. 163, 591–607.

Andersson, A.K., Flodström, M., Sandler, S., 2001. Cytokine-induced inhibition of insulin release from mouse pancreatic β-cells deficient in inducible nitric oxide synthase. Biochem. Biophys. Res. Commun. 281, 396–403.

Arkin, J.M., Alsdorf, R., Bigornia, S., et al., 2008. Relation of cumulative weight burden to vascular endothelial dysfunction in obesity. Am. J. Cardiol. 101, 98–101.

Ayyobi, A.F., Brunzell, J.D., 2003. Lipoprotein distribution in the metabolic syndrome, type 2 diabetes mellitus, and familial combined hyperlipidemia. Am. J. Cardiol. 92, 27J–33J.

Banerjee, J., Nema, V., Dhas, Y., Mishra, N., 2017. Role of microRNAs in type 2 diabetes and associated vascular complications. Biochimie 139, 9–19.

Basta, G., Schmidt, A.M., DeCaterina, R., 2004. Advanced glycation end products and vascular inflammation: implications for accelerated atherosclerosis in diabetes. Cardiovasc. Res. 63, 582–592.

Biessels, G.J., van der Heide, L.P., Kamal, A., Bleys, R.L., Gispen, W.H., 2002. Ageing and diabetes: implications for brain function. Eur. J. Pharmacol. 441, 1–14.

Bhaskar, K., Miller, M., Chludzinski, A., Herrup, K., Zagorski, M., Lamb, B.T., 2009. The PI3K-Akt-mTOR pathway regulates Abeta oligomer induced neuronal cell cycle events. Mol. Neurodegener. 4, 14.

Bloom, G.S., 2014. Amyloid-β and Tau: the trigger and bullet in Alzheimer disease pathogenesis. JAMA Neurol. 71, 505–508.

Bookchin, R.M., Gallop, P.M., 1968. Structure of hemoglobin AIc: nature of the N-terminal beta chain blocking group, Biochem. Biophys. Res. Commun., 32. pp. 86–93.

Borger, E., Aitken, L., Du, H., Zhang, W., Gunn-Moore, F.J., Yan, S.S., 2013. Is amyloid binding alcohol dehydrogenase a drug target for treating Alzheimer's disease? Curr. Alzheimer Res. 10, 21–29.

Boulton, A.J.M., Vinik, A.I., Arezzo, J.C., Bril, V., Feldman, E.L., et al., 2005. Diabetic neuropathies: a statement by the american diabetes association, Diabetes Care, 28. pp. 956–962.

Broad, A., Kirby, J.A., Jones, D.E.J., 2007. Toll-like receptor interactions: tolerance of MyD88-dependent cytokines but enhancement of MyD88-independent interferon-β production. Immunology 120, 103–111.

Brownlee, M., 2001. Biochemistry and molecular cell biology of diabetic complications. Nature 414, 813–820.

Brownlee, M., 2005. The pathobiology of diabetic complications: a unifying mechanism. Diabetes 54, 1615–1625.

Bruce, K.D., Byrne, C.D., 2009. The metabolic syndrome: common origins of a multifactorial disorder. Postgrad. Med. J. 85, 614–621.

Bruning, J.C., Gautam, D., Burks, D.J., Gillette, J., Schubert, M., Orban, P.C., et al., 2000. Role of brain insulin receptor in control of body weight and reproduction. Science 5487, 2122–2125.

Bryan, S., Baregzay, B., Spicer, D., Singal, P.K., Khaper, N., 2013. Redox-inflammatory synergy in the metabolic syndrome. Can. J. Physiol. Pharmacol. 91, 22–30.

Busquets, O., Ettcheto, M., Pallàs, M., et al., 2017. Long-term exposition to a high fat diet favors the appearance of β-amyloid depositions in the brain of C57BL/6J mice. A potential model of sporadic Alzheimer's disease. Mech. Ageing Dev. 162, 38–45.

Caccamo, A., Majumder, S., Richardson, A., Strong, R., Oddo, S., 2010. Molecular interplay between mammalian target of rapamycin (mTOR), amyloid-beta, and tau: effects on cognitive impairments. J. Biol. Chem. 285, 13107–13120.

Cai, Z., Zhao, B., Li, K., et al., 2012. Mammalian target of rapamycin: a valid therapeutic target through the autophagy pathway for Alzheimer's disease? J. Neurosci. Res. 90, 1105–1118.

Cameron, A.J., Shaw, J.E., Zimmet, P.Z., 2004. The metabolic syndrome: prevalence in worldwide populations. Endocrinol. Metab. Clin. North Am. 33, 351–375.

Caron, A., Lee, S., Elmquist, J.K., Gautron, L., 2018. Leptin and brain–adipose crosstalks. Nat. Rev. Neurosci. 19, 153–165.

Carvalho, C., Santos, M.S., Oliveira, C.R., Moreira, P.I., 2015. Alzheimer's disease and type 2 diabetes-related alterations in brain mitochondria, autophagy and synaptic markers. Biochim. Biophys. Acta 1852, 1665–1675.

Ceriello, A., 1993. Coagulation activation in diabetes mellitus: the role of hyperglycaemia and therapeutic prospects. Diabetologia 3, 1119–1125.

Cha, S.H., Wolfgang, M., Tokutake, Y., Chohnan, S., Lane, M.D., 2008. Differential effects of central fructose and glucose on hypothalamic malonyl-CoA and food intake. Proc. Natl. Acad. Sci. U.S.A. 105, 16871–16875.

Chen, T.J., Wang, D.C., Chen, S.S., 2009. Amyloid-beta interrupts the PI3K-Akt-mTOR signaling pathway that could be involved in brain-derived neurotrophic factor-induced Arc expression in rat cortical neurons. J. Neurosci. Res. 87, 2297–2307.

Cho, N.H., Shaw, J.E., Karuranga, S., Huang, Y., da Rocha Fernandes, J.D., et al., 2018. IDF Diabetes atlas: Global estimates of diabetes prevalence for 2017 and projections for 2045. Diabetes Res. Clin. Pract. 138, 271–281.

d'Abramo, C., Ricciarelli, R., Pronzato, M.A., Davies, P., 2006. Troglitazone, a peroxisome proliferator-activated receptor-gamma agonist, decreases tau phosphorylation in CHOtau4R cells. J. Neurochem. 98, 1068–1077.

Dal Col, J., Dolcetti, R., 2008. GSK-3beta inhibition: at the crossroad between Akt and mTOR constitutive activation to enhance cyclin D1 protein stability in mantle cell lymphoma. Cell Cycle 7, 2813–2816.

Damjanac, M., Rioux Bilan, A., Paccalin, M., et al., 2008. Dissociation of Akt/PKB and ribosomal S6 kinase signaling markers in a transgenic mouse model of Alzheimer's disease. Neurobiol. Dis. 29, 354–367.

Daniels, L.B., Maisel, A.S., 2015. Cardiovascular biomarkers and sex: the case for women. Nat. Rev. Cardiol. 12, 588–596.

De Felice, F.G., Ferreira, S.T., 2014. Inflammation, defective insulin signaling, and mitochondrial dysfunction as common molecular denominators connecting type 2 diabetes to Alzheimer disease. Diabetes 63, 2262–2272.

Desroches, S., Lamarche, B., 2007. The evolving definitions and increasing prevalence of the metabolic syndrome. Appl. Physiol. Nutrit. Metab. 32, 23–32.

Dietrich, M.O., Spuch, C., Antequera, D., Rodal, I., de Yebenes, J.G., Molina, et al., 2007. Megalin mediates the transport of leptin across the blood-CSF barrier. Neurobiol. Aging 29, 902–912.

Dolan, R.J., 2002. Emotion, cognition, and behavior. Science 298, 1191–1194.

Dun, X.P., Parkinson, D., 2017. Role of Netrin-1 signaling in nerve regeneration. Int. J. Mol. Sci. 18, 491.

Ebert, M.S., Sharp, P.A., 2012. Roles for microRNAs in conferring robustness to biological processes. Cell. 149, 505–524.

Eckel, R.H., Grundy, S.M., Zimmet, P.Z., 2005. The metabolic syndrome. Lancet 365, 1415–1428.

Erol, A., 2007. Insulin resistance is an evolutionarily conserved physiological mechanism at the cellular level for protection against increased oxidative stress. Bioessays 29, 811–818.

Fang, C.L., Wu, W.H., Liu, Q., Sun, X., Ma, Y., Zhao, Y.F., et al., 2010. Dual functions of beta-amyloid oligomer and fibril in Cu(II)-induced H_2O_2 production. Regul. Pept. 163, 1–6.

Farooqui, A.A., 2013. Metabolic Syndrome: An Important Risk Factor for Stroke, Alzheimer, and Depression. Spinger, New York.

Farooqui, A.A., Farooqui, T., 2013. Contribution of phospholipid, sphingolipid, and cholesterol-derived lipid mediators in the pathogenesis of metabolic syndrome and neurological disorders. In: Farooqui, T., Farooqui, A.A. (Eds.), Metabolic Syndrome and Neurological Disorders. Wiley Blackwell, Ames, Iowa.

Farooqui, A.A., Farooqui, T., Panza, F., Frisardi, V., 2012. Metabolic syndrome as a risk factor for neurological disorders. Cell. Mol. Life Sci. 69, 741–762.

Fernández-Sánchez, A., Madrigal-Santillán, E., Bautista, M., Esquivel-Soto, J., Morales-González, A., et al., 2011. Inflammation, oxidative stress, and obesity. Int. J. Mol. Sci. 12, 3117–3132.

Ferreira, I.L., Resende, R., Ferreiro, E., Rego, A.C., Pereira, C.F., 2010. Multiple defects in energy metabolism in Alzheimer's disease. Curr. Drug Targets 11, 1193–1206.

Fewlass, D.C., Noboa, K., Pi-Sunyer, F.X., Johnston, J.M., Yan, S.D., Tezapsidis, N., 2004. Obesity-related leptin regulates Alzheimer's Aβ. FASEB J. 18, 1870–1878.

Finucane, F.M., Luan, J., Wareham, N.J., Sharp, S.J., O'Rahilly, S., Balkau, B., et al., 2009. Correlation of the leptin:adiponectin ratio with measures of insulin resistance in nondiabetic individuals. Diabetologia 52, 2345–2349.

Fitzgerald, K.A., Rowe, D.C., Barnes, B.J., Caffrey, D.R., Visintin, A., Latz, E., et al., 2003. LPS-TLR4 signaling to IRF-3/7 and NF-κB involves the toll adapters TRAM and TRIF. J. Exp. Med. 198, 1043–1055.

Folli, F., Corradi, D., Fanti, P., Davalli, A., Paez, A., Giaccari, A., et al., 2011. The role of oxidative stress in the pathogenesis of type 2 diabetes mellitus micro- and macrovascular complications: avenues for a mechanistic-based therapeutic approach. Curr. Diabetes Rev. 7, 313–324.

Fournet, M., Bonté, F., Desmoulière, A., 2018. Glycation damage: a possible hub for major pathophysiological disorders and aging. Aging Dis. 9, 880–900.

Frimat, M., Daroux, M., Litke, R., et al., 2017. Kidney, heart and brain: three organs targeted by ageing and glycation. Clin. Sci. (Lond.) 131, 1069–1092.

Frisardi, V., Solfrizzi, V., Seripa, D., Capurso, C., Santamato, A., Sancarlo, D., et al., 2010. Metabolic-cognitive syndrome: a cross-talk between metabolic syndrome and Alzheimer's disease. Ageing Res. Rev. 9, 399–417.

Gan, D., 2003. Diabetes Atlas, 2nd ed. International Diabetes Federation, Brussels.

Gao, S., Zhang, X., Qin, Y., et al., 2016. Dual actions of Netrin-1 on islet insulin secretion and immune modulation. Clin. Sci. 130, 1901–1911.

Geng, J., Wang, L., Zhang, L., Qin, C., Song, Y., Ma, Y., et al., 2018. Blood-brain barrier disruption induced cognitive impairment is associated with increase of inflammatory cytokine. Front. Aging Neurosci. 10, 129.

Giacco, F., Brownlee, M., 2010. Oxidative stress and diabetic complications. Circ. Res. 107, 1058–1070.

Godzik, L., Rusinek, H., Kamer, A., Pirraglia, E., Tsui, W., Mosconi, L., et al., 2016. Effects of vascular risk factors, statins, and antihypertensive drugs on PiB deposition in cognitively normal subjects. Alzheimers Dement. (Amst.) 2, 95–104.

Gokce, N., Vita, J.A., McDonnell, M., et al., 2005. Effect of medical and surgical weight loss on endothelial vasomotor function in obese patients. Am. J. Cardiol. 95, 266–268.

Griffin, R.J., Moloney, A., Kelliher, M., Johnston, J.A., Ravid, R., Dockery, P., et al., 2005. Activation of Akt/PKB, increased phosphorylation of Akt substrates and loss and altered distribution of Akt and PTEN are features of Alzheimer's disease pathology. J. Neurochem. 93, 105–117.

Grimmer, T., Faust, M., Auer, F., Alexopoulos, P., Forstl, H., Henriksen, G., et al., 2012. White matter hyperintensities predict amyloid increase in Alzheimer's disease. Neurobiol. Aging. 33, 2766–2773.

Grundy, S.M., Cleeman, J.I., Daniels, S.R., Donato, K.A., Eckel, R.H., Franklin, B.A., et al., 2005. Diagnosis and management of the metabolic syndrome: an American Heart Association/National Heart, Lung, and Blood Institute Scientific Statement. Circulation 112, 2735–2752.

Haan, M.N., 2006. Therapy insight: type 2 diabetes mellitus and the risk of late-onset Alzheimer's disease. Nat. Clin. Pract. Neurol. 2, 159–166.

Haataja, L., Gurlo, T., Huang, C.J., Butler, P.C., 2008. Islet amyloid in type 2 diabetes, and the toxic oligomer hypothesis. Endocr. Rev. 29, 303–316.

Haffner, S.J., Cassells, H., 2003. Hyperglycemia as a cardiovascular risk factor. Am. J. Med. 115 (Suppl. 8A), 6S–11S.

Hao, K., Di Narzo, A.F., Ho, L., Luo, W., Li, S., et al., 2015. Shared genetic etiology underlying Alzheimer's disease and type 2 diabetes. Mol. Aspects Med. 43-44, 66–76.

Hayden, M.S., Ghosh, S., 2008. Shared principles in NF-kappaB signaling. Cell. 132, 344–362.

He, P., Zhong, Z., Lindholm, K., Berning, L., Lee, W., Lemere, C., et al., 2007. Deletion of tumor necrosis factor death receptor inhibits amyloid β generation and prevents learning and memory deficits in Alzheimer's mice. J. Cell. Biol. 178, 829–841.

Holland, W.L., Miller, R.A., Wang, Z.V., Sun, K., Barth, B.M., et al., 2011. Receptor-mediated activation of ceramidase activity initiates the pleiotropic actions of adiponectin. Nat. Med. 17, 55–63.

Hotamisligil, G.S., 2006. Inflammation and metabolic disorders. Nature 444, 860–867.
Hotamisligil, G.S., 2017. Foundations of immunometabolism and implications for metabolic health and disease. Immunity 47, 406–420.
Howard, B.V., Rodriguez, B.L., Bennett, P.H., et al., 2002. Diabetes and cardiovascular disease. Circulation 105, 132–137.
Hsu, P.P., Kang, S.A., Rameseder, J., Zhang, Y., Ottina, K.A., Lim, D., et al., 2011. The mTOR-regulated phosphoproteome reveals a mechanism of mTORC1-mediated inhibition of growth factor signaling. Science 332, 1317–1322.
Huang, X., Liu, G., Guo, J., Su, Z., 2018. The PI3/Akt pathway in obesity and type 2 diabetes. Int. J. Biol. Sci. 14, 1483–1496.
Hughes, T.M., Lopez, O.L., Evans, R.W., Kamboh, M.I., Williamson, J.D., Klunk, W. E., et al., 2014. Markers of cholesterol transport are associated with amyloid deposition in the brain. Neurobiol. Aging. 35, 802–807.
Hwu, C.M., Liou, T.L., Hsiao, L.C., Lin, M.W., 2009. Prehypertension is associated with insulin resistance. Q. J. Med. 102, 705–711.
International Diabetes Federation, 2015. IDF Diabetes Atlas, 7th ed. International Diabetes Federation, Brussels, Belgium. Available from: http://www.diabetesatlas.org.
Iqbal, K., Liu, F., Gong, C.X., Alonso, A.C., Grundke-Iqbal, I., 2009. Mechanisms of tau-induced neurodegeneration. Acta Neuropathol. 118, 53–69.
Iwabu, M., Yamauchi, T., Okada-Iwabu, M., Sato, K., Nakagawa, T., Funata, M., et al., 2010. Adiponectin and AdipoR1 regulate PGC-1alpha and mitochondria by Ca^{2+} and AMPK/SIRT1. Nature 464, 1313–1319.
Janson, T., Laedtke, J.E., Parisi, P., O'Brien, R.C., Petersen, P.C., et al., 2004. Butler Increased risk of type 2 diabetes in Alzheimer disease. Diabetes 53, 474–481.
Kampoli, A.M., Tousoulis, D., Briasoulis, A., Latsios, G., Papageorgiou, N., et al., 2011. Potential pathogenic inflammatory mechanisms of endothelial dysfunction induced by type 2 diabetes mellitus. Curr. Pharm. Des. 17, 4147–4158.
Kaushik, S., Cuervo, A.M., 2012. Chaperone-mediated autophagy: a unique way to enter the lysosome world. Trends Cell Biol. 22, 407–417.
Kautzky-Willer, A., Harreiter, J., Pacini, G., 2016. Sex and gender differences in risk, pathophysiology and complications of type 2 diabetes mellitus. Endocr. Rev. 37, 278–316.
Kawai, T., Akira, S., 2010. The role of pattern-recognition receptors in innate immunity: update on Toll-like receptors. Nat. Immunol. 11, 373–384.
Kawasaki, E., Gill, R.G., Eisenbarth, G.S., 1999. Type I diabetes mellitus. In: Eisenbarth, G.S. (Ed.), Endocrine and Organ Specific Autoimmunity. Landes Bioscience, Austin, TX, pp. 149–182.
Kikuchi, S., Shinpo, K., Takeuchi, M., Yamagishi, S., Makita, Z., Sasaki, N., et al., 2003. Glycation—a sweet tempter for neuronal death. Brain Res. Brain Res. Rev. 41, 306–323.
Kiliaan, A.J., Arnoldussen, I.A., Gustafson, D.R., 2014. Adipokines: a link between obesity and dementia? Lancet Neurol. 13, 913–923.
Kim, W., Hudson, B.I., Moser, B., Guo, J., Rong, L.L., Lu, Y., et al., 2005. Receptor for advanced glycation end products and its ligands: a journey from the complications of diabetes to its pathogenesis. Ann. N. Y. Acad. Sci. 1043, 553–561.
Kolovou, G.D., Anagnostopoulou, K.K., Salpea, K.D., Mikhailidis, D.P., 2007. The prevalence of metabolic syndrome in various populations. Am. J. Med. Sci. 333, 362–371.
Koulis, C., Watson, A.M., Gray, S.P., Jandeleit-Dahm, K.A., 2015. Linking RAGE and Nox in diabetic micro- and macrovascular complications. Diabetes Metab. 41, 272–281.
Krishna, C.S., Srikanta, S., 2015. Type 1 diabetes pathogenesis – prevention. Ind. J. Endocrinol. Metab. 19 (Suppl. 1), S58–S63.

Kumar, A., Harris, T.E., Keller, S.R., Choi, K.M., Magnuson, M.A., Lawrence Jr., J.C., 2008. Muscle-specific deletion of rictor impairs insulin-stimulated glucose transport and enhances Basal glycogen synthase activity. Mol. Cell. Biol. 28, 61–70.
Lakka, H.M., Laaksonen, D.E., Lakka, T.A., Niskanen, L.K., Kumpusalo, E., Tuomilehto, J., et al., 2002. The metabolic syndrome and total and cardiovascular disease mortality in middle-aged men. JAMA 288, 2709–2716.
Langbaum, J.B., Chen, K., Launer, L.J., Fleisher, A.S., Lee, W., Liu, X., et al., 2012. Blood pressure is associated with higher brain amyloid burden and lower glucose metabolism in healthy late middle-age persons. Neurobiol. Aging. 33, 827 e811–e829.
Laughlin, G.A., Barrett-Connor, E., Cummins, K.M., Daniels, L.B., Wassel, C.L., Ix, J.H., 2013. Sex-specific association of fetuin-A with type 2 diabetes in older community-dwelling adults: the Rancho Bernardo Study. Diabetes Care 36, 1994–2000.
Lee, H.J., Ryu, J.M., Jung, Y.H., Lee, S.J., Kim, J.Y., Lee, S.H., et al., 2016. High glucose upregulates BACE1-mediated Aβ production through ROS-dependent HIF-1α and LXRα/ABCA1-regulated lipid raft reorganization in SK-N-MC cells. Sci. Rep. 6, 36746.
Leung, P.S., 2016. The potential protective action of vitamin D in hepatic insulin resistance and pancreatic islet dysfunction in type 2 diabetes mellitus. Nutrients 8, pii: E147.
Levin, B.E., Kang, L., Sanders, N.M., Dunn-Meynell, A.A., 2006. Role of neuronal glucosensing in the regulation of energy homeostasis. Diabetes 55 (Suppl. 2), S122–S130.
Levitt, N.S., Lambert, E.V., Woods, D., Hales, C.N., Andrew, R., Seckl, J.R., 2000. Impaired glucose tolerance and elevated blood pressure in low birth weight, nonobese, young south African adults: early programming of cortisol axis. J. Clin. Endocrinol. Metab. 85, 4611–4618.
Li, Y.W., Aronow, W.S., 2011. Diabetes mellitus and cardio vascular disease. J. Clin. Exp. Cardiol. 2, 114.
Liu, A.K., Chang, R.C.-C., Pearce, R.K.B., Gentleman, S.M., 2015. Nucleus basalis of Meynert revisited: anatomy, history and differential involvement in Alzheimer's and Parkinson's disease. Acta Neuropathol. 129, 527–540.
Loerch, P.M., Lu, T., Dakin, K.A., Vann, J.M., Isaacs, A., et al., 2008. Evolution of the aging brain transcriptome and synaptic regulation. PLoS One 3, e3329.
López-Jaramillo, P., Gómez-Arbeláez, D., López-López, J., López-López, C., Martínez-Ortega, J., Gómez-Rodríguez, A., et al., 2014. The role of leptin/adiponectin ratio in metabolic syndrome and diabetes. Horm. Mol. Biol. Clin. Investig. 18, 37–45.
Lu, T., Pan, Y., Kao, S.Y., Li, C., Kohane, I., Chan, J., et al., 2004. Gene regulation and DNA damage in the ageing human brain. Nature 429, 883–891.
Ma, Y.Q., Wu, D.K., Liu, J.K., 2013. mTOR and tau phosphorylated proteins in the hippocampal tissue of rats with type 2 diabetes and Alzheimer's disease. Mol. Med. Rep. 7, 623–627.
Madonna, R., De Caterina, R., 2011. Cellular and molecular mechanisms of vascular injury in diabetes—part II: cellular mechanisms and therapeutic targets. Vascul. Pharmacol. 54, 75–79.
Maiese, K., Chong, Z.Z., Wang, S., Shang, Y.C., 2012. Oxidant stress and signal transduction in the nervous system with the PI 3-K, Akt, and mTOR cascade. Int. J. Mol. Sci. 13, 13830–138366.
Manigrasso, M.B., Juranek, J., Ramasamy, R., Schmidt, A.M., 2014. Unlocking the biology of RAGE in diabetic microvascular complications. Trends Endocrinol. Metab. 25, 15–22.

Mao, X., Kikani, C.K., Riojas, R.A., Langlais, P., Wang, L., Ramos, F.J., et al., 2006. APPL1 binds to adiponectin receptors and mediates adiponectin signalling and function. Nat. Cell Biol. 8, 516–523.

Marzella, L., Ahlberg, J., Glaumann, H., 1981. Autophagy, heterophagy, microautophagy and crinophagy as the means for intracellular degradation. Virchows Arch. B Cell Pathol. Incl. Mol. Pathol. 36, 219–234.

Marzban, L., Park, K., Verchere, C.B., 2003. Islet amyloid polypeptide and type 2 diabetes. Exp. Gerontol. 38, 347–351.

McCoy, M.K., Martinez, T.N., Ruhn, K.A., Szymkowski, D.E., Smith, C.G., Botterman, B.R., et al., 2006. Blocking soluble tumor necrosis factor signaling with dominant-negative tumor necrosis factor inhibitor attenuates loss of dopaminergic neurons in models of Parkinson's disease. J. Neurosci. 26, 9365–9375.

Méndez, J.D., Xie, J., Aguilar-Hernández, M., Méndez-Valenzuela, V., 2010. Trends in advanced glycation end products research in diabetes mellitus and its complications. Mol. Cell. Biochem. 341, 33–41.

Miele, C., Riboulet, A., Maitan, M.A., Oriente, F., Romano, C., Formisano, P., et al., 2003. Human glycated albumin affects glucose metabolism in L6 skeletal muscle cells by impairing insulin-induced receptor substrate (IRS) signaling through a protein kinase C alpha-mediated mechanism. J. Biol. Chem. 278, 47376–47387.

Modena, M.G., Bonetti, L., Coppi, F., Bursi, F., Rossi, R., 2002. Prognostic role of reversible endothelial dysfunction in hypertensive postmenopausal women. J. Am. Coll. Cardiol. 40, 505–510.

Morrison, J.H., Baxter, M.G., 2012. The aging cortical synapse: hallmarks and implications for cognitive decline. Nat. Rev. Neurosci. 13, 240–250.

Moseley, K.F., 2012. Type 2 diabetes and bone fractures. Curr. Opin. Endocrinol. Diabetes Obes. 19, 128–135.

Muoio, D.M., Newgard, C.B., 2008. Molecular and metabolic mechanisms of insulin resistance and β-cell failure in type 2 diabetes. Nat. Rev. Mol. Cell Biol. 9, 193–205.

Naitoh, T., Kitahara, M., Tsuruzoe, N., 2001. Tumor necrosis factor alpha is induced through phorbol ester- and glycated human albumin-dependent pathway in THP-1 cells. Cell. Signal. 13, 331–334.

Ng, R.C., Cheng, O.Y., Jian, M., Kwan, J.S., Ho, P.W., Cheng, K.K., et al., 2016. Chronic adiponectin deficiency leads to Alzheimer's disease-like cognitive impairments and pathologies through AMPK inactivation and cerebral insulin resistance in aged mice. Mol. Neurodegener. 11, 71.

Nishikawa, T., Edelstein, D., Du, X.L., Yamagishi, S., Matsumura, T., et al., 2000. Normalizing mitochondrial superoxide production blocks three pathways of hyperglycaemic damage. Nature 404, 787–790.

Nolan, C.J., Damm, P., Prentki, M., 2011. Type 2 diabetes across generations: from pathophysiology to prevention and management. Lancet 378, 169–181.

Nowotny, K., Jung, T., Hohn, A., Weber, D., Grune, T., 2015. Advanced glycation end products and oxidative stress in type 2 diabetes mellitus. Biomolecules 5, 194–222.

Nuzzo, D., Picone, P., Baldassano, S., et al., 2015. Insulin resistance as common molecular denominator linking obesity to Alzheimer's disease. Curr. Alzheimer Res. 12, 723–735.

Odegaard, J.I., Chawla, A., 2013. Pleiotropic actions of insulin resistance and inflammation in metabolic homeostasis. Science 339, 172–177.

Oddo, S., 2012. The role of mTOR signaling in Alzheimer disease. Front. Biosci. (Schol. Ed.) 4, 941–952.

Orr, M.E., Salinas, A., Buffenstein, R., Oddo, S., 2014. Mammalian target of rapamycin hyperactivity mediates the detrimental effects of a high sucrose diet on Alzheimer's disease pathology. Neurobiol. Aging. 35, 1233–1242.

Övünç Hacıhamdioğlu, D., Hacıhamdioğlu, B., Altun, D., Müftüoğlu, T., Karademir, F., Süleymanoğlu, S., 2016. Urinary Netrin-1: a new biomarker for the early diagnosis of renal damage in obese children. J. Clin. Res. Pedia Endocrinol. 8, 282–287.
Pan, A., Wang, Y., Talaei, M., Hu, F.B., Wu, T., 2015. Relation of active, passive, and quitting smoking with incident type 2 diabetes: a systematic review and meta-analysis. Lancet Diabetes Endocrinol 3, 958–967.
Papon, M.A., El, K.N., Marcouiller, F., Julien, C., Morin, F., Bretteville, A., et al., 2013. Deregulation of protein phosphatase 2A and hyperphosphorylation of tau protein following onset of diabetes in NOD mice. Diabetes 62, 609–617.
Papatheodorou, K., Papanas, N., Banach, M., Papazoglou, D., Edmonds, M., 2016. Complications of diabetes 2016. J. Diabetes Res. 2016, 6989453.
Parillo, M., Riccardi, G., 2004. Diet composition and the risk of type 2 diabetes: epidemiological and clinical evidence. Br. J. Nutr. 92, 7–19.
Paschou, S.A., Papadopoulou-Marketou, N., Chrousos, G.P., Kanaka-Gantenbein, C., 2018. On type 1 diabetes mellitus pathogenesis. Endocr. Connect. 7, R38–R46.
Pei, J.J., Hugon, J., 2008. mTOR-dependent signalling in Alzheimer's disease. J. Cell. Mol. Med. 12 (6B), 2525–2532.
Pilch, P.F., Lee, J., 2004. Insulin receptor family. In: Lennarz, W., Lane, M.D. (Eds.), Encyclopedia of Biological Chemistry, Vol. 2. Elsevier Science, San Diego, pp. 436–440.
Planel, E., Tatebayashi, Y., Miyasaka, T., Liu, L., Wang, L., Herman, M., et al., 2007. Insulin dysfunction induces in vivo tau hyperphosphorylation through distinct mechanisms. J. Neurosci. 27, 13635–13648.
Platt, T.L., Beckett, T.L., Kohler, K., Niedowicz, D.M., Murphy, M.P., 2016. Obesity, diabetes, and leptin resistance promote tau pathology in a mouse model of disease. Neuroscience 315, 162–174.
Portet, F., Brickman, A.M., Stern, Y., Scarmeas, N., Muraskin, J., Provenzano, F.A., et al., 2012. Metabolic syndrome and localization of white matter hyperintensities in the elderly population. Alzheimers Dement. 8, S88–S95. e81.
Prada, P.O., Hirabara, S.M., de Souza, C.T., et al., 2007. L-glutamine supplementation induces insulin resistance in adipose tissue and improves insulin signalling in liver and muscle of rats with diet-induced obesity. Diabetologia 50, 1949–1959.
Pugliese, A., 2016. Insulitis in the pathogenesis of type 1 diabetes. Pediatr. Diabetes. 17 (Suppl. 22), 31–36.
Qaseem, A., Snow, V., Cross Jr, J., Forciea, M.A., Hopkins Jr, R., Shekelle, P., et al., 2008. Current pharmacologic treatment of dementia: a clinical practice guideline from the American College of Physicians and the American Academy of Family Physicians. Ann. Intern. Med. 148, 370–378.
Qu, Z.S., Li, L., Sun, X.J., Zhao, Y.W., Zhang, J., Geng, Z., et al., 2014. Glycogen synthase kinase-3 regulates production of amyloid-beta peptides and tau phosphorylation in diabetic rat brain. Sci. World J. 2014, 878123.
Ramirez, S.H., Potula, R., Fan, S., Eidem, T., Papugani, A., Reichenbach, N., et al., 2009. Methamphetamine disrupts blood-brain barrier function by induction of oxidative stress in brain endothelial cells. J. Cereb. Blood Flow Metab. 29, 1933–1945.
Ranganathan, P., Jayakumar, C., Navankasattusas, S., Li, D.Y., Kim, I.M., Ramesh, G., 2014. UNC5B receptor deletion exacerbates tissue injury in response to AKI. J. Am. Soc. Nephrol. 25, 239–249.
Reed, B., Villeneuve, S., Mack, W., DeCarli, C., Chui, H.C., Jagust, W., 2014. Associations between serum cholesterol levels and cerebral amyloidosis. JAMA Neurol. 71, 195–200.
Ren, J., Pulakat, L., Whaley-Connell, A., Sowers, J.R., 2010. Mitochondrial biogenesis in the metabolic syndrome and cardiovascular disease. J. Mol. Med. (Berl.) 88, 993–1001.

Ridderstråle, M., Groop, L., 2009. Genetic dissection of type 2 diabetes. Mol. Cell. Endocrinol. 297, 10–17.
Ruckdeschel, K., Pfaffinger, G., Haase, R., Sing, A., Weighardt, H., Häcker, G., et al., 2004. Signaling of apoptosis through TLRs critically involves toll/IL-1 receptor domain-containing adapter inducing IFN-β, but not MyD88, in bacteria-infected murine macrophages. J. Immunol. 173, 3320–3328.
Sadanand, S., Balachandar, R., Bharath, S., 2016. Memory and executive functions in persons with type 2 diabetes: a meta-analysis. Diabetes Metab. Res. Rev. 32, 132–142.
Saltiel, A.R., Olefsky, J.M., 2017. Inflammatory mechanisms linking obesity and metabolic disease. J. Clin. Invest. 127, 1–4.
Schedin-Weiss, S., Winblad, B., Tjernberg, L.O., 2014. The role of protein glycosylation in Alzheimer disease. FEBS J. 281, 46–62.
Schmidt, A.M., 2017. 2016 ATVB Plenary Lecture: receptor for advanced glycation endproducts and implications for the pathogenesis an treatment of cardiometabolic disorders: spotlight on the macrophage. Arterioscler. Thromb. Vasc. Biol. 37, 613–621.
Schuh, A.F., Rieder, C.M., Rizzi, L., Chaves, M., Roriz-Cruz, M., 2011. Mechanisms of brain aging regulation by insulin: implications for neurodegeneration in late-onset Alzheimer's disease. ISRN Neurol. 2011, 306905.
Sciacqua, A., Candigliota, M., Ceravolo, R., et al., 2003. Weight loss in combination with physical activity improves endothelial dysfunction in human obesity. Diabetes Care 26, 1673–1678.
Selkoe, D.J., 2008. Soluble oligomers of the amyloid β-protein impair synaptic plasticity and behavior. Behav. Brain Res. 192, 106–113.
Serrano-Pozo, A., Frosch, M.P., Masliah, E., Hyman, B.T., 2011. Neuropathological alterations in Alzheimer disease. Cold Spring Harb. Perspect. Med. 1, a006189.
Settembre, C., Fraldi, A., Medina, D.L., Ballabio, A., 2013. Signals from the lysosome: a control centre for cellular clearance and energy metabolism. Nat. Rev. Mol. Cell Biol. 14, 283–296.
Shaw, J.E., Sicree, R.A., Zimmet, P.Z., 2010. Global estimates of the prevalence of diabetes for 2010 and 2030. Diabetes Res. Clin. Pract. 87, 4–14.
Slipczuk, L., Bekinschtein, P., Katche, C., Cammarota, M., Izquierdo, I., Medina, J.H., 2009. BDNF activates mTOR to regulate GluR1 expression required for memory formation. PLoS One 4, e6007.
Son, S.M., Song, H., Byun, J., et al., 2012. Altered APP processing in insulin-resistant conditions is mediated by autophagosome accumulation via the inhibition of mammalian target of rapamycin pathway. Diabetes 61, 3126–3138.
Song, F., Schmidt, A.M., 2012. Glycation and insulin resistance: novel mechanisms and unique targets? Arterioscler. Thromb. Vasc. Biol. 32, 1760–1765.
Spanswick, D., Smith, M.A., Mirshamsi, S., Routh, V.H., Ashford, M.L., 2000. Insulin activates ATP-sensitive K + channels in hypothalamic neurons of lean, but not obese rats. Nat. Neurosci. 3, 757–758.
Spilman, P., Podlutskaya, N., Hart, M.J., et al., 2010. Inhibition of mTOR by rapamycin abolishes cognitive deficits and reduces amyloid-beta levels in a mouse model of Alzheimer's disease. PLoS One 5, e9979.
Sriram, K., Matheson, J.M., Benkovic, S.A., Miller, D.B., Luster, M.I., O'Callaghan, J.P., 2002. Mice deficient in TNF receptors are protected against dopaminergic neurotoxicity: implications for Parkinson's disease. FASEB J 16, 1474–1476.
Stadlmayr, A., Aigner, E., Huber-Schönauer, U., et al., 2015. Relations of vitamin D status, gender and type 2 diabetes in middle-aged Caucasians. Acta Diabetol. 52, 39–46.
Steen, E., Terry, B.M., Rivera, E.J., Cannon, J.L., Neely, T.R., Tavares, R., et al., 2005. Impaired insulin and insulin-like growth factor expression and signaling mechanisms in Alzheimer's disease–is this type 3 diabetes? J. Alzheimers Dis. 7, 63–80.

Stumvoll, M., Goldstein, B.J., Van Haeften, T.W., 2005. Type 2 diabetes: principles of pathogenesis and therapy. Lancet 365, 1333–1346.

Styskal, J., van Remmen, H., Richardson, A., Salmon, A.B., 2012. Oxidative stress and diabetes: what can we learn about insulin resistance from antioxidant mutant mouse models? Free Radic. Biol. Med. 52, 46–58.

Sun, K.L.W., Correia, J.P., Kennedy, T.E., 2011. Netrins: versatile extracellular cues with diverse functions. Development 138, 2153–2169.

Takeda, K., Akira, S., 2004. TLR signaling pathways. Semin. Immunol. 16, 3–9.

Takeda, K., Kaisho, T., Akira, S., 2003. Toll-like receptors. Annu. Rev. Immunol. 21, 335–376.

Tang, X., Luo, Y.-X., Chen, H.-Z., Liu, D.P., 2014. Mitochondria, endothelial cell function, and vascular diseases. Front. Physiol. 5, 175.

Tesseur, I., Zou, K., Esposito, L., Bard, F., Berber, E., Can, J.V., et al., 2006. Deficiency in neuronal TGF-β signaling promotes neurodegeneration and Alzheimer's pathology. J. Clin. Invest. 116, 3060–3069.

Toledo, J.B., Toledo, E., Weiner, M.W., Jack Jr, C.R., Jagust, W., Lee, V.M., et al., 2012. Alzheimer's disease neuroimaging initiative: cardiovascular risk factors, cortisol, and amyloid-beta deposition in Alzheimer's disease neuroimaging initiative. Alzheimer Dement. 8, 483–489.

Treviño, S., Vázquez-Roque, R.A., López-López, G., et al., 2017. Metabolic syndrome causes recognition impairments and reduced hippocampal neuronal plasticity in rats. J. Chem. Neuroanat. 82, 65–75.

Tu, S., Okamoto, S., Lipton, S.A., Xu, H., 2014. Oligomeric Aβ-induced synaptic dysfunction in Alzheimer's disease. Mol. Neurodegener. 9, 48.

Tu, T., Zhang, C., Yan, H., et al., 2015. CD146 acts as a novel receptor for netrin-1 in promoting angiogenesis and vascular development. Cell Res. 25, 275–287.

Umegaki, H., 2014. Type 2 diabetes as a risk factor for cognitive impairment: current insights. Clin. Interv. Aging. 9, 1011–1019.

Unger, J.W., Moss, A.M., Livingston, J.N., 1991. Immunohistochemical localization of insulin receptors and phosphotyrosine in the brainstem of the adult rat. Neuroscience 42, 853–861.

Vaishya, S., Sarwade, R.D., Seshadri, V., 2018. MicroRNA, proteins, and metabolites as novel biomarkers for prediabetes, diabetes, and related complications. Front. Endocrinol. (Lausanne). 9, 180.

Vorhees, C.V., Williams, M.T., 2006. Morris water maze: procedures for assessing spatial and related forms of learning and memory. Nat. Protoc. 1, 848–858.

Uppu, R.M., Parinandi, N.L., 2011. Insulin sensitization and resistance interrelationship revisited with a quantitative molecular model approach. J. Diabetes Metab. 2, 106e.

Viola, K.L., Klein, W.L., 2015. Amyloid β oligomers in Alzheimer's disease pathogenesis, treatment, and diagnosis. Acta Neuropathol. 129, 183–206.

Waterson, M.J., Horvath, T.L., 2015. Neuronal regulation of energy homeostasis: beyond the hypothalamus and feeding. Cell. Metab. 22, 962–970.

Webster, S.J., Bachstetter, A.D., Nelson, P.T., Schmitt, F.A., Van Eldik, L.J., 2014. Using mice to model Alzheimer's dementia: an overview of the clinical disease and the preclinical behavioral changes in 10 mouse models. Front. Genet. 5, 88.

Wendt, T., Bucciarelli, L., Qu, W., Lu, Y., Yan, S.F., et al., 2002. Receptor for advanced glycation endproducts (RAGE) and vascular inflammation: insights into the pathogenesis of macrovascular complications in diabetes. Curr. Atheroscler. Rep. 4, 228–237.

Whaley-Connell, A., Sowers, J.R., 2012. Oxidative stress in the cardiorenal metabolic syndrome. Curr. Hypertens. Rep. 14, 360–365.

Whincup, P.H., Kaye, S.J., Owen, C.G., Huxley, R., Cook, D.G., Anazawa, S., et al., 2008. Birth weight and risk of type 2 diabetes: a systematic review. JAMA 300, 2886–2897.

Wilcox, G., 2005. Insulin and insulin resistance. Clin. Biochem. Rev. 26, 19–39.
Wild, S., Roglic, G., Green, A., et al., 2004. Global prevalence of diabetes: estimates for the year 2000 and projections for 2030. Diabetes Care 27, 1047–1053.
Wrighten, S.A., Piroli, G.G., Grillo, C.A., Reagan, L.P., 2009. A look inside the diabetic brain: contributors to diabetes-induced brain aging. Biochim. Biophys. Acta 1792, 444–453.
Wyss-Coray, T., 2006. Inflammation in Alzheimer disease: driving force, bystander or beneficial response? Nat. Med. 12, 1005–1015.
Wyss-Coray, T., Lin, C., Yan, F., Yu, G., Rohde, M., McConlogue, L., et al., 2001. TGF-b1 promotes microglial amyloid-b clearance and reduces plaque burden in transgenic mice. Nat. Med. 7, 612–618.
Xu, W.L., Qiu, C.X., Wahlin, A., Winblad, B., Fratiglioni, L., 2004. Diabetes mellitus and risk of dementia in the Kungsholmen project: a 6-year follow-up study. Neurology 63, 1181–1186.
Yamamoto, M., Sato, S., Hemmi, H., Hoshino, K., Kaisho, T., Sanjo, H., et al., 2003. Role of adaptor TRIF in the MyD88-independent toll-like receptor signaling pathway. Science 301, 640–643.
Yamauchi, T., Kadowaki, T., 2013. Adiponectin receptor as a key player in healthy longevity and obesity-related diseases. Cell Metabol. 17, 185–196.
Yamauchi, T., Kamon, J., Ito, Y., Tsuchida, A., Yokomizo, T., Kita, S., et al., 2003. Cloning of adiponectin receptors that mediate antidiabetic metabolic effects. Nature 423, 762–769.
Yang, Y., Song, W., 2013. Molecular links between Alzheimer's disease and diabetes mellitus. Neuroscience 250, 140–150.
Yankner, A., Lu, T., Bruce, L.P., 2008. The aging brain. Annu. Rev. Pathol. Mech. Dis. 3, 41–66.
Yimer, E.M., Zewdie, K.A., Hishe, H.Z., 2018. Netrin as a novel biomarker and its therapeutic implications in diabetes mellitus and diabetes-associated complications. J. Diabetes Res. 2018, 8250521.
Zemke, D., Azhar, S., Majid, A., 2007. The mTOR pathway as a potential target for the development of therapies against neurological disease. Drug. News Perspect. 20, 495–499.
Zhang, Y., Sun, X., Icli, B., Feinberg, M.W., 2017. Emerging roles for microRNAs in diabetic microvascular disease: novel targets for therapy. Endocr. Rev. 38, 145–168.
Zheng, H., Zheng, Y., Zhao, L., Chen, M., Bai, G., Hu, Y., et al., 2017. Cognitive decline in type 2 diabetic db/db mice may be associated with brain region-specific metabolic disorders. Biochim. Biophys. Acta 1863, 266–273.
Zick, Y., 2005. Ser/Thr phosphorylation of IRS proteins: a molecular basis for insulin resistance. Sci. STKE 2005, pe4.
Zimmermann, R., Panzenböck, U., Wintersperger, A., Levak-Frank, S., Graier, W., Glatter, O., et al., 2001. Lipoprotein lipase mediates the uptake of glycated LDL in fibroblasts, endothelial cells, and macrophages. Diabetes 50, 1643–1653.
Zlokovic, B.V., 2011. Neurovascular pathways to neurodegeneration in Alzheimer's disease and other disorders. Nat. Rev. Neurosci. 12, 723–738.

Further reading

Alonso, A.D., Di Clerico, J., Li, B., Corbo, C.P., Alaniz, M.E., Grundke-Iqbal, I., et al., 2010. Phosphorylation of tau at Thr212, Thr231, and Ser262 combined causes neurodegeneration. J. Biol. Chem. 285, 30851–30860.

Costa, R.O., Ferreiro, E., Martins, I., Santana, I., Cardoso, S.M., et al., 2012. Amyloid beta-induced ER stress is enhanced under mitochondrial dysfunction conditions. Neurobiol. Aging. 33 (824), e825–816.

Ghemrawi, R., Pooya, S., Lorentz, S., Gauchotte, G., Arnold, C., Gueant, J.L., et al., 2013. Decreased vitamin B12 availability induces ER stress through impaired SIRT1-deacetylation of HSF1. Cell Death Dis. 4, e553.

International Diabetes Federation, 2013. IDF Diabetes Atlas, 6th ed International Diabetes Federation, Brussels, Belgium. Available from: http://www.idf.org/diabetesatlas.

Kissebah, A.H., Vydelingum, N., Murray, M., Evans, D.J., Hartz, A.J., Kalkhoff, R.K., 1982. Relation of body fat distribution to metabolic complications of obesity. Metab. J. Clin. Endocrinol. 54, 254–260.

Li, Y., Xu, S., Giles, A., Nakamura, K., Lee, J.W., Hou, X., et al., 2011. Hepatic overexpression of SIRT1 in mice attenuates endoplasmic reticulum stress and insulin resistance in the liver. FASEB J. 25, 1664–1679.

Stumvoll, M., Goldstein, B.J., van Haeften, T.W., 2005. Type 2 diabetes: principles of pathogenesis and therapy. Lancet 365, 1333–1346.

Takeuchi, M., Yamagishi, S., 2008. Possible involvement of advanced glycation end-products (AGEs) in the pathogenesis of Alzheimer's disease. Curr. Pharm. Des. 14, 973–978.

Xing, Z., Gauldie, J., Cox, G., Baumann, H., Jordana, M., Lei, X.F., et al., 1998. IL-6 is an antiinflammatory cytokine required for controlling local or systemic acute inflammatory responses. J. Clin. Invest. 101, 311–320.

CHAPTER 3

Insulin resistance and heart disease

Introduction

Cardiovascular disease (CVD) or heart disease is a chronic inflammatory atherosclerotic condition, which is primarily associated with the large- and the medium-sized arteries. CVD is a major health problem in the United States. Approximately 6 million Americans suffer from CVD, with 700,000 new cases diagnosed every year. The likelihood of having CVD increases with age reaching 10 per 1000 population in individuals older than 65 years of age (Lloyd-Jones et al., 2010; Writing Group Members et al., 2016; CDC N, 2015; Benjamin et al., 2017). In the United States, CVD accounts for ~600,000 deaths (25%) each year (CDC N, 2015; Benjamin et al., 2017), and after a continuous decline over the last five decades, its incidences are increasing again (Roth et al., 2015). Advances in CVD treatments have significantly reduced the death from the CVD. However, in spite of these advances, 5-year mortality from CVD remains at about 50% in both men and women (Roger et al., 2004) indicating that CVD is the leading cause of morbidity and mortality worldwide. Multiple conventional risk factors have been identified for CVD. These factors include age, sex, genetic predisposition, diet, and lifestyle. These factors lead to hypertension, dyslipidemia, advanced age, and endothelial dysfunction (Mendelsohn and Larrick, 2013). Other important risk factors for CVD are sedentary lifestyle, oxidative stress, insulin resistance, obesity, type 2 diabetes, and smoking (Fig. 3.1) (NDIC, 2014; Petersen and Shulman, 2018). Moreover, it is widely accepted that excessive dietary intake of saturated fats and cholesterol plays a role in the onset and development of CVD through changes in plasma low-density lipoprotein-cholesterol (LDL-c) (Das and Das, 2010). However, nonlipid risk factors can also contribute to the development of CVD. About one half of the deaths due to this condition occur in individuals with normal cholesterol levels (Castelli, 1998; Das and Das, 2010). Among above risk factors, insulin resistance in the myocardium plays a prominent role in the pathogenesis

Insulin Resistance as a Risk Factor in Visceral and Neurological Disorders
DOI: https://doi.org/10.1016/B978-0-12-819603-8.00003-1

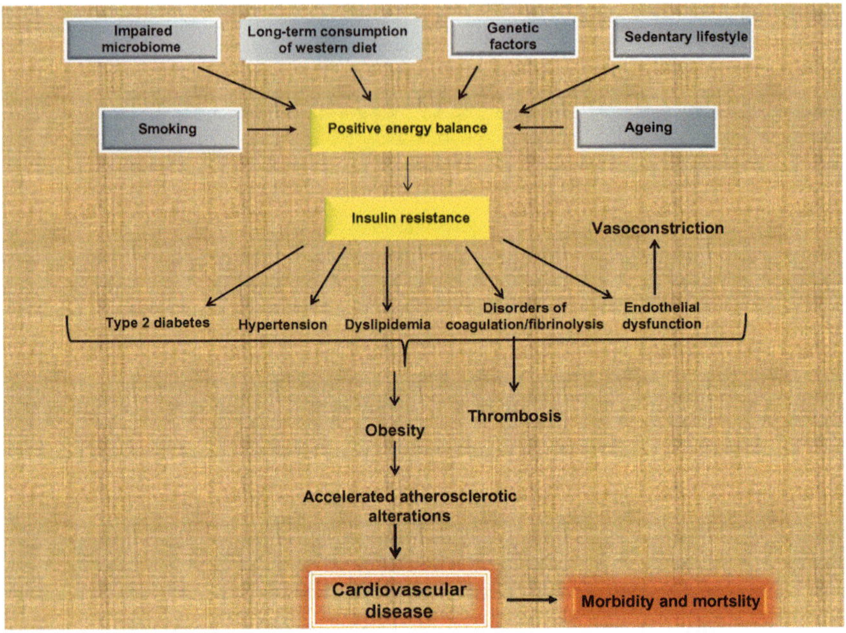

Figure 3.1 Factors associated with insulin resistance and relationship with obesity and heart disease.

of CVD at least through three different mechanisms: (1) signal transduction alterations, (2) impaired regulation of substrate metabolism, and (3) altered delivery of substrates to the myocardium. Seventy percent of adult Americans not only have insulin resistance, but are overweight or obese. The prevalence of insulin resistance and visceral obesity stands at 60% and continues to rise. Less than 20% of American adults exercise sufficiently, and over 65% engage in no vigorous physical activity. Among adults, 20% have diabetes, 38% have hypertension, 36% have prehypertension, 40% are prediabetes, 12% are both prediabetes and prehypertension, and 15% of the population with either diabetes, hypertension, or dyslipidemia are undiagnosed. Half of adults have at least two cardiovascular risk factors. Not even 1% of the population attains ideal cardiovascular health. Up to 65% of patients do not have their conventional risk biomarkers under control. Only 30% of higher risk patients with CVD address and achieve aggressive LDL targets. Of those patients with multiple risk factors, fewer than 10% have all of them adequately controlled. Even when patients are titrated to evidence-based targets, about 70% of cardiac events remain unaddressed.

Based on the above information, it is proposed that sedentary lifestyle (consistently low levels of physical activity) and overnutrition may lead induction of oxidative stress, endothelial dysfunction, and onset of low-grade inflammation especially in aged animals and humans. Among these process, vascular endothelial dysfunction is widely recognized as a hallmark of a variety of cardiovascular-related diseases such as hypertension, atherosclerosis, and metabolic disorders (Reho and Rahmouni, 2017; Davignon and Ganz, 2004). A myriad of signaling pathways have been established that underlie these deficits in endothelial-mediated relaxation including oxidative stress, inflammation, glucose/insulin metabolism, and the renin–angiotensin system with considerable complexity arising in the crosstalk between these pathways and the resulting integration of cardiovascular responses. A key integrating molecular hub for these signaling pathways in the vascular system is the mammalian target of rapamycin complex 1 (mTORC1) pathway. Previous studies have demonstrated that mTORC1 activity in vascular endothelial cells is elevated in response to angiotensin II, glucose (Fan et al., 2017), insulin (Kim et al., 2012), and oxidative stress (Rajapakse et al., 2011) suggesting a potential role in mediating the vascular functional effects of these signaling pathways. However, the overall contribution of mTORC1-dependent activity in the regulation of these vascular endothelial signaling pathways and ultimately vascular function remains unclear.

The consumption of high-calorie diet and sedentary lifestyle induce coronary atherosclerosis are responsible for the vast majority of the CVD events associated with heart disease in humans. The pathophysiology of CVD is very complex due to the involvement of multiple factors (Peluso et al., 2012; Touyz, 2004; Rodrigo et al., 2011). Ischemia/reperfusion injury also contributes to the pathogenesis of CVD. The redox imbalance during ischemia/reperfusion injury triggers the activity of a number of signaling pathways associated with reactive oxygen species (ROS) and reactive nitrogen species (RNSs) (Elahi et al., 2009). As mentioned earlier, ROS not only include oxygen radicals such as superoxide and hydroxyl radicals but also nonradical derivatives of O_2, including H_2O_2 and ozone (O_3) in cells. RNSs include nitric oxide (NO), nitrogen dioxide (NO_2^-), peroxynitrite ($OONO^-$), dinitrogen trioxide (N_2O_3), and nitrous acid (HNO_2). High levels of ROS and RNS impair beta-cell function and exacerbate insulin resistance in type 2 diabetes, which often coexists in patients with atherosclerosis and CVD (Wellen and Hotamisligil, 2005). It is also reported that in cardiac surgery with extracorporeal circulation, electrical and structural myocardial remodeling due to the excessive production of these ROSs may lead to the development of arrhythmias such as atrial fibrillation (Rodrigo et al., 2008).

Insulin signaling and nitric oxide production in cardiovascular diseases

Ingestion of food promotes the release of insulin from the pancreas. This anabolic hormone promotes the uptake of circulating glucose in heart tissue and myocardium by binding to an insulin receptor. At the molecular level, the binding of insulin to its receptors activates receptor autophosphorylation, which triggers a downstream signaling cascade through the phosphorylation of tyrosine residues of the insulin receptor substrates (IRSs; IRS-1 or IRS-2), followed by phosphorylation of phosphatidylinositol 3-kinase (PtdIns 3K), phosphoinositide-dependent kinase-1 (Akt; Akt1 and Akt2), protein kinase C (PKC), forkhead box O (FOXO), and mammalian target of rapamycin (mTOR) (Fig. 3.2) (Janus et al., 2016). These events result in an increased translocation of the glucose transporter 4 (GLUT4) to the membrane, thus facilitating glucose uptake (Bogan, 2012). After uptake, free glucose is rapidly phosphorylated to glucose 6-phosphate, which subsequently enters different metabolic pathways (Zimmer, 1996). In the cardiomyocytes, insulin promotes glucose and fatty acid uptake, but inhibits the use of fatty acids as an energy source. As a result of insulin resistance, the pancreas attempts to compensate by secreting increasing amounts of insulin, resulting in hyperinsulinemia and insulin resistance (Reaven, 2012). It is becoming increasingly evident that a strong correlation occurs between insulin resistance and risk of developing CVD (Gast et al., 2012). Molecular mechanisms, which contribute to the association between insulin resistance and CVD (Reaven, 2012; Bornfeldt and Tabas, 2011; Davidson and Parkin, 2009; Laakso and Kuusisto, 2014), include the development of atherosclerosis, vascular function, hypertension, and macrophage accumulation (Laakso and Kuusisto, 2014).

Metabolic diseases such as obesity, insulin resistance, and type 2 diabetes are all linked to CVDs such as cardiac hypertrophy and heart failure. Diabetic cardiomyopathy, in particular, is characterized by structural and functional alterations in the heart muscle of people with type 2 diabetes. This finally leads to heart failure. Several mechanisms have been involved in the pathogenesis of diabetic cardiomyopathy, such as alterations in myocardial energy metabolism and calcium signaling. Metabolic

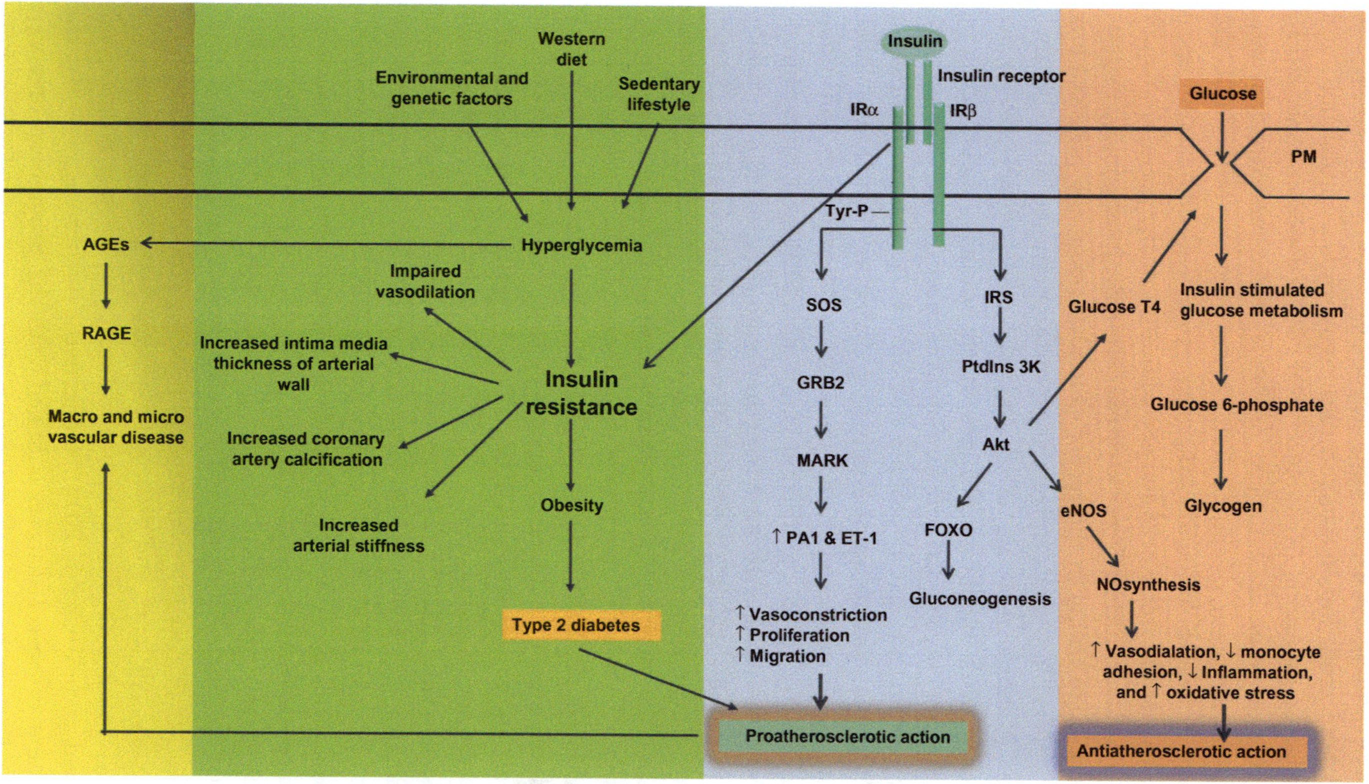

Figure 3.2 Glucose transport and insulin signaling in the heart.

disturbances during diabetic cardiomyopathy are characterized by increased lipid oxidation, intramyocardial triglyceride accumulation, and reduced glucose utilization (Kauppinen et al., 2013; Palomer et al., 2013). Overall changes result in enhanced oxidative stress, mitochondrial dysfunction, and apoptosis of the cardiomyocytes. On the other hand, the progression of heart failure and cardiac hypertrophy usually entails a local rise in cytokines in cardiac cells and the activation of the proinflammatory transcription factor nuclear factor kappa B (NF-κB). Interestingly, increasing evidences are arising in the recent years that point to a potential link between chronic low-grade inflammation in the heart and metabolic dysregulation (Kauppinen et al., 2013; Palomer et al., 2013). CVDs are accompanied with significant changes in myocardial insulin signaling, which influences myocardial structure and function. Hence, insulin resistance in heart disease may contribute not only to adverse left ventricular remodeling, but also to mitochondrial dysfunction (Riehle and Abel, 2016). In experimental studies, induction of insulin resistance, inflammation (Packer, 2018), vasoconstriction (Li et al., 2013), and antinatriuretic effects (Manhiani et al., 2011) are associated with undesirable and harmful changes in patients with heart disease. As mentioned in Chapter 2, Insulin resistance, diabetes, and metabolic syndrome, type 2 diabetes is a metabolic disease that occurs frequently mainly in the middle-aged and elderly period and accounts for above 90% of diabetes. Diabetes often leads to CVD (Morel et al., 2010; Zhang et al., 2018). The morbidity and mortality rates of type 2 diabetes complicated with coronary heart disease (CHD) are increasing. Approximately 70% of type 2 diabetes eventually die of cardiovascular and cerebrovascular diseases, and CHD is responsible for 50% of deaths (Gaede et al., 2008). Occurrence and development mechanism of CHD in patients with type 2 diabetes is not clear yet. Increasing number of studies are conducted on insulin resistance, oxidative stress, disorder of lipid metabolism, endothelial dysfunction, hypertension, and inflammatory reaction (Santi et al., 2015; Chazova et al., 2007). It is becoming increasingly evident that insulin resistance is a common pathophysiological change of metabolic syndrome, type 2 diabetes, and atherosclerosis. When insulin resistance occurs, the concentration of inflammatory cytokines and inflammatory-sensitive proteins increases in the body, leading to the inflammatory reaction in the arterial wall (Mahmoud and Al-Ozairi, 2013). Collective evidence suggests that the development of type 2 diabetes is linked with the progression of CVDs including heart failure. The underlying mechanisms involved in both conditions influence the structure, function, and metabolism of the heart tissue. Thus many diabetic patients also develop diabetic cardiomyopathy. This condition defined by ventricular dysfunction in the absence of coronary atherosclerosis and hypertension (Ryden et al., 2013). In the early stages, diabetic cardiomyopathy is usually asymptomatic. The earliest manifestations of diabetic cardiomyopathy include decrease in left ventricle compliance, impaired early diastolic filling, and prolonged isovolumetric relaxation (Jia et al., 2016). Furthermore, an increase in interstitial and perivascular fibrosis also occurs in the diabetic heart, which is different from the fibrosis observed following myocardial infarction and with coronary artery disease or hypertension (Kasznicki and Drzewoski, 2014).

It is well known that branched chained amino acids (BCAAs) play an important role in progression of type 2 diabetes and whole-body insulin resistance including cardiac insulin resistance (Newgard et al., 2009; Sun et al., 2016; Karwi et al., 2019; Wang et al., 2016). The molecular mechanisms associated with BCAAs-mediated insulin resistance are not fully understood. However, recent studies have indicated that BCAAs induce whole-body insulin resistance by increasing muscle BCAA oxidation, and decreasing glucose and fatty acid oxidation (Newgard et al., 2009). It has also been shown that in the heart that accumulation of BCAAs occurs in conjunction with a decrease, rather than an increase, in BCAA oxidation in obese mice (Fillmore et al., 2018). In addition, the accumulation of BCAAs may not only activate mTOR (Lynch and Adams, 2014), but also impair insulin signaling, due to phosphorylation of IRS-1 via directly activating p70S6K1 (Newgard et al., 2009; Dodd and Tee, 2012; Han et al., 2012; Wolfson et al., 2015).

Hypertension also plays an important role in the pathogenesis of heart disease. Many factors contribute to the pathophysiology of hypertension such as upregulation of the renin–angiotensin–aldosterone system, activation of the sympathetic nervous system (SNS), perturbed G protein-coupled receptor signaling, inflammation, and altered T-cell function. These processes are linked to increased production of ROS, decrease in NO production, and reduction in antioxidant capacity in the cardiovascular system (Fig. 3.3). Although oxidative stress may not be solely associated with the etiology of hypertension, it amplifies blood pressure elevation in the presence of other prohypertensive factors may contribute to hypertension. In the cardiovascular system, ROS play an important physiological role in controlling endothelial function, vascular tone, and cardiac function, along with pathophysiological changes in inflammation,

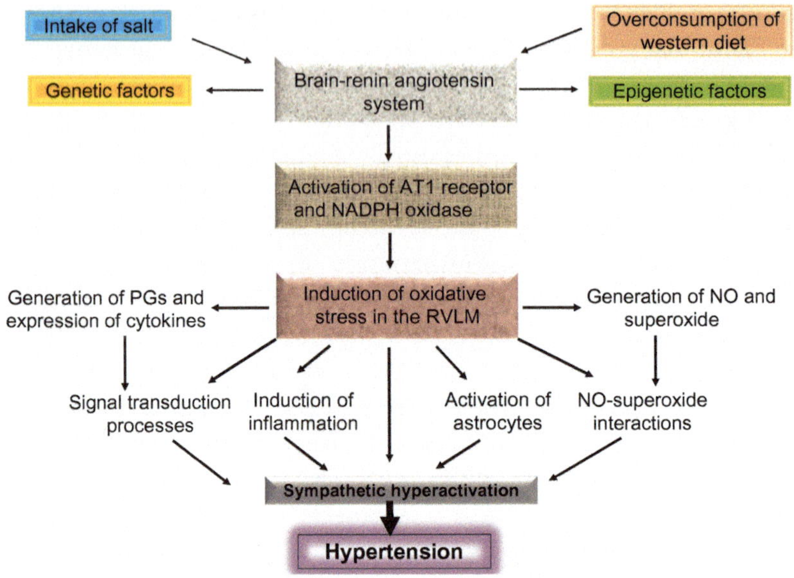

Figure 3.3 Hypothetical diagram showing the concept regulation of sympathetic nerve activity via brain renin–angiotensin system and oxidative stress.

hypertrophy, proliferation, apoptosis, migration, fibrosis, angiogenesis, and rarefaction, all of which are important processes contributing to endothelial dysfunction and cardiovascular remodeling in hypertension. A major source for cardiovascular ROS is the activation of nicotinamide adenine dinucleotide phosphate (NADPH) oxidase. Other sources include mitochondrial dysfunction and uncoupled NO synthase (NOS). Although convincing data from animal studies support a causative role for oxidative stress in the pathogenesis of hypertension, there is still no solid evidence that oxidative stress causes hypertension in humans. However, biomarkers of excess ROS are increased in patients with hypertension and oxidative damage is important in the molecular mechanisms associated with cardiovascular and renal injury in hypertension (Montezano and Touyz, 2012). In addition, activation of the SNS also plays an important role in the pathogenesis of hypertension (Fig. 3.3). Thus many studies have indicated that oxidative stress is not only supported by angiotensin II type 1 (AT1) receptor, but also promoted by NADPH oxidase and SNS activation in the autonomic brain regions. In hypertensive rats, the induction of oxidative stress in the rostral ventrolateral medulla (RVLM) promotes sympathoexcitation through the involvement of NO and inflammation (Kishi and Hirooka, 2012; Kishi, 2013; Hirooka, 2011; Guyenet et al., 2018).

Environmental factors such as intake of high salt intake and consumption of high-calorie diet increase the oxidative stress in the brain in the RVLM. This process promotes activation of the central sympathetic outflow and increases the risk of hypertension (Kishi and Hirooka, 2012; Kishi, 2013; Hirooka, 2011; Guyenet et al., 2018; Farooqui, 2015). This hypothesis is supported by the observation that several orally administered AT1 receptor blockers induce sympathoinhibition via reduction of oxidative stress through the inhibition of central AT1 receptor (Kishi and Hirooka, 2012; Kishi, 2013; Hirooka, 2011). Similarly, hyperglycemia in diabetic patients may not only lead to an increase in the production of ROS, but also may also contribute to other structural changes including cardiac hypertrophy, which may be increased by hyperinsulinemia due to insulin resistance (Lehrke and Marx, 2017; Nishikawa and Araki, 2007). Oxidative stress plays a major role in downstream diabetic complications and may also be involved in gene activation and remodeling of the myocardium due to cell death mediated by ROS (Lehrke and Marx, 2017). Abnormalities in adipokine (adiponectin, leptin, apelin, and adipsin) secretion link type 2 diabetes with heart failure. This may also exert a multitude of downstream pathophysiological effects in these pathological conditions (Dunmore and Brown, 2013). Abnormalities in insulin also mediates a change in the energy substrate of the heart by increasing the supply of pyruvate (Hermann et al., 1999; Hasenfuss et al., 2002). In the diabetic heart, there is a shift in metabolic substrates, with a decrease in glucose availability (Rijzewijk et al., 2010) and an increase in fatty acid oxidation to meet energetic demands (Lehrke and Marx, 2017). In early stages of heart failure, there is a metabolic substrate shift from fatty acids to glucose oxidation, but in later more advanced stages, insulin resistance may develop and a decrease in glucose utilization may occur (Neubauer, 2007; Lehrke and Marx, 2017) and several studies have indicated that long-term induction of insulin resistance may be an independent predictor of heart failure and worsening prognosis of heart disease (Doehner et al., 2005).

Insulin signaling in vasculature

The vascular endothelium forms the innermost layer of the blood vessel and produces and releases a variety of vasoactive substances and growth factors to regulate vascular homeostasis and angiogenesis. Endothelium not only provides an anticoagulant surface, regulates fluid, and molecule traffic between blood and tissues, but also contributes to vascular homeostasis and repair, and plays vital role in vascular tone and blood flow regulation. Recent studies have indicated that at the homeostatic stage, vascular endothelium is in a tightly regulated by a balance between pro- and antioxidants, vasodilators and vasoconstrictors, pro- and antiinflammatory molecules, and pro- and antithrombotic signals. An unhealthy or dysfunctional endothelium loses its balance and displays prooxidant, vasoconstrictor, proinflammatory, and prothrombotic properties. One hallmark of vascular endothelial dysfunction is impaired endothelial dependent dilation, which is predictive of future CVD events (Donato et al., 2018).

The vascular actions of insulin are complex. Insulin signaling within the vascular endothelium involves the binding of insulin to its receptor. This process promotes Akt-dependent phosphorylation and activation of endothelial NOS (eNOS). This enzyme acts on L-arginine to produce L-citrulline and NO. Production of basal levels of NO by endothelial cells regulates vasomotor tone and preserves the nonthrombogenic behavior of the vascular lining. The synthesis of NO is not only stimulated by receptor-dependent agonists (acetylcholine, bradykinin), but also by nonreceptor-dependent agonists (calcium ionophores). Collective evidence suggests that NO is a multifunctional signaling molecule that controls vascular tone, inhibits platelet function, prevents adhesion of leukocytes, reduces proliferation of the intima, and acts as a vasodilator. NO is also involved in the maintenance of metabolic and cardiovascular homeostasis (Fig. 3.4). The enhancement of arginase activity and the increase in asymmetric dimethylarginine and hyperhomocysteinemia levels also contribute to the pathogenesis of atherosclerosis by intervening NO bioavailability in human beings (Chen et al., 2018). Under normal conditions, NO diffuses into the vascular smooth muscle cells (VSMCs) and activates guanylate cyclase, which leads to cyclic guanosine monophosphate-mediated vasodilation. Shear stress is a key activator of eNOS under physiological circumstances and this helps adapt organ perfusion to changes in cardiac output. Other signaling molecules such as bradykinin, adenosine, vascular endothelial growth factor (expressed in response to hypoxia), and serotonin (released during platelet aggregation) can also activate eNOS (Taddei et al., 1992). In contrast, in type 2 diabetes, obesity, chronic kidney disease, CVD, peripheral arterial disease, the bioavailability of NO is markedly decreased. An enhanced inactivation and/or reduction in the synthesis of NO is seen as an important risk

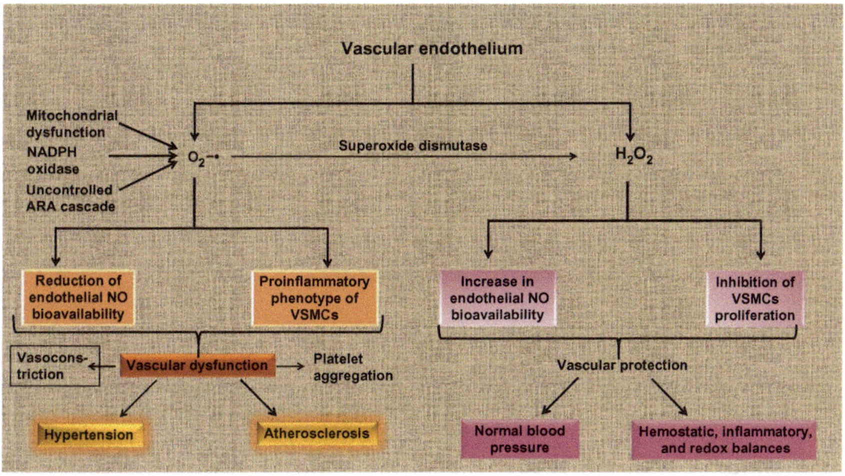

Figure 3.4 Events contributing to vascular dysfunction and vascular protection in endothelial cells.

factors for CVD. This condition is called as endothelial dysfunction. Endothelial dysfunction promotes vasospasm, thrombosis, vascular inflammation, and proliferation of VSMCs (Forstermann et al., 2017). Vascular oxidative stress with an increased production of ROS plays a major role in the pathogenesis of vascular dysfunction (Fig. 3.5). Oxidative stress is mainly caused by an imbalance between the activity of endogenous prooxidative enzymes (such as NADPH oxidase, the mitochondrial respiratory chain, and uncontrolled arachidonic acid cascade) and antioxidative enzymes (such as superoxide dismutase, glutathione peroxidase, heme oxygenase, thioredoxin peroxidase/peroxiredoxin, catalase, and paraoxonase) in favor of the former. These effects of NO are antagonized by the activation of the Ras—RAF—MAPK pathway, which contribute to the stimulation of cell growth and differentiation and increases the synthesis and secretion of the potent vasoconstrictor—endothelin-1 (ET-1) (Steinberg et al., 1994; Zeng and Quon, 1996). In addition, increased ROS levels reduce the amount of bioactive NO by chemical inactivation to form toxic $ONOO^-$. Peroxynitrite-in turn-can "uncouple" endothelial NOS to become a dysfunctional superoxide-generating enzyme that contributes to vascular oxidative stress. Oxidative stress and endothelial dysfunction are known to promote atherogenesis (Forstermann et al., 2017). Converging evidence suggests that consequences of endothelial dysfunction result in a range from proatherosclerotic molecular events to

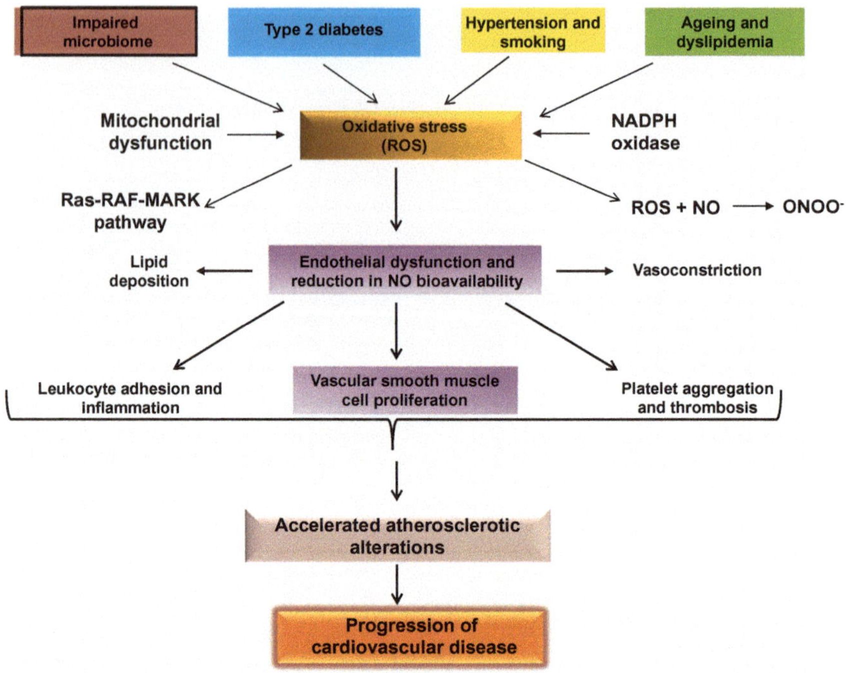

Figure 3.5 Progression from risk factors to atherosclerosis and cardiovascular disease mediated by oxidative stress and endothelial dysfunction. The early detection of endothelial dysfunction is a critical point in the prevention of atherosclerosis and cardiovascular disease because this dysfunction can be an initial reversible step in the process of atherosclerosis.

heart attack. Proatherosclerotic molecular events include increased lipid permeability and the promotion of oxidative and inflammatory environments within atheromatous plaques that favor plaque rupture and prothrombotic events, as seen in the acute coronary syndrome. Endothelial function therefore represents an integrated index of both the overall cardiovascular risk factor burden and the sum of vasculoprotective factors in a given individual. Given its role in the atherosclerotic process, it is not surprising that many studies demonstrate a prognostic role for endothelial dysfunction, as measured both in the coronary arteries and the circulation (Penny et al., 2001).

In type 2 diabetes, selective inhibition of the PtdIns 3K–Akt–eNOS pathway, together with compensatory hyperinsulinemia leads to unmasking and stimulation of the MAPK-mediated production of ET-1 (Potenza et al., 2005; Marasciulo et al., 2006), and vascular smooth muscle proliferation, which not only promote the formation of atherosclerotic plaque, but also the induction of peripheral artery disease (Beckman et al., 2002; Wang et al., 2003). Enhancement in endothelial function and secretion of ET-1, along with heightened sympathetic activity may represent key contributing factors in enhanced vasoconstriction of small diameter arteries and arterioles in the insulin-resistant state, thereby increasing systemic vascular resistance to blood flow and elevating arterial blood pressure. In addition, insulin-resistance elevates blood glucose levels and promote the formation of ROS and advanced glycation end products (AGEs). These metabolites interact with proteins and lipids leading not only to cross-linking of collagen and elastin fibers, but also loss of vascular compliance (i.e., arterial stiffening) (Schleicher et al., 1997; Sell and Monnier, 2012). Hyperglycemia accelerates formation of AGEs, which accumulate in the extracellular matrix of vessels and contribute to vascular damage in diabetes (Vlassara and Uribarri, 2014). AGEs stimulate production of ROS, which in turn further enhance AGE formation. AGEs are also antigenic and hence induce immune responses (Vlassara and Uribarri, 2014). In addition to AGEs, dicarbonyl methylglyoxal, a by-product of glycolysis, accumulates in tissues and contributes to diabetes-associated vascular damage (Nigro et al., 2017). Drugs, such as aminoguanidine, vitamins, angiotensin-converting enzyme inhibitors, angiotensin-II receptor blockers, statins, and metformin inhibit AGE formation. Alagebrium, an AGE breaker not only reduces levels of AGEs, but also decreases arterial stiffness.

High levels of ROS in endothelium activates NF-κB, which migrates to the nucleus where it upregulates the expression of proinflammatory cytokines and chemokines, such as tumor necrosis factor-beta (TNF-β), interleukin-1β (IL-1β), interleukin-6 (IL-6), and monocyte chemoattractant protein-1 and adhesion molecules, such as vascular cell adhesion molecule-1 (VCAM-1) and intercellular adhesion molecule-1 (ICAM-1), which are required for the adhesion of leukocytes to the endothelial surface (Dessì et al., 2013). The early phase of atherogenesis is characterized by the attraction/adherence of monocytes to the vascular endothelium and their migration into the vessel wall. The expression of cellular adhesion molecules promotes the adhesion of leukocytes to the vascular endothelium and is induced by inflammatory factors, including IL-1 and tumor necrosis factor-α (TNF-α) (Willerson and Ridker, 2004). In particular, VCAM-1 binds specifically to those classes of leukocytes found in nascent atheroma (i.e., monocytes and T lymphocytes). Both macrophages and endothelial cells produce ICAM-1 in response to inflammatory cytokines (e.g., IL-1, TNF-α, interferon-γ), whereas VCAM-1 expression is mainly restricted to endothelial cells. Furthermore, the progressive accumulation of macrophages and their uptake of oxidized LDL ultimately lead to the generation of foam cells and initiates fatty streaks.

These processes are also accompanied by increase in circulating acute phase protein called C-reactive protein (CRP). In atherosclerotic plaques, CRP is present with monocytes and lipoproteins (Torzewski et al., 2000). CRP contributes to phagocytosis, which clears necrotic tissues in the atherosclerotic plaques and perpetuates inflammatory response (Zakynthinos and Pappa, 2009; Sproston and Ashworth, 2018). There are indications that persons with no manifestations of vascular disease and elevated CRP have a three- to fourfold increased relative risk of myocardial infarction (Mach et al., 1997). Furthermore, baseline levels of CRP are elevated in patients with unstable angina and are associated with an unfavorable short-term prognosis. CRP levels may be a valid prognostic marker for differentiation between patients with unstable angina and chronic stable angina; however, they fail to differentiate patients with stable coronary artery disease from patients with acute coronary syndrome (Kaptoge et al., 2010). CRP also plays important roles in host responses to infection including the complement pathway, apoptosis, phagocytosis, NO release, and the production of cytokines, particularly IL-6 and TNF-α (Sproston and Ashworth, 2018). The endothelial expression of these factors contributes to the development of inflammation within the arterial wall and promotes the formation of atherosclerotic plaques, which are composed of VSMCs, monocyte/macrophages, T lymphocytes, and other inflammatory cells, in addition to intra- and extracellular lipid and cellular debris (Li et al., 1993). In addition, endothelial dysfunction contributes to impairment in insulin action by altering the transcapillary passage of insulin to target tissues (Cersosimo and DeFronzo, 2006). Reduced expansion of the capillary network, with attenuation of microcirculatory blood flow to metabolically active tissues, contributes to the impairment of insulin-stimulated glucose and lipid metabolism. Vascular injury caused by increased levels of proinflammatory eicosanoids, and oxidative stress to the vessel wall triggers inflammatory reactions and responses by releasing more chemoattractants, cytokines, and chemokines, which worsen the insulin resistance and endothelial dysfunction (Cersosimo and DeFronzo, 2006). Elevated levels of circulating lipids and lipid mediators in LDL have significant effects on the development of plaques (Libby et al., 2011) supporting the view that the migration of LDL into the vessel wall with subsequent oxidation and endothelial dysfunction are key processes initiating atherogenesis.

As mentioned above, aging is an important risk factor for CVD. This is largely because of dysfunctional arteries (Donato et al., 2018), which are composed of multiple concentric layers, each with has a distinct composition and function. The intima is the innermost artery layer and is a composite of two layers. The luminal layer, known as the basal lamina, is comprised of a thin basement membrane with a proteoglycan rich matrix and small amounts of collagen (Donato et al., 2018). The second intimal layer is musculoelastic. It is composed of elastin fibers, individual smooth muscle cells, and collagen. The intima is separated from the media, the inner layer of the artery. It is composed of lamellar units that are composites of elastin fibers, circumferentially oriented VSMC layers, collagen fibers, and a glycosaminoglycan viscoelastic gel, commonly referred to as a "ground substance" (Stary et al., 1992). It is interesting to note that elastin within the internal elastic lamina is oriented longitudinally in the direction of luminal blood flow, while in the media it is oriented circumferentially (Farand et al., 2007). Stiffening of the large elastic arteries is an emerging and independent predictor of CVD (Mattace-Raso et al., 2006; Mitchell et al., 2010; Willum Hansen et al., 2006). In CVD, much of the arterial stiffening can be attributed to structural remodeling within the artery (Diez, 2007). The structural alterations within the intimal, medial, and/or adventitial layers of the artery involves the increased deposition of collagen (a protein providing mechanical stiffness to the artery), reductions in elastin (a protein, which provides elasticity to arteries and veins), and greater cross-linking by AGEs (Zieman et al., 2007; Steppan et al., 2012; Fleenor et al., 2012a,b; Sindler et al., 2011). As mentioned in Chapter 2, Insulin resistance, diabetes, and metabolic syndrome, AGEs interact with two main types of cell surface receptors: (1) scavenger receptors, which remove and degrade AGEs, and (2) receptors for AGEs (RAGEs), which trigger-specific cellular signaling responses on AGE binding. RAGE is a member of the immunoglobulin family and interacts many ligands including AGEs, high mobility group protein B1, S100 calcium binding proteins (including calgranulin), amyloid-b-protein, and amphotericin. RAGEs are not only expressed in endothelial cells and VSMCs, but also on neurons, glial cells, and macrophages (Wendt et al., 2002). AGE–RAGE signaling involves NF-kB, mitogen-activated protein kinases (MAPK; ERK1/2, p38MAPK), and NADPH oxidases. AGE–RAGE signaling also induces expression of vascular adhesion molecule-1, selectin, vascular endothelial growth factor, and proinflammatory cytokines (IL-1β, IL-6, and TNF-α) (Manigrasso et al., 2014) (Fig. 3.6). CVD-related risk factors (hypertension, type 2 diabetes, dyslipidemia, renal disease, and smoking) are associated with increased arterial stiffness (Benetos et al., 2002). Stiffness of large elastic arteries (aorta and carotid arteries) is commonly assessed regionally and locally by aortic pulse-wave velocity and carotid artery compliance, respectively (Mitchell et al., 2010; Tanaka et al., 2000). Large elastic arteries are known to stiffen with advancing age (Mitchell et al., 2010), which is associated with a higher risk of CVD-related mortality (Willum Hansen et al., 2006; Ben-Shlomo et al., 2014). Increase in vascular smooth muscle tone and structural components of the arterial wall are thought to be the predominate factors contributing to the increase in stiffness (Lakatta and Levy, 2003; Najjar et al., 2005), driven by reductions in NO bioavailability and increases in oxidative stress and inflammation (Fleenor, 2012). The efficacy of blood pressure treatment and differences in efficacy between different types of antihypertensive agents has been used to measure and decrease arterial stiffness (Ong et al., 2011; Van Bortel et al., 2011).

It is well known that higher physical activity in late-life, and habitual physical activity from mid-life to late-life lower central arterial stiffness and pressure pulsatility in a large population-based sample of community-dwelling older adults (Tanaka et al., 2018). Furthermore, exercise also produce cardiovascular and cerebrovascular changes and muscular fitness by elevating energy consumption, improving insulin sensitivity, increasing blood flow, strengthening the immune system, reducing inflammation, promoting sleep, and controlling weight (Fig. 3.7). The molecular mechanisms underlying abovementioned processes are not fully understood. However, it is becoming increasingly evident that exercise not only improves the dyslipidemic profile by raising high-density lipoprotein-cholesterol (HDL-c) and lowering triglycerides in the body (Lakka and Laaksonen, 2007), but also increases expression of GLUT4 and other proteins involved in insulin signaling and glucose metabolism (Houmard et al., 1993). An important factor, which correlates with CVD risk is HDL particle size. Hence, an increase in the size of LDL and HDL particles and a decrease in very low-density lipoprotein (VLDL) particle size, rather than HDL levels per se, upon exercise may provide protection from CVD risk (Kraus et al., 2002). Exercise also increases angiogenesis by upregulating the expression of VEGF (Louissaint et al., 2002), BDNF, apolipoprotein E (Schuit et al., 2001), catechol-O-methyltransferase

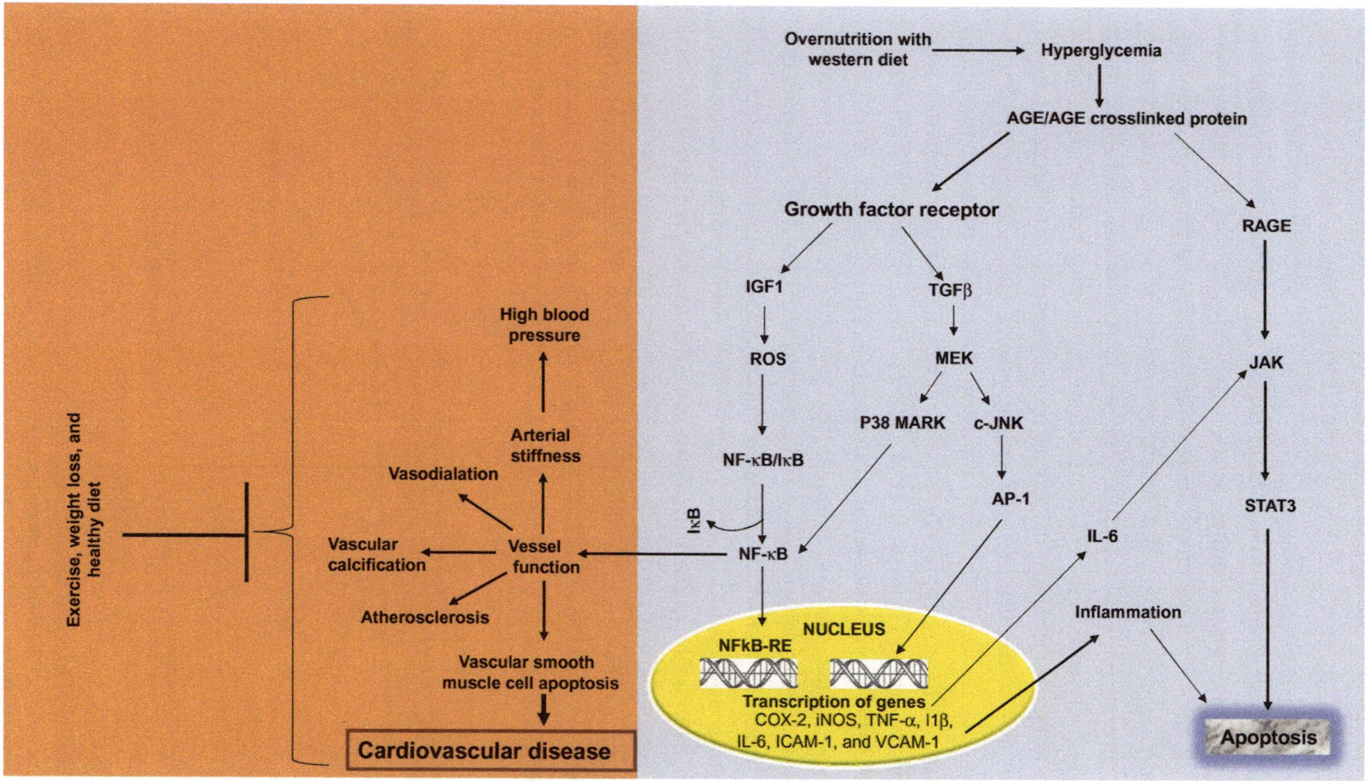

Figure 3.6 Interactions of AGEs with their receptors and downstream signaling in vascular system.
AP-1, Activator protein-1; *c-JNK*, cjun N-terminal kinase; *ERK*, extracellular signal-related protein kinase; *IGF-1*, insulin like growth factor 1; *IL-6*, interleukin-6; *JAK*, janus kinase; *MAPK*, mitogen-activated protein kinase; *NF-κB*, nuclear factor kappa B; *p38MARK*, p38 mitogen-activated protein kinase; *RAGE*, receptor for AGEs; *STAT3*, signal transducers and activator of transcription 3; *TGF-β*, transforming growth factor β.

(Stroth et al., 2010), endorphins, and NO and suppressing atherogenesis (Fig. 3.7) (Camargo et al., 2013). Production of metabolites by exercise increases cardiac output and lowers heart rate, and prevents cardiac hypertrophy. In addition, these molecules improve the cardiovascular status by decreasing atherosclerotic plaques, aortic valve calcification, increasing vascular smooth muscles relaxation, decreasing high blood pressure, reducing body weight, improving insulin sensitivity, helping in glycemic control, improving lipoprotein-lipid profiles, preventing obesity, and decreasing risk of type 2 diabetes (Fig. 3.8). All these processing reduce the risk of CVD. In addition to changes in plasma lipids, exercise can directly impact the homeostasis of the arterial wall to antagonize the progression of atherosclerotic disease and thereby contribute to the well-documented reduction in CVD in people with active lifestyles, when compared with sedentary individuals (Thompson et al., 2003). Even in people with symptomatic coronary artery disease, an increase in regular physical activity can improve VO_2 max and, at high doses (~2200 kcal/week), promote regression of atherosclerotic lesions (Hambrecht et al., 1993). Collective evidence suggests moderate exercise alters plasma lipid profile and increases HDL concentration and particle size. An important question is how much exercise is optimal for cardiovascular health benefit. Moderate exercise is beneficial for cardiac health. However, too much exercise may produce detrimental effects (O'Keefe et al., 2012). The cardioprotective effect of exercise can be explained by the hormesis, which is defined as a biphasic dose response whereby moderate exercise stimulates resistance to

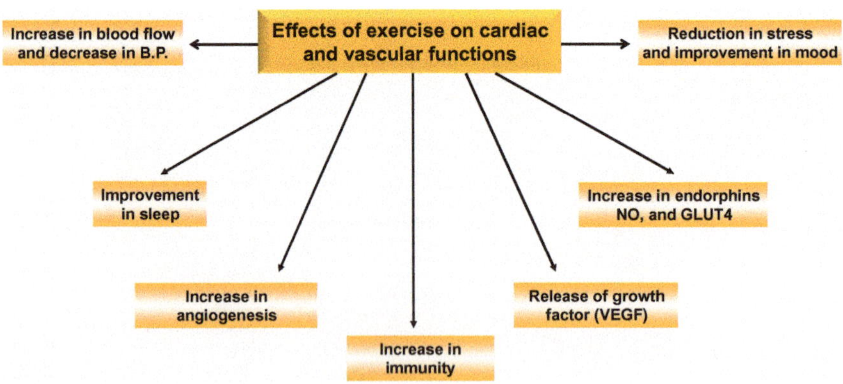

Figure 3.7 Effects of exercise on cardiovascular function.
B.P., Blood pressure; *GLUT4*, glucose transporter; *NO*, nitric oxide; *VEGF*, vascular endothelial growth factor.

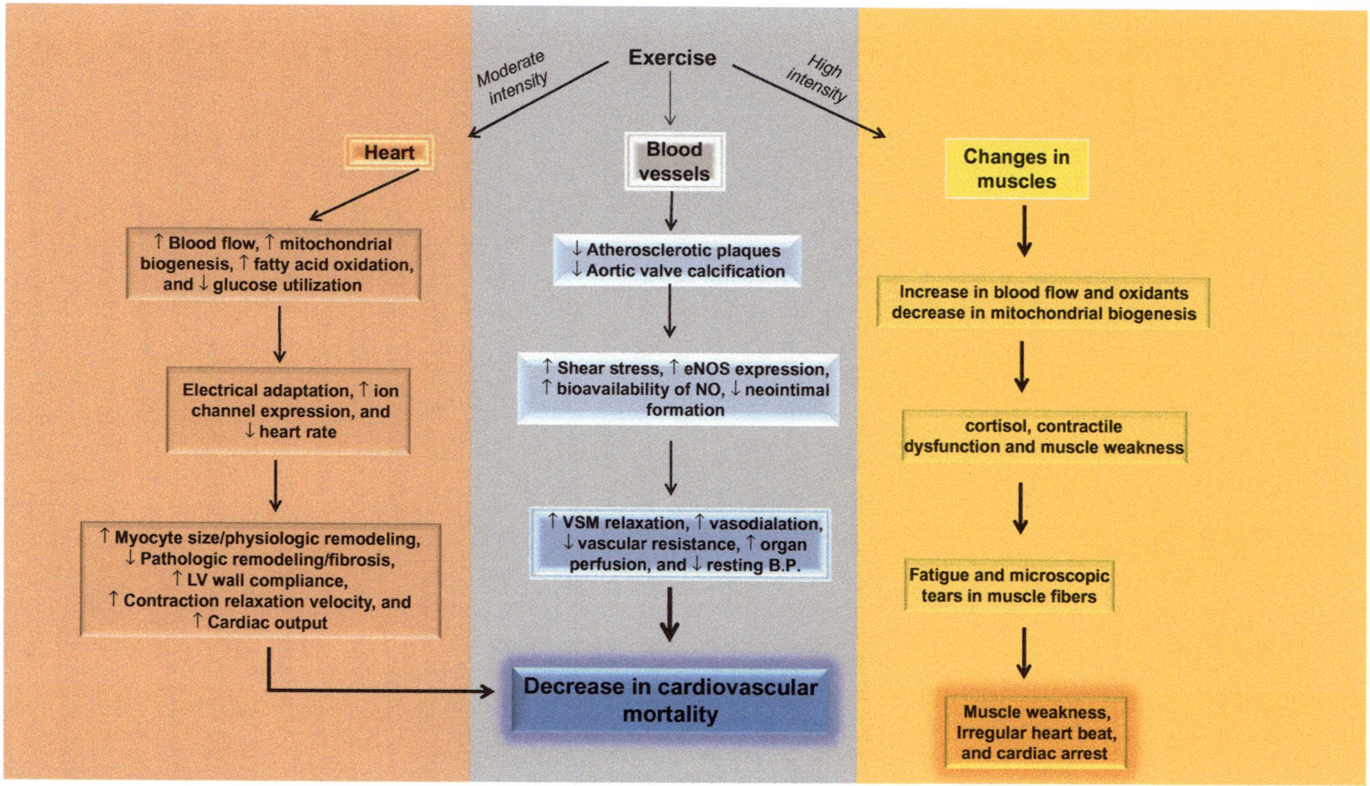

Figure 3.8 Effects of exercise on cardiovascular and muscular functions.
eNOS, Endothelial nitric oxide synthase; *LV*, left ventricle; *VSM*, vascular smooth muscle.

stress and improves biological fitness, while too much exercise induces damage and inhibits function. This concept of hormesis can be extended to the effect of exercise on ROS and RNS production in muscles (Radak et al., 2005, 2008). ROS and RNS generated during exercise can cause oxidative and nitrosative stress and damage to skeletal muscle, heart muscle, and muscles of other organs, and neurochemical changes in the brain. It is well known that at low levels ROS and RNS serve as the chemical agents (mediators) for maintaining cellular milieu and transferring messages from one subcellular organelle to another to conduct physiological functions. Several studies have indicated that individuals who completed at least 25 marathons over a period of 25 years have higher than expected levels of coronary artery calcification and calcified coronary plaque volume when compared with sedentary individuals (Roberts et al., 2017; Laddu et al., 2017; Aengevaeren et al., 2017). This is tempting to speculate that more studies are needed to identify the mechanisms that impart cardiovascular benefits of exercise in order to develop more effective exercise regimens, test the interaction of exercise with diet, and develop pharmacological interventions for those unwilling or unable to exercise.

It has been hypothesized that early hyperglycemia not only leads to marked increase in AGE formation, but also facilitates the induction of oxidative stress. Over time, mitochondrial respiratory chain proteins become increasingly glycated and mitochondrial DNA damage occurs leading to a self-perpetuating cycle of AGE formation and oxidative stress independently of hyperglycemia (Testa et al., 2017). AGEs are known to play an important role in the pathogenesis of large elastic artery stiffness with advancing age (Steppan et al., 2012; Fleenor et al., 2012a; Kass et al., 2001). Although it is well known that AGEs contribute to arterial stiffening, less in known about mechanisms leading to the arterial stiffness. However, it is proposed that glycation of structural extracellular proteins in the myocardial matrix may not only increase the myocardial stiffness, but may contribute to impaired relaxation and diastolic dysfunction (Candido et al., 2003). In addition to glycated collagen, glycation of elastin and laminin in basement membrane have also been shown to impair endothelial cell adhesion and migration by disrupting cell attachment sites (Haitoglou et al., 1992). These alterations in cell–matrix interactions are associated with a reduction in stress-induced NO production by endothelial cells and impaired vasodilation.

The RAGE gene promoter region contains an NF-κβ binding domain, suggesting that expression of RAGE may be upregulated as part of the inflammatory response. This creates a positive feedback loop, as activation of RAGE by AGEs leads to a series of phosphorylation reactions, including MAPK activation, and results in translocation of NF-κβ to the cell nucleus and enhanced expression of additional proinflammatory cytokines and proteins including ras, IL-6, TNF-α, TGF-β, and vascular adhesion molecules (VCAM-1, ICAM-1, ET-1) (Fig. 3.5) (Ramasamy et al., 2005, 2009; Fishman et al., 2018). RAGE activation also enhances activity of the JAK/STAT3 signaling pathway and upregulation of interferon responsive genes (Ott et al., 2014). Treatment with AGEs cross-linking breaking compounds, such as Alagebrium attenuates arterial stiffening in older adults (Kass et al., 2001) and rodents (Steppan et al., 2012). Moreover, treatment of biologically active AGEs ex vivo produces greater mechanical stiffness in cultured aortic rings from young mice further supporting a role of AGEs in arterial stiffening (Fleenor et al., 2012a,b). Collectively, these studies indicate that enhanced cardiovascular stiffness is a precursor to atherosclerosis, systolic hypertension, cardiac diastolic dysfunction, and impairment of coronary and cerebral flow. It is also reported that in aged obese subject, stiffness of endothelial and VSMCs, extracellular matrix remodeling, perivascular adipose tissue inflammation, and immune cell dysfunction contribute to the development of arterial stiffness. Enhanced endothelial cortical stiffness decreases endothelial generation of NO, and increased oxidative stress promotes destruction of NO. In addition, advancing age is associated with marked increase in oxidative stress (Fleenor et al., 2012a,b) and inflammatory (Zou et al., 2006; Csiszar et al., 2008; Ungvari et al., 2010) signaling cascades. These processes in turn may contribute to arterial stiffening in CVD development.

Molecular mechanism of atherosclerosis

Atherosclerosis is a multifactorial inflammatory disease of the arteries characterized by lipid accumulation within the artery walls. In humans, atherosclerotic plaques are usually found in the aorta, the coronary arteries, and cerebral arteries, but also in peripheral arteries. Advanced atherosclerotic plaques grow large to block blood flow resulting in various CVDs including CHD, angina, carotid artery disease, peripheral artery disease, and chronic kidney disease. The onset of atherosclerosis involves a network of vascular wall cells and mediators (Steinberg, 2002), in which macrophages play critical roles by producing proinflammatory factors and via transition to lipid-laden foam cells that initiate the formation of atherosclerotic lesion (Yan and Hansson, 2007). The pathogenesis of atherosclerosis involves multiple pathways, with reactions occurring continuously rather than in discrete steps, each one molecularly complex, involving intersecting pathways, and rich crosstalk horizontally and vertically (Libby, 2012; Quillard and Libby, 2012; Wang and Bennett, 2012). Chronic inflammation is closely associated with the initiation to progression of atherosclerosis. Atherosclerosis can be assessed by monitoring the arterial stiffness, which can be monitored using pulse-wave velocity, the cardio-ankle vascular index, the ankle-brachial index, pulse pressure, the augmentation index, flow-mediated dilation, carotid intima media thickness, and arterial stiffness index-β. Arterial stiffness is generally considered an independent predictor of cardiovascular and cerebrovascular diseases (Saiki et al., 2016). Role of AGEs in the formation of glycated collagen, glycated elastin, and glycated laminin and the induction of arterial stiffness has been described earlier (Haitoglou et al., 1992). The atherosclerotic process is initiated when lipid-containing lipoproteins accumulate on the intima and activate the endothelium (Kuiper et al., 2007). Inflammatory responses are characterized by the recruitment of circulating leukocytes and the production of growth factors which encourage cell migration and proliferation (Kuiper et al., 2007; Mallika et al., 2007). Animal model studies have indicated that the retention/accumulation of serum LDL on intima and sedentary lifestyle is the crucial factor for the initiation and progression of atherosclerosis (Lichtenstein et al., 2006). The delivery and retention of lipoproteins appear to be dependent on lipoprotein concentration, lipoprotein size, and the integrity of the endothelium (Skålén et al., 2002; Leitinger, 2003). Indeed, modification of retained lipoproteins contribute to the release of phospholipids and phospholipid-derived lipid mediators that can activate endothelium (Leitinger, 2003). It is the basis for increased expression of proinflammatory cytokines, chemokines, and adhesion molecules by endothelial cells. This ultimately leads to the loss of the morphofunctional integrity of the endothelium. The accumulation of lipids and lipid-derived lipid mediators (prostaglandins, leukotrienes, thromboxanes) within the artery wall is a major factor for the initiation of inflammatory process in the artery. The attachment of white blood cells initiates the more production of chemokines and promotes its migration into the subendothelial space (Ross, 1999; Kuiper et al., 2007). Monocytes enter the intima and differentiate into macrophages and express at high level scavenger receptors and toll-like receptors. Scavenger receptors may contribute to internalize apoptotic cell fragments, oxidized LDL particles, and other dendrites. Lipid deposition into macrophages contribute to foam cell (a hallmark of atheromata) formation. Phagocytic caveolae, which are sites strongly associated with cholesterol homeostasis and signal transduction, contain the structural protein, caveolin-1. Increased activity of caveolin-1, mediated by activation of transcription factor early growth response 1, promotes monocyte to macrophage differentiation, a possible critical step in atherogenesis (Fu et al., 2012). Suppression of monocyte/macrophage caveolin-1 is known to inhibit foam cell formation and impede atherogenesis. Toll-like receptors initiate a signal cascade that leads to macrophage activation and production of inflammatory cytokines, proteases, and cytotoxic radical molecules (Ross, 1999; Kuiper et al., 2007). Also, T-lymphocytes infiltrate the atherosclerotic lesions. These T-cells recognize antigens presented to them by activated macrophages. Activated T cells therefore differentiate mainly into T-helper 1 cells and begin producing interferon-γ, which in turn increases the process of antigen presentation by macrophages to lymphocytes and stimulates synthesis of other cytokines like TNF and IL-1. All these cytokines stimulate the production of many other inflammatory and cytotoxic molecules thus increasing the burden of the inflammatory reaction.

Another important factor in the development of heart disease is the involvement of dysbiosis, a process associated with changes in the composition of gut microbiota. Dysbiosis is linked with the pathogenesis of many conditions including atherosclerosis, hypertension, heart failure, chronic kidney disease, obesity, and type 2 diabetes (Saad et al., 2016). Recent studies have indicated that crosstalk between gut microbiota and host intestinal tract not only involves multiple overlapping pathways, including the autonomic, neuroendocrine, and immune systems, but also bacterial metabolites and neuromodulatory molecules (Quigley, 2017). This crosstalk produces a variety of effects in the host. For example, generation of bioactive metabolites [trimethylamine/trimethylamine *N*-oxide (TMAO), short-chain fatty acids, and primary and secondary bile acids] that have impact on host physiology (Quigley, 2017; Moludi et al., 2018; Bu and Wang, 2018). The significance of short-chain fatty acids pathway has been described in Chapter 1, Insulin resistance and obesity. TMAO pathway may be involved in the etiology of hypertension, atherosclerosis, coronary artery disease, diabetes, and renal failure. On the contrary, a number of studies have shown protective functions of TMAO, such as stabilization of proteins and protection of cells from osmotic and hydrostatic stresses (Nowiński and Ufnal, 2018). In addition to abovementioned metabolism-dependent pathways, metabolism-independent processes have also been suggested to potentially contribute to CVD pathogenesis. For example, heart failure—associated splanchnic circulation congestion, bowel wall edema, and impaired intestinal barrier function are thought to result in bacterial translocation, the presence of bacterial products in the systemic circulation and heightened inflammatory state. These processes may also contribute to further progression of heart failure and atherosclerosis (Tang et al., 2017). A better understanding of the pathogenesis of atherosclerosis can be obtained from the following sites (http://www.ncbl.nlm.nih.gov/; http://www.pubmedcentral.nih.gov/). It is now well established that atherosclerosis is the culprit behind coronary artery disease, cerebral vascular diseases, and peripheral vascular diseases. The pathogenic mechanisms of above conditions are complex and major players include lipoprotein transport, homocysteine metabolism, induction of inflammation, and infection in pathogenesis of coronary artery disease.

Carbohydrate metabolism, insulin resistance, and heart disease

Insulin resistance is a condition in which insulin receptor signaling in cells is impaired. Insulin resistance is caused by the exposure of insulin receptor to high levels of insulin. Insulin resistance downregulates insulin receptors via a negative homeostatic mechanism, or disruption of signaling molecules downstream of the insulin receptor, such as IRS-1 and IRS-2. It is characterized by defects in uptake and oxidation of glucose, a decrease in glycogen synthesis, and, to a lesser extent, the ability to suppress lipid oxidation (Petersen and Shulman, 2018). Many studies have indicated that free fatty acids are the predominant substrate for ATP synthesis in the adult myocardium, but the cardiac metabolic network is highly flexible. It can use other substrates, such as glucose, lactate, or amino acids. Many studies have indicated that effects of insulin resistance in different tissues depend on their physiological as well as metabolic functions. Due to their high metabolic demand, insulin resistance has significant effects on skeletal muscle, adipocytes, and liver tissue, which are the main targets of intracellular glucose transport as well as glucose and lipid metabolism (Dimitriadis et al., 2011). Skeletal muscle and adipocytes accounts for about 60%—70% and 10% of insulin-stimulated glucose uptake, respectively, via the GLUT4 receptors. Insulin resistance not only contributes to impaired glycogen synthesis and protein degradation in skeletal muscles, but also inhibits lipoprotein lipase activity in adipocytes leading to an increased release of free fatty acids and inflammatory cytokines such as IL-6, TNF-α, and leptin. Additionally, the liver accounts for 30% of insulin-stimulated glucose disposal and insulin resistance leads to impaired glucose output and fatty acid metabolism leading to increased diacylglycerol (DAG), triacylglycerol (TAG) content, and VLDL secretion from liver (Wilcox, 2005; Petersen and Shulman, 2018). As mentioned above, insulin resistance promotes endothelial cell dysfunction by lowering the production of NO from endothelial cells and increasing the release of procoagulant factors leading to platelet aggregation. In an insulin-resistant state, the PtdIns 3K pathway is affected whereas the MAP kinase pathway is intact, which causes mitogenic effect of insulin in endothelial cells leading to atherosclerosis (Wu and Meininger, 2009; Wang et al., 2003). During insulin resistance, metabolic changes can promote the development of CVD. For instance, insulin resistance can induce an imbalance in glucose metabolism leading to chronic hyperglycemia, which in turn triggers oxidative stress and causes an inflammatory response leading to cell damage (Ginsberg, 2000; Petersen and Shulman, 2018).

Fatty acids metabolism, insulin resistance, and heart disease

Dietary fat contains monounsaturated, polyunsaturated, trans and saturated fatty acids. The metabolism of these fatty acids markedly effects human health. Monounsaturated fatty acids (MUFAs; extra virgin olive oil) are known to reduce LDL cholesterol levels in the blood and lower the risk of heart disease and stroke. Olive oil also provide nutrients like vitamin E and phytochemicals like tyrosol, hydroxy-tyrosol, oleuropein, and oleocanthal, which are important antioxidants. Polyunsaturated fatty acids (PUFAs) have more than one unsaturated carbon bond in the molecule. They are found in soybean, corn, peanut, and sunflower oils. They produce beneficial effects in heart when eaten in moderation and when used to replace saturated fat and trans fatty acids in the diet. They also decrease LDL cholesterol. Among PUFAs, consumption of excess of arachidonic acid results in production of high levels of proinflammatory eicosanoids (prostaglandins, leukotrienes, and thromboxanes). Docosahexaenoic acid (DHA) is enriched in fish oil. Oxidation of DHA results in formation of docosanoids (D-series resolvins, protectins, and maresins) (Farooqui, 2012). DHA-derived lipid mediators produce antioxidants, antiinflammatory, and antiapoptotic effects (Farooqui, 2012). The risk of diabetes is inversely associated with a higher intake of PUFA, but no associations have been found with MUFA or saturated fatty acids (Salmerón et al., 2001).

Trans fatty acids are geometric isomers of monounsaturated and PUFAs having at least one carbon—carbon double bond with hydrogens on opposite sides of the double bond (trans configuration). Trans fatty acids are known to increase total and LDL cholesterol and decrease the HDL cholesterol (Mensink and Katan, 1990) supporting the view that intake of trans fatty acids increases the risk of CVD by increasing the ratio of LDL cholesterol to HDL cholesterol (Willett et al., 1993). Furthermore, a high intake of trans fat is associated with a higher risk of both diabetes (Salmerón et al., 2001; Iqbal, 2014). Similarly, saturated fatty acids increase LDL cholesterol and increase the risk for heart disease (Mensink, 2016; Houston, 2018). Thus a high intake of saturated fatty acids is associated with an increased risk of cardiovascular events (Zong et al., 2016) and reducing the dietary content of saturated fatty acids is emphasized in CVD prevention (Briggs et al., 2017). Collective evidence suggests that the type of oil consumed is related to risk factors for noncommunicable diseases. Among various dietary oils, sesame oil plays a supportive role in hypertensive treatment (Sankar et al., 2006), and vegetable oils rich in PUFA not only decrease diastolic blood pressure (Rao et al., 1981), but also lower fasting plasma glucose levels (Trevisan et al., 1990).

As mentioned in Chapter 2, Insulin resistance, diabetes, and metabolic syndrome, overnutrition is a major cause of insulin resistance. Insulin resistance due to overnutrition has been best characterized in the liver and adipocytes. Increase in food intake impairs fatty acid oxidation with redirects the long-chain acyl coenzyme As (CoAs) to pathways that favor the synthesis of DAG, ceramide, and TAG (Fig. 3.9) (Chavez and Summers, 2003; Boden et al., 2001). These lipid metabolites contribute to insulin resistance by interfering with insulin signaling. There is also an increase in the concentration of malonyl CoA by insulin (Bandyopadhyay et al., 2006). In addition, insulin is known to inhibit the expression of beta-oxidation enzymes in the hepatocyte. Accumulation of fat in the liver results in not only liver steatosis, but also in imbalance in free fatty acid availability, and impairment in the oxidative capacity of mitochondria causing mitochondrial dysfunction and further accumulation of lipid and lipid mediators in the cell (Muoio and Newgard, 2008). At the molecular level, excessive concentrations of DAG interfere with the phosphorylation of IRS proteins and activation of PtdIns 3K. This effect is mediated by serine kinases in the PKC family that phosphorylate IRS proteins at serine

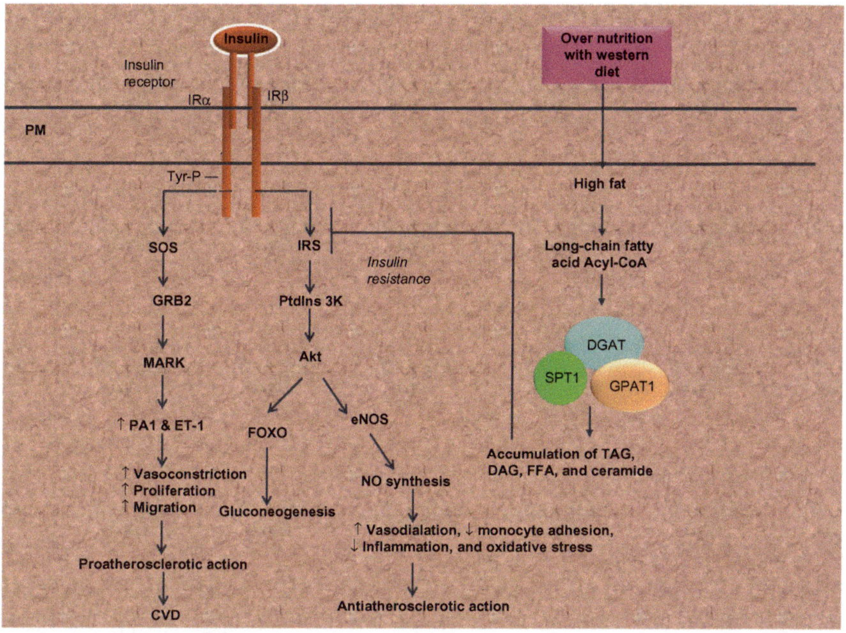

Figure 3.9 Triacylglycerol, diacylglycerol, free fatty acid, and ceramide-mediated changes in insulin signaling.
Aβ, Beta amyloid; *Akt*, serine/threonine protein kinase; *APP*, amyloid precursor protein; *DGAT*, diacylglycerol acyltransferase; *FFA*, free fatty acid; *FOXO*, forkhead box O; *GPAT1*, glycerol-3-phosphate acyltransferase; *GSK-3*, glycogen synthase kinase 3; *IRS*, insulin receptor substrate; *LCA-CoA*, long-chain fatty acid coenzyme A; *PtdIns 3K*, phosphatidylinositol 3 kinase; *SPT1*, serine palmitoyltransferase; ↑, increase; and ↓, decrease.

residues impeding their ability to activate insulin signaling through PtdIns 3K (Fig. 3.8). This effect not only contributes to insulin resistance, but also prevents GLUT4 translocation (Roden, 2004). In adipose tissue, this interference with insulin signaling results in reduced lipoprotein lipase function and increased activity of hormone-sensitive lipase leading to further increases in circulating free fatty acids (Petersen and Shulman, 2006; Frayn, 2002; Samuel et al., 2010).

Insulin resistance–mediated changes in systemic lipid metabolism may result in the development of dyslipidemia and the well-known lipid triad: (1) high levels of plasma TAGs, (2) low levels of HDL, and (3) the appearance of small dense LDLs. This triad, along with endothelial dysfunction may further induce aberrant insulin signaling, contributing to atherosclerotic plaque formation. Insulin resistance and the cardiac metabolic changes promote myocardial damage by at least three different mechanisms: (1) alterations in signal transduction processes, (2) impairment in regulation of substrate metabolism, (3) changes in the delivery of substrates to the myocardium, and (4) reduction in HDL in the insulin-resistant state may increase in activity of cholesteryl ester transfer protein. The transfer of cholesteryl esters from HDL to TAG-rich lipoproteins (Borggreve et al., 2003). This suppresses and reverses cholesterol transport from the arterial wall and promotes atherosclerotic plaque formation.

Biochemical links between insulin resistance and heart disease

It is well known that under physiological conditions, insulin stimulates the use of metabolic substrates in multiple tissues including heart, skeletal muscle, liver, and adipose tissue. In the cardiomyocytes, insulin promotes glucose and fatty acid uptake, but blocks the use of fatty acids as an energy source. As a result of insulin resistance, the pancreas attempts to compensate by secreting increasing amounts of insulin, resulting in hyperinsulinemia. Insulin resistance contributes to CVD not only via the development of atherosclerosis and endothelial dysfunction, but also through the induction of hypertension and macrophage accumulation (Bornfeldt and Tabas, 2011; Davidson and Parkin, 2009; Laakso and Kuusisto, 2014; Ormazabal et al., 2018). These processes are supported not only by alterations in signal transduction processes, but also by alterations delivery of substrates to the myocardium.

Contribution of lipid mediators in heart disease

Lipid mediators are chemical messengers that are synthesized and release in response to endothelial and myocytic cell's stimulation from membrane phospholipids, sphingolipid, and cholesterol. Lipid mediators play important roles in internal and external communication and modulate cellular responses such as the differentiation, adhesion, migration, inflammation, and oxidative stress. These processes are modulated by eicosanoids (prostaglandins, leukotrienes, thromboxanes, and lipoxins) and docosanoids (E and D series resolvins, protectins, and maresins). These lipid mediators are generated from membrane phospholipids, sphingolipid, and cholesterol by the action of phospholipases A_2, cyclooxygenases, and lipoxygenases on ARA, DHAs, and eicosapentaenoic acids (Fig. 3.10), respectively

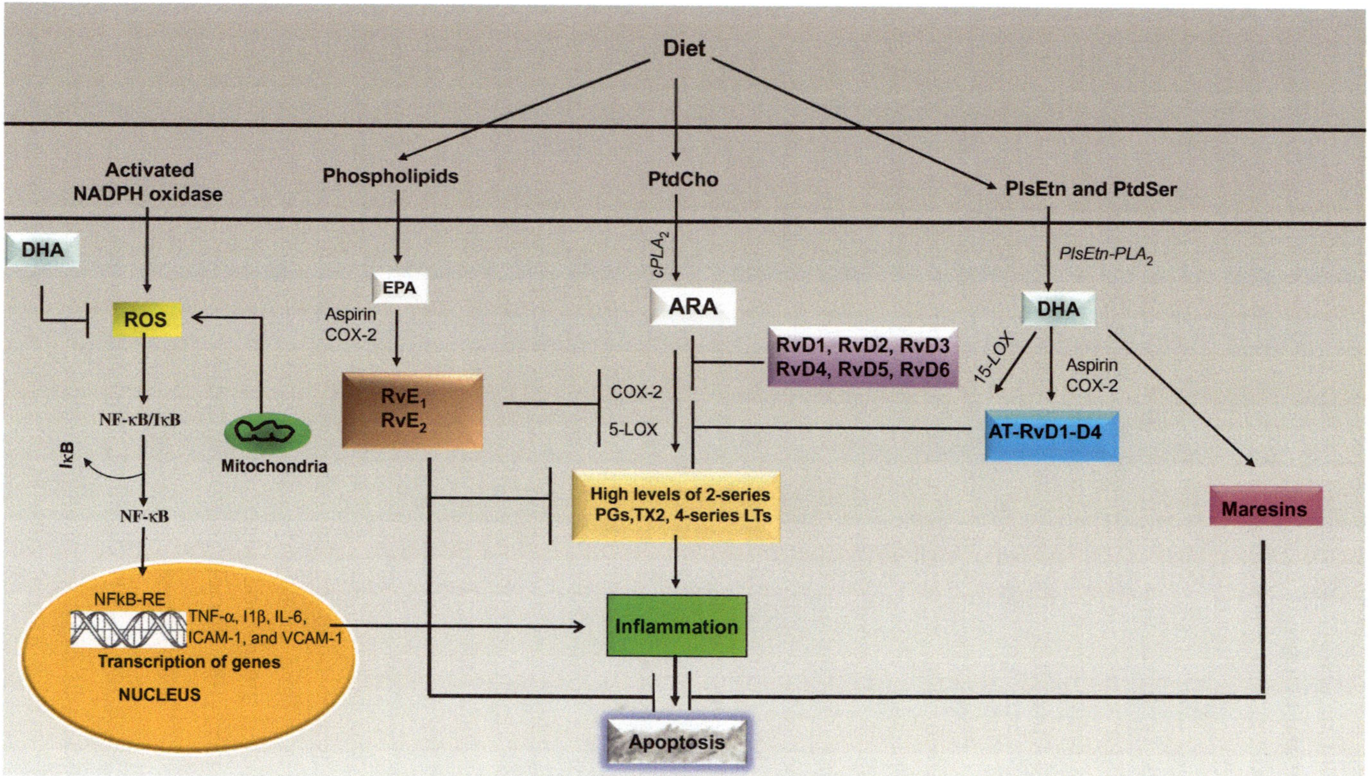

Figure 3.10 Regulation of oxidative stress, inflammation, and apoptotic cell death by docosanoids.
ARA, Arachidonic acid; *AT-RvD1-4*, aspirin-triggered D series resolvins; *COX-2*, cyclooxygenase-2; *cPLA*$_2$, cytosolic phospholipase A_2; *DHA*, docosahexaenoic acid; *EPA*, eicosapentaenoic acid; *5-LOX*, 5-lipoxygenase; *PlsEtn*, plasmalogen; *PlsEtn-PLA*$_2$, plasmalogen-selective PLA_2; *PtdCho*, phosphatidylcholine; *RvE*$_1$, resolving E_1; *RvE*$_2$, resolving E_2; *RvD1-RvD6*, resolving D series.

(Lavie et al., 2009; Serhan et al, 2015). Under normal conditions, low levels of lipid mediators are needed for signal transduction, gene expression, and neural cell proliferation and differentiation resulting in endothelial and myocytic cell survival but high levels of lipid mediators may cause cell death through the induction of oxidative stress, neuroinflammation, and apoptosis (Wu et al., 2017; Pirault and Bäck, 2018). Levels of lipid mediators in cardiovascular system and heat tissue are partly regulated by diet. The high intake of food enriched in ARA (vegetable oils) elevates not only oxidative stress, but also increases levels of proinflammatory eicosanoids, and upregulates the expression of proinflammatory cytokines and chemokines (Fig. 3.9). ARA and ARA-derived metabolites have prothrombotic, proaggregatory, and proinflammatory properties, and unresolved inflammation is central to the pathophysiology of commonly occurring vascular diseases such as atherosclerosis, aneurysm, and deep vein thrombosis. These diseases are responsible for considerable morbidity and mortality. In contrast, diet enriched in DHA (fish and fish oil) produces docosanoids, which not only downregulate proinflammatory cytokines but also induce antiinflammatory, antithrombotic, antiarrhythmic, hypolipidemic, and vasodilatory effects (Elajami et al., 2016). The resolution of vascular inflammation is an important driver of vessel wall remodeling and functional recovery. DHA-derived docosanoids not only orchestrate key cellular processes driving resolution, but also promote rapid return to homeostasis. At present, the threshold concentrations of ARA-derived lipid mediators, which facilitate endothelial dysfunction and myocytic cell death are not known. Similarly, levels of DHA-derived lipid mediators, which protect from myocytic cell death is not known. In cardiovascular system, endothelial and myocytic death not only depends upon elevated levels of lipid mediators, but also on crosstalk (interplay) among glycerophospholipid-, glycosphingolipid-, and cholesterol-derived lipid mediators. Studies on lipid-derived mediators fall in a fast-paced research area related to cardiovascular system and provide opportunities for target-based therapeutic intervention using inhibitors of lipid mediator synthesizing enzymes (Pirault and Bäck, 2018).

Insulin resistance, heart disease, and Alzheimer's disease

Epidemiologic and clinico-pathologic studies have indicated that there is considerable overlap in molecular events among insulin resistance, type 2 diabetes, cerebrovascular lesions, CVD, and AD. These conditions occur in middle-aged to elderly populations. Onset of type 2 diabetes, CVD, and AD occurs in human subjects with increasing age, and all these conditions are the leading causes of death. The primary causes of CVD are CHD, hypertension, stroke, and heart failure. These diseases are frequently interconnected and share an underlying pathology of atherosclerosis. Vascular pathology in aging heart and brain includes cardio- and cerebral amyloid angiopathy, multiple or recurrent microhemorrhages and ischemic infarcts promoting onset hypertension and increased risk for mild cognitive impairment (Graff-Radford et al., 2007). It should be noted that in heart disease, nerve growth factor receptor (tyrosine kinase receptor 1) regulates the survival and maintenance of sympathetic neurons that innervate the heart and is a crucial factor in myocardial ischemia, a main cause of sudden death in insulin resistance and type 2 diabetes (Ieda et al., 2006). Similarly, in AD, the nerve growth factor receptor is downregulated suggesting the induction of imbalance in NGF/NTRK1 signaling (Hock et al., 2000). The risk for dementia and heart disease is very high in type 2 diabetic patients along with severe systolic hypertension. In addition, the degree of CAD is independently associated with the cardinal neuropathological lesions of AD (Hulette and Welsh-Bohmer, 2007). Consequently, the association of heart failure with an increased risk of AD suggests that the general advice to keep the heart healthy may contribute to healthy brain (Qiu et al., 2006). The strong association of cardiovascular risk factors with AD suggests that these diseases share some biologic pathways. Based on the above information, it can be suggested that the use of cardiovascular therapies may prove useful for the treatment or prevention of AD and dementia (de la Torre, 2009).

Conclusion

Insulin resistance is identified as an impaired biologic response to insulin stimulation of target tissues, primarily liver, muscle, and adipose tissue. Insulin resistance is primarily an acquired condition related to excess body fat and genetic causes. It impairs glucose disposal, resulting in a compensatory increase in beta-cell insulin production and hyperinsulinemia. The metabolic consequences of insulin resistance can result in hyperglycemia, hypertension, dyslipidemia, visceral adiposity, hyperuricemia, elevated inflammatory markers, endothelial dysfunction, and a prothrombic state. Progression of insulin resistance can lead to the development of type 2 diabetes and CVD. In young adults, insulin resistance modulates endothelial dysfunction. In contrast, among elderly humans advancing age is independently associated with the development of vascular endothelial dysfunction and CVD. The endothelial dysfunction is caused by the reduction in NO bioavailability downstream of endothelial oxidative stress and inflammation that can be further modulated by traditional CVD risk factors in older adults. Greater endothelial oxidative stress with aging is a result of augmented production from the intracellular enzymes NADPH oxidase and uncoupled eNOS, as well as from mitochondrial respiration in the absence of appropriate increases in antioxidant defenses as regulated by relevant transcription factors, such as FOXO. The increased oxidative stress in patients with type 2 diabetes and CVD is a consequence of many abnormalities, including hyperglycemia, insulin resistance, hyperinsulinemia, and dyslipidemia, each of which is associated with overproduction of mitochondrial superoxide in endothelial cells of large and small vessels as well as the abnormalities in myocardial metabolism. In addition, oxidative stress is also supported by (1) the polyol pathway flux, (2) increased formation of AGEs, (3) increased expression of the RAGE, (4) activation of PKC isoforms, and (5) overactivity of the hexosamine pathway. Furthermore, in diabetic patients with heart disease, oxidative stress–mediated DNA damage produces poly (ADP-ribose) polymerase overactivation and reduces activity of glyceraldehyde 3-phosphate dehydrogenase, a factor common to all the five above-mentioned pathways. Inflammation is supported by the activation of NF-κB, a critical inflammatory transcription factor located in the cytoplasm. NF-κB is sensitive not only to insulin resistance and age-related endothelial redox change, but also to expression of proinflammatory cytokines. This can further suppress endothelial function. This can further suppress endothelial function. Converging evidence suggests that the leading cause of death in people with insulin resistance and type 2 diabetes with heart and blood vessel disease (70%–80%). The risk of heart and blood vessel disease is eight times higher in patients with insulin resistance and type 2 diabetes. Patients with type 2 diabetes without previous heart attacks are at higher risk of a heart attack as compared with nondiabetic patients who have previously had a heart attack.

References

Aengevaeren, V.L., Mosterd, A., Braber, T.L., Prakken, N.H.J., Doevendans, P.A., Grobbee, D.E., et al., 2017. Relationship between lifelong exercise volume and coronary atherosclerosis in athletes. Circulation 136, 138–148.

Bandyopadhyay, G.K., Yu, J.G., Ofrecio, J., Olefsky, J.M., 2006. Increased malonyl-CoA levels in muscle from obese and type 2 diabetic subjects lead to decreased fatty acid oxidation and increased lipogenesis; thiazolidinedione treatment reverses these defects. Diabetes 55, 2277–2285.

Beckman, J.A., Creager, M.A., Libby, P., 2002. Diabetes and atherosclerosis: epidemiology, pathophysiology, and management. JAMA 287, 2570–2581.

Benetos, A., Waeber, B., Izzo, J., Mitchell, G., Lawrence, R., Roland, A., et al., 2002. Influence of age, risk factors, and cardiovascular and renal disease on arterial stiffness: clinical applications. Am. J. Hypertens. 15, 1101–1108.

Benjamin, E.J., Blaha, M.J., Chiuve, S.E., Cushman, M., Das, S.R., Deo, R., et al., 2017. Heart disease and stroke statistics-2017 update: a report from the American Heart Association. Circulation 135, e146–603.

Ben-Shlomo, Y., Spears, M., Boustred, C., May, M., Anderson, S.G., Benjamin, E.J., et al., 2014. Aortic pulse wave velocity improves cardiovascular event prediction: an individual participant meta-analysis of prospective observational data from 17,635 subjects. J. Am. Coll. Cardiol. 63, 636–646.

Boden, G., Lebed, B., Schatz, M., Homko, C., Lemieux, S., 2001. Effects of acute changes of plasma free fatty acids on intramyocellular fat content and insulin resistance in healthy subjects. Diabetes 50, 1612–1617.

Bogan, J.S., 2012. Regulation of glucose transporter translocation in health and diabetes. Annu. Rev. Biochem. 81, 507–532.

Borggreve, S.E., De Vries, R., Dullaart, R.P., 2003. Alterations in high-density lipoprotein metabolism and reverse cholesterol transport in insulin resistance and type 2 diabetes mellitus: role of lipolytic enzymes, lecithin: cholesterol acyltransferase and lipid transfer proteins. Eur. J. Clin. Invest. 33, 1051–1069.

Bornfeldt, K.E., Tabas, I., 2011. Insulin resistance, hyperglycemia, and atherosclerosis. Cell. Metab. 14, 575–585.

Briggs, M.A., Petersen, K.S., Kris-Etherton, P.M., 2017. Saturated fatty acids and cardiovascular disease: replacements for saturated fat to reduce cardiovascular risk. Healthcare 5, pii: E29.

Bu, J., Wang, Z., 2018. Cross-talk between gut microbiota and heart via the routes of metabolite and immunity. Gastroenterol. Res. Pract. 2018, 6458094.

Candido, R., Forbes, J.M., Thomas, M.C., Thallas, V., Dean, R.G., Burns, W.C., et al., 2003. A breaker of advanced glycation end products attenuates diabetes-induced myocardial structural changes. Circ. Res. 92, 785–792.

Camargo, L.H., Alves, F.H., Biojones, C., Correa, F.M., Resstel, L.B., Crestani, C.C., 2013. Involvement of N-methyl-D-aspartate glutamate receptor and nitric oxide in cardiovascular responses to dynamic exercise in rats. Eur. J. Pharmacol. 713, 16–24.

Castelli, W.P., 1998. The new pathophysiology of coronary artery disease. Am. J. Cardiol. 82, 60T–65T.

CDC N, 2015. Underlying Cause of Death 1999-2013 on CDC WONDER Online Database. Data are From the Multiple Cause of Death Files, 1999-2013, as Compiled From Data Provided by the 57 Vital Statistics Jurisdictions Through the Vital Statistics Cooperative Program (accessed February 3, 2015).

Cersosimo, E., DeFronzo, R.A., 2006. Insulin resistance and endothelial dysfunction: the road map to cardiovascular diseases. Diabetes Metab. Res. Rev. 22, 423–436.

Chavez, J.A., Summers, S.A., 2003. Characterizing the effects of saturated fatty acids on insulin signaling and ceramide and diacylglycerol accumulation in 3T3-L1 adipocytes and C2C12 myotubes. Arch. Biochem. Biophys. 419, 101–109.

Chazova, T.E., Masenko, V.P., Zykov, K.A., Golitsyna, T.Iu, 2007. The role of inflammation factors in development of acute coronary syndrome in patients with type 2 diabetes mellitus and impaired glucose tolerance. Ter. Arkh. 79, 60–64.

Chen, J.Y., Ye, Z.X., Wang, X.F., Chang, J., Yang, M.W., Zhong, H.H., et al., 2018. Nitric oxide bioavailability dysfunction involves in atherosclerosis. Biomed. Pharmacother. 97, 423–428.

Csiszar, A., Wang, M., Lakatta, E.G., Ungvari, Z., 2008. Inflammation and endothelial dysfunction during aging: role of NF-κB. J. Appl. Physiol. 105, 1333–1341.
Das, M., Das, D.K., 2010. Resveratrol and cardiovascular health. Mol. Aspects. Med. 31, 503–512.
Davidson, J.A., Parkin, C.G., 2009. Is hyperglycemia a causal factor in cardiovascular disease? Does proving this relationship really matter? Yes. Diabetes Care 32 (Suppl. 2), S331–S333.
Davignon, J., Ganz, P., 2004. Role of endothelial dysfunction in atherosclerosis. Circulation 109, III27–III32.
de la Torre, J.C., 2009. Cerebrovascular and cardiovascular pathology in Alzheimer's disease. Int. Rev. Neurobiol. 84, 35–48.
Dessì, M., Noce, A., Bertucci, P., Manca di Villahermosa, S., Zenobi, R., Castagnola, V., et al., 2013. Atherosclerosis, dyslipidemia, and inflammation: the significant role of polyunsaturated fatty acids. ISRN Inflamm. 2013, 191823.
Diez, J., 2007. Arterial stiffness and extracellular matrix. Adv. Cardiol. 44, 76–95.
Dimitriadis, G., Mitrou, P., Lambadiari, V., Maratou, E., Raptis, S.A., 2011. Insulin effects in muscle and adipose tissue. Diabetes Res. Clin. Pract. 93, S52–S59.
Dodd, K.M., Tee, A.R., 2012. Leucine and mTORC1: a complex relationship. Am. J. Physiol. Endocrinol. Metab. 302, E1329–E1342.
Doehner, W., Rauchhaus, M., Ponikowski, P., Godsland, I.F., von Haehling, S., Okonko, D.O., et al., 2005. Impaired insulin sensitivity as an independent risk factor for mortality in patients with stable chronic heart failure. J. Am. Coll. Cardiol. 46, 1019–1026.
Donato, A.J., Machin, D.R., Lesniewski, L.A., 2018. Mechanisms of dysfunction in the aging vasculature and role in age-related disease. Circ. Res. 123, 825–848.
Dunmore, S.J., Brown, J.E., 2013. The role of adipokines in beta-cell failure of type 2 diabetes. J. Endocrinol. 216, T37–T45.
Elahi, M.M., Cagampang, F.R., Mukhtar, D., Anthony, F.W., Ohri, S.K., Hanson, M.A., 2009. Long-term maternal high-fat feeding from weaning through pregnancy and lactation predisposes offspring to hypertension, raised plasma lipids and fatty liver in mice. Br. J. Nutr. 102, 514–519.
Elajami, T.K., Colas, R.A., Dalli, J., Chiang, N., Serhan, C.N., Welty, F.K., 2016. Specialized proresolving lipid mediators in patients with coronary artery disease and their potential for clot remodeling. FASEB J. 30, 2792–2801.
Fan, W., Han, D., Sun, Z., Ma, S., Gao, L., Chen, J., et al., 2017. Endothelial deletion of mTORC1 protects against hindlimb ischemia in diabetic mice via activation of autophagy, attenuation of oxidative stress and alleviation of inflammation. Free Radic. Biol. Med. 108, 725–740.
Farand, P., Garon, A., Plante, G.E., 2007. Structure of large arteries: orientation of elastin in rabbit aortic internal elastic lamina and in the elastic lamellae of aortic media. Microvasc. Res. 73, 95–99.
Farooqui, A.A., 2012. n-3 fatty acid-derived lipid mediators in the brain: new weapons against oxidative stress and inflammation. Curr. Med. Chem. 19, 532–543.
Farooqui, A.A., 2015. High Calorie Diet and the Human Brain: Metabolic Consequences of Long-Term Consumption. Springer International Publishing, Switzerland.
Fillmore, N., Wagg, C.S., Zhang, L., Fukushima, A., Lopaschuk, G.D., 2018. Cardiac branched-chain amino acid oxidation is reduced during insulin resistance in the heart. Am. J. Physiol. Endocrinol. Metab. 315, E1046–E1052.
Fishman, S.L., Sonmez, H., Basman, C., Singh, V., Poretsky, L., 2018. The role of advanced glycation end-products in the development of coronary artery disease in patients with and without diabetes mellitus: a review. Mol. Med. 24, 59.
Fleenor, B.S., 2012. Large elastic artery stiffness withaging: novel translational mechanisms and interventions. Aging Dis. 4, 76–83.

Fleenor, B.S., Seals, D.R., Zigler, M.L., Sindler, A.L., 2012a. Superoxide-lowering therapy with TEMPOL reverses arterial dysfunction with aging in mice. Aging Cell. 11, 269–276.

Fleenor, B.S., Sindler, A.L., Eng, J.S., Nair, D.P., Dodson, R.B., Seals, D.R., 2012b. Sodium nitrite de-stiffening of large elastic arteries with aging: role of normalization of advanced glycation end-products. Exp. Gerontol. 47, 588–594.

Forstermann, U., Xia, N., Li, H., 2017. Roles of vascular oxidative stress and nitric oxide in the pathogenesis of atherosclerosis. Circ. Res. 120, 713-120.

Frayn, K.N., 2002. Adipose tissue as a buffer for daily lipid flux. Diabetologia 45, 1201–1210.

Fu, Y., Moore, X.L., Lee, M.K.S., et al., 2012. Caveolin-1 plays a critical role in the differentiation of monocytes into macrophages. Arterioscler. Thromb. Vasc. Biol. 32, e117–e125.

Gaede, P., Lund-Andersen, H., Parving, H.H., Pedersen, O., 2008. Effect of a multifactorial intervention on mortality in type 2 diabetes. N. Engl. J. Med. 358, 580–591.

Gast, K.B., Tjeerdema, N., Stijnen, T., Smit, J.W., Dekkers, O.M., 2012. Insulin resistance and risk of incident cardiovascular events in adults without diabetes: meta-analysis. PLoS One 7, e52036.

Ginsberg, H.N., 2000. Insulin resistance and cardiovascular disease. J. Clin. Invest. 106, 453–458.

Graff-Radford, N.R., Crook, J.E., Lucas, J., Boeve, B.F., Knopman, D.S., Ivnik, R.J., 2007. Association of low plasma Abeta42/Abeta40 ratios with increased imminent risk for mild cognitive impairment and Alzheimer disease. Arch. Neurol. 64, 354–362.

Guyenet, P.G., Stornetta, R.L., Holloway, B.B., Souza, G.M.P.R., Abbott, S.B.G., 2018. Rostral ventrolateral medulla and hypertension. Hypertension 72, 559–566.

Haitoglou, C.S., Tsilibary, E.C., Brownlee, M., Charonis, A.S., 1992. Altered cellular interactions between endothelial cells and nonenzymatically glucosylated laminin/type IV collagen. J. Biol. Chem. 267, 12404–12407.

Hambrecht, R., Niebauer, J., Marburger, C., Grunze, M., Kalberer, B., Hauer, K., et al., 1993. Various intensities of leisure time physical activity in patients with coronary artery disease: effects on cardiorespiratory fitness and progression of coronary atherosclerotic lesions. J. Am. Coll. Cardiol. 22, 468–477.

Han, J.M., Jeong, S.J., Park, M.C., Kim, G., Kwon, N.H., Kim, H.K., et al., 2012. Leucyl-tRNA synthetase is an intracellular leucine sensor for the mTORC1-signaling pathway. Cell. 149, 410–424.

Hasenfuss, G., Maier, L.S., Hermann, H.P., Luers, C., Hunlich, M., Zeitz, O., et al., 2002. Influence of pyruvate on contractile performance and Ca(2 +) cycling in isolated failing human myocardium. Circulation 105, 194–199.

Hermann, H.P., Pieske, B., Schwarzmuller, E., Keul, J., Just, H., Hasenfuss, G., 1999. Haemodynamic effects of intracoronary pyruvate in patients with congestive heart failure: an open study. Lancet 353, 1321–1323.

Hirooka, Y., 2011. Oxidative stress in the cardiovascular center has a pivotal role in the sympathetic activation in hypertension. Hypertens. Res. 34, 407–412.

Hock, C., Heese, K., Hulette, C., Rosenberg, C., Otten, U., 2000. Region-specific neurotrophin imbalances in Alzheimer disease: decreased levels of brain-derived neurotrophic factor and increased levels of nerve growth factor in hippocampus and cortical areas. Arch. Neurol. 57, 846–851.

Houmard, J.A., Shinebarger, M.H., Dolan, P.L., Leggettfrazier, N., Bruner, R.K., Mccammon, M.R., et al., 1993. Exercise training increases GLUT-4 protein concentration in previously sedentary middle-aged men. Am. J. Physiol. 264, E896–E901.

Houston, M., 2018. The relationship of saturated fats and coronary heart disease: fa(c)t or fiction? A commentary. Ther. Adv. Cardiovasc. Dis. 12, 33–37.

Hulette, C.M., Welsh-Bohmer, K., 2007. Coronary artery disease is associated with Alzheimer disease neuropathology in APOE4 carriers. Neurology 68, 471.
Ieda, M., Kanazawa, H., Ieda, Y., Kimura, K., Matsumura, K., Tomita, Y., et al., 2006. Nerve growth factor is critical for cardiac sensory innervation and rescues neuropathy in diabetic hearts. Circulation 114, 2351–2363.
Iqbal, M.P., 2014. Trans fatty acids—a risk factor for cardiovascular disease. Pak. J. Med. Sci. 30, 194–197.
Janus, A., Szahidewicz-Krupska, E., Mazur, G., Doroszko, A., 2016. Insulin resistance and endothelial dysfunction constitute a common therapeutic target in cardiometabolic disorders. Mediators Inflamm. 2016, 3634948.
Jia, G., DeMarco, V.G., Sowers, J.R., 2016. Insulin resistance and hyperinsulinaemia in diabetic cardiomyopathy. Nat. Rev. Endocrinol. 12, 144–153.
Kaptoge, S., Di Angelantonio, E., Lowe, G., Pepys, M.B., Thompson, S.G., Collins, R., et al., 2010. C-reactive protein concentration and risk of coronary heart disease, stroke, and mortality: An individual participant meta-analysis. Lancet 375, 132–140.
Karwi, Q.G., Zhang, L., Wagg, C.S., Wang, W., Ghandi, M., Thai, D., et al., 2019. Targeting the glucagon receptor improves cardiac function and enhances insulin sensitivity following a myocardial infarction. Cardiovasc. Diabetol. 18, 1.
Kass, D.A., Shapiro, E.P., Kawaguchi, M., Capriotti, A.R., Scuteri, A., deGroof, R.C., et al., 2001. Improved arterial compliance by a novel advanced glycation end-product crosslink breaker. Circulation 104, 1464–1470.
Kasznicki, J., Drzewoski, J., 2014. Heart failure in the diabetic population—pathophysiology, diagnosis and management. Arch. Med. Sci. 10, 546–556.
Kauppinen, A., Suuronen, T., Ojala, J., Kaarniranta, K., Salminen, A., 2013. Antagonistic crosstalk between NF-κB and SIRT1 in the regulation of inflammation and metabolic disorders. Cell. Signal. 25, 1939–1948.
Kim, J.A., Jang, H.J., Martinez-Lemus, L.A., Sowers, J.R., 2012. Activation of mTOR/p70S6 kinase by ANG II inhibits insulin-stimulated endothelial nitric oxide synthase and vasodilation. Am. J. Physiol. Endocrinol. Metab. 302, E201–E208.
Kishi, T., 2013. Regulation of the sympathetic nervous system by nitric oxide and oxidative stress in the rostral ventrolateral medulla: 2012 Academic Conference Award from the Japanese Society of Hypertension. Hypertens. Res. 36, 845–851.
Kishi, T., Hirooka, Y., 2012. Oxidative stress in the brain causes hypertension via sympathoexcitation. Front. Physiol. 3, 335.
Kraus, W.E., Houmard, J.A., Duscha, B.D., Knetzger, K.J., Wharton, M.B., McCartney, J.S., et al., 2002. Effects of the amount and intensity of exercise on plasma lipoproteins. N. Engl. J. Med. 347, 1483–1492.
Kuiper, J., van Puijvelde, G.H., van Wanrooij, E.J., et al., 2007. Immunomodulation of the inflammatory response in atherosclerosis. Curr. Opin. Lipidol. 18, 521–526.
Laakso, M., Kuusisto, J., 2014. Insulin resistance and hyperglycaemia in cardiovascular disease development. Nat. Rev. Endocrinol. 10, 293–302.
Laddu, D.R., Rana, J.S., Murillo, R., Sorel, M.E., Quesenberry Jr, C.P., Allen, N.B., et al., 2017. 25-Year physical activity trajectories and development of subclinical coronary artery disease as measured by coronary artery calcium: the Coronary Artery Risk Development in Young Adults (CARDIA) study. Mayo Clin. Proc. 92, 1660–1670.
Lakatta, E.G., Levy, D., 2003. Arterial and cardiac aging: major shareholders in cardiovascular disease enterprises: Part II: the aging heart in health: links to heart disease. Circulation 107, 346–354.
Lakka, T.A., Laaksonen, D.E., 2007. Physical activity in prevention and treatment of the metabolic syndrome. Appl. Physiol. Nutr. Metab. 32, 76–88.
Lavie, C.J., Milani, R.V., Mehra, M.R., Ventura, H.O., 2009. Omega-3 polyunsaturated fatty acids and cardiovascular diseases. J. Am. Coll. Cardiol. 54, 585–594.

Lehrke, M., Marx, N., 2017. Diabetes mellitus and heart failure. Am. J. Cardiol. 120, S37–S47.
Leitinger, N., 2003. Oxidized phospholipids as modulators of inflammation in atherosclerosis. Curr. Opin. Lipidol. 14, 421–430.
Li, H., Cybulsky, M.I., Gimbrone, M.A., Libby, P., 1993. An atherogenic diet rapidly induces VCAM-1, a cytokine-regulatable mononuclear leukocyte adhesion molecule, in rabbit aortic endothelium. Arterioscler. Thromb. 13, 197–204.
Li, Q., Park, K., Li, C., Rask-Madsen, C., Mima, A., Qi, W., et al., 2013. Induction of vascular insulin resistance and endothelin-1 expression and acceleration of atherosclerosis by the overexpression of protein kinase C-beta isoform in the endothelium. Circ. Res. 113, 418–427.
Libby, P., 2012. Inflammation in atherosclerosis. Arterioscler. Thromb. Vasc. Biol. 32, 2045–2051.
Libby, P., Ridker, P.M., Hansson, G.K., 2011. Progress and challenges in translating the biology of atherosclerosis. Nature 473, 317–325.
Lichtenstein, A.H., Appel, L.J., Brands, M., Carnethon, M., Daniels, S., Franch, H.A., et al., 2006. Summary of american heart association diet and lifestyle recommendations revision 2006. Arterioscler. Thromb. Vasc. Biol. 26, 2186–2191.
Lloyd-Jones, D., Adams, R.J., Brown, T.M., Carnethon, M., Dai, S., De Simone, G., et al., 2010. Executive summary: heart disease and stroke statistics-2010 update: a report from the American Heart Association. Circulation 121, 948–954.
Louissaint Jr, A., Rao, S., Leventhal, C., Goldman, S.A., 2002. Coordinated interaction of neurogenesis and angiogenesis in the adult songbird brain. Neuron 34, 945–960.
Lynch, C.J., Adams, S.H., 2014. Branched-chain amino acids in metabolic signalling and insulin resistance. Nat. Rev. Endocrinol. 10, 723.
Mach, F., Lovis, C., Gaspoz, J.M., Unger, P.F., Bouillie, M., Urban, P., et al., 1997. C-reactive protein as a marker for acute coronary syndromes. Eur. Heart J. 18, 1897–1902.
Mahmoud, F., Al-Ozairi, E., 2013. Inflammatory cytokines and the risk of cardiovascular complications in type 2 diabetes. Dis. Markers 35, 235–241.
Mallika, V., Goswami, B., Rajappa, M., 2007. Atherosclerosis pathophysiology and the role of novel risk factors: a clinicobiochemical perspective. Angiology 58, 513–522.
Manhiani, M.M., Cormican, M.T., Brands, M.W., 2011. Chronic sodium-retaining action of insulin in diabetic dogs. Am. J. Physiol. Renal. Physiol. 300, F957–F965.
Manigrasso, M.B., Juranek, J., Ramasamy, R., Schmidt, A.M., 2014. Unlocking the biology of RAGE in diabetic microvascular complications. Trends Endocrinol. Metab. 25, 15–22.
Marasciulo, F.L., Montagnani, M., Potenza, M.A., 2006. Endothelin-1: the yin and yang on vascular function. Curr. Med. Chem. 13, 1655–1665.
Mattace-Raso, F.U.S., van der Cammen, T.J.M., Hofman, A., van Popele, N.M., Bos, M. L., Schalekamp, M.A.D.H., et al., 2006. Arterial stiffness and risk of coronary heart disease and stroke: the Rotterdam Study. Circulation 113, 657–663.
Mendelsohn, A.R., Larrick, J.W., 2013. Dietary modification of the microbiome affects risk for cardiovascular disease. Rejuvenation Res. 16, 241–244.
Mensink, R., 2016. Effects of Saturated Fatty Acids on Serum Lipids and Lipoproteins: A Systematic Review and Regression Analysis. World Health Organization, Geneva, Switzerland.
Mensink, R.P., Katan, M.B., 1990. Effect of dietary trans fatty acids on high-density and low-density lipoprotein cholesterol levels in healthy subjects. N. Engl. J. Med. 323, 439–445.
Mitchell, G.F., Hwang, S.-J., Vasan, R.S., Larson, M.G., Pencina, M.J., Hamburg, N.M., et al., 2010. Arterial stiffness and cardiovascular events: the Framingham Heart Study. Circulation 121, 505–511.

Moludi, J., Alizadeh, M., Lotfi Yagin, N., Pasdar, Y., Nachvak, S.M., Abdollahzad, H., et al., 2018. New insights on atherosclerosis: a cross-talk between endocannabinoid systems with gut microbiota. J. Cardiovasc. Thorac. Res. 10, 129–137.

Montezano, A.C., Touyz, R.M., 2012. Molecular mechanisms of hypertension—reactive oxygen species and antioxidants: a basic science update for the clinician. Can. J. Cardiol. 28, 288–295.

Morel, O., Kessler, L., Ohlmann, P., Bareiss, P., 2010. Diabetes and the platelet: toward new therapeutic paradigms for diabetic atherothrombosis. Atherosclerosis. 212, 367–376.

Muoio, D.M., Newgard, C.B., 2008. Mechanisms of disease: molecular and metabolic mechanisms of insulin resistance and beta-cell failure in type 2 diabetes. Nat. Rev. Mol. Cell Biol. 9, 193–205.

Najjar, S.S., Scuteri, A., Lakatta, E.G., 2005. Arterial aging: is it an immutable cardiovascular risk factor? Hypertension 46, 454–462.

NDIC (National Diabetes Information Clearinghouse), 2014. Diabetes, heart disease, and stroke. <http://diabetes.niddk.nih.gov/dm/pubs/stroke/index.aspx>.

Neubauer, S., 2007. The failing heart—an engine out of fuel. N. Engl. J. Med. 356, 1140–1151.

Newgard, C.B., An, J., Bain, J.R., Muehlbauer, M.J., Stevens, R.D., Lien, L.F., et al., 2009. A branched-chain amino acid-related metabolic signature that differentiates obese and lean humans and contributes to insulin resistance. Cell. Metab. 9, 311–326.

Nigro, C., Leone, A., Raciti, G.A., et al., 2017. Methylglyoxal-glyoxalase 1 balance: the root of vascular damage. Int. J. Mol. Sci. 18, 188–202.

Nishikawa, T., Araki, E., 2007. Impact of mitochondrial ROS production in the pathogenesis of diabetes mellitus and its complications. Antioxid. Redox. Signal. 9, 343–353.

Nowiński, A., Ufnal, M., 2018. Trimethylamine N-oxide: a harmful, protective or diagnostic marker in lifestyle diseases? Nutrition 46, 7–12.

O'Keefe, J.H., Patil, H.R., Lavie, C.J., Magalski, A., Vogel, R.A., McCullough, P.A., 2012. Potential adverse cardiovascular effects from excessive endurance exercise. Mayo Clin. Proc. 87, 587–595.

Ong, K.T., Delerme, S., Pannier, B., Safar, M.E., Benetos, A., Laurent, S., et al., 2011. Aortic stiffness is reduced beyond blood pressure lowering by short-term and long-term antihypertensive treatment: a meta-analysis of individual data in 294 patients. J. Hypertens. 29, 1034–1042.

Ormazabal, V., Nair, S., Elfeky, O., Aguayo, C., Salomon, C., Zuñiga, F.A., 2018. Association between insulin resistance and the development of cardiovascular disease. Cardiovasc. Diabetol. 17, 122.

Ott, C., Jacobs, K., Haucke, E., Navarrete Santos, A., Grune, T., Simm, A., 2014. Role of advanced glycation end products in cellular signaling. Redox Biol. 2, 411–429.

Packer, M., 2018. Potentiation of insulin signaling contributes to heart failure in type 2 diabetes: a hypothesis supported by both mechanistic studies and clinical trials. JACC Basic Transl. Sci. 3, 415–419.

Palomer, X., Salvadó, L., Barroso, E., Vázquez-Carrera, M., 2013. An overview of the crosstalk between inflammatory processes and metabolic dysregulation during diabetic cardiomyopathy. Int. J. Cardiol. 168, 3160–3172.

Peluso, I., Morabito, G., Urban, L., Ioannone, F., Serafini, M., 2012. Oxidative stress in atherosclerosis development: the central role of LDL and oxidative burst. Endocr. Metab. Immune Disorders Drug Targets 12, 351–360.

Penny, W.F., Ben-Yehuda, O., Kuroe, K., Long, J., Bond, A., Bhargava, V., et al., 2001. Improvement of coronary artery endothelial dysfunction with lipid-lowering therapy: heterogeneity of segmental response and correlation with plasma-oxidized low-density lipoprotein. J. Am. Coll. Cardiol. 37, 766–774.

Petersen, K.F., Shulman, G.I., 2006. Etiology of insulin resistance. Am. J. Med. 119 (5 Suppl. 1), S10–S16.
Petersen, M.C., Shulman, G.I., 2018. Mechanisms of insulin action and insulin resistance. Physiol. Rev. 98, 2133–2223.
Pirault, J., Bäck, M., 2018. Lipoxin and resolvin receptors transducing the resolution of inflammation in cardiovascular disease. Front. Pharmacol. 9, 1273.
Potenza, M.A., Marasciulo, F.L., Chieppa, D.M., Brigiani, G.S., Formoso, G., et al., 2005. Insulin resistance in spontaneously hypertensive rats is associated with endothelial dysfunction characterized by imbalance between NO and ET-1 production. Am. J. Physiol. Heart Circ. Physiol. 289, H813–H822.
Qiu, C., Winblad, B., Marengoni, A., Klarin, I., Fastbom, J., Fratiglioni, L., 2006. Heart failure and risk of dementia and Alzheimer disease: a population-based cohort study. Arch. Intern. Med. 166, 1003–1008.
Quigley, E.M.M., 2017. Microbiota-brain-gut axis and neurodegenerative diseases. Curr. Neurol. Neurosci. Rep. 17, 94.
Quillard, T., Libby, P., 2012. Molecular imaging of atherosclerosis for improving diagnostic and therapeutic development. Circ. Res. 111, 231–244.
Radak, Z., Chung, H.Y., Goto, S., 2005. Exercise and hormesis: oxidative stress-related adaptation for successful aging. Biogerontology 6, 71–75.
Radak, Z., Chung, H.Y., Koltai, E., Taylor, A.W., Goto, S., 2008. Exercise, oxidative stress and hormesis. Ageing Res. Rev. 7, 34–42.
Rajapakse, A.G., Yepuri, G., Carvas, J.M., Stein, S., Matter, C.M., et al., 2011. Hyperactive S6K1 mediates oxidative stress and endothelial dysfunction in aging: inhibition by resveratrol. PLoS One. 6, e19237.
Ramasamy, R., Vannucci, S.J., Yan, S.S., Herold, K., Yan, S.F., Schmidt, A.M., 2005. Advanced glycation end products and RAGE: a common thread in aging, diabetes, neurodegeneration, and inflammation. Glycobiology 15, 16R–28R.
Ramasamy, R., Yan, S.F., Schmidt, A.M., 2009. RAGE: therapeutic target and biomarker of the inflammatory response—the evidence mounts. J. Leukoc. Biol. 86, 505–512.
Rao, R.H., Rao, U.B., Srikantia, S.G., 1981. Effect of polyunsaturate-rich vegetable oils on blood pressure in essential hypertension. Clin. Exp. Hypertens. 3, 27–38.
Reaven, G., 2012. Insulin resistance and coronary heart disease in nondiabetic individuals. Arterioscler. Thromb. Vasc. Biol. 32, 1754–1759.
Reho, J.J., Rahmouni, K., 2017. Oxidative and inflammatory signals in obesity-associated vascular abnormalities. Clin. Sci. (Lond.) 131, 1689–1700.
Riehle, C., Abel, E.D., 2016. Insulin signaling and heart failure. Circ. Res. 118, 1151–1169.
Rijzewijk, L.J., Jonker, J.T., van der Meer, R.W., Lubberink, M., de Jong, H.W., et al., 2010. Effects of hepatic triglyceride content on myocardial metabolism in type 2 diabetes. J. Am. Coll. Cardiol. 56, 225–233.
Roberts, W.O., Schwartz, R.S., Garberich, R.F., Carlson, S., Knickelbine, T., Schwartz, J.G., et al., 2017. Fifty men, 3510 marathons, cardiac risk factors, and coronary artery calcium scores. Med. Sci. Sports Exerc. 49, 2369–2373.
Roden, M., 2004. How free fatty acids inhibit glucose utilization in human skeletal muscle. News Physiol. Sci. 19, 92–96.
Rodrigo, R., Cereceda, M., Castillo, R., Asenjo, R., Zamorano, J., Araya, J., et al., 2008. Prevention of atrial fibrillation following cardiac surgery: basis for a novel therapeutic strategy based on non-hypoxic myocardial preconditioning. Pharmacol. Therapeut. 118, 104–127.
Rodrigo, R., González, J., Paoletto, F., 2011. The role of oxidative stress in the pathophysiology of hypertension. Hypertension Res. 34, 431–440.

Roger, V.L., Weston, S.A., Redfield, M.M., Hellermann-Homan, J.P., Killian, J., Yawn, B.P., et al., 2004. Trends in heart failure incidence and survival in a community-based population. JAMA 292, 344–350.
Ross, R., 1999. Atherosclerosis: an inflammatory disease. N. Engl. J. Med. 340, 115–126.
Roth, G.A., Forouzanfar, M.H., Moran, A.E., Barber, R., Nguyen, G., et al., 2015. Demographic and epidemiologic drivers of global cardiovascular mortality. N. Engl. J. Med. 372, 1333–1341.
Ryden, L., Grant, P.J., Anker, S.D., Berne, C., Cosentino, F., Danchin, N., et al., 2013. ESC guidelines on diabetes, pre-diabetes, and cardiovascular diseases developed in collaboration with the EASD: the task force on diabetes, pre-diabetes, and cardiovascular diseases of the European Society of Cardiology (ESC) and developed in collaboration with the European Association for the Study of Diabetes (EASD). Eur. Heart J. 34, 3035–3087.
Saad, M.J., Santos, A., Prada, P.O., 2016. Linking gut microbiota and inflammation to obesity and insulin resistance. Physiology (Bethesda). 31, 283–293.
Saiki, A., Sato, Y., Watanabe, R., Watanabe, Y., Imamura, H., Yamaguchi, T., et al., 2016. The role of a novel arterial stiffness parameter, cardio-ankle vascular index (CAVI), as a surrogate marker for cardiovascular diseases. J. Atheroscler. Thromb. 23, 155–168.
Salmerón, J., Hu, F.B., Manson, J.E., Stampfer, M.J., Colditz, G.A., Rimm, E.B., et al., 2001. Dietary fat intake and type 2 diabetes in women. Am. J. Clin. Nutr. 73, 1019–1026.
Samuel, V.T., Petersen, K.F., Shulman, G.I., 2010. Lipid-induced insulin resistance: unravelling the mechanism. Lancet 375, 2267–2277.
Santi, D., Giannetta, E., Isidori, A.M., Vitale, C., Aversa, A., Simoni, M., 2015. Therapy of endocrine disease. Effects of chronic use of phosphodiesterase inhibitors on endothelial markers in type 2 diabetes mellitus: a meta-analysis. Eur. J. Endocrinol. 172, R103–R114.
Serhan, C.N., Dalli, J., Colas, R.A., Winkler, J.W., Chiang, N., 2015. Protectins and maresins: new pro-resolving families of mediators in acute inflammation and resolution bioactive metabolome. Biochim. Biophys. Acta 1851, 397–413.
Sankar, D., Rao, M.R., Sambandam, G., Pugalendid, K.V., 2006. Effect of sesame oil on diuretics or ß-blockers in the modulation of blood pressure, anthropometry, lipid profile, and redox status. Yale J. Biol. Med. 79, 19–26.
Schleicher, E.D., Wagner, E., Nerlich, A.G., 1997. Increased accumulation of the glycoxidation product N(epsilon)-(carboxymethyl)lysine in human tissues in diabetes and aging. J. Clin. Invest. 99, 457–468.
Schuit, A.J., Feskens, E.J.M., Launer, L.J., Kromhout, D., 2001. Physical activity and cognitive decline, the role of the apolipoprotein e4 allele. Med. Sci. Sports Exer. 33, 772–777. 2001.
Sell, D.R., Monnier, V.M., 2012. Molecular basis of arterial stiffening: role of glycation—a mini-review. Gerontology 58, 227–237.
Sindler, A.L., Fleenor, B.S., Calvert, J.W., Marshall, K.D., Zigler, M.L., Lefer, D.J., et al., 2011. Nitrite supplementation reverses vascular endothelial dysfunction and large elastic artery stiffness with aging. Aging Cell. 10, 429–437.
Skålén, K., Gustafsson, M., Rydberg, E.K., et al., 2002. Subendothelial retention of atherogenic lipoproteins in early atherosclerosis. Nature 417, 750–754.
Sproston, N.R., Ashworth, J.J., 2018. Roles of c-reactive protein at sites of inflammation and infection. Front. Immunol. 9, 754.
Stary, H.C., Blankenhorn, D.H., Chandler, A.B., Glagov, S., Insull Jr, J., Richardson, M., et al., 1992. A definition of the intima of human arteries and of its atherosclerosis-prone regions. A report from the Committee on Vascular Lesions of the Council on Arteriosclerosis. Circulation 85, 391–405.

Steinberg, D., 2002. Atherogenesis in perspective: hypercholesterolemia and inflammation as partners in crime. Nat. Med. 8, 1211–1217.

Steinberg, H.O., Brechtel, G., Johnson, A., Fineberg, N., Baron, A.D., 1994. Insulin-mediated skeletal muscle vasodilation is nitric oxide dependent. A novel action of insulin to increase nitric oxide release. J. Clin. Invest. 94, 1172–1179.

Steppan, J., Tran, H., Benjo, A.M., Pellakuru, L., Barodka, V., Ryoo, S., et al., 2012. Alagebrium in combination with exercise ameliorates age-associated ventricular and vascular stiffness. Exp. Gerontol. 47, 565–572.

Stroth, S., Reinhardt, R.K., Thöne, J., Hille, K., Schneider, M., Härtel, S., et al., 2010. Impact of aerobic exercise training on cognitive functions and affect associated to the COMT polymorphism in young adults. Neurobiol. Learn. Memory 2010 (94), 364–372.

Sun, H., Olson, K.C., Gao, C., Prosdocimo, D.A., Zhou, M., Wang, Z., et al., 2016. Catabolic defect of branched-chain amino acids promotes heart failure. Circulation 133, 2038–2049.

Taddei, S., Virdis, A., Mattei, P., Arzilli, F., Salvetti, A., 1992. Endothelium-dependent forearm vasodilation is reduced in normotensive subjects with familial history of hypertension. J. Cardiovasc. Pharmacol. 20 (Suppl. 12), S193–S195.

Tanaka, H., Dinenno, F.A., Monahan, K.D., Clevenger, C.M., DeSouza, C.A., Seals, D. R., 2000. Aging, habitual exercise, and dynamic arterial compliance. Circulation 102, 1270–1275.

Tanaka, H., Palta, P., Folsom, A.R., Meyer, M.L., Matsushita, K., Evenson, K.R., et al., 2018. Habitual physical activity and central artery stiffening in older adults: the Atherosclerosis Risk in Communities study. J. Hypertens. 36, 1889–1894.

Tang, W.H., Kitai, T., Hazen, S.L., 2017. Gut microbiota in cardiovascular health and disease. Circ. Res. 120, 1183–1196.

Testa, R., Bonfigli, A.R., Prattichizzo, F., La Sala, L., De Nigris, V., Ceriello, A., 2017. The "metabolic memory" theory and the early treatment of hyperglycemia in prevention of diabetic complications. Nutrients 9, 437–445.

Thompson, P.D., Buchner, D., Pina, I.L., Balady, G.J., Williams, M.A., Marcus, B.H., et al., 2003. Exercise and physical activity in the prevention and treatment of atherosclerotic cardiovascular disease: a statement from the Council on Clinical Cardiology (Subcommittee on Exercise, Rehabilitation, and Prevention) and the Council on Nutrition, Physical Activity, and Metabolism (Subcommittee on Physical Activity). Circulation 107, 3109–3016.

Torzewski, M., Rist, C., Mortensen, R.F., Zwaka, T.P., Bienek, M., Waltenberger, J., et al., 2000. C-reactive protein in the arterial intima: role of C-reactive protein receptor-dependent monocyte recruitment in atherogenesis. Arterioscler. Thromb. Vasc. Biol. 20, 2094–2099.

Touyz, R.M., 2004. Reactive oxygen species, vascular oxidative stress, and redox signaling in hypertension: what is the clinical significance? Hypertension 44, 248–252.

Trevisan, M., Krogh, V., Freudenheim, J., Blake, A., Muti, P., Panico, S., et al., 1990. Consumption of olive oil, butter, and vegetable oils and coronary heart disease risk factors. JAMA 263, 688–692.

Ungvari, Z., Kaley, G., de Cabo, R., Sonntag, W.E., Csiszar, A., 2010. Mechanisms of vascular aging: new perspectives. J. Gerontol. A Biol. Sci. Med. Sci. 65, 1028–1041.

Van Bortel, L.M., De Backer, T., De Buyzere, M., 2011. How to treat arterial stiffness beyond blood pressure lowering? J. Hypertens. 29, 1051–1053.

Vlassara, H., Uribarri, J., 2014. Advanced glycation end products (AGE) and diabetes: cause, effect, or both? Curr. Diab. Rep. 14, 453.

Wang, J.C., Bennett, M., 2012. Aging and atherosclerosis. Mechanisms, functional consequences, and potential therapeutics for cellular senescence. Circ. Res. 111, 245–259.

Wang, C.C., Gurevich, I., Draznin, B., 2003. Insulin affects vascular smooth muscle cell phenotype and migration via distinct signaling pathways. Diabetes 52, 2562–2569.
Wang, W., Zhang, F., Xia, Y., Zhao, S., Yan, W., Wang, H., et al., 2016. Defective branched chain amino acid catabolism contributes to cardiac dysfunction and remodeling following myocardial infarction. Am. J. Physiol. Heart Circ. Physiol. 311, H1160–H1169.
Wellen, K.E., Hotamisligil, G.S., 2005. Inflammation, stress, and diabetes. J. Clin. Invest. 115, 1111–1119.
Wendt, T., Bucciarelli, L., Qu, W., Lu, Y., Yan, S.F., et al., 2002. Receptor for advanced glycation endproducts (RAGE) and vascular inflammation: insights into the pathogenesis of macrovascular complications in diabetes. Curr. Atheroscler. Rep. 4, 228–237.
Wilcox, G., 2005. Insulin and insulin resistance. Clin. Biochem. Rev. 26, 19–39.
Willerson, J.T., Ridker, P.M., 2004. Inflammation as a cardiovascular risk factor. Circulation 109, II-2–II-10.
Willett, W.C., Stampfer, M.J., Manson, J.E., Colditz, G.A., Speizer, F.E., Rosner, B.A., et al., 1993. Intake of trans fatty acids and risk of coronary heart disease among women. Lancet 341, 581–585.
Willum Hansen, T., Staessen, J.A., Torp-Pedersen, C., Rasmussen, S., Thijs, L., Ibsen, H., et al., 2006. Prognostic value of aortic pulse wave velocity as index of arterial stiffness in the general population. Circulation 113, 664–670.
Wolfson, R.L., Chantranupong, L., Saxton, R.A., Shen, K., Scaria, S.M., Cantor, J.R., et al., 2015. Sestrin2 is a leucine sensor for the mTORC1 pathway. Science 351, 43–48.
Writing Group Members, Mozaffarian, D., Benjamin, E.J., Go, A.S., Arnett, D.K., Blaha, M.J., et al., 2016. Executive summary: heart disease and stroke statistics—2016 update: a report from the American Heart Association. Circulation 133, 447–454.
Wu, G., Meininger, C.J., 2009. Nitric oxide and vascular insulin resistance, BioFactors (Oxford, England), 35. pp. 21–27.
Wu, B., Mottola, G., Schaller, M., Upchurch Jr, G.R., Conte, M.S., 2017. Resolution of vascular injury: specialized lipid mediators and their evolving therapeutic implications. Mol. Aspects Med. 58, 72–82.
Yan, Z.Q., Hansson, G.K., 2007. Innate immunity, macrophage activation, and atherosclerosis. Immunol. Rev. 219, 187–203.
Zakynthinos, E., Pappa, N., 2009. Inflammatory biomarkers in coronary artery disease. J. Cardiol. 53, 317–333.
Zeng, G., Quon, M.J., 1996. Insulin-stimulated production of nitric oxide is inhibited by wortmannin. Direct measurement in vascular endothelial cells. J. Clin. Invest. 98, 894–898.
Zhang, J., Jin, J., Liu, J., He, Y., Zhang, P., Ye, W., et al., 2018. A study of the correlation of insulin resistance and leptin with inflammatory factors and vascular endothelial injury in T2DM patients with CHD. Exp. Ther. Med. 16, 265–269.
Zieman, S.J., Vojtech, M., Clattenburg, L., Corretti, M.C., Capriotti, A.R., Gerstenblith, G., et al., 2007. Advanced glycation endproduct crosslink breaker (alagebrium) improves endothelial function in patients with isolated systolic hypertension. J. Hypertens. 25, 577–583.
Zimmer, H.G., 1996. Regulation of and intervention into the oxidative pentose phosphate pathway and adenine nucleotide metabolism in the heart. Mol. Cell. Biochem. 160–161, 101–109.
Zong, G., Li, Y., Wanders, A.J., Alssema, M., Zock, P.L., Willett, W.C., et al., 2016. Intake of individual saturated fatty acids and risk of coronary heart disease in US men and women: two prospective longitudinal cohort studies. BMJ 355, i5796.
Zou, Y., Yoon, S., Jung, K.J., Kim, C.H., Son, T.G., Kim, M.S., et al., 2006. Upregulation of aortic adhesion molecules during aging. J. Gerontol. A Biol. Sci. Med. Sci. 61, 232–244.

Further reading

Ayyadhury, S., Heese, K., 2007. Neurotrophins—more than neurotrophic. Curr. Immunol. Rev. 3, 189–215.

Heese, K., Akatsu, H., 2006. Alzheimer's disease—an interactive perspective. Curr. Alzheimer Res. 3, 109–121.

Lesniewski, L.A., Durrant, J.R., Connell, M.L., Henson, G.D., Black, A.D., Donato, A.J., et al., 2011. Aerobic exercise reverses arterial inflammation with aging in mice. Am. J. Physiol. Heart Circ. Physiol. 301, H1025–H1032.

Polychronopoulos, E., Pounis, G., Bountziouka, V., Zeimbekis, A., Tsiligianni, I., Qira, B.E., et al., 2010. Dietary meat fats and burden of cardiovascular disease risk factors, in the elderly: a report from the MEDIS study. Lipids Health Dis. 9, 30.

Serhan, C.N., 2014. Pro-resolving lipid mediators are leads for resolution physiology. Nature 510, 92–101.

Zieman, S.J., Melenovsky, V., Kass, D.A., 2005. Mechanisms, pathophysiology, and therapy of arterial stiffness. Arterioscler. Thromb. Vasc. Biol. 25, 932–943.

CHAPTER 4

Insulin resistance and sleep apnea

Introduction

Sleep is a complex and essential biological process, which is essential for human life regardless of age, sex, or ethnic origin. Although sleep is a vital process, we still do not fully understand why do we sleep? What induces sleep? What induces wakefulness, and how many hours are needed to achieve "restorative sleep"? Sleep is known to modulate learning and memory processing, cellular repair, and brain development (Tononi and Cirelli, 2006; Dinges, 2006). Behaviorally, sleep is characterized by four criteria: (1) reduction in motor activity, (2) decrease in response to stimulation, (3) induction of stereotypic postures (in humans, e.g., lying down with eyes closed), and (4) relatively easy reversibility (distinguishing it from coma, hibernation, and estivation) (Rechtschaffen and Siegel, 2000). Sleep consists of two different stages: non−rapid eye movement (non-REM) sleep and rapid eye movement (REM) sleep (Rechtschaffen and Siegel, 2000). These stages can be measured by polysomnography: non-REM consists of three stages, N1, N2, and N3—and REM sleep. As non-REM sleep becomes deeper, there is slowing of electroencephalographic frequencies and increasing synchronization of cortical neuronal activity. At its deepest point—stage N3 or slow wave sleep (SWS)—the neuronal synchrony appears as large slow wave activity on the electroencephalogram (Berry et al., 2012). SWS is considered a marker of the sleep homeostat as it rebounds during recovery sleep after prolonged wakefulness (Dijk et al., 1990). REM sleep is characterized by cortical desynchrony and REMs and is associated with dream states (V).

Many genes are affected by insufficient sleep and abnormal circadian rhythms. These genes include *PER1*, *PER2*, *PER3*, *CRY2*, *CLOCK*, *NR1D1*, *NR1D2*, *RORA*, *DEC1*, and *CSNK1E*. Other genes that modulate sleep homeostasis *IL6*, *STAT3*, *KCNV2*, and *CAMK2D* are also affected by insufficient sleep (Möller-Levet et al., 2013). These genes not only modulate chromatin remodeling, but also regulate immune and stress

Insulin Resistance as a Risk Factor in Visceral and Neurological Disorders
DOI: https://doi.org/10.1016/B978-0-12-819603-8.00004-3

responses. On the basis of gene expression studies, it is proposed that insufficient sleep affects the human blood transcriptome, disrupts its circadian regulation, and intensifies the effects of acute total sleep deprivation. In addition, age-related sleep disruption promotes significant downstream consequences for brain and body health. Sleep in young adults supports every major physiological system within the body, including immune, metabolic, thermoregulatory, endocrine, and cardiovascular function (Irwin, 2015); and numerous cognitive and affective neural processes, such as learning and memory, emotional regulation, attention, motivation, decision making, and motor control (Walker, 2009). It is well established that pattern of sleep changes across the lifespan, from infancy through old age in humans (Ohayon et al., 2004). Older people experience less sleep, more frequent awakenings, a reduction of SWS (Dijk et al., 1999), and blunting of the amplitudes of circadian rhythms such as core body temperature (Carrier et al., 1996) and activity (Hu et al., 2009). Recent studies have indicated that sleep plays important roles in brain plasticity, memory, and learning (Diekelmann and Born, 2010). In addition, sleep facilitates the consolidation of new information or abilities learned the previous day (Peigneux et al., 2004). The characteristics of sleep also predict the ability to learn new material after the sleep episode (Lafortune et al., 2014), and total sleep deprivation impedes subsequent encoding in episodic memory (Yoo et al., 2007). As mentioned above, disturbed sleep is common in old age and in patients with neurodegenerative disorders, such as Alzheimer's disease (AD) and other dementing disorders. More than 60% of AD patients develop sleep disturbance, which often occurs at the early stages of the disease or even before the onset of major cognitive decline (Guarnieri et al., 2012). Impaired sleep in these patients has been attributed to the progression of AD pathology to brain regions that regulate the sleep—wake or circadian rhythm (Ju et al., 2014). It is not known how and when sleep problems manifest and involved in earlier stages of the dementia. Collective evidence suggests that normal sleep is important for many aspects of mammalian metabolism and homeostasis such as relaxing the physiological stress, which leads to neuroprotection in central nervous system and cardio-protection in cardiovascular system.

In humans, sleep and sleep deprivation is accompanied by changes in metabolism (Buxton et al., 2012) and hormone secretion leading to and dysregulated physical and psychological feelings (Van Cauter et al., 2008; Hanlon and Van Cauter, 2011). For example, the unfavorable consequences of obstructive sleep apnea (OSA) include abnormal arterial blood

Table 4.1 Effect of sleep on hormones.

Hormone	Level during daytime	Levels during night	References
Prolactine	Low	High	Gronfier and Brandenberger (1998) and Carley and Farabi (2016)
Growth hormone	Low	High	Gronfier and Brandenberger (1998) and Carley and Farabi (2016)
Cortisol	Low	High	Gronfier and Brandenberger (1998) and Carley and Farabi (2016)
TSH	Low	High	Gronfier and Brandenberger (1998) and Carley and Farabi (2016)
Melatonin			Carley and Farabi (2016)

gas components and decreased parasympathetic and increased sympathetic activity, all of which are harmful for cardiovascular system (Yu et al., 2017). Hormones, which are influenced by sleep, include adrenocortotropic hormone, cortisol, and melatonin; some are strongly influenced by sleep, such as thyroid-stimulating hormone (TSH) prolactin, and growth hormone (Table 4.1) (Gronfier and Brandenberger, 1998; Carley and Farabi, 2016). These hormonal changes contribute to the activation of sympathetic nervous system and increase in the release of catecholamine resulting in glycogenesis and decrease in insulin sensitivity (Aurora and Punjabi, 2013). In addition, there is stimulation of the hypothalamic–pituitary–adrenal axis (HPA axis). This leads to increase in levels of cortisol which can further impair glucose metabolism by decreasing insulin release (Punjabi and Polotsky, 2005; Kent et al., 2015). Collective evidence suggests that sleep plays a vital role in the normal homeostasis of glucose metabolism and insulin sensitivity. Sleep loss is considered as a novel risk factor for insulin resistance, type 2 diabetes, metabolic syndrome (MetS), heart disease, and stroke (Fig. 4.1) (Farré et al., 2018; Beaudin et al., 2017; Gabryelska et al., 2018). Sleep loss, whether voluntary or disease related, affects millions of individuals in our modern society. Over the past few decades, the average sleep duration of Americans has decreased by 1.5–2 hours (Spiegel et al., 2005; Cooper et al., 2018). Interestingly, the trends in increased obesity and diabetes seem to mirror the time period for the increase in sleep loss. The molecular mechanism

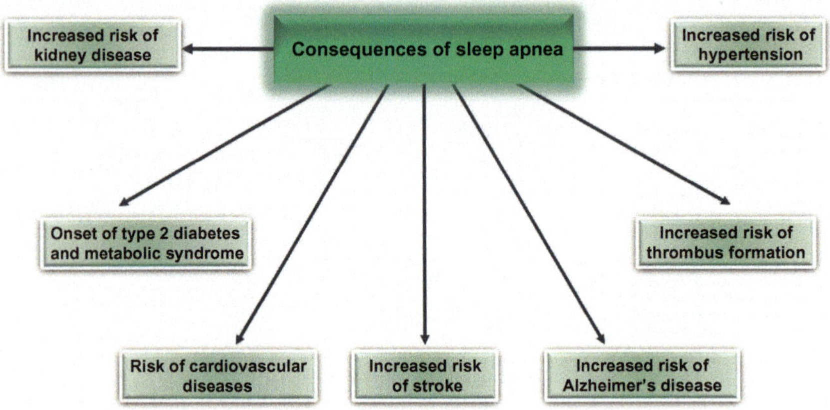

Figure 4.1 Metabolic consequences of sleep apnea.

which links sleep loss to heart disease and stroke are not fully understood. However, it is proposed that β-cell dysfunction, insulin resistance, type 2 diabetes, and MetS may contribute to cardiovascular and cerebrovascular complications in the pathogenesis of heart disease and stroke (Bonsignore and Zito, 2008). Emerging evidence indicates that metabolic disorders can exacerbate OSA, creating a bidirectional relationship between OSA and metabolic physiology (Minakawa et al., 2019).

Obstructive sleep apnea

OSA is a common chronic sleep disorder found in the general population. It is characterized by recurrent pauses in breathing during sleep, which not only leads to hypercapnia, arousals, reductions in intrathoracic pressure, and sleep fragmentation in middle age men. The clinical manifestations of OSA include snoring, choking/gasping episodes, excessive daytime sleepiness, nonrestorative sleep, nocturia, long duration of sleep, morning headaches, loss of libido, irritability, cognitive impairment, and decreased in concentration and memory (Epstein et al., 2009). OSA is also characterized by collapse of upper airways and intermittent hypoxia along with unstable breathing. Upper airway is known to have little rigid support and is largely dependent on neuromuscular control for the maintenance of patency. Anatomically OSA patients are obese men, who have small pharyngeal airway. In middle age human, OSA can be improved by weight loss (Somers et al., 2008). In children, OSA may be caused by enlarged tonsils and adenoids

(Schwab et al., 2003; Calvin et al., 2009). Awake OSA patients have greater airway resistance, which activates mechanoreceptors to trigger reflex pharyngeal dilator muscle activity, thus maintaining the efficacy and patency of airway (Schwab et al., 2003; Fogel et al., 2001). During sleep, dilator muscle activity is diminished, producing pharyngeal narrowing and intermittent collapse of the upper airway (Fogel et al., 2003; Calvin et al., 2009). This can lead to a combination of hypopneas, or reduction in airflow associated with a fall in oxygen saturation, or apneas, or complete cessation of airflow. The number of apneas and hypopneas per hour of sleep is termed the apnea–hypopnea index (AHI) and has been used as a marker of OSA severity. The diagnosis of OSA can be made when the AHI is >5 in a patient with excessive daytime sleepiness (Sleep-related breathing disorders in adults, 1999; Calvin et al., 2009). As mentioned above, apneas and hypopneas lead to hypoxia and hypercapnia, which stimulate ventilatory drive and ultimately arousal from sleep and restoration of airway patency. The intermittent hypoxia can be severe, with arterial oxygen saturation dropping to <60% in some patients and associated with a disruption of normal autonomic and hemodynamic responses to sleep (Somers et al., 1993a), including increased sympathetic activity to peripheral blood vessels leading to vasoconstriction and acute increases in blood pressure (BP; Somers et al., 1989a,b, 2008). In the general population, risk factors for OSA are age, male gender, body mass index (BMI), MetS, high BP, and a variety of craniofacial and oropharyngeal features (Fig. 4.2) (Young et al., 2002; Dempsey et al., 2010;

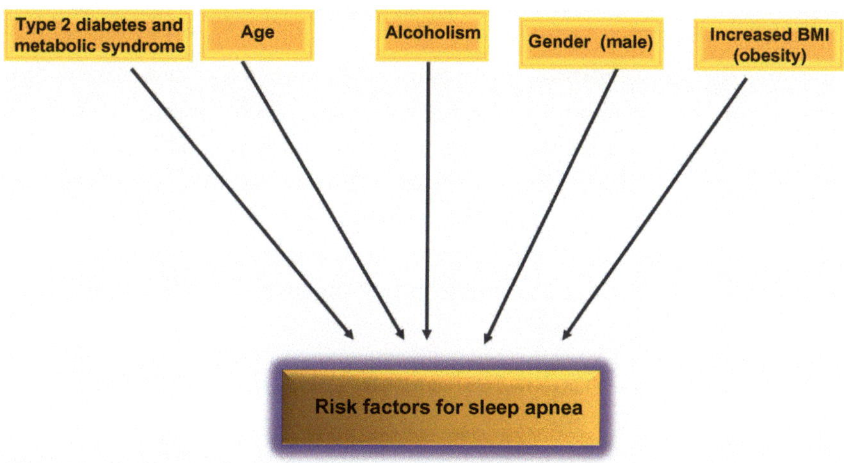

Figure 4.2 Risk factors for sleep apnea.

Cairns et al., 2016). Epidemiological studies indicate that OSA is a common disorder affecting about 34% of men and 17% of women (Peppard et al., 2013), and prevalence is rapidly rising due to the strong association of OSA with obesity, with at least 60% of OSA patients being obese (Peppard et al., 2000; Young et al., 2008). Similarly, OSA patients also have hypertension (Javaheri et al., 2017). The prevalence of hypertension varies considerably affecting 24%–26% of men and 17%–28% of women between 30 and 70 years of age (Peppard et al., 2013; Kearney et al., 2005; Javaheri et al., 2017).

Biochemical changes in obstructive sleep apnea

In OSA patients, chronic intermittent hypoxia induces obesity, dyslipidemia, insulin resistance, and pancreatic endocrine dysfunction. Recent reports provide new insights in possible mechanisms by which intermittent hypoxia affects lipid and glucose metabolism. Chronic intermittent hypoxia induces dyslipidemia by upregulating lipid biosynthesis in the liver, increasing adipose tissue lipolysis with subsequent free fatty acid (FFA) flux to the liver, and by inhibiting lipoprotein clearance. Chronic intermittent hypoxia affects glucose metabolism by inducing sympathetic activation, elevating systemic inflammation, increasing counter-regulatory hormones and fatty acids, and causing direct pancreatic beta-cell injury. Intermittent hypoxia also contributes to oxidative stress and inflammation (Fig. 4.3) (Schulz et al., 2014; Olea et al., 2014; Murphy et al., 2017; He et al., 2014), overactivation of sympathetic nervous system (Iiyori et al., 2007), tissue hypoxia (Almendros et al., 2011; Sherwani et al., 2013), and sleep fragmentation. As mentioned above, intermittent hypoxia is also influenced by several sleep hormones including cortisol, melatonin, TSH, prolactin, and growth hormone (Gronfier and Brandenberger, 1998; Carley and Farabi, 2016). These hormonal changes contribute to the activation of sympathetic nervous system and increase in the release of catecholamine resulting in glycogenesis and decrease in insulin sensitivity (Aurora and Punjabi, 2013). These changes are also associated with hypertension, breathing abnormalities, insulin resistance, and ventricular hypertrophy (Spiegel et al., 2004, 2005).

Oxidative stress, insulin resistance, and obstructive sleep apnea

In OSA patients, chronic intermittent hypoxia potentiates the carotid body (CB) chemosensory discharge leading to sympathetic overflow,

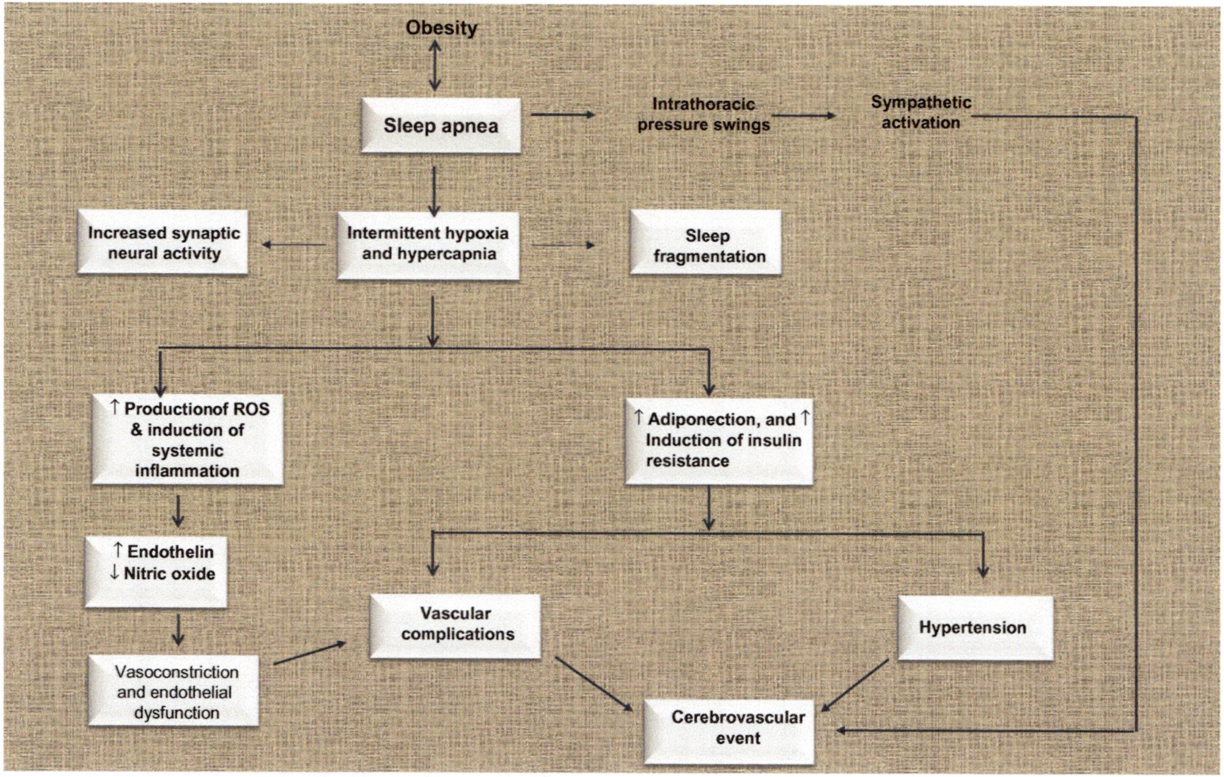

Figure 4.3 Biochemical changes associated with sleep apnea.
ROS, Reactive oxygen species; ↑, increase; and ↓, decrease.

autonomic dysfunction, and hypertension along with the formation of reactive oxygen species (ROS). These metabolites are formed from the reduction of molecular oxygen or by oxidation of water to yield products such as superoxide anion ($O_2^{\cdot -}$), hydrogen peroxide (H_2O_2), and hydroxyl radical ($^{\bullet}OH$) (Farooqui, 2014) resulting in the induction of oxidative stress, which may contribute the development of cardiovascular disease, vascular injury, and endothelial dysfunction. In low amounts, ROS contribute to a number of physiological processes that produce desired cellular responses. However, large quantities of ROS in a biological system are associated with cellular damage to lipids, membrane proteins, and DNA. Hyperglycemia causes an increase in oxidative stress markers such as membrane lipid peroxidation (Farooqui, 2013). The degree of lipid peroxidation is directly proportional to the glucose concentrations in diabetic patients. These processes occur via several well-studied mechanisms, including increased polyol pathway flux, increased intracellular formation of advanced glycation end products, activation of protein kinase C, or overproduction of superoxide by the mitochondrial electron transport chain (Farooqui, 2013, 2014).

It is well known that hypoxia-inducible factor (HIF)-1 is the main transcriptional regulator of the hypoxia response. Genetic inhibition of HIF-1α in adipocytes protects mice from high-fat diet (HFD)-induced adipose tissue inflammation and restores insulin sensitivity not only in the adipose tissue but also in liver and skeletal muscle (Lee et al., 2014). Intermittent hypoxia increases the intracellular Ca^{2+} levels, which induce the increased expression of HIF-1α and a decrease in the levels of HIF-2α (Fig. 4.4) (Semenza and Prabhakar, 2018). Intermittent hypoxia also elevates HIF-1α-dependent expression of Nox2, encoding for the prooxidant enzyme nicotinamide adenine dinucleotide phosphate oxidase 2, and reduces HIF-2α-dependent expression of Sod2, encoding the antioxidant enzyme superoxide dismutase 2. These changes in gene expression promote persistently elevated ROS levels in the CB, brainstem, and adrenal medulla that are required for the development of hypertension and breathing abnormalities. The ROS generated by dysregulated HIF activity in the CB results in oxidation and inhibition of hemoxygenase 2, and the resulting reduction in the levels of carbon monoxide leads to increased hydrogen sulfide production, triggering glomus cell depolarization. These findings support the view that the pathophysiology of OSA involves the dysregulation of O_2-regulated transcription factors, gasotransmitters, and sympathetic outflow that affects BP and breathing (Semenza and Prabhakar, 2018).

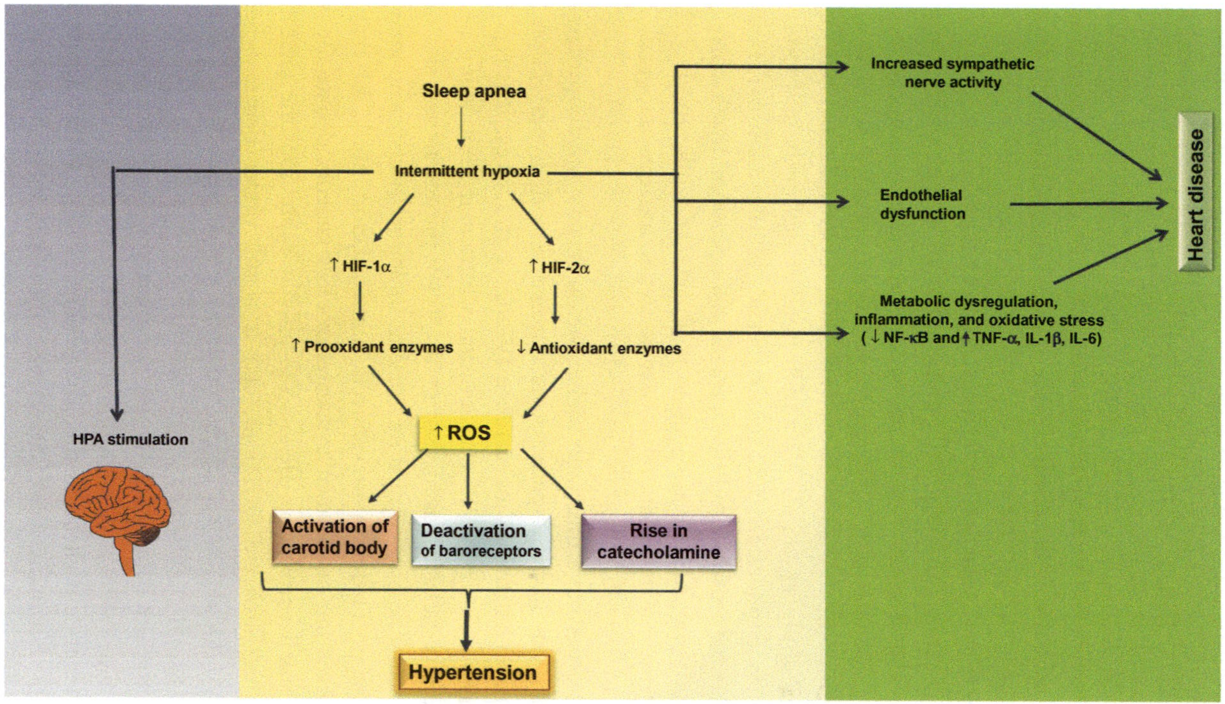

Figure 4.4 Effect of intermittent hypoxia on the pathogenesis of cardiovascular diseases.
HIF-1α, Hypoxia-inducible factor-1α; *HIF-2α*, hypoxia-inducible factor-2α; *IL-1β*, interleukin-1β; IL-6, interleukin-6; *TNF-α*, tumor necrosis factor-α; ↑, increase; and ↓, decrease.

High ROS production has been implicated in the development of insulin resistance and chronic diabetic complications in the brain. In visceral organs, chronic exposure to high glucose concentration can increase the metabolic flux in mitochondria through the hexosamine pathway, leading to more ROS production; this may perturb insulin secretion and β-cell survival through multiple mechanisms. Intermittent hypoxia in OSA contributes to the development of hypertension by increasing oxidative stress. Oxidative stress results in the production of ROS and reduces circulating nitric oxide (NO), a key endothelial-derived molecule that mediates control of vascular tone. Impaired NO release from endothelial cells leads to vasoconstriction and is regarded as an initiator and promoter of cardiovascular disease in patients with OSA. In this context, several studies have documented increased markers of oxidative stress in OSA patients when compared to controls, including superoxide levels in blood neutrophils (Schulz et al., 2000) and 8-isoprostane levels in blood and exhaled breath condensate (Carpagnano et al., 2003; Alonso-Fernandez et al., 2009). One study found that exhaled 8-isoprostane levels correlated positively with the AHI (Carpagnano et al., 2003) while another study also found lower nitrate and nitrite levels (reflecting lower overall NO production) in the OSA group (Alonso-Fernandez et al., 2009).

Recent studies have indicated higher serum matrix metalloproteinase-9 (MMP-9) levels in OSA patients are correlated with hypertension and left ventricular hypertrophy occurrence (Wang et al., 2018). It is reported that the activation of MMP-9 precedes left ventricular remodeling in rats with hypertensive heart failure (Sakata et al., 2004); the researchers also demonstrated that angiotensin-converting enzyme inhibitor (ACEI) can attenuate MMP-9 activity. Therefore preventively administration of ACEI may be helpful in preventing cardiovascular diseases of OSA patients, despite the need for further clinical verification. Besides the protein expression, the gene level of MMP-9 also attracted attentions, and it is still under debate whether MMP-9 polymorphisms affect OSAS occurrence and onset of its complications (Cao et al., 2015; Yalcinkaya et al., 2015).

Importantly, the role of MMP-9 is not restricted to cardiovascular disorders; it is also associated with the progression of microalbuminuria in noninsulin-dependent diabetes mellitus (Ebihara et al., 1998). It is interesting to note that many OSA patients are obese and suffer from diabetes. This is tempting to speculate that systematically exploring the clinical significance of serum MMP-9 may be important for elucidating OSA complication mechanisms.

Obstructive sleep apnea–mediated changes in lipid metabolism

In adipose tissue, inflammation and insulin resistance promote lipolysis with increased and persistent FFA flux into the systemic circulation which is linked to the development of hepatic and muscle insulin resistance and β-cell dysfunction (Cooke et al., 2016). OSA impairs the clearance of triglyceride-rich lipoproteins from human plasma, a process, which is reversed using continuous positive airway pressure (CPAP). This is a well-established treatment for OSA (Lin et al., 2015). The hydrolysis of triglyceride-rich lipoprotein into fatty acids is catalyzed by enzyme lipoprotein lipase (LPL). This is a rate limiting enzyme, which is downregulated by hypoxia (Yao et al., 2013; Drager et al., 2012). LPL contributes to the hydrolysis of triglyceride from chylomicrons and very low-density lipoproteins into fatty acids, which are taken up by peripheral tissues, including the heart, muscle, and adipose tissues after food intake. In humans and mice, deficiency of LPL results in severe hypertriglyceridemia (Weinstock et al., 1995). Because of the critical role that LPL plays in lipoprotein metabolism and tissue-specific substrate delivery and utilization, LPL activity is carefully orchestrated in a tissue-specific manner to meet the energy demands of various tissues at different nutritional statuses. LPL activity is regulated by various factors, including angiopoietin-like (ANGPTL) proteins, such as ANGPTL3, 4, and 8 (Haller et al., 2017; Zhang, 2016). ANGPTLs are a family of proteins composed of eight members and have been associated with various metabolic pathways, including insulin resistance, oxidative stress, and dyslipidemia (Zhang, 2016). ANGPTL4 and 8 are two members of this family and have been shown to play a role in lipid metabolism through regulating plasma lipid levels by inhibiting LPL activity (Abid et al., 2016; Santulli, 2014). ANGPTL4 is a ubiquitously expressed protein that has been shown to be transcriptionally regulated by hypoxia (Hu et al., 2016; Babapoor-Farrokhran et al., 2015). Moreover, ANGPTL8 has been shown to regulate the activity of LPL through its interaction with ANGPTL3 (Quagliarini et al., 2012). Both our and other research groups have shown that ANGPTL8 is associated with insulin resistance and that its levels are increased in obesity, metabolic disease, and type 2 diabetes (Abu-Farha et al., 2016, 2017). Considering that OSA is a known risk factor for dyslipidemia and the roles of ANGPTL4 and 8 in regulating the level of lipids in plasma as well as their involvement in metabolic-related pathways, it has been hypothesized that ANGPTL4 and 8 levels may be dysregulated in OSA.

Inflammation, insulin resistance, and obstructive sleep apnea

Intermittent hypoxic episodes in OSA are accompanied by repetitive oxygen desaturation and resaturation. This process may contribute to the development of inflammation. Systemic inflammation is strongly linked to the pathogenesis of atherosclerosis and hypertension; however, its role in OSA is confounded by the presence of obesity, a chronic inflammatory state. In human and rodents, the onset of OSA is also accompanied by the stimulation of toll-like receptors (TLR), and in particular the TLR4. TLRs are a family of pattern-recognition receptors that play a critical role in the innate immune system by activating proinflammatory signaling pathways in response to microbial pathogens (Farooqui, 2014). At the molecular level, this receptor is linked with nuclear factor-κB (NF-κB). High levels of ROS activate NF-κB and facilitate its translocation to the nucleus, where it binds to NFr-κB-response element and promote the expression of proinflammatory cytokines (TNF-α, IL-2, IL-4, IL-5, IL-6, IL-8, and IFN-γ), chemokines (monocyte chemotactic protein-1), and adhesion molecules [intercellular adhesion molecule-1 (ICAM-1), vascular cell adhesion molecule-1 and E-selectin] in specific brain regions which regulate sleep (Fig. 4.5) (Irwin et al., 2008; Brandt et al., 2004; Ramesh et al., 2007). The induction of inflammation plays important roles in atherogenesis, arterial thrombus formation, and induction of heart disease (Rifai and Ridker, 2002). As mentioned above, intermittent apnea-related hypoxia and postapneic reoxygenation probably contribute not only to more ROS production, but also further increase in inflammatory mediators, triggering upper airway and systemic inflammation. Upper airway inflammation, aggravated by mechanical injury caused by repeated pharyngeal collapse, increases airway obstruction. The systemic inflammatory process also increases release of oxygen-free radicals beyond physiological antioxidant capacity, generating oxidative stress. Various diseases and/or anatomical conditions of the upper airways play a significant role in the ethiopathogenesis of OSA (Salamanca et al., 2013; Passàli et al., 2014). In addition, OSA also promotes many other biochemical changes such as reduction in glucose tolerance, activation of the sympathetic nervous system, reduction in leptin levels, and increases in the secretion of

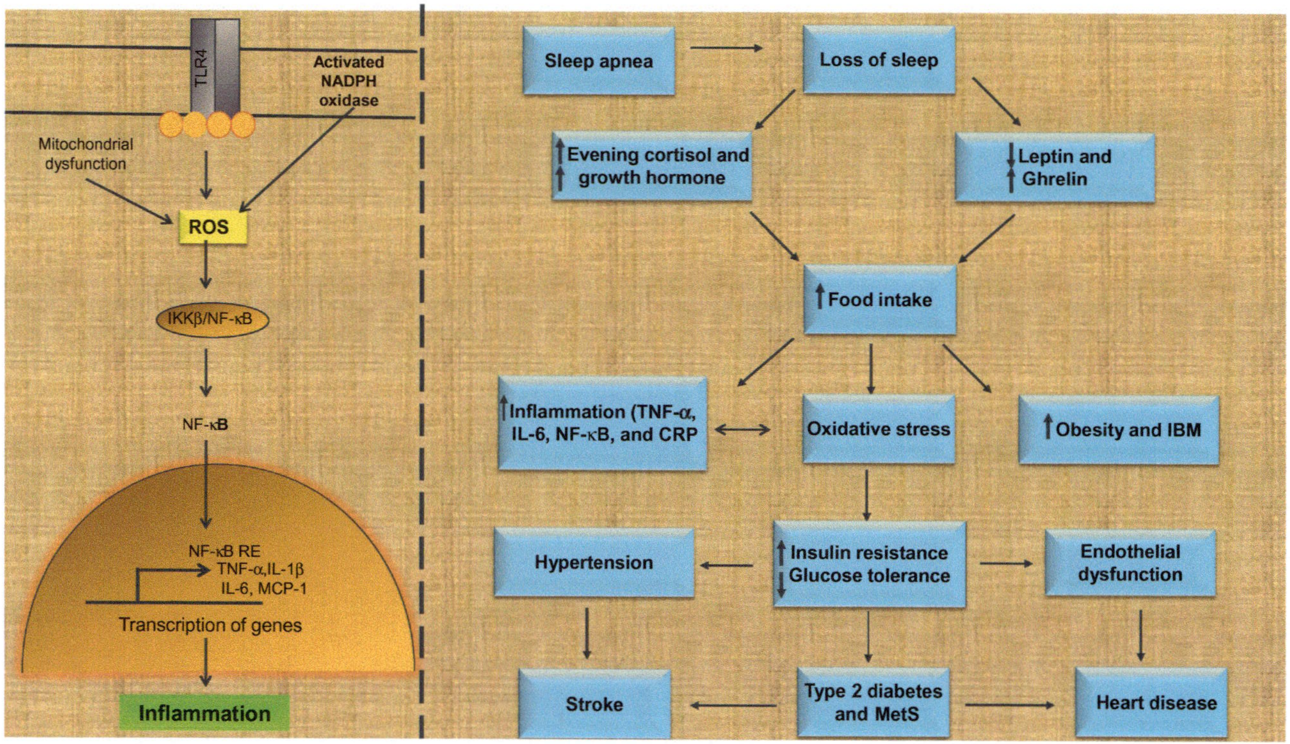

Figure 4.5 Relationship among sleep, insulin resistance, and neurological disorders.
↑, Increase; ↓, decrease.

catecholamines by the adrenal medulla resulting in increase in BP, and breathing abnormalities (Spiegel et al., 2004, 2005). These changes contribute to MetS, a multifactorial condition, which is a risk factor for stroke, AD, and depression (Farooqui, 2018).

Recent studies have indicated that in OSA patients, intermittent hypoxia promotes systemic inflammation through the involvement of platelets. Blood platelets support inflammation not only by transcellular production of bioactive lipids, but also by delivering both specific enzymes and substrate molecules (Gabryelska et al., 2018). Despite their lack of nucleus, platelets synthetize proteins in a stimuli-dependent manner. Atherosclerosis and consequent cardiovascular complications result from disruption in homeostasis of both of the platelet roles: blood coagulation and inflammatory processes modulation. In OSA patients, platelet blood count test can be used as markers of cardiovascular comorbidity (Gabryelska et al., 2018). Persistent activation of platelets results in enhanced spontaneous aggregability and alterations in cytokine production. Furthermore, platelet–lymphocyte ratio has been suggested to be an independent marker for cardiovascular disease in OSA syndrome and continuous positive air pressure therapy impacts platelet parameters and phenotype (Gabryelska et al., 2018).

Biomarkers for obstructive sleep apnea

The biomarkers are metabolites whose concentration, presence, and activity are closely associated with the pathogenesis of the disease processes. Biomarkers can be detected in patients with the disease not only for early detection of the disease (preclinical stage) and monitoring the disease progression, but also for following the treatment response more sensitively and objectively. At the present time, very little information is available on an ideal and specific biomarker of OSA. The discovery of an ideal and specific biomarker will not only improve the differential diagnosis of OSA, but also track its progression, and measure the efficacy of treatment. The main problem in developing an ideal biomarker for OSA has been the slow understanding of pathogenesis of OSA. The ideal biomarker of OSA should be highly sensitive and specific for OSA, should be dose–responsive and correlate to severity of disease, and should be involved in an important causal pathway, so that changes in the biomarker levels reliably predict improvements in the outcome (Shih and Malhotra, 2011). Several different OSA biomarkers have been described over the

Table 4.2 Potential biomarkers for sleep apnea.

Potential marker	Sensitivity (%)	Specificity
HbA1c + CRP + EPO	81	60
HbA1c	38	88
CRP	52	77
EPO	66	58
BMI	64	78
Neck circumference	44	93

Source: Summarized from Papanas, N., Steiropoulos, P., Nena, E., et al., 2009. HbA1C is associated with severity of obstructive sleep apnea hypopnea syndrome in nondiabetic men. Vasc. Health Risk. Manag. 5, 751–756; Canto, Gde L., Pachêco-Pereira, C., Aydinoz, S., Major, P.W., Flores-Mir, C., Gozal, D., 2015. Biomarkers associated with obstructive sleep apnea: a scoping review. Sleep. Med. Rev. 23, 28–45; Fleming, W.E., Ferouz-Colborn, A., Samoszuk, M.K., Azad, A., Lu, J., Riley, J.S., et al., 2016. Blood biomarkers of endocrine, immune, inflammatory, and metabolic systems in obstructive sleep apnea. Clin. Biochem., 49, 854–861; Fleming, W.E., Holty, J.C., Bogan, R.K., Hwang, D., Ferouz-Colborn, A.S., Budhiraja, R., et al., 2018. Use of blood biomarkers to screen for obstructive sleep apnea. Nat. Sci. Sleep 10, 159–167; Abrams B., 2018. Comment on: use of blood biomarkers to screen for obstructive sleep apnea. Nat. Sci. Sleep 10, 313–315.

last 20 years (Table 4.2). They include glycated hemoglobin, C-reactive protein, erythropoietin, TNF-α, IL-6, IL-10, uric acid, and isoprotane (Papanas et al., 2009; Canto Gde et al., 2015; Fleming et al., 2016, 2018; Abrams, 2018). These biomarkers do not fulfill all requirements of an ideal biomarkers. For future studies on OSA biomarkers, it is important have a complete understanding of how OSA biomarkers change over time throughout the progression of the disease. To determine and evaluate whether a treatment is working, it is of utmost importance to choose biomarkers that are most relevant and specific to OSA. The biomarkers that are most dynamic (in terms of their rate of change over time) will likely not be the same ones at various stages of OSA. In addition, multiple studies have indicated that there is a clear, positive association between obesity and AHI (Gami et al., 2003). More specifically, increased visceral obesity (Vgontzas and Kales, 1999) and neck circumference (Davies et al., 1992) have been linked to OSA. The size and adiposity of upper airway structures have also been shown to be important and OSA patients have larger tongues with increased adiposity on volumetric magnetic resonance analysis (Punjabi, 2008). Increased abdominal visceral adiposity decreases lung volumes, including the functional residual volume, which reduces traction on the pharynx and has been hypothesized to lead to increased pharyngeal collapsibility and, thus, OSA (Gami et al., 2003). Collective evidence suggests that the pathophysiology of OSA is fundamentally based on mechanisms whereby collapsing forces within the upper airways exceed the ability of dilating muscles to maintain airway patency leading to complete or partial upper airway occlusion. It is now generally accepted that obesity is an important prerequisite for OSA, the hypothesized mechanisms by which obesity contributes to OSA vary widely. Indeed, the relative contributions of an individuals' physical weight versus an individuals' metabolic physiology in the development of OSA is actively investigated by researchers (Phillips et al., 2016).

Obstructive sleep apnea and heart disease

As mentioned above, obesity is a major risk factor for OSA, which is characterized by periodic upper airway occlusion during sleep and high levels of leptin in plasma. The neuromuscular activity that maintains upper airway patency during sleep and the anatomy of upper airway are key factors involved in the pathogenesis of sleep disorders. Leptin has been implicated in the pathogenesis of OSA not only through the central regulation of upper airway patency and diaphragmatic control, but also via modulation of sleep architecture, ventilatory function, and hypercapnic ventilatory responses. Experimental data from animal models show that genetic forms of leptin deficiency and leptin resistance in mice contribute to higher occurrence of flow-limited breathing and pharyngeal collapsibility (Pho et al., 2016; Yao et al., 2016). However, human studies do not consistently support the data from the animal models. The effects of leptin on the pathophysiology of OSA in humans have also been complicated by leptin's diurnal variation, adiposity, age, and gender. Therefore more studies are needed on the involvement of leptin in the pathogenesis of OSA in humans. OSA not only contribute to metabolic alterations such as insulin resistance and type 2 diabetes, but is also associated with wide array of cardiovascular abnormalities including atherosclerosis, coronary artery disease, congestive heart failure, arrhythmias, infarction, atrial fibrillation, hypertension, and stroke (Farré et al., 2018; Beaudin et al., 2017; Gabryelska et al., 2018). These consequences of OSA can be attributed to oxidative stress, inflammation, and sympathetic overflow induced by intermittent hypoxia. In addition, other factors, which contribute to cardiovascular abnormalities, are sleep fragmentation and comorbid metabolic diseases (Somers et al., 2008; Dempsey et al., 2010; Iturriaga et al., 2016). OSA patients also show increased sympathetic, vasopressor and ventilatory responses to hypoxia, attributed to a potentiated hypoxic peripheral chemoreflex (Somers et al., 2008; Dempsey et al., 2010). The enhanced CB chemosensory discharge induced by intermittent hypoxia is linked with local oxidative stress and increased endothelin-1 (ET-1) levels in the CB (Rey et al., 2006; Del Rio et al., 2016). The enhanced CB chemosensory discharge plays a crucial role in the onset and progression of hypertension induced by intermittent hypoxia. Collective evidence suggests that CB plays a crucial role in the onset and progression of the hypertension induced by intermittent hypoxia during OSA.

The typical pattern of chronic intermittent hypoxia in OSA with repetitive short cycles of desaturation followed by rapid reoxygenation is strikingly different from the chronic sustained hypoxia, which is associated with other chronic respiratory or cardiac conditions. Notably, intermittent hypoxia itself is considered as a "double-edged sword" and there is increasing evidence that short exposures to mild intermittent hypoxia may lead to adaptive responses through preconditioning effects and, hence, may be cardioprotective for patients with mild OSA (Lavie, 2015). In contrast, severe intermittent hypoxia due to induction of oxidative stress and inflammation may be very harmful for the heart function due to the development of vascular dysfunction, atherosclerotic changes, and formation of plaques. Thus, in C57BL/6J wild-type mice, the induction of intermittent hypoxia for 2 weeks produces vascular remodeling in the aorta leading to increase in intima-media thickness (Arnaud et al., 2011). However, atherosclerotic response to intermittent hypoxia in this mouse strain only occurs after 12 weeks and following the consumption of a high-cholesterol diet, suggesting that the intermittent-induced atherogenic process is amplified in the presence of other risk factors (Savransky et al., 2007). These findings have been reproduced in atherosclerosis-prone apolipoprotein E deficient ($ApoE^{-/-}$) mice. These studies have indicated that intermittent hypoxia induces vascular functional changes which not only precede the atherosclerotic process, but also regulate endothelium-dependent impairment of vasodilation and increase vasoconstrictor responses (Dematteis et al., 2008; Phillips et al., 2004). ET-1 appears to play a major role in this process. Intermittent hypoxia increases plasma and tissue levels of ET-1 in animals and treatment with the dual endothelin-receptor antagonist bosentan prevents the intermittent hypoxia-induced vascular remodeling (Branco et al., 2016; Lefebvre et al., 2006). The detrimental cardiovascular effects of chronic intermittent hypoxia also extend to the heart. In rat model of sleep apnea, intermittent hypoxia increases the susceptibility rats to myocardial infarction, to enlarge infarct size and to enhance the incidence of ischemic arrhythmias (Belaidi et al., 2009; Morand et al., 2018). Beyond these acute consequences, intermittent hypoxia has also shown to lead to cardiac remodeling with biventricular hypertrophy, cardiac fibrosis, dilatation and decreased stroke volume and hence, lending mechanistic insight into the development of heart failure in OSA (Hayashi et al., 2011; Li et al., 2016).

Mechanisms contributing to increased risk of developing above cardiovascular complications in OSA are complex and intertwine with each other. Best established among them are: increased sympathetic activation, altered vascular regulation, endothelial dysfunction, arterial hypertension, oxidative stress, and chronic systemic inflammation (Somers et al., 2008). OSA-induced oxidative stress and inflammation not only impair vascular function and structure, but also contribute to the development of atherosclerotic lesions. High levels of ROS in endothelial, vascular smooth muscle, and adventitial cells are known to initiate atherogenic processes (Harrison et al., 2003). ROS-activated proinflammatory transcription factors, such as NF-κB, stimulate the production of inflammatory cytokines that cause proliferation of vascular smooth muscle cells in the intimal layer (Brown, 2008) and adhesion of leukocytes to the endothelium (Adams et al., 2000). ROS promote insulin resistance through the activation of redox-sensitive signaling pathways, such as p38 mitogen-activated protein kinase (p38MAPK), c-Jun N-terminal kinase (JNK), IkB kinase (IKK), and extracellular signal-regulated kinase (ERK), all of which increase serine phosphorylation of insulin receptor substrate proteins and impaired insulin signaling (Evans et al., 2002; Nishikawa and Araki, 2007). ROS also interferes with PtdIns 3K/AKT signaling to decrease both NOS expression and glucose metabolism, but the MAPK pathway is stimulated. Insulin resistance results in compensated hyperinsulinemia. All these effects not only contribute to disruption of insulin-stimulated glucose metabolism and increased risk of CVD, but also atherosclerosis in type 2 diabetics mellitus and OSA (DeFronzo, 2010). The pathogenesis of OSA-mediated atherosclerosis is complicated. It involves changes in hormonal secretion, and a proinflammatory state. Oxidative stress induces inflammation; inflammation promotes ROS production. Oxidative stress and inflammation impair pancreatic β-cell activity and insulin sensitivity. These effects form a vicious cycle that increases the complexity of the causes of OSA-accelerated atherosclerosis.

Number of studies have shown increased sympathetic activation following intermittent hypoxia, both in animal and human models of OSA leading to hypertension (Beaudin et al., 2017; Foster et al., 2007). OSA patients have increased sympathetic traffic to peripheral blood vessels and cardiac sympathetic drive (Narkiewicz et al., 1998). It leads to upregulation of renin–angiotensin–aldosterone pathway and downregulation of NO synthesis (Prabhakar et al., 2001). About 69% of patients with coronary artery disease have OSA (Cepeda-Valery et al., 2014). CPAP, the gold standard therapy for OSA, has not only been shown to reduce cardiovascular mortality and improve cardiovascular risk factors including BP, lipid profile, and insulin sensitivity (Anandam et al., 2013; Nadeem et al., 2014; Yang et al., 2013), but also reduces sympathetic activity as well as increases arterial baroreflex sensitivity (Maser et al., 2008; Marrone et al., 2011).

Obstructive sleep apnea and hypertension

OSA may also increase the risk of CVD through reduction in cerebral blood flow, impaired endothelial function, increased platelet activation, inflammation, and oxidative stress related to the intermittent hypoxemia–reoxygenation and arousals associated with increased sympathetic tone (Ramos et al., 2015). The pathophysiology of hypertension in OSA is complex and multifactorial. Many factors including sympathetic tone, peripheral vasoconstriction, increased renin–angiotensin–aldosterone activity, and altered baroreceptor reflexes are involved in the pathophysiology of OSA. Key factors relating to OSA that promote the potential mechanisms of BP elevations include intermittent hypoxia and recurring microarousals from sleep, both of which are central features of severe sleep-disordered breathing events. Intermittent hypoxia has long been recognized as one of the mechanisms in the development of OSA-related hypertension, with early animal studies showing a robust association between intermittent hypoxia as a marker of OSA and both acute and persistent increases in BP (Fletcher, 2001). However, recent studies indicate that fragmented sleep also represents a distinct trigger factor for elevated BP in that sleep fragmentation and frequent arousals in patients with periodic leg movements have also been found to be associated with elevated BP (Koo et al., 2015). During OSA, episodes increase in catecholamine levels are associated with increase in heart rate and BP that is most prominent during the postapneic hyperventilation, soaring as high as 240/130 mmHg (Somers et al., 1993b). This increase in BP is associated with cardiovascular events such as coronary spasm, angina, and arrhythmias

(Okcay et al., 2008). Collective evidence suggests that there is a close association between OSA and hypertension (Peppard et al., 2000; Lavie et al., 2000; Somers et al., 2008). Furthermore, OSA is also linked with numerous other comorbidities such as insulin resistance, type 2 diabetes, and cardiovascular diseases, and this information continues to grow (Somers et al., 2008). On the basis of this information, it is now believed that intermittent hypoxia during OSA is a major risk factor for type 2 diabetes, MetS, systemic arterial hypertension, coronary artery disease, heart failure, and stroke (McNicholas and Bonsigores, 2007; Bradley and Floras, 2003). Processes, which link OSA with cardiovascular disease, include hypoxemia, hypercapnia and rapid recurrent changes in intrathoracic pressure, triggering a wide variety of autonomic and hemodynamic responses (Bradley and Floras, 2003; Lanfranchi and Somers, 2001). Collective evidence suggests that the link between OSA and hypertension has been well established. However, the mechanisms underlying the onset and progression of the arterial BP elevation are not fully understood. It has been proposed that chronic intermittent hypoxia not only produces oxidative stress, inflammation, and sympathetic overflow, but also endothelial dysfunction and hypertension (Lévy et al., 2008; Garvey et al., 2009; Iturriaga et al., 2009). In addition, new evidences suggest that the CB is involved in generation of autonomic and cardiovascular and ventilatory alterations elicited by chronic intermittent hypoxemia (CIH) (Iturriaga et al., 2014, 2017; Prabhakar et al., 2015). The cycles of hypoxia–reoxygenation produce oxidative stress in the CB and enhance its chemosensory responsiveness to hypoxia. The enhanced CB chemosensory drive leads to sympathetic hyperactivation of the sympathoadrenal axis and the renin–angiotensin system (Iturriaga et al., 2009; Prabhakar et al., 2012).

Obstructive sleep apnea and metabolic syndrome

MetS is a multifactorial condition characterized by a variety of symptoms such as obesity with abundant visceral fat, dyslipidemia, insulin resistance, hypertension, fatty liver disease, atherosclerosis, and type 2 diabetes (Fig. 4.6). The prevalence of MetS among adults in the United States is estimated to be 25%; however, prevalence increases with age such that an estimated 40% of adults aged 65 years and older have the condition (Scuteri et al., 2005). Like the prevalence of diabetes, the prevalence of MetS has increased considerably from 2001 to 2013 due to long-term

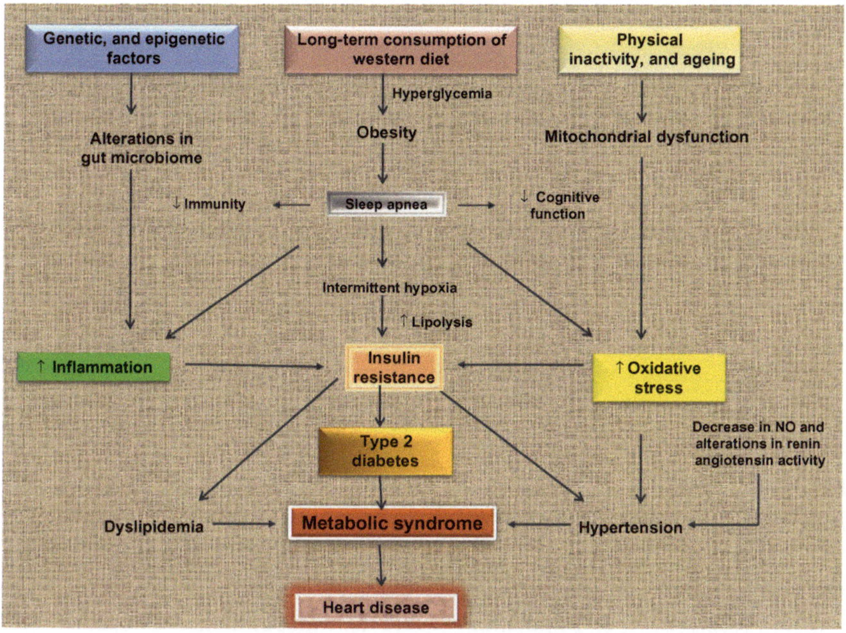

Figure 4.6 Contribution of sleep apnea and insulin resistance in the pathogenesis of heart disease.

consumption of western diet and lack of physical activity (Khosravi-Boroujeni et al., 2017). These findings have been attributed to aging, life-style changes, population growth, obesity, and decline in physical activity. Central obesity has been labeled as a critical component of the MetS.

It is becoming increasingly evident that there is a strong correlation between MetS and OSA. Thus sleep disruption can contribute not only to autonomic imbalance, but also to insulin resistance (Coughlin et al., 2004; Levy et al., 2011). Moreover, chronic sleep deprivation resulting from sleep-disordered breathing or behavioral causes can lead to excessive daytime sleepiness and lethargy, which in turn can contribute to increasing obesity (Tasali and Van Cauter, 2002). The complex nature of this multifactorial system makes it difficult to establish the causal links that connect one factor to another. There is significant evidence for the correlation between MetS and OSA. It is clear that OSA independently leads to insulin resistance, a major component of type 2 diabetes and MetS pathogenesis (Fig. 4.6) (Nannapaneni et al., 2013). MetS patients show an increase in epinephrine, norepinephrine, and/or cortisol leading to an increase in gluconeogenesis and a decrease in skeletal muscle uptake of glucose (Trombetta et al., 2013).

This is due to the induction of intermittent hypoxemia in OSA patients. This process may lead to sympathetic excitation, decrease in insulin sensitivity, and reduction in glucose uptake resulting in stimulation of hepatic gluconeogenesis. As mentioned above, increase in ROS levels results in activation of NF-κB with subsequent production of proinflammatory mediators such as TNF-α, IL-1β, IL-6, and interleukin 8 (IL-8) (Ryan, 2017). In addition, in MetS and OSA patients, oxidative stress is linked to inflammatory processes with sympathetic signaling (Gileles-Hillel et al., 2017). Obesity, on the other hand, leads to hypertrophied, dysfunctional adipocytes (Gami et al., 2003). These adipocytes attract cytotoxic T cells (CD8 + T) as well as macrophages with the latter produce proinflammatory cytokines like IL-6, TNF-α along with expression of inducible nitric oxide synthase (iNOS) (Gileles-Hillel et al., 2017; Gami et al., 2003). The production of FFAs by lipolysis not only leads to impairment in insulin-signaling pathway, but also induction of insulin resistance with remarkable metabolic consequences through the involvement of HIF-1α (Ryan, 2017). Converging evidence suggests that in MetS patients, the prevalence of moderate to severe OSA is very high (~60%). In this population, OSA is independently associated with increased glucose and triglyceride levels as well as markers of inflammation, arterial stiffness, and atherosclerosis. A recent randomized, controlled, crossover studies have shown that effective treatment of OSA with CPAP for 3 months significantly reduces several components of the MetS, including BP, triglyceride levels, and visceral fat. It is also proposed that OSA may be linked to different components of MetS through intermittent hypoxemia causing oxidative stress and visceral adipose tissue inflammation (Ryan, 2017).

Several investigators have indicated that development of obesity is more crucial for the induction of OSA than either age or gender (Jalilolghadr et al., 2016). For every percent in weight reduction, there is a 3% reduction in the AHI (Ip et al., 2002). Increase in BMI also correlates with an increase in incidences of OSA (Young et al., 2005). For the onset of MetS, both abdominal and central obesities are important. Central obesity is linked to higher leptin production with leptin resistance leading to increase probability of developing OSA. Metabolic abnormalities increase the chance of upper airway collapsibility. In MetS patients, neck circumference is a better predictor of OSA than general obesity. Central obesity impact is greater on the upper airway function when compared to peripheral obesity, as stated earlier (Young et al., 2005). Patients with obesity and OSA have approximately a 67% more total neck fat compared to the normal person. This leads to a smaller upper airway area and greater compression on said airway while sleeping. Central obesity is more closely related to fat depositions in the neck, unlike peripheral obesity. This leads to a more notable narrowing of the upper airway while at sleep (Mortimore et al., 1998). Several mechanisms may explain the observed associations between OSA and vascular disease in MetS patients. Hyperglycemia and OSA result in the activation of Aldose reductase, protein kinase C, advanced glycation end products, and increased oxidative and nitrosative stress that can lead to insulin resistance, increased inflammation, cellular and endothelial dysfunction, and vascular disease (Fig. 4.6) (Allahdadi et al., 2008; Jelic et al., 2010). In recent studies also indicate that OSA may be related to increased oxidative stress, nitrosative stress, and impaired microvascular complications in patients with MetS (Tahrani et al., 2012, 2013). Collective evidence suggests that induction of several processes is common between OSA and MetS. These processes include similar changes in upper way pathophysiology, peaking of both pathological conditions at the middle age, and induction of oxidative stress and inflammation.

Obstructive sleep apnea and stroke

Stroke is caused by the disruption of the blood flow into a brain region and the consequent oxygen and nutrient deprivation, resulting in cell death and severe brain damage. It is the fifth most common cause of death and the most frequent cause of permanent disability in adults worldwide. Ischemic stroke, which is caused by the obstruction of a blood vessel by a thrombus, represents 87% of all strokes (Benjamin et al., 2018). Stroke-mediated injury impairs the neurologic functions by breaking down the cellular and subcellular integrity not only through the induction of excitotoxic injury, but also due to reduction in ATP synthesis and marked decrease in glucose and oxygen levels. This injury is not only mediated by the overactivation of glutamatergic receptors, large Ca^{2+}-influx, and alterations in redox status, but also by induction of oxidative stress and neuroinflammation (Farooqui, 2018). These processes are mediated by the stimulation of phospholipases A_2, C, and D (PLA_2, PLC, and PLD), calcium/calmodulin-dependent kinases, MAPKs such as ERK, p38, and JNK, NOSs, calpains, calcineurin, and endonucleases. Stimulation of these enzymes bring them in contact with appropriate substrates and modulates cell survival/degeneration through the generation of proinflammatory cytokines and chemokine by injured neurons (Farooqui, 2018). These processes promote the expression of cell adhesion molecules, P-selectin,

E-selectin, and ICAM-1 on the endothelial cell surface. These molecules promote the infiltration of leukocytes into the brain parenchyma and facilitate the clearance of debris in the infarct area. Moreover, endothelial cells increase the expression of chemokines in order to guide leukocytes to the site of injury (Mizuma and Yenari, 2017). However, there is evidence showing that beside their beneficial role, infiltrating immune cells also impair the ischemic brain by producing cytotoxic mediators that can extend the inflammatory response and increase brain damage (Simats et al., 2016; Farooqui, 2018). Traditional risk factors such as age, male sex, ethnicity, hypertension, and atrial fibrillation explain 60%–80% of the risk of stroke. OSA may lead to stroke through its associations with potent vascular risk factors, such as hypertension, diabetes mellitus, obesity, and atrial fibrillation. OSA may also increase stroke and CVD risk through reduction in cerebral blood flow, altered cerebral autoregulation, impaired endothelial function, increased platelet activation, systemic inflammation, and oxidative stress related to the intermittent hypoxemia–reoxygenation and arousals associated to increased sympathetic tone (Fig. 4.7) (Ramos et al., 2015). Furthermore, chronic intermittent hypoxia markedly increases ET-1 in cerebral blood vessels. Treatment with agonist of endothelin type A receptor reduces oxidative stress. This effect can be abolished by neocortical application of the endothelin type A receptor antagonist BQ123 supporting the view that intermittent hypoxia alters key regulatory mechanisms of the cerebral circulation through involvement of ET-1, and increase in ET-1 levels may increase stroke risk in patients with sleep apnea by reducing cerebrovascular reserves and increasing the brain's susceptibility to cerebral ischemia (Capone et al., 2012). It should be noted that patients with acute ischemic stroke and OSA have worse outcomes than those without OSA, including increased risk of early neurological worsening, mortality, decreased functional recovery, nonfatal cardiovascular events, longer hospital stays, delirium, and depressed mood (Sahlin et al., 2008; Sandberg et al., 2001). Because early neurologic worsening is likely a reflection of injury in the ischemic penumbra, some investigators have proposed that early identification and treatment of OSA may prevent clinical deterioration (Sahlin et al., 2008; Sandberg et al., 2001). Furthermore, the consequences of untreated sleep disorders (cognitive dysfunction, altered mood, sleepiness, and fatigue) may impede stroke rehabilitation, lengthen hospital stay, and influence stroke outcomes and stroke recurrence (Hermann and Bassetti, 2009). Following stroke-mediated injury, induction of oxidative stress and inflammation may

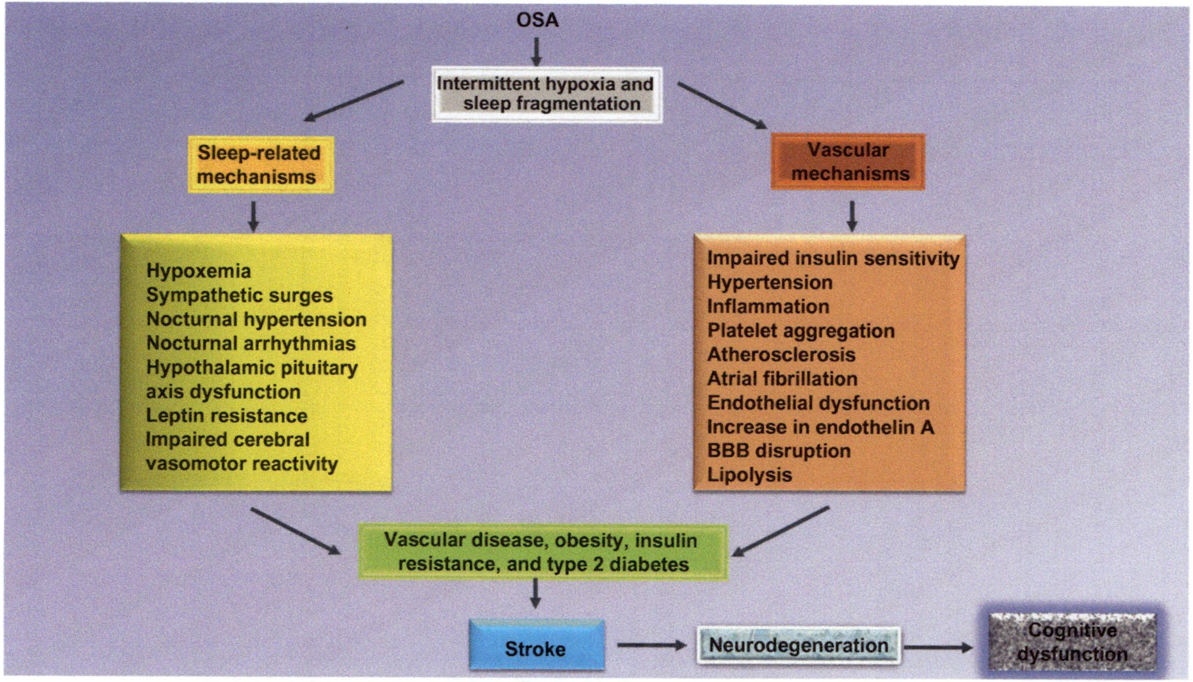

Figure 4.7 Contribution of sleep apnea and insulin resistance in the pathogenesis of stroke.

contribute to free radical generation, release of proinflammatory (TNF-α, IL1β, IL6, and ICAM) and prothrombotic mediators (TNF-α and platelet activating factor) resulting in enhanced activity of thrombocytes and coagulation factors leading to hypercoagulable state, and progression of atherosclerosis. These processes may trigger atherothrombotic and embolic strokes. Onset of stroke is also promoted by surges in sympathetic nervous system activity, alterations in BP, and development of insulin resistance. Thus OSA may increase risk factors for stroke and directly contribute to pathophysiological stresses implicated in stroke. It is proposed that variability in OSA, which includes a relatively high prevalence in young African Americans (Rosen et al., 2003) and in older women (Redline et al., 2003), roughly parallels disparities in stroke prevalence in the population. This observation supports the plausibility that unrecognized OSA-hypopnea may explain a portion of the population variability in stroke. The treatment of OSA has been considered an important intervention for reducing the morbidity and mortality associated with stroke and CVD (Barbe et al., 2012). However, treatment of OSA has not consistently reduced cardiovascular risk (McEvoy et al., 2016); results partly explained by methodological limitations, such as suboptimal adherence to positive airway pressure therapy. Importantly, some limitations may lie on the need to better identify the subset of individuals who respond more favorably to OSA therapy, with the potential of a greater reduction in adverse outcomes. Recent studies by Chen et al. (2015) have indicated that the severity of OSA is positively associated with the total antioxidant capacity (TAC) and CRP in ischemic stroke subjects. Specifically, the results of the study show that the levels of CRP and TAC are positively correlated with the oxyhemoglobin desaturation index. Furthermore, the TAC levels are negatively correlated with mean arterial oxygen saturation in stroke survivors with severe OSA (Chen et al., 2015). However, what has been lacking to date is examination of the relationships among the biomarkers, sleep, and functional and health-related outcomes after stroke rehabilitation. Thus more studies are needed on the involvement of OSA in the stroke injury.

Obstructive sleep apnea and its relationship with various diseases

Many cross-sectional and longitudinal studies have indicated that intermittent hypoxia in OSA as an independent risk factor for the development of a variety of adverse metabolic disease states, including obesity, hypertension, insulin resistance, type 2 diabetes, nonalcoholic fatty liver disease, dyslipidemia, atherosclerosis, and early mortality (Buxton et al., 2010; Buxton and Marcelli, 2010; Grandner et al., 2010). In addition, sleep disruption is tightly associated with an increased susceptibility to a broad range of visceral (cardiovascular), psychiatric (depression), and neurodegenerative diseases [stroke, AD, Parkinson's disease (PD), and dementia]. Among these diseases, sleep disturbances may precede the onset of more typical symptoms, in some cases by decades (Sterniczuk et al., 2013; Lim et al., 2013). The most striking example is REM behavior disorder (RBD), a condition in which normal muscle paralysis is lost during REM sleep. Over 80% of all RBD patients will eventually develop PD or another synucleinopathy, often decades later (Schenck et al., 2013). Mouse models of AD, PD, and Huntington's disease pathology also exhibit sleep abnormalities (Sterniczuk et al., 2013). Sleep deprivation increases cerebrospinal fluid (CSF) markers of neuronal injury and alters plasma markers of inflammation in humans (Benedict et al., 2014; Frey et al., 2007), and induces the unfolded protein response in the brain of mice, indicating endoplasmic reticulum stress and potential neuronal injury (Naidoo et al., 2008). Thus inadequate sleep could prime the brain for neurodegeneration by promoting processes such as inflammation and synaptic damage which exert pathogenic effects across diseases. Studies on the effect of sleep deprivation on memory indicate that the hippocampus is particularly vulnerable to sleep loss. The molecular signaling pathways that modulate changes in synaptic efficacy observed after sleep deprivation are not fully understood. However, multiple studies have indicated that changes in *N*-methyl-D-aspartate and gamma-aminobutyric acid (GABA) receptors signaling may create the molecular milieu conducive to both functional and structural synaptic plasticity changes during LTP induction and attenuation (Poe et al., 2010; Havekes et al., 2012).

Nasal CPAP has long been the mainstay of therapy for OSA, but definitive studies demonstrating the efficacy of CPAP in improving metabolic outcomes, or in reducing incident disease burden, are lacking. Despite its low compliance (Hong et al., 2017), CPAP treatment is known to improve BP (Somers et al., 1995), attenuates heart failure (Arias et al., 2005), improves cardiac function (Otto et al., 2007), and can significantly reduce mortality due to cardiovascular diseases (Partinen et al., 1988). CPAP can also improve AHI (Issa and Sullivan, 1986) and blood oxygenation in individuals presenting predominantly with central sleep apnea (Bradley et al., 2005). Recent studies have indicated that in most randomized clinical trials, CPAP treatment improves short-term insulin resistance (Salord et al., 2016). However, the impact of CPAP on long-term insulin resistance is not known (Gileles-Hillel et al., 2016). In non-obese, HFD fed rats, treatment with metformin not only increases insulin sensitivity, but also prevents the development of OSA independently of body weight (Ramadan et al., 2007). These observations are tempting to speculate that future studies on effects and management of OSA particularly with respect to limiting OSA-related metabolic dysfunction should be performed on targeting mechanisms of OSA-induced beneficial and adverse health outcomes.

Obstructive sleep apnea and Alzheimer's disease

AD, the most common form of dementia, is a complex disease characterized by the accumulation of extracellular β-amyloid (Aβ) plaques (senile plaques) and intracellular neurofibrillary tangles (NFTs) composed of Tau amyloid fibrils (Hardy, 2009) leading to the loss of synapses and degeneration of neurons in cortical and subcortical areas and hippocampus (Fig. 4.8) resulting in a loss of cognitive brain functions, along with progressive impairment of activities of daily living. Multiple converging pathways are known to influence the relationship among sleep, circadian disruption, and AD. These mechanisms include changes in HPA axis, activation of microglia, levels of melatonin, levels of orexin, and ERK/MARK signaling pathways (Fig. 4.8).

Cognitive performance varies throughout the 24-hour cycle day and is high during the day about 24 hours after waking except for a dip in the afternoon and low at night (Burke, 2015). Several cognitive processes are affected by circadian timing (Krishnan and Lyons, 2015). The suprachiasmatic nucleus controls cognitive function indirectly through its effects on sleep and wakefulness and perhaps through more direct effects (Wright et al., 2012). Circadian misalignment, where the timing of daily activities—such as sleep and wake—are not aligned to the endogenous circadian timing, results in impaired cognitive function. Chronic jet lag leads to cognitive performance and is associated with increased cortisol levels (Cho, 2000). Experimental jet lag in rodents result in long-term cognitive deficits (Gibson, 2010). Thus when circadian rhythm is misaligned and circadian amplitude is lowered in old age and dementia patients, it also coupled with cognitive impairment (Smarr, 2014). Cognitive function is

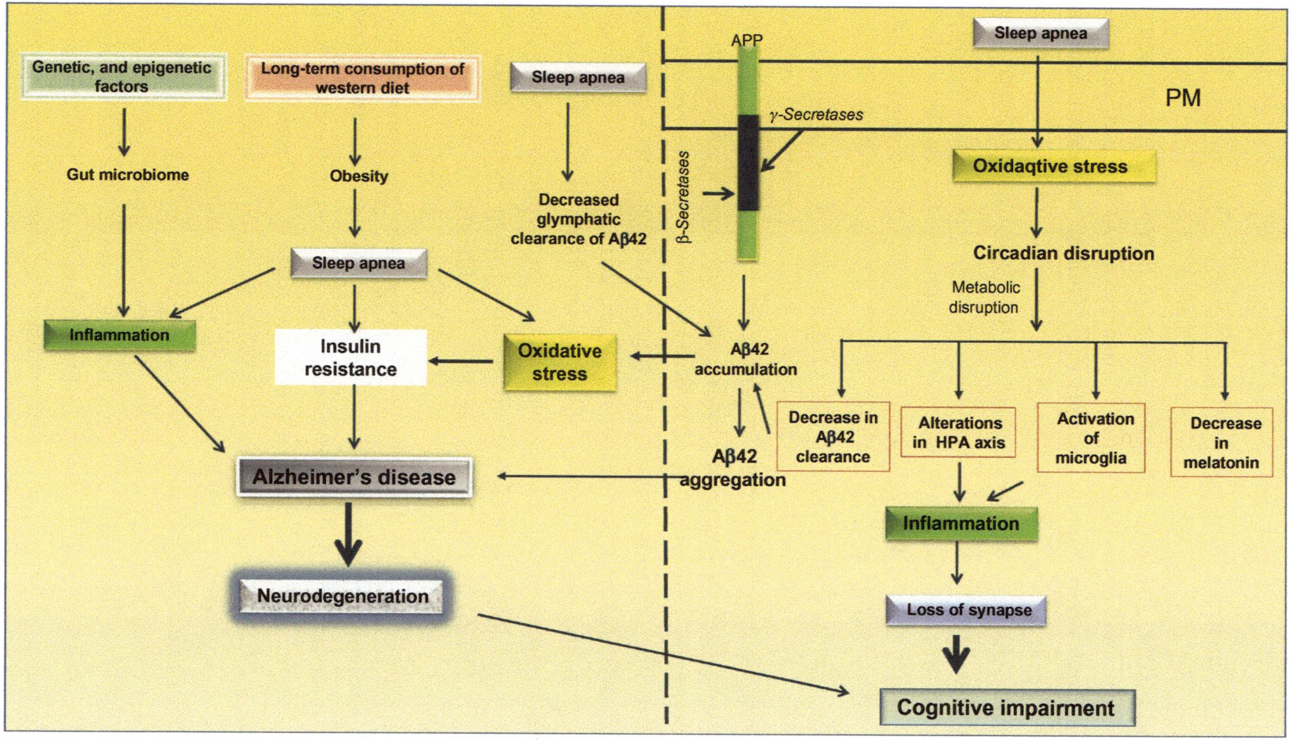

Figure 4.8 Contribution of sleep apnea and insulin resistance in the pathogenesis of Alzheimer's disease. Aβ, β-amyloid; PM, plasma membrane.

composed of multiple cognitive domains including memory, calculation, spatial orientation, structure ability and executive ability, language comprehension and expression (Dolan, 2002). How these factors ultimately contribute to memory impairments and cognitive deficits that clinically characterize the disease remains elusive (Hardy, 2009; Farooqui, 2017; Selkoe and Hardy, 2016). Although progressive deteriorations of memory, language, and intellect are the classic hallmarks of AD, sleep disturbance is a common and often highly disruptive behavioral symptom associated with AD. Epidemiological studies have reported that up to 45% of patients with AD have sleep disturbances (McCurry et al., 2000). These neurobehavioral symptoms may appear at an early stage of the AD process but seem to be usually correlated to a more severe cognitive decline (McCurry et al., 2000). The major risk factors for the development of AD are aging and a family history of the disease. More than 95% of all AD cases occur in individuals over the age of 60 years and are defined as sporadic AD or late-onset AD. Less than 5% of cases are defined as familial AD and can be seen in patients as young as 30 years of age (Farooqui, 2017). In over 90% of cases, AD begins after the age of 65 as a sporadic form of dementia (Ashraf et al., 2014). Many cases of AD are found to coexist with other chronic diseases like type 2 diabetes, MetS, and cardiovascular diseases (Ashraf et al., 2014; Farooqui, 2017).

Production and accumulation of Aβ in the brain is closely linked to the 24-hour sleep–wake cycle, with high extracellular levels during wakefulness and low extracellular levels during sleep (Kang et al., 2009; Roh et al., 2012). A major driver and modulator of Aβ production appears to be neuronal activity, which is higher during wakefulness as compared with sleep. This hypothesis is supported by the observation that unilateral vibrissal stimulation increases, while unilateral vibrissal removal decreases, interstitial fluid (ISF) levels of Aβ in the contralateral barrel cortex of transgenic mice (Tg2576) (Bero et al., 2011). In humans, ISF Aβ concentrations have been shown to increase in patients with acute brain damage. The improvement in neurological status results in decrease in ISF Aβ concentrations (Brody et al., 2008).

Recent studies have indicated that there is an important relationship between disrupted sleep and brain glymphatic system in AD (Mendelsohn and Larrick, 2013; Lee et al., 2015; O'Donnell et al., 2015; Krueger et al., 2016). Glymphatic system consists of paravascular channels located around blood vessels of the brain. CSF flows along paraarterial space, reaches the capillary bed, and penetrates into the brain parenchyma, where it gets mixed with ISF. After the collection of metabolic waste, CSF moves to paravenous space and then enters cervical lymphatic vessels (Ratner et al., 2015). Converging evidence suggests that an important function of the glymphatic system is the removal of metabolites and neurotoxic compounds, including soluble Aβ from the CNS parenchyma (Kyrtsos and Baras, 2015; Bakker et al., 2016; Simon and Iliff, 2016). Microdialysis studies have indicated that more than half of Aβ can be removed from the brain through the glymphatic system (Iliff et al., 2012). It seems that sleep may control activities and function of glymphatic system during natural sleep. Moreover, there is a marked increase of the brain's interstitial space as compared with wakefulness due to the shrinkage of astroglial cells (Mendelsohn and Larrick, 2013; Xie et al., 2013; Kress et al., 2014; O'Donnell et al., 2015). The enlargement of the extracellular space accelerates clearance processes. It has been found that in mice, the clearance of the Aβ during sleep is twofold faster than during wakefulness (Xie et al., 2013). Another other animal study has indicated that the speed of clearance through the glymphatic system depends on the body posture (Lee et al., 2015). The glymphatic transport is the most efficient in the lateral position, which is the most common during sleep.

Microdialysis on interstitial concentration of Aβ in mice has indicated that wakefulness correlates with Aβ levels (Kang et al., 2009), and synaptic activity is associated with increased release of Aβ (Roh et al., 2012). A reason for this increase in Aβ with higher synaptic activity during wakefulness may be due to increase in endocytosis (Cirrito et al., 2008), which is essential for Aβ production (Rajendran et al., 2006; Rajendran and Paolicelli, 2018; Ben Halima et al., 2016). Another explanation for this observation is that sleep increases Aβ clearance (Xie et al., 2013) and that sleep deprivation affects both Aβ production and clearance, thus mechanistically explaining the higher amyloid deposition in sleep-deprived patients (Osorio et al., 2011; Minakawa et al., 2017, 2019). Sleep also affects microglial activity (Ingiosi et al., 2013; Bellesi et al., 2017) and impaired sleep may affect microglial clearance of Aβ. Indeed, addition of orexin, the crucial molecule that regulates the sleep—wake cycle in mammals, to BV-2 microglia cells impaired microglial phagocytic activity and Aβ clearance (An et al., 2017; Xie et al., 2013). On the basis of above studies, it is proposed that production and accumulation of Aβ may be linked with the sleep—wake cycle in the development and progression of AD (Fig. 4.8) (Lim et al., 2014; Musiek and Holtzman, 2016). The molecular mechanism contributing to Aβ production with sleep—wake cycle is not fully understood. However, stress-mediated activation of HPA axis promotes AD pathogenesis by promoting the cleavage of the amyloid precursor protein (APP) by β-secretase enzyme 1 (BACE1) followed by γ-secretase (Park, 2015). This pathway leads to the production of Aβ42, a synaptotoxic peptide, which is the main constituent of neuritic plaques. The production of Aβ42 initiates a cascade of processes resulting in accumulating intracellular tau and ultimately neuronal demise (Selkoe and Hardy, 2016; Farooqui, 2017). In addition, based on animal model studies, it is proposed that astrocytes play a central role in linking sleep—wake cycle with the pathogenesis of AD through the involvement of astrocyte—neuron—lactate—shuttle (ANLS) (Pellerin and Magistretti, 2012; Vanderheyden et al., 2018) and the disruption of ANLS neurometabolic coupling by astrocytes may be associated with sleep, Aβ, and AD pathophysiology.

On the basis of Aβ clearance studies, it is reported that the levels of Aβ are altered in both early and late forms of AD (Tarasoff-Conway et al., 2015). It is proposed that there is a link between impaired glymphatic system function and onset of AD. Several studies have indicated that here is a diurnal oscillation of the Aβ level in the brain ISF (Musiek, 2015) and endogenous neuronal activity and function not only influences the regional concentration of the Aβ in the ISF (Bero et al., 2011), but also reduces neuronal activity in some stages of sleep. This may also cause the oscillations of the Aβ concentrations. SWS with periodic neuronal hyperpolarization and diminished neuronal firing in some brain regions can be associated with decreased Aβ production (Musiek, 2015). Thus alterations in sleep quality may contribute to the onset and progression of the AD both through impaired glymphatic clearance and disturbances in the Aβ production in case of disordered SWS.

As mentioned above, an important factor contribute to the sleep—wake cycle is hypothalamic neuropeptide orexin-A (hypocretin 1). Levels of this neuropeptides are increased in during wakefulness (Kiyashchenko et al., 2002). In transgenic APPswe (Tg2576) mice, intracerebroventricular administration of orexin at the beginning of the light (i.e., inactive) period can acutely increase both wakefulness and Aβ levels in ISF. Conversely, intracerebroventricular treatment over 24 hours with a dual orexin receptor antagonist (almorexant) decreases Aβ ISF levels supporting the role of orexin in Aβ accumulation (Kang et al., 2009). Moreover, daily treatment with almorexant for 8 weeks reduces the formation of Aβ plaques in several brain regions in APPswe/PS1dE9 mice

(Kang et al., 2009). These basic and translational studies are supported by epidemiological studies showing that deficient or fragmented sleep in cognitively normal individuals is a risk factor for the future development of symptomatic AD (Sterniczuk et al., 2013; Lim et al., 2013; Roh et al., 2014), and that people with amyloid plaque pathology develop detectable declines in sleep efficiency prior to the onset of cognitive symptoms (Ju et al., 2013). Human studies have also shown a correlation between concentrations of orexin in the CSF and that of tau protein, another pathologic hallmark of AD and the primary constituent of NFTs. High concentrations of CSF orexin-A are associated with increased amounts of phophorylated tau (Osorio et al., 2016), but not with Aβ levels in CSF (Wennstrom et al., 2012). Another study has indicated that in amyloid transgenic mice, knocking out (APP/PS1dE9/OR$^{-/-}$) the orexin gene results in decreased wakefulness and a subsequent reduction in amyloid pathology (Roh et al., 2014). These findings have not been confirmed in humans, where studies on orexin remain controversial. This suggests that a critical understanding of the molecular and cellular mechanisms responsible for this feedback relationship between sleep, Aβ, and human AD pathology is still lacking, and more studies are urgently required on this important topic.

Obstructive sleep apnea and depression

As mentioned above, OSA is a sleep-related breathing disorders characterized by repetitive, partial, or complete collapse of the upper airway during sleep, causing impaired gaseous exchange, and sleep disturbance. Several cross-sectional studies have indicated that OSA contribute to increased levels of total cholesterol, LDL, and triacylglycerol, whereas others report no such relationships (Tsioufis et al., 2007; Tokuda et al., 2008; Drager et al., 2010). In contrast, depression is one of the most common neuropsychiatric disorder by behavioral changes such as sleep disturbances, psychomotor retardation or agitation, fatigue, feelings of worthlessness or guilt, and psychomotor changes leading to diminished cognitive functioning, loss of energy, concentration difficulties/indecisiveness, irritability, and low self-esteem (Davidson et al., 2002). The pathogenesis of depression involves alterations in levels of neurotransmitters, neuropeptide, and cytokines (dopamine, norepinephrine, serotonin, vasopressin, and inflammatory cytokines), decrease in magnesium, defects in neurogenesis, decrease in synaptic plasticity, mitochondrial dysfunction, and alterations in redox imbalance (Fig. 4.9) (Sullivan et al., 2000; Anisman and

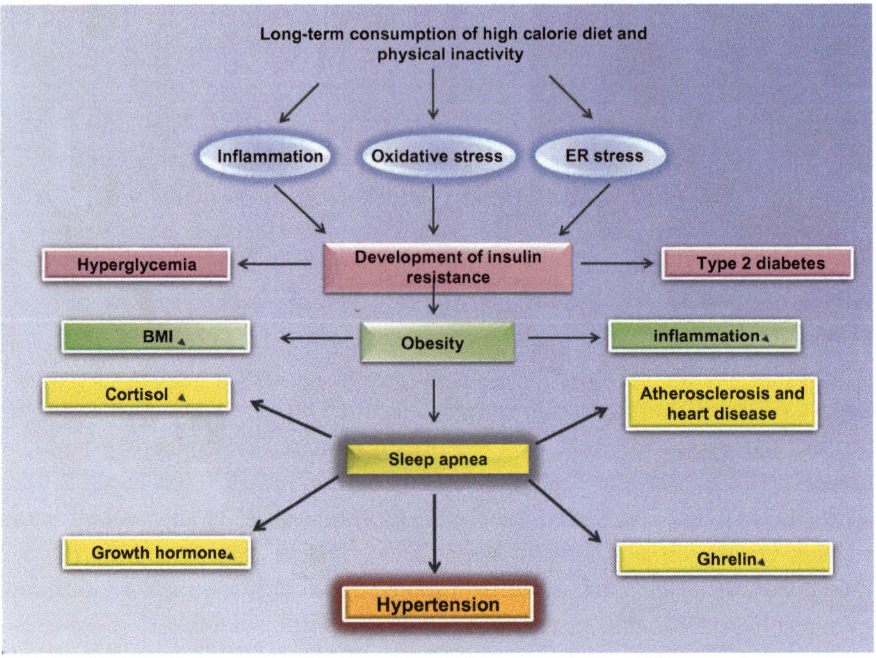

Figure 4.9 Contribution of sleep apnea and insulin resistance in depression.

Matheson, 2005; Wurtman, 2005; Dowlati et al., 2010). The relationship between depression and OSA is very complex. The linkage between OSA and depression may not only involve neurochemical changes in sleep fragmentation and intermittent hypoxia and hypoxemia, but also alterations in levels of serotonin, norepinephrine, and GABA (Ejaz et al., 2011). Some studies have reported significantly higher rates of depression among OSA patients (Sharafkhaneh et al., 2005; Ejaz et al., 2011). The underlying mechanisms supporting the association between OSA and depression are not fully understood. However, several possibilities have been reported. First, sleep fragmentation or oxygen desaturation during sleep in OSA patients may impact the presentation of mood symptoms, although the results regarding this possibility are mixed. In a randomized controlled trial, hypoxia in OSA was shown to potentially associate with depression, as a significant reduction of depressive symptoms was observed for patients with OSA receiving oxygen therapy (Bardwell et al., 2007). This report is supported by a recent imaging study, which indicate that hypoxemia linked with OSA may play a part in affecting mood (Schroder and O'Hara, 2005). Second, it is possible that there is an overlap in signaling pathway involving proinflammatory markers and neurotransmitters associated with the pathogenesis of OSA and depression. Thus OSA and depression are associated not only with increased levels of proinflammatory cytokines (IL-6 and TNF-α) (Vgontzas et al., 2000; Irwin and Miller, 2007), but also with alterations in several excitatory and inhibitory neurotransmitters (serotonin, norepinephrine, and GABA), which may also be involved in both the sleep–wake cycle and regulation of depression. Involvement of insulin resistance–related risk factors (e.g., obesity, cardiovascular disease, and MetS) is another plausible explanation. Furthermore, both OSA and depression are associated with increased morbidity and mortality, and decreased quality of life. Finally, as depression was prevalently observed in patients with chronic medical diseases (Katon et al., 2007), OSA may decrease patients' quality of life, further leading to other chronic diseases including depression (Harris et al., 2009).

Conclusion

OSA is a sleep disorder characterized by repetitive upper airway collapse that leads to cycles of intermittent hypoxia, sleep fragmentation, and/or intrathoracic pressure swings. OSA is associated glucose intolerance, insulin resistance, and type 2 diabetes, independently of other confounding factors, including age, obesity, and central adiposity. Pathogenesis of OSA involves exposure to intermittent hypoxia, which not only contribute to compromised whole-body glucose homeostasis, glucose tolerance, and muscle glucose uptake, but also induce insulin resistance by impairing secretion insulin from pancreatic β cells. Multiple mechanisms have been identified by which sleep apnea adversely affects cardiovascular and cerebrovascular structure and function. Epidemiological studies have indicated that OSA is associated with increases in the incidence and progression of not only type 2 diabetes, MetS, coronary heart disease, heart failure, and atrial fibrillation, but also stroke and AD. Oxidative stress, inflammation, and increase in sympathetic activity are always associated with the pathogenesis of sleep apnea, but their relative contribution in the pathogenesis sleep apnea is not known.

References

Abid, K., Trimeche, T., Mili, D., Msolli, M.A., Trabelsi, I., Nouira, S., et al., 2016. ANGPTL4 variants E40K and T266M are associated with lower fasting triglyceride levels and predicts cardiovascular disease risk in Type 2 diabetic Tunisian population. Lipids Health Dis. 15, 63.

Abrams, B., 2018. Comment on: use of blood biomarkers to screen for obstructive sleep apnea. Nat. Sci. Sleep 10, 313–315.

Abu-Farha, M., Al Khairi, I., Cherian, P., Chandy, B., Sriraman, D., Alhubail, A., et al., 2016. Increased ANGPTL3:4 and ANGPTL8/betatrophin expression levels in obesity and T2D. Lipids Health Dis. 15, 181.

Abu-Farha, M., Abubaker, J., Tuomilehto, J., 2017. ANGPTL8 (betatrophin) role in diabetes and metabolic diseases. Diabetes Metab. Res. Rev. 33, e2919.

Adams, M.R., Kinlay, S., Blake, G.J., Orford, J.L., Ganz, P., Selwyn, A.P., 2000. Atherogenic lipids and endothelial dysfunction: mechanisms in the genesis of ischemic syndromes. Annu. Rev. Med. 51, 149–167.

Allahdadi, K.J., Duling, L.C., Walker, B.R., Kanagy, N.L., 2008. Eucapnic intermittent hypoxia augments endothelin-1 vasoconstriction in rats: role of PKCδ. Am. J. Physiol. Heart Circ. Physiol. 294, H920–H927.

Almendros, I., Farre, R., Planas, A.M., et al., 2011. Tissue oxygenation in brain, muscle, and fat in a rat model of sleep apnea: differential effect of obstructive apneas and intermittent hypoxia. Sleep 34, 1127–1133.

Alonso-Fernandez, A., Garcia-Rio, F., Arias, M.A., et al., 2009. Effects of CPAP on oxidative stress and nitrate efficiency in sleep apnoea: a randomised trial. Thorax 64, 581–586.

An, H., Cho, M.H., Kim, D.H., Chung, S., Yoon, S.Y., 2017. Orexin impairs the phagocytosis and degradation of amyloid-beta fibrils by microglial cells. J. Alzheimer Dis. 58, 253–261.

Anandam, A., Patil, M., Akinnusi, M., Jaoude, P., El-Solh, A.A., 2013. Cardiovascular mortality in obstructive sleep apnoea treated with continuous positive airway pressure or oral appliance: an observational study. Respirology 18, 1184–1190.

Anisman, H., Matheson, K., 2005. Stress, depression, and anhedonia: caveats concerning animal models. Neurosci. Biobehav. Rev. 29, 525–546.

Arias, M.A., Garcia-Rio, F., Alonso-Fernandez, A., Mediano, O., Martinez, I., Villamor, J., 2005. Obstructive sleep apnea syndrome affects left ventricular diastolic function: effects of nasal continuous positive airway pressure in men. Circulation 112, 375–383.

Arnaud, C., Beguin, P.C., Lantuejoul, S., Pepin, J.L., Guillermet, C., Pelli, G., et al., 2011. The inflammatory preatherosclerotic remodeling induced by intermittent hypoxia is attenuated by RANTES/CCL5 inhibition. Am. J. Respir. Crit. Care Med. 184, 724–731.

Ashraf, G.M., Greig, N.H., Khan, T.A., Hassan, I., Tabrez, S., Shakil, S., 2014. Protein misfolding and aggregation in Alzheimer's disease and type 2 diabetes mellitus. CNS Neurol. Disord. Drug Targets 13, 1280–1293.

Aurora, R.N., Punjabi, N.M., 2013. Obstructive sleep apnoea and type 2 diabetes mellitus: a bidirectional association. Lancet Respir. Med. 1, 329–338.

Babapoor-Farrokhran, S., Jee, K., Puchner, B., Hassan, S.J., Xin, X., Rodrigues, M., et al., 2015. Angiopoietin-like 4 is a potent angiogenic factor and a novel therapeutic target for patients with proliferative diabetic retinopathy. Proc. Natl. Acad. Sci. U.S.A. 112, E3030–E3039.

Bakker, E.N., Bacska, B.J., Arbel-Ornath, M., Aldea, R., Bedussi, B., Morris, A.W., et al., 2016. Lymphatic clearance of the brain: perivascular, paravascular and significance for neurodegenerative diseases. Cell. Mol. Neurobiol. 36, 181–194.

Barbe, F., Duran-Cantolla, J., Sanchez-de-la-Torre, M., Martinez-Alonso, M., Carmona, C., Barcelo, A., et al., 2012. Effect of continuous positive airway pressure on the incidence of hypertension and cardiovascular events in nonsleepy patients with obstructive sleep apnea: a randomized controlled trial. JAMA 307 (20), 2161–2168.

Beaudin, A.E., Waltz, X., Hanly, P.J., Poulin, M.J., 2017. Impact of obstructive sleep apnoea and intermittent hypoxia on cardiovascular and cerebrovascular regulation. Exp. Physiol. 102, 743–763.

Belaidi, E., Joyeux-Faure, M., Ribuot, C., Launois, S.H., Levy, P., Godin-Ribuot, D., 2009. Major role for hypoxia inducible factor-1 and the endothelin system in promoting myocardial infarction and hypertension in an animal model of obstructive sleep apnea. J. Am. Coll. Cardiol. 53, 1309–1317.

Bellesi, M., de Vivo, L., Chini, M., Gilli, F., Tononi, G., Cirelli, C., 2017. Sleep loss promotes astrocytic phagocytosis and microglial activation in mouse cerebral cortex. J. Neurosci. 37, 5263–5273.

Benedict, C., Cedernaes, J., Giedraitis, V., Nilsson, E.K., Hogenkamp, P.S., Vågesjö, E., et al., 2014. Acute sleep deprivation increases serum levels of neuron-specific enolase (NSE) and S100 calcium binding protein B (S-100B) in healthy young men. Sleep 37, 195–198.

Ben Halima, S., Mishra, S., Raja, K.M., Willem, M., Baici, A., Simons, K., et al., 2016. Specific inhibition of beta-secretase processing of the Alzheimer disease amyloid precursor protein. Cell Rep. 14, 2127–2141.

Benjamin, E.J., Virani, S.S., Callaway, C.W., et al., 2018. Heart disease and stroke statistics—2018 update: a report from the American Heart Association. Circulation 137, e67–e492.

Bero, A.W., Yan, P., Roh, J.H., Cirrito, J.R., Stewart, F.R., Raichle, M.E., et al., 2011. Neuronal activity regulates the regional vulnerability to amyloid-beta deposition. Nat. Neurosci. 14, 750–756.

Berry, R.B., et al.,American Academy of Sleep Medicine 2012. The AASM Manual for the Scoring of Sleep and Associated Events: Rules, Terminology and Technical Specifications. American Academy of Sleep Medicine, Darien, IL.

Bonsignore, M.R., Zito, A., 2008. Metabolic effects of the obstructive sleep apnea syndrome and cardiovascular risk. Arch. Physiol. Biochem. 114, 255e60–255e60.

Bradley, T.D., Floras, J.S., 2003. Sleep apnea and heart failure. Part I: Obstructive sleep apnea. Circulation 107, 1671–1678.

Bradley, T.D., Logan, A.G., Kimoff, R.J., Series, F., Morrison, D., Ferguson, K., et al., Investigators C 2005. Continuous positive airway pressure for central sleep apnea and heart failure. N. Engl. J. Med. 353, 2025–2033.

Bardwell, W.A., Norman, D., Ancoli-Israel, S., Loredo, J.S., Lowery, A., Lim, W., et al., 2007. Effects of 2-week nocturnal oxygen supplementation and continuous positive airway pressure treatment on psychological symptoms in patients with obstructive sleep apnea: a randomized placebo-controlled study. Behav. Sleep. Med. 5, 21–38.

Brandt, J.A., Churchill, L., Rehman, A., Ellis, G., Memet, S., Israel, A., et al., 2004. Sleep deprivation increases the activation of nuclear factor kappa B in lateral hypothalamic cells. Brain Res. 1004, 91–97.

Branco, T., Tozer, A., Magnus, C.J., Sugino, K., Tanaka, S., Lee, A.K., et al., 2016. Near-perfect synaptic integration by Nav1.7 in hypothalamic neurons regulates body weight. Cell 165, 1749–1761.

Brody, D.L., Magnoni, S., Schwetye, K.E., Spinner, M.L., Esparza, T.J., Stocchetti, N., et al., 2008. Amyloid-beta dynamics correlate with neurological status in the injured human brain. Science 321, 1221–1224.

Brown, N.J., 2008. Aldosterone and vascular inflammation. Hypertension 51, 161–167.

Burke, T.M., 2015. Sleep inertia, sleep homeostatic and circadian influences on higher-order cognitive functions. J. Sleep. Res. 24, 364–371.

Buxton, O.M., Marcelli, E., 2010. Short and long sleep are positively associated with obesity, diabetes, hypertension, and cardiovascular disease among adults in the United States. Soc. Sci. Med. 71, 1027–1036.

Buxton, O.M., Ravlova, M., Reid, E.W., Wang, W., Simonson, D.C., Adler, G.K., 2010. Sleep restriction for one week reduces insulin sensitivity in healthy men. Diabetes 59, 2126–2133.

Buxton, O.M., Cain, S.W., O'Connor, S.P., Porter, J.H., Duffy, J.F., Wang, W., et al., 2012. Adverse metabolic consequences in humans of prolonged sleep restriction combined with circadian disruption. Sci. Transl. Med. 14, 129ra43.

Cairns, J., Freire-Pritchett, P., Wingett, S.W., Várnai, C., Dimond, A., Plagnol, V., et al., 2016. CHiCAGO: robust detection of DNA looping interactions in Capture Hi-C data. Genome Biol. 17, 127.

Calvin, A.D., Albuquerque, F.N., Lopez-Jimenez, F., Somers, V.K., 2009. Obstructive sleep apnea, inflammation, and the metabolic syndrome. Metab. Syndr. Relat. Disord. 7, 271–278.

Canto Gde, L., Pachêco-Pereira, C., Aydinoz, S., Major, P.W., Flores-Mir, C., Gozal, D., 2015. Biomarkers associated with obstructive sleep apnea: a scoping review. Sleep Med. Rev. 23, 28–45.

Cao, C., Wu, B., Wu, Y., et al., 2015. Functional polymorphisms in the promoter region of MMP-2 and MMP-9 and susceptibility to obstructive sleep apnea. Sci. Rep. 5, 8966.

Capone, C., Faraco, G., Coleman, C., Young, C.N., Pickel, V.M., Anrather, J., et al., 2012. Endothelin 1-dependent neurovascular dysfunction in chronic intermittent hypoxia. Hypertension 60, 106–113.

Carley, D.W., Farabi, S.S., 2016. Physiology of Sleep. Diabetes Spectr. 29, 5–9.

Carpagnano, G.E., Kharitonov, S.A., Resta, O., Foschino-Barbaro, M.P., Gramiccioni, E., Barnes, P.J., 2003. 8-Isoprostane, a marker of oxidative stress, is increased in exhaled breath condensate of patients with obstructive sleep apnea after night and is reduced by continuous positive airway pressure therapy. Chest 124, 1386–1392.

Carrier, J., Monk, T.H., Buysse, D.J., Kupfer, D., 1996. Amplitude reduction of the circadian temperature and sleep rhythms in the elderly. Chronobiol. Int. 13, 373–386.

Cepeda-Valery, B., Chaudhry, K., Slipczuk, L., Pressman, G.S., Figueredo, V.M., Lavie, C.J., et al., 2014. Association between obesity and severity of coronary artery disease at the time of acute myocardial infarction: another piece of the puzzle in the "obesity paradox". Int. J. Cardiol. 176, 247–249.

Chen, C.Y., Chen, C.L., Yu, C.C., Chen, T.T., Tseng, S.T., Ho, C.H., 2015. Association of inflammation and oxidative stress with obstructive sleep apnea in ischemic stroke patients. Sleep Med. 16, 113–118.

Cho, K., 2000. Chronic jet lag produces cognitive deficits. J. Neurosci. 20, RC66.

Cirrito, J.R., Kang, J.E., Lee, J., Stewart, F.R., Verges, D.K., Silverio, L.M., et al., 2008. Endocytosis is required for synaptic activity-dependent release of amyloid-beta in vivo. Neuron 58, 42–51.

Cooke, A.A., Connaughton, R.M., Lyons, C.L., McMorrow, A.M., Roche, H.M., 2016. Fatty acids and chronic low-grade inflammation associated with obesity and the metabolic syndrome. Eur. J. Pharmacol. 785, 207–214.

Cooper, C.B., Neufeld, E.V., Dolezal, B.A., Martin, J.L., 2018. Sleep deprivation and obesity in adults: a brief narrative review. BMJ Open Sport Exerc. Med. 4, e000392.

Coughlin, S.R., Mawdsle, L., Mugarza, J.A., Calverley, P.M.A., Wilding, J.P.H., 2004. Obstructive sleep apnoea is independently associated with an increased prevalence of metabolic syndrome. Eur. Heart J. 25, 735–741.

Davies, R.J., Ali, N.J., Stradling, J.R., 1992. Neck circumference and other clinical features in the diagnosis of the obstructive sleep apnoea syndrome. Thorax 47, 101–105.

Davidson, R.J., Pizzagalli, D., Nitschike, J.B., Putnam, K., 2002. Depression: perspectives from affective neuroscience. Annu. Rev. Psychol. 53, 545–574.
DeFronzo, R.A., 2010. Insulin resistance, lipotoxicity, type 2 diabetes and atherosclerosis: the missing links. The Claude Bernard Lecture 2009. Diabetologia 53, 1270–1287.
Del Rio, R., Andrade, D.C., Lucero, C., Arias, P., Iturriaga, R., 2016. Carotid body ablation abrogates hypertension and autonomic alterations induced by intermittent hypoxia in rats. Hypertension 68436–68445.
Dematteis, M., Julien, C., Guillermet, C., Sturm, N., Lantuejoul, S., Mallaret, M., et al., 2008. Intermittent hypoxia induces early functional cardiovascular remodeling in mice. Am. J. Respir. Crit. Care Med. 177, 227–235.
Dempsey, J.A., Veasey, S.C., Morgan, B.R., O'Donnell, C.P., 2010. Pathophysiology of sleep apnea. Physiol. Rev. 90, 47–112. 2010.
Diekelmann, S., Born, J., 2010. The memory function of sleep. Nat. Rev. Neurosci. 11, 114–126.
Dijk, D.J., Brunner, D.P., Borbely, A.A., 1990. Time course of EEG power density during long sleep in humans. Am. J. Physiol. 258 (3 Pt 2), R650–R661.
Dijk, D.J., Duffy, J.F., Riel, E., Shanahan, T.L., Czeisler, C.A., 1999. Ageing and the circadian and homeostatic regulation of human sleep during forced desynchrony of rest, melatonin and temperature rhythms. J. Physiol. 516, 611–627.
Dinges, D.F., 2006. The state of sleep deprivation: from functional biology to functional consequences. Sleep Med. Rev. 10, 303–305.
Dolan, R.J., 2002. Emotion, cognition, and behavior. Science 298, 1191–1194.
Dowlati, Y., Herrmann, N., Swardfager, W., Liu, H., Sham, L., et al., 2010. A meta-analysis of cytokines in major depression. Biol. Psychiatry 67, 446–457.
Drager, L.F., Jun, J.C., Polotsky, V.Y., 2010. Metabolic consequences of intermittent hypoxia: relevance to obstructive sleep apnea. Best Pract. Res. Clin. Endocrinol. Metab. 24, 843–851.
Drager, L.F., Li, J., Shin, M.K., Reinke, C., Aggarwal, N.R., Jun, J.C., et al., 2012. Intermittent hypoxia inhibits clearance of triglyceride-rich lipoproteins and inactivates adipose lipoprotein lipase in a mouse model of sleep apnoea. Eur. Heart J. 33, 783–790.
Ebihara, I., Nakamura, T., Shimada, N., Koide, H., 1998. Increased plasma metalloproteinase-9 concentrations precede development of microalbuminuria in non-insulin-dependent diabetes mellitus. Am. J. Kidney Dis. 32, 544–550.
Ejaz, S.M., Khawaja, I.S., Bhatia, S., Hurwitz, T.D., 2011. Obstructive sleep apnea and depression: a review. Innov. Clin. Neurosci. 8, 17–25.
Epstein, L.J., Kristo, D., Strollo Jr, P.J., Friedman, N., Malhotra, A., Patil, S.P., et al., 2009. Clinical guideline for the evaluation, management and long-term care of obstructive sleep apnea in adults. J. Clin. Sleep Med. 5, 263–276.
Evans, J.L., Goldfine, I.D., Maddux, B.A., Grodsky, G.M., 2002. Oxidative stress and stress-activated signaling pathways: a unifying hypothesis of type 2 diabetes. Endocr. Rev. 23, 599–622.
Farooqui, A.A., 2013. Metabolic Syndrome. Springer-Science -Business Media, New York.
Farooqui, A.A., 2014. Inflammation and Oxidative Stress in Neurological Disorders. Springer International Publishing, Switzerland.
Farooqui, A.A., 2017. Neurochemical Aspects of Alzheimer's Disease. Academic Press, San Diego, CA.
Farooqui, A.A., 2018. Ischemic and Traumatic Brain and Spinal Cord Injuries. Academic Press, San Diego, CA.
Farré, N., Otero, J., Falcones, B., Torres, M., Jorba, I., Gozal, D., et al., 2018. Intermittent hypoxia mimicking sleep apnea increases passive stiffness of myocardial extracellular matrix. A multiscale study. Front. Physiol. 9, 1143.

Fleming, W.E., Ferouz-Colborn, A., Samoszuk, M.K., Azad, A., Lu, J., Riley, J.S., et al., 2016. Blood biomarkers of endocrine, immune, inflammatory, and metabolic systems in obstructive sleep apnea. Clin. Biochem. 49, 854–861.

Fleming, W.E., Holty, J.C., Bogan, R.K., Hwang, D., Ferouz-Colborn, A.S., Budhiraja, R., et al., 2018. Use of blood biomarkers to screen for obstructive sleep apnea. Nat. Sci. Sleep. 10, 159–167.

Fletcher, E.C., 2001. Physiological consequences of intermittent hypoxia: systemic blood pressure. J. Appl. Physiol. 90, 1600–1605.

Fogel, R.B., Malhotra, A., Pillar, G., et al., 2001. Genioglossal activation in patients with obstructive sleep apnea versus control subjects. Mechanisms of muscle control. Am. J. Respir. Crit. Care Med. 164, 2025–2030.

Fogel, R.B., Trinder, J., Malhotra, A., et al., 2003. Within-breath control of genioglossal muscle activation in humans: effect of sleep-wake state. J. Physiol. 550, 899–910.

Foster, G.E., Poulin, M.J., Hanly, P.J., 2007. Intermittent hypoxia and vascular function: implications for obstructive sleep apnoea. Exp. Physiol 92, 51–65.

Frey, D.J., Fleshner, M., Wright Jr., K.P., 2007. The effects of 40 hours of total sleep deprivation on inflammatory markers in healthy young adults. Brain Behav. Immun. 21, 1050–1057.

Gabryelska, A., èukasik, Z.M., Makowska, J.S., Białasiewicz, P., 2018. Obstructive sleep apnea: from intermittent hypoxia to cardiovascular complications via blood platelets. Front. Neurol. 9, 635.

Gami, A.S., Caples, S.M., Somers, V.K., 2003. Obesity and obstructive sleep apnea. Endocrinol. Metab. Clin. North Am. 32, 869–894.

Garvey, J.F., Taylor, C.T., McNicholas, W.T., 2009. Cardiovascular disease in obstructive sleep apnoea syndrome: the role of intermittent hypoxia and inflammation. Eur. Respir. J. 33, 1195–1205.

Gibson, E.M., 2010. Experimental 'jet lag' inhibits adult neurogenesis and produces long-term cognitive deficits in female hamsters. PLoS One 5, e15267.

Gileles-Hillel, A., Kheirandish-Gozal, L., Gozal, D., 2016. Biological plausibility linking sleep apnoea and metabolic dysfunction. Nat. Rev. Endocrinol. 12, 290–298.

Gileles-Hillel, A., Almendros, I., Khalyfa, A., Nigdelioglu, R., Qiao, Z., Hamanaka, R.B., et al., 2017. Prolonged exposures to intermittent hypoxia promote visceral white adipose tissue inflammation in a murine model of severe sleep apnea: effect of normoxic recovery. Sleep 40.

Grandner, M.A., Hale, L., Moore, M., Patel, N.P., 2010. Mortality associated with short sleep duration: the evidence, the possible mechanisms, and the future. Sleep Med. Rev. 14, 191–203.

Gronfier, C., Brandenberger, G., 1998. Ultradian rhythms in pituitary and adrenal hormones: their relations to sleep. Sleep Med. Rev. 2, 17–29.

Guarnieri, B., Adorni, F., Musicco, M., Appollonio, I., Bonanni, E., Caffarra, P., et al., 2012. Prevalence of sleep disturbances in mild cognitive impairment and dementing disorders: a multicenter italian clinical cross-sectional study on 431 patients. Dement. Geriatr. Cogn. Disord. 33, 50–58.

Haller, J.F., Mintah, I.J., Shihanian, L.M., Stevis, P., Buckler, D., Alexa-Braun, C.A., et al., 2017. ANGPTL8 requires ANGPTL3 to inhibit lipoprotein lipase and plasma triglyceride clearance. J. Lipid Res. 58, 1166–1173.

Hanlon, E.C., Van Cauter, E., 2011. Quantification of sleep behavior and of its impact on the cross-talk between the brain and peripheral metabolism. Proc. Natl. Acad. Sci. U.S.A. 14, 15609–15616.

Hardy, J., 2009. The amyloid hypothesis for Alzheimer's disease: a critical reappraisal. J. Neurochem. 110, 1129–1134.

Harris, M., Glozier, N., Ratnavadivel, R., Grunstein, R.R., 2009. Obstructive sleep apnea and depression. Sleep Med. Rev. 13, 437–444.

Harrison, D., Griendling, K.K., Landmesser, U., Hornig, B., Drexler, H., 2003. Role of oxidative stress in atherosclerosis. Am. J. Cardiol. 91, 7A–11A.

Havekes, R., Vecsey, C.G., Abel, T., 2012. The impact of sleep deprivation on neuronal and glial signalling pathways important for memory and synaptic plasticity. Cell Signal. 24, 1251–1260.

Hayashi, T., Yoshioka, T., Hasegawa, K., Miyamura, M., Mori, T., Ukimura, A., et al., 2011. Inhalation of hydrogen gas attenuates left ventricular remodeling induced by intermittent hypoxia in mice. Am. J. Physiol. Heart Circ. Physiol. 301, H1062–H1069.

He, Q., Yang, Q.C., Zhou, Q., et al., 2014. Effects of varying degrees of intermittent hypoxia on proinflammatory cytokines and adipokines in rats and 3T3-L1 adipocytes. PLoS One 9, e86326.

Hermann, D.M., Bassetti, C.L., 2009. Sleep-related breathing and sleep-wake disturbances in ischemic stroke. Neurology 73, 1313–1322.

Hong, S.O., Chen, Y.F., Jung, J., Kwon, Y.D., Liu, S.Y.C., 2017. Hypoglossal nerve stimulation for treatment of obstructive sleep apnea (OSA): a primer for oral and maxillofacial surgeons. Maxillofac. Plast. Reconstr. Surg. 39, 27.

Hu, K., Van Someren, E.J., Shea, S.A., Scheer, F.A., 2009. Reduction of scale invariance of activity fluctuations with aging and Alzheimer's disease: Involvement of the circadian pacemaker. Proc. Natl. Acad. Sci. U.S.A. 106, 2490–2494.

Hu, K., Babapoor-Farrokhran, S., Rodrigues, M., Deshpande, M., Puchner, B., Kashiwabuchi, F., et al., 2016. Hypoxia-inducible factor 1 upregulation of both VEGF and ANGPTL4 is required to promote the angiogenic phenotype in uveal melanoma. Oncotarget 7, 7816–7828.

Iiyori, N., Alonso, L.C., Li, J., et al., 2007. Intermittent hypoxia causes insulin resistance in lean mice independent of autonomic activity. Am. J. Respir. Crit. Care Med. 175, 851–857.

Iliff, J.J., Wang, M., Liao, Y., Plogg, B.A., Peng, W., Gundersen, G.A., et al., 2012. A paravascular pathway facilitates CSF flow through the brain parenchyma and the clearance of interstitial solutes, including amyloid β. Sci. Transl. Med. 4, 147ra111.

Ingiosi, A.M., Opp, M.R., Krueger, J.M., 2013. Sleep and immune function: glial contributions and consequences of aging. Curr. Opin. Neurobiol. 23, 806–811.

Ip, M.S., Lam, B., Ng, M.M., Lam, W.K., Tsang, K.W., Lam, K.S., 2002. Obstructive sleep apnea is independently associated with insulin resistance. Am. J. Respir. Crit. Care Med. 165, 670–676.

Irwin, M.R., 2015. Why sleep is important for health: a psychoneuroimmunology perspective. Annu. Rev. Psychol. 66, 143–172.

Irwin, M.R., Miller, A.H., 2007. Depressive disorders and immunity: 20 years of progress and discovery. Brain Behav. Immun. 21, 374–383.

Irwin, M.R., Wang, M., Ribeiro, D., Cho, H.J., Olmstead, R., Breen, E.C., et al., 2008. Sleep loss activates cellular inflammatory signaling. Biol. Psychiatry 64, 538–540.

Issa, F.G., Sullivan, C.E., 1986. Reversal of central sleep apnea using nasal CPAP. Chest 90, 165–171.

Iturriaga, R., Moya, E.A., Del Rio, R., 2009. Carotid body potentiation induced by intermittent hypoxia: implications for cardiorespiratory changes induced by sleep apnoea. Clin. Exp. Pharmacol. Physiol. 36, 1197–1204.

Iturriaga, R., Andrade, D.C., Del Rio, R., 2014. Enhanced carotid body chemosensory activity and the cardiovascular alterations induced by intermittent hypoxia. Front. Physiol. 5, 468.

Iturriaga, R., Del Rio, R., Idiaquez, J., Somers, V.K., 2016. Carotid body chemoreceptors, sympathetic neural activation, and cardiometabolic disease. Biol. Res. 49, 13.

Iturriaga, R., Oyarce, M.P., Dias, A.C.R., 2017. Role of carotid body in intermittent hypoxia-related hypertension. Curr. Hypertens. Rep. 19, 38.
Jalilolghadr, S., Yazdi, Z., Mahram, M., Babaei, F., Esmailzadehha, N., Nozari, H., et al., 2016. Sleep architecture and obstructive sleep apnea in obese children with and without metabolic syndrome: a case control study. Sleep Breath. 20, 845−851.
Javaheri, S., Barbe, F., Campos-Rodriguez, F., et al., 2017. Sleep apnea: types, mechanisms, and clinical cardiovascular consequences. J. Am. Coll. Cardiol. 69, 841−858.
Jelic, S., Lederer, D.J., Adams, T., et al., 2010. Vascular inflammation in obesity and sleep apnea. Circulation 121, 1014−1021.
Ju, Y.E., McLeland, J.S., Toedebusch, C.D., Xiong, C., Fagan, A.M., Duntley, S.P., et al., 2013. Sleep quality and preclinical Alzheimer disease. JAMA Neurol. 70, 587−593.
Ju, Y.-E.S., Lucey, B.P., Holtzman, D.M., 2014. Sleep and Alzheimer disease pathology—a bidirectional relationship. Nat. Rev. Neurol. 10, 115−119.
Kang, J.E., Lim, M.M., Bateman, R.J., Lee, J.J., Smyth, L.P., Cirrito, J.R., et al., 2009. Amyloid-beta dynamics are regulated by orexin and the sleep-wake cycle. Science 326, 1005−1007.
Katon, W., Lin, E.H., Kroenke, K., 2007. The association of depression and anxiety with medical symptom burden in patients with chronic medical illness. Gen. Hosp. Psychiatry 29, 147−155.
Kearney, P.M., Whelton, M., Reynolds, K., Muntner, P., Whelton, P.K., et al., 2005. Global burden of hypertension: analysis of worldwide data. Lancet 365, 217−223.
Kent, B.D., Mc Nicholas, W.T., Ryan, S., 2015. Insulin resistance, glucose intolerance and diabetes mellitus in obstructive sleep apnoea. J. Thorac. Dis. 7, 1343−1357.
Khosravi-Boroujeni, H., Sarrafzadegan, N., Sadeghi, M., Roohafza, H., Talaei, M., Ng, S. K., et al., 2017. Secular trend of metabolic syndrome and its components in a cohort of Iranian adults from 2001 to 2013. Metab. Syndr. Relat. Disord. 15, 137−144.
Kiyashchenko, L.I., Mileykovskiy, B.Y., Maidment, N., Lam, H.A., Wu, M.F., John, J., et al., 2002. Release of hypocretin (orexin) during waking and sleep states. J. Neurosci. Off. J. Soc. Neurosci. 22, 5282−5286.
Koo, B.B., Sillau, S., Dean, D.A., et al., 2015. Periodic limb movements during sleep and prevalent hypertension in the multi-ethnic study of atherosclerosis. Hypertension 65, 70−77.
Kress, B.T., Iliff, J.J., Xia, M., Wang, M., Wei, H.S., Zeppenfeld, D., et al., 2014. Impairment of paravascular clearance pathways in the aging brain. Ann. Neurol. 76, 845−861.
Krishnan, H.C., Lyons, L.C., 2015. Synchrony and desynchrony in circadian clocks: impacts on learning and memory. Learn. Mem. 22, 426−437.
Krueger, J.M., Frank, M.G., Wisor, J.P., Roy, S., 2016. Sleep function: toward elucidating an enigma. Sleep Med. Rev. 28, 46−54.
Kyrtsos, C.R., Baras, J.S., 2015. Modeling the role of the glymphatic pathway and cerebral blood vessel properties in Alzheimer's disease pathogenesis. PLoS One 10, e0139574.
Lafortune, M., Gagnon, J.F., Martin, N., Latreille, V., Dubé, J., Bouchard, M., et al., 2014. Sleep spindles and rapid eye movement sleep as predictors of next morning cognitive performance in healthy middle-aged and older participants. J. Sleep. Res. 23, 159−167.
Lanfranchi, P., Somers, V.K., 2001. Ostructive sleep apnea and vascular disease. Respir. Res. 2, 315−319.
Lavie, L., 2015. Oxidative stress in obstructive sleep apnea and intermittent hypoxia—revisited—the bad ugly and good: implications to the heart and brain. Sleep Med. Rev. 20, 27−45.

Lavie, P., Herer, P., Hoffstein, V., 2000. Obstructive sleep apnoea syndrome as a risk factor for hypertension: population study. Br. Med. J. 320, 479–482.
Lee, Y.S., Kim, J.W., Osborne, O., Oh da, Y., Sasik, R., Schenk, S., et al., 2014. Increased adipocyte O_2 consumption triggers HIF-1alpha, causing inflammation and insulin resistance in obesity. Cell 157, 1339–1352.
Lee, H., Xie, L., Yu, M., Kang, H., Feng, T., Deane, R., et al., 2015. The effect of body posture on brain glymphatic transport. J. Neurosci. 35, 11034–11044.
Lefebvre, B., Godin-Ribuot, D., Joyeux-Faure, M., et al., 2006. Functional assessment of vascular reactivity after chronic intermittent hypoxia in the rat. Respir. Physiol. Neurobiol. 150, 278–286.
Lévy, P., Pépin, J.L., Arnaud, C., Tamisier, R., Borel, J.C., Dematteis, M., et al., 2008. Intermittent hypoxia and sleep-disordered breathing: current concepts and perspectives. Eur. Respir. J. 32, 1082–1095.
Levy, P., Tamisier, R., Minville, C., Launois, S., Pepin, J.-L., 2011. Sleep apnoea syndrome in 2011: current concepts and future directions. Eur. Respir. Rev. 20, 134–146.
Li, S., Feng, J., Wei, S., Qian, X., Cao, J., Chen, B., 2016. Delayed neutrophil apoptosis mediates intermittent hypoxia-induced progressive heart failure in pressure-overloaded rats. Sleep Breath. 20, 95–102.
Lim, A.S., Kowgier, M., Yu, L., Buchman, A.S., Bennett, D.A., 2013. Sleep fragmentation and the risk of incident Alzheimer's disease and cognitive decline in older persons. Sleep 36, 1027–1032.
Lim, M.M., Gerstner, J.R., Holtzman, D.M., 2014. The sleep–wake cycle and Alzheimer's disease: what do we know? Neurodegener. Dis. Manag. 4, 351–362.
Lin, M.T., Lin, H.H., Lee, P.L., Weng, P.H., Lee, C.C., Lai, T.C., et al., 2015. Beneficial effect of continuous positive airway pressure on lipid profiles in obstructive sleep apnea: a meta-analysis. Sleep Breath. 19, 809–817.
Marrone, O., Salvaggio, A., Bue, A.L., Bonanno, A., Riccobono, L., Insalaco, G., et al., 2011. Blood pressure changes after automatic and fixed CPAP in obstructive sleep apnea: relationship with nocturnal sympathetic activity. Clin. Exp. Hypertens. 33, 373–380.
Maser, R.E., Lenhard, M.J., Rizzo, A.A., Vasile, A.A., 2008. Continuous positive airway pressure therapy improves cardiovascular autonomic function for persons with sleep-disordered breathing. Chest 133, 86–91.
McCurry, S.M., Reynolds, C.F., Ancoli-Israel, S., Teri, L., Vitiello, M.V., 2000. Treatment of sleep disturbance in Alzheimer's disease. Sleep Med. Rev. 4, 603e28.
McEvoy, R.D., Antic, N.A., Heeley, E., et al., 2016. CPAP for prevention of cardiovascular events in obstructive sleep apnea. N. Engl. J. Med. 375, 919–931.
McNicholas, W.T., Bonsigore, M.R., 2007. Sleep apnoea as an independent risk factor for cardiovascular disease: current evidence, basic mechanisms and research priorities. Eur. Respir. J. 29, 156–178.
Mendelsohn, A.R., Larrick, J.W., 2013. Sleep facilitates clearance of metabolites from the brain: glymphatic function in aging and neurodegenerative diseases. Rejuvenation Res. 16, 518–523.
Minakawa, E.N., Miyazaki, K., Maruo, K., Yagihara, H., Fujita, H., Wada, K., et al., 2017. Chronic sleep fragmentation exacerbates amyloid beta deposition in Alzheimer's disease model mice. Neurosci. Lett. 653, 362–369.
Minakawa, E.N., Wada, K., Nagai, Y., 2019. Sleep disturbance as a potential modifiable risk factor for Alzheimer's disease. Int. J. Mol. Sci. 20. pii: E803.
Mizuma, A., Yenari, M.A., 2017. Anti-inflammatory targets for the treatment of reperfusion injury in stroke. Front. Neurol. 8, 467.
Möller-Levet, C.S., Archer, S.N., Bucca, G., Laing, E.E., Slak, A., Kabiljo, R., et al., 2013. Effects of insufficient sleep on circadian rhythmicity and expression amplitude

of the human blood transcriptome. Proc. Natl. Acad. Sci. U S A 110, E1132–1141.
Morand, J., Arnaud, C., Pepin, J.L., Godin-Ribuot, D., 2018. Chronic intermittent hypoxia promotes myocardial ischemia-related ventricular arrhythmias and sudden cardiac death. Sci. Rep. 8, 2997.
Mortimore, I.L., Marshall, I., Wraith, P.K., Sellar, R.J., Douglas, N.J., 1998. Neck and total body fat deposition in nonobese and obese patients with sleep apnea compared with that in control subjects. Am. J. Respir. Crit. Care Med. 157, 280–283.
Murphy, A.M., Thomas, A., Crinion, S.J., et al., 2017. Intermittent hypoxia in obstructive sleep apnoea mediates insulin resistance through adipose tissue inflammation. Eur. Respir. J. 49. pii: 1601731.
Musiek, E.S., 2015. Circadian clock disruption in neurodegenerative diseases: cause and effect? Front. Pharmacol. 6, 29.
Musiek, E.S., Holtzman, D.M., 2016. Mechanisms linking circadian clocks, sleep, and neurodegeneration. Science 354, 1004–1008.
Nadeem, R., Singh, M., Nida, M., et al., 2014. Effect of CPAP treatment for obstructive sleep apnea hypopnea syndrome on lipid profile: a meta-regression analysis. J. Clin. Sleep Med. 10, 1295–1302.
Naidoo, N., Ferber, M., Master, M., Zhu, Y., Pack, A.I., 2008. Aging impairs the unfolded protein response to sleep deprivation and leads to proapoptotic signaling. J. Neurosci. 28, 6539–6548.
Nannapaneni, S., Ramar, K., Surani, S., 2013. Effect of obstructive sleep apnea on type 2 diabetes mellitus: A comprehensive literature review. World J. Diabetes. 4, 238–244.
Narkiewicz, K., van de Borne, P.J., Cooley, R.L., Dyken, M.E., Somers, V.K., 1998. Sympathetic activity in obese subjects with and without obstructive sleep apnea. Circulation 98, 772–776.
Nishikawa, T., Araki, E., 2007. Impact of mitochondrial ROS production in the pathogenesis of diabetes mellitus and its complications. Antioxid. Redox. Signal. 9, 343–353.
O'Donnell, J., Ding, F., Nedergaard, M., 2015. Distinct functional states of astrocytes during sleep and wakefulness: is norepinephrine the master regulator? Curr. Sleep Med. Rep. 1, 1–8.
Ohayon, M.M., Carskadon, M.A., Guilleminault, C., Vitiello, M.V., 2004. Meta-analysis of quantitative sleep parameters from childhood to old age in healthy individuals: developing normative sleep values across the human lifespan. Sleep 27, 1255–1273.
Okcay, A., Somers, V.K., Caples, S.M., 2008. Obstructive sleep apnea and hypertension. J. Clin. Hypertens 10, 549–555.
Olea, E., Agapito, M.T., Gallego-Martin, T., et al., 2014. Intermittent hypoxia and diet-induced obesity: effects on oxidative status, sympathetic tone, plasma glucose and insulin levels, and arterial pressure. J. Appl. Physiol. (1985) 117, 706–719.
Osorio, R.S., Pirraglia, E., Agüera-Ortiz, L.F., During, E.H., Sacks, H., Ayappa, I., et al., 2011. Greater risk of Alzheimer's disease in older adults with insomnia. J. Am. Geriatr. Soc. 59, 559–562.
Osorio, R.S., Ducca, E.L., Wohlleber, M.E., Tanzi, E.B., Gumb, T., Twumasi, A., et al., 2016. Orexin-A is associated with increases in cerebrospinal fluid phosphorylated-tau in cognitively normal elderly subjects. Sleep 39, 1253–1260.
Otto, M.E., Belohlavek, M., Romero-Corral, A., Gami, A.S., Gilman, G., Svatikova, A., et al., 2007. Comparison of cardiac structural and functional changes in obese otherwise healthy adults with versus without obstructive sleep apnea. Am. J. Cardiol. 99, 1298–1302.
Papanas, N., Steiropoulos, P., Nena, E., et al., 2009. HbA_{1C} is associated with severity of obstructive sleep apnea hypopnea syndrome in nondiabetic men. Vasc. Health Risk. Manag. 5, 751–756.

Park, H.J., 2015. The stress response neuropeptide CRF increases amyloid-beta production by regulating gamma-secretase activity. EMBO J. 34, 1674–1686.
Partinen, M., Jamieson, A., Guilleminault, C., 1988. Long-term outcome for obstructive sleep apnea syndrome patients. Mortality. Chest 94, 1200–1204.
Passàli, D., Tatti, P., Toraldo, M., et al., 2014. OSAS and metabolic diseases: Round Table, 99th SIO National Congress, Bari 2012. Acta Otorhinolaryngol. Ital. 34, 158–166.
Peigneux, P., Laureys, S., Fuchs, S., Collette, F., Perrin, F., Reggers, J., et al., 2004. Are spatial memories strengthened in the human hippocampus during slow wave sleep? Neuron 44, 535–545.
Pellerin, L., Magistretti, P.J., 2012. Sweet sixteen for ANLS. J. Cereb. Blood Flow Metab. 32, 1152–1166.
Peppard, P.E., Young, T., Palta, M., Dempsey, J., Skatrud, J., 2000. Longitudinal study of moderate weight change and sleep-disordered breathing. JAMA 284, 3015–3021.
Peppard, P.E., Young, T., Barnet, J.H., Palta, M., Hagen, E.W., Hla, K.M., 2013. Increased prevalence of sleep-disordered breathing in adults. Am. J. Epidemiol. 177, 1006–1014.
Phillips, S.A., Olson, E.B., Morgan, B.J., Lombard, J.H., 2004. Chronic intermittent hypoxia impairs endothelium-dependent dilation in rat cerebral and skeletal muscle resistance arteries. Am. J. Physiol. Heart Circ. Physiol. 286, H388–H393.
Phillips, C.L., Hoyos, C.M., Yee, B.J., Grunstein, R.R., 2016. CrossTalk opposing view: sleep apnoea causes metabolic syndrome. J. Physiol. 594, 4691–4694.
Pho, H., Hernandez, A.B., Arias, R.S., et al., 2016. The effect of leptin replacement on sleep-disordered breathing in the leptin-deficient ob/ob mouse. J. Appl. Physiol. 120, 78–86.
Poe, G.R., Walsh, C.M., Bjorness, T.E., 2010. Cognitive neuroscience of sleep. Prog. Brain Res. 185, 1–19.
Prabhakar, N.R., Fields, R.D., Baker, T., Fletcher, E.C., 2001. Intermittent hypoxia: cell to system. Am. J. Physiol. Lung Cell. Mol. Physiol. 281, L524–L528.
Prabhakar, N.R., Kumar, G.K., Peng, Y.J., 2012. Sympatho-adrenal activation by chronic intermittent hypoxia. J. Appl. Physiol. 113, 1304–1310.
Prabhakar, N.R., Peng, Y.J., Kumar, G.K., Nanduri, J., 2015. Peripheral chemoreception and arterial pressure responses to intermittent hypoxia. Compr. Physiol. 5, 561–577.
Punjabi, N.M., 2008. The epidemiology of adult obstructive sleep apnea. Proc. Am. Thorac. Soc. 5, 136–143.
Punjabi, N.M., Polotsky, V.Y., 2005. Disorders of glucose metabolism in sleep apnea. J. Appl. Physiol. 99, 1998–2007.
Quagliarini, F., Wang, Y., Kozlitina, J., Grishin, N.V., Hyde, R., Boerwinkle, E., et al., 2012. Atypical angiopoietin-like protein that regulates ANGPTL3. Proc. Natl. Acad. Sci. U.S.A. 109, 19751–19756.
Rajendran, L., Paolicelli, R.C., 2018. Microglia-mediated synapse loss in Alzheimer's disease. J. Neurosci. 38, 2911–2919.
Rajendran, L., Honsho, M., Zahn, T.R., Keller, P., Geiger, K.D., Verkade, P., et al., 2006. Alzheimer's disease beta-amyloid peptides are released in association with exosomes. Proc. Natl. Acad. Sci. U.S.A. 103, 11172–11177.
Ramadan, W., Dewasmes, G., Petitjean, M., Wiernsperger, N., Delanaud, S., Geloen, A., et al., 2007. Sleep apnea is induced by a high-fat diet and reversed and prevented by metformin in non-obese rats. Obesity (Silver Spring, MD) 15, 1409–1418.
Ramesh, V., Thatte, H.S., McCarley, R.W., Basheer, R., 2007. Adenosine and sleep deprivation promote NF-kappaB nuclear translocation in cholinergic basal forebrain. J. Neurochem. 100, 1351–1363.
Ramos, A.R., Seixas, A., Dib, S.I., 2015. Obstructive sleep apnea and stroke: links to health disparities. Sleep Health 1, 244–248.

Ratner, V., Zhu, L., Kolesov, I., Nedergaard, M., Benveniste, H., Tannenbaum, A., 2015. Optimal-mass-transfer-based estimation of glymphatic transport in living brain. Proc. SPIE Int. Soc. Opt. Eng. 9413, 94131J.

Rechtschaffen, A., Siegel, J., 2000. Sleep and dreaming. In: Kandel, E.R., Schwartz, J.H., Jessell, T.M. (Eds.), Principles of Neural Science, fourth ed. McGraw-Hill, New York, pp. 936–947.

Redline, S., Schluchter, M.D., Larkin, E.K., Tishler, P.V., 2003. Predictors of longitudinal change in sleep-disordered breathing in a nonclinic population. Sleep 26, 703–709.

Rey, S., Del Rio, R., Iturriaga, R., 2006. Role of endothelin-1 on the enhanced carotid body activity induced by chronic intermittent hypoxia. Adv. Exp. Med. Biol. 580, 345–350.

Rifai, N., Ridker, P.M., 2002. Inflammatory markers and coronary heart disease. Curr. Opin. Lipidol. 13, 3383–3389.

Roh, J.H., Huang, Y., Bero, A.W., Kasten, T., Stewart, F.R., Bateman, R.J., et al., 2012. Disruption of the sleep-wake cycle and diurnal fluctuation of beta-amyloid in mice with Alzheimer's disease pathology. Sci. Transl. Med 4, 150ra22.

Roh, J.H., Finn, M.B., Stewart, F.R., Mahan, T.E., Cirrito, J.R., Heda, A., et al., 2014. Potential role of orexin and sleep modulation in the pathogenesis of Alzheimer's disease. J. Exp. Med. 211, 2487–2496.

Rosen, C.L., Larkin, E.K., Kirchner, H.L., et al., 2003. Prevalence and risk factors for sleep-disordered breathing in 8- to 11-year-old children: association with race and prematurity. J. Pediatr. 142, 383–389.

Ryan, S., 2017. Adipose tissue inflammation by intermittent hypoxia: mechanistic link between obstructive sleep apnoea and metabolic dysfunction. J. Physiol. 595, 2423–2430.

Sahlin, C., Sandberg, O., Gustafson, Y., et al., 2008. Obstructive sleep apnea is a risk factor for death in patients with stroke: a 10-year follow-up. Arch. Intern. Med. 168, 297–301.

Sakata, Y., Yamamoto, K., Mano, T., et al., 2004. Activation of matrix metalloproteinases precedes left ventricular remodeling in hypertensive heart failure rats: its inhibition as a primary effect of angiotensin-converting enzyme inhibitor. Circulation 109, 2143–2149.

Salamanca, F., Costantini, F., Bianchi, A., et al., 2013. Identification of obstructive sites and patterns in obstructive sleep apnoea syndrome by sleep endoscopy in 614 patients. Acta Otorhinolaryngol. Ital. 33, 261–266.

Salord, N., Fortuna, A.M., Monasterio, C., Gasa, M., Perez, A., Bonsignore, M.R., et al., 2016. A randomized controlled trial of continuous positive airway pressure on glucose tolerance in obese patients with obstructive sleep apnea. Sleep 39, 35–41.

Sandberg, O., Franklin, K.A., Bucht, G., Gustafson, Y., 2001. Sleep apnea, delirium, depressed mood, cognition, and ADL ability after stroke. J. Am. Geriatr. Soc. 49, 391–397.

Santulli, G., 2014. Angiopoietin-like proteins: a comprehensive look. Front. Endocrinol. 5, 4.

Savransky, V., Nanayakkara, A., Li, J., Smith, P.L., Rodriguez, A., Polotsky, V.Y., et al., 2007. Chronic intermittent hypoxia induces atherosclerosis. Am. J. Respir. Crit. Care Med. 175, 1290–1297.

Schenck, C.H., Boeve, B.F., Mahowald, M.W., 2013. Delayed emergence of a parkinsonian disorder or dementia in 81% of older men initially diagnosed with idiopathic rapid eye movement sleep behavior disorder: a 16-year update on a previously reported series. Sleep Med. 14, 744–748.

Schroder, C.M., O'Hara, R., 2005. Depression and obstructive sleep apnea (OSA). Ann. Gen. Psychiatry 4, 13.

Schulz, R., Mahmoudi, S., Hattar, K., et al., 2000. Enhanced release of superoxide from polymorphonuclear neutrophils in obstructive sleep apnea. Impact of continuous positive airway pressure therapy. Am. J. Respir. Crit. Care Med. 162 (2 Pt 1), 566–570.

Schulz, R., Murzabekova, G., Egemnazarov, B., et al., 2014. Arterial hypertension in a murine model of sleep apnea: role of NADPH oxidase 2. J. Hypertens. 32, 300–305.

Schwab, R.J., Pasirstein, M., Pierson, R., et al., 2003. Identification of upper airway anatomic risk factors for obstructive sleep apnea with volumetric magnetic resonance imaging. Am. J. Respir. Crit. Care Med. 168, 522–530.

Scuteri, A., Morrell, C., Najjar, S., Lakatta, E., 2005. The metabolic syndrome in older individuals: prevalence and prediction of cardiovascular events: the Cardiovascular Health Study. Diabetes Care 28, 882–887.

Selkoe, D., Hardy, J., 2016. The amyloid hypothesis of Alzheimer's disease at 25 years. EMBO Mol. Med. 8, 595–608.

Semenza, G.L., Prabhakar, N.R., 2018. The role of hypoxia-inducible factors in carotid body (patho) physiology. J. Physiol. 596, 2977–2983.

Sharafkhaneh, A., Giray, N., Richardson, P., Young, T., Hirshkowitz, M., 2005. Association of psychiatric disorders and sleep apnea in a large cohort. Sleep 28, 1405–1411.

Sherwani, S.I., Aldana, C., Usmani, S., et al., 2013. Intermittent hypoxia exacerbates pancreatic beta-cell dysfunction in a mouse model of diabetes mellitus. Sleep 36, 1849–1858.

Shih, J.L., Malhotra, A., 2011. Could vitamins be helpful to patients with sleep apnea? Chest 139, 237–238.

Simats, A., García-Berrocoso, T., Montaner, J., 2016. Neuroinflammatory biomarkers: from stroke diagnosis and prognosis to therapy. Biochim. Biophys. Acta Mol. Basis Dis. 1862, 411–424.

Simon, M.J., Iliff, J.J., 2016. Regulation of cerebrospinal fluid (CSF) flow in neurodegenerative, neurovascular and neuroinflammatory disease. Biochim. Biophys. Acta 1862, 442–451.

No author listed. 1999. Sleep-related breathing disorders in adults: recommendations for syndrome definition and measurement techniques in clinical research. The Report of an American Academy of Sleep Medicine Task Force. Sleep 22, 667–689.

Smarr, B.L., 2014. A time to remember: the role of circadian clocks in learning and memory. Behav. Neurosci. 128, 283–303.

Somers, V.K., Mark, A.L., Zavala, D.C., Abboud, F.M., 1989a. Contrasting effects of hypoxia and hypercapnia on ventilation and sympathetic activity in humans. J. Appl. Physiol. 67, 2101–2106.

Somers, V.K., Mark, A.L., Zavala, D.C., Abboud, F.M., 1989b. Influence of ventilation and hypocapnia on sympathetic nerve responses to hypoxia in normal humans. J. Appl. Physiol. 67, 2095–2100.

Somers, V.K., Dyken, M.E., Mark, A.L., Abboud, F.M., 1993a. Sympathetic-nerve activity during sleep in normal subjects. N. Engl. J. Med. 328, 303–307.

Somers, V.K., Dyken, M.E., Skinner, J.L., 1993b. Autonomic and hemodynamic responses and interactions during the Mueller maneuver in humans. J. Auton. Nerv. Syst. 44, 253–259.

Somers, V.K., Dyken, M.E., Clary, M.P., Abboud, F.M., 1995. Sympathetic neural mechanisms in obstructive sleep apnea. J. Clin. Invest. 96, 1897–1904.

Somers, V.K., White, D.P., Amin, R., et al., 2008. Sleep apnea and cardiovascular disease: an American Heart Association/American College of Cardiology Foundation Scientific Statement from the American Heart Association Council for High Blood Pressure Research Professional Education Committee, Council on Clinical

Cardiology, Stroke Council, and Council on Cardiovascular Nursing. J. Am. Coll. Cardiol. 52, 686–717.
Spiegel, K., Leproult, R., L'Hermite-Baleriaux, M., Copinschi, G., Penev, P.D., Van Cauter, E., 2004. Leptin levels are dependent on sleep duration: relationships with sympathovagal balance, carbohydrate regulation, cortisol, and thyrotropin. J. Clin. Endocrinol. Metab. 89, 5762–5771.
Spiegel, K., Knutson, K., Leproult, R., Tasali, E., Cauter, E.V., 2005. Sleep loss: a novel risk factor for insulin resistance and Type 2 diabetes. J. Appl. Physiol. 99, 2008–2019.
Sterniczuk, R., Theou, O., Rusak, B., Rockwood, K., 2013. Sleep disturbance is associated with incident dementia and mortality. Curr. Alzheimer Res. 10, 767–775.
Sullivan, P.F., Neale, M.C., Kendler, K.S., 2000. Genetic epidemiology of major depression: review and meta-analysis. Am. J. Psychiatry 157, 1552–1562.
Tahrani, A.A., Ali, A., Raymond, N.T., et al., 2012. Obstructive sleep apnea and diabetic neuropathy: a novel association in patients with type 2 diabetes. Am. J. Respir. Crit. Care Med. 186, 434–441.
Tahrani, A.A., Ali, A., Raymond, N.T., et al., 2013. Obstructive sleep apnea and diabetic nephropathy: a cohort study. Diabetes Care 36, 3718–3737.
Tarasoff-Conway, J.M., Carare, R.O., Osorio, R.S., Glodzik, L., Butler, T., Fieremans, E., et al., 2015. Clearance systems in the brain-implications for Alzheimer disease. Nat. Rev. Neurol. 11, 457–470.
Tasali, E., Van Cauter, E., 2002. Sleep-disordered breathing and the current epidemic of obesity: consequence or contributing factor? Am. J. Respir. Crit. Care Med. 165, 562–563.
Tokuda, F., Sando, Y., Matsui, H., Koike, H., Yokoyama, T., 2008. Serum levels of adipocytokines, adiponectin and leptin, in patients with obstructive sleep apnea syndrome. Intern. Med. 47, 1843–1849.
Tononi, G., Cirelli, C., 2006. Sleep function and synaptic homeostasis. Sleep Med. Rev. 10, 49–62.
Trombetta, I.C., Maki-Nunes, C., Toschi-Dias, E., Alves, M.J., Rondon, M.U., Cepeda, F.X., et al., 2013. Obstructive sleep apnea is associated with increased chemoreflex sensitivity in patients with metabolic syndrome. Sleep 36, 41–49.
Tsioufis, C., Thomopoulos, K., Dimitriadis, K., Amfilochiou, A., Tousoulis, D., Alchanatis, M., et al., 2007. The incremental effect of obstructive sleep apnoea syndrome on arterial stiffness in newly diagnosed essential hypertensive subjects. J. Hypertens. 25, 141–146.
Van Cauter, E., Spiegel, K., Tasali, E., Leproult, R., 2008. Metabolic consequences of sleep and sleep loss. Sleep Med. 14 (Suppl. 1), S23–S28.
Vanderheyden, W.M., Lim, M.M., Musiek, E.S., Gerstner, J.R., 2018. Alzheimer's disease and sleep-wake disturbances: amyloid, astrocytes, and animal models. J. Neurosci. 38, 2901–2910.
Vgontzas, A.N., Kales, A., 1999. Sleep and its disorders. Ann. Rev. Med. 50, 387–400.
Vgontzas, A.N., Papanicolaou, D.A., Bixler, E.O., et al., 2000. Sleep apnea and daytime sleepiness and fatigue: relation to visceral obesity, insulin resistance, and hypercytokinemia. J. Clin. Endocrinol. Metab. 85, 1151–1158.
Walker, M.P., 2009. The role of sleep in cognition and emotion. Ann. N. Y. Acad. Sci. 1156, 168–197.
Wang, S., Li, S., Wang, B., Liu, J., Tang, Q., 2018. Matrix metalloproteinase-9 is a predictive factor for systematic hypertension and heart dysfunction in patients with obstructive sleep apnea syndrome. Biomed. Res. Int. 1569701. 2018.
Weinstock, P.H., Bisgaier, C.L., Aalto-Setala, K., Radner, H., Ramakrishnan, R., Levak-Frank, S., et al., 1995. Severe hypertriglyceridemia, reduced high density lipoprotein, and neonatal death in lipoprotein lipase knockout mice. Mild hypertriglyceridemia with impaired very low-density lipoprotein clearance in heterozygotes. J. Clin. Invest. 96, 2555–2568.

Wennstrom, M., Londos, E., Minthon, L., Nielsen, H.M., 2012. Altered CSF orexin and α-synuclein levels in dementia patients. J. Alzheimer Dis. 29, 125–132.
Wright, K.P., Lowry, C.A., Lebourgeois, M.K., 2012. Circadian and wakefulness-sleep modulation of cognition in humans. Front. Mol. Neurosci. 5, 50.
Wurtman, R.J., 2005. Genes, stress, and depression. Metabolism 54 (Suppl. 1), 16–19.
Xie, L., Kang, H., Xu, Q., Chen, M.J., Liao, Y., Thiyagarajan, M., et al., 2013. Sleep drives metabolite clearance from the adult brain. Science 342, 373–377.
Yalcinkaya, M., Erbek, S.S., Babakurban, S.T., et al., 2015. Lack of association of matrix metalloproteinase-9 promoter gene polymorphism in obstructive sleep apnea syndrome. J. Craniomaxillofac. Surg. 43, 1099–1103.
Yang, D., Liu, Z., Yang, H., Luo, Q., 2013. Effects of continuous positive airway pressure on glycemic control and insulin resistance in patients with obstructive sleep apnea: a meta-analysis. Sleep Breath. 17, 33–38.
Yao, Q., Shin, M.K., Jun, J.C., Hernandez, K.L., Aggarwal, N.R., Mock, J.R., et al., 2013. Effect of chronic intermittent hypoxia on triglyceride uptake in different tissues. J. Lipid Res. 54, 1058–1065.
Yao, Q., Pho, H., Kirkness, J., et al., 2016. Localizing effects of leptin on upper airway and respiratory control during sleep. Sleep 39, 1097–1106.
Yoo, S.S., Hu, P.T., Gujar, N., Jolesz, F.A., Walker, M.P., 2007. A deficit in the ability to form new human memories without sleep. Nat. Neurosci. 10, 385–392.
Young, T., Shahar, E., Nieto, F.J., Redline, S., Newman, A.B., Gottlieb, D.J., et al., 2002. Predictors of sleep-disordered breathing in community-dwelling adults: the sleep heart health study. Arch. Intern. Med. 162, 893–900.
Young, T., Peppard, P.E., Taheri, S., 2005. Excess weight and sleep-disordered breathing. J. Appl. Physiol. (1985) 99, 1592–1599.
Young, T., Finn, L., Peppard, P.E., et al., 2008. Sleep disordered breathing and mortality: eighteen-year follow-up of the Wisconsin sleep cohort. Sleep 31, 1071–1078.
Yu, J., Zhou, Z., McEvoy, R.D., et al., 2017. Association of positive airway pressure with cardiovascular events and death in adults with sleep apnea: a systematic review and meta-analysis. J. Am. Med. Assoc. 318, 156–166.
Zhang, R., 2016. The ANGPTL3-4-8 model, a molecular mechanism for triglyceride trafficking. Open Biol. 6, 150272.

Further reading

Benedict, C., Byberg, L., Cedernaes, J., Hogenkamp, P.S., Giedratis, V., Kilander, L., et al., 2015. Self-reported sleep disturbance is associated with Alzheimer's disease risk in men. Alzheimer Dement. 11, 1090–1097.
Briancon-Marjollet, A., Monneret, D., Henri, M., Hazane-Puch, F., Pepin, J.L., Faure, P., et al., 2016. Endothelin regulates intermittent hypoxia-induced lipolytic remodelling of adipose tissue and phosphorylation of hormone-sensitive lipase. J. Physiol. 594, 1727–1740.
Castaneda, A., Jauregui-Maldonado, E., Ratnani, I., Varon, J., Surani, S., 2018. Correlation between metabolic syndrome and sleep apnea. World J. Diabetes 9, 66–71.
Feigin, V.L., Forouzanfar, M.H., Krishnamurthi, R., Mensah, G.A., Connor, M., et al., 2014. Global and regional burden of stroke during 1990-2010: findings from the Global Burden of Disease Study 2010. Lancet 383, 245–254.
Mozaffarian, D., Benjamin, E.J., Go, A.S., Arnett, D.K., Blaha, M.J., et al., 2015. Heart disease and stroke statistics--2015 update: a report from the American Heart Association. Circulation 131, e29–322.

Reimer, M.A., Flemons, W.W., 2003. Quality of life in sleep disorders. Sleep Med. Rev. 7, 335–349.

Sterniczuk, R., Dyck, R.H., Laferla, F.M., Antle, M.C., 2010. Delayed emergence of a parkinsonian disorder or dementia in 81% of older men initially diagnosed with idiopathic rapid eye movement sleep behavior disorder: a 16-year update on a previously reported series. Brain Res. 1348, 139–148.

Wei, Q., Bian, Y., Yu, F., et al., 2016. Chronic intermittent hypoxia induces cardiac inflammation and dysfunction in a rat obstructive sleep apnea model. J. Biomed. Res. 30, 490–495.

CHAPTER 5

Insulin resistance and stroke

Introduction

As mentioned in Chapter 2, Insulin resistance, diabetes, and metabolic syndrome, insulin resistance is defined as a decrease in tissue response to insulin stimulation. It may be caused by the inability of insulin receptors to bind insulin or faulty activation of the proximal insulin signaling cascade. Insulin resistance is characterized by hyperglycemia, hyperinsulinemia, defects in uptake and oxidation of glucose, decrease in glycogen synthesis, and ability to suppress lipid oxidation (Morino et al., 2006; Thaler and Schwartz, 2010; Thaler et al., 2012). These processes disturb the physiological function of several vital organs via the impairment of insulin signaling and the disturbance of intracellular signaling transduction. The mechanisms contributing to insulin resistance are not fully elucidated. However, mitochondrial dysfunction is closely associated with the pathogenesis of insulin resistance. To this end, mitochondria play a pivotal role in insulin signaling (Cheng et al., 2010). As mentioned in Chapter 1, Insulin resistance and obesity, insulin binds with its receptor, mediating the activation of cellular glucose uptake through glucose transporters. Following uptake, glucose is converted to pyruvate by the glycolytic process and these pyruvates are then converted to Acetyl-CoA, a substrate of the Krebs cycle, by glucose oxidation (Schenk et al., 2008; Montgomery and Turner, 2015). In addition, insulin stimulates the uptake of cellular fatty acids into the cells and the fatty acids are further converted to fatty acyl-CoA (Schenk et al., 2008). Fatty acyl-CoA can either be converted into several lipid products, including diacylglycerol (DAG), triacylglycerol (TAG), and ceramide or be directly transported to mitochondria to induce mitochondrial β-oxidation, resulting in the production of acetyl-CoA and CO_2 in the Krebs cycle (Schenk et al., 2008; Montgomery and Turner, 2015). Generation of long-chain fatty acids, DAG, TAG, and ceremide promotes the induction of insulin resistance. In addition, genetic and environmental factors and complex interactions among inflammation, endoplasmic reticulum stress (ERS), oxidative stress, and dysregulated autophagy also play important roles in the pathogenesis of insulin

Insulin Resistance as a Risk Factor in Visceral and Neurological Disorders
DOI: https://doi.org/10.1016/B978-0-12-819603-8.00005-5

resistance. Several studies have also indicated that increased levels of adipocytokines and genetic alterations in insulin signaling may also contribute to the pathogenesis of insulin resistance (Shoelson et al., 2006; Schenk et al., 2008; Hotamisligil, 2008; Brüning et al., 1997). Recently, stress-activated c-Jun N-terminal kinase (JNK) has been recognized as a central mediator of insulin resistance. JNK is a member of the mitogen-activated protein kinases (MAPK) family. It differs from classical MAPK such as ERK in the fact that JNK activity is more potently induced in response to cellular stress than to mitogens (Chang and Karin, 2001). JNK not only mediates and modulates effects of stress on insulin resistance through inhibitory phosphorylation of insulin receptor substrate, but also promotes plaque instability in the arterial wall. The suppression of the JNK pathway has been shown to improve insulin resistance and glucose tolerance supporting the view that JNK may serve as a crucial link between stress and metabolic diseases and can be used as a promising therapeutic target (Solinas and Becattini, 2016). Over-nutrition (long-term consumption of western diet), changes in gut microbiota composition, and physical inactivity may also underlie the development of insulin resistance and type 2 diabetes (McAuley and Mann, 2006; Weickert, 2012; Farooqui, 2015). Thus long-term consumption of energy-dense/high-fat and high-carbohydrate diets and physical inactivity is strongly and positively associated with obesity, which negatively effects insulin sensitivity, particularly when the excess of body fat is located in abdominal region and around vasculature (Fig. 5.1). Accumulation of intraabdominal fat mass is the most important cause of insulin resistance and type 2 diabetes. Simply being overweight ($BMI > 25\ kg/m^2$) raises the risk of developing type 2 diabetes by a factor of 3 (Brancati et al., 1999). It is known since decades that this effect can be effectively reversed by reduction of excess body weight (Newburgh, 1972) in obese patients with poorly controlled type 2 diabetes even modest weight loss, if maintained, markedly reduces plasma glucose concentrations and improves markers of glucose metabolism (Redmon et al., 2005). Therefore appropriate dietary measures and exercise as part of a healthy lifestyle are known to substantially reduce insulin resistance and risk of type 2 diabetes (Maggio and Pi-Sunyer, 1997).

The link among consumption of high-fat diet, obesity, and insulin resistance is not limited to the high-energy content of fatty foods, but also stimulation of storage mechanisms. Adipose tissue is an important energy sink. It not only stores the energy, but also controls energy expenditure, appetite regulation, and glucose regulation. Fat in abdominal and around

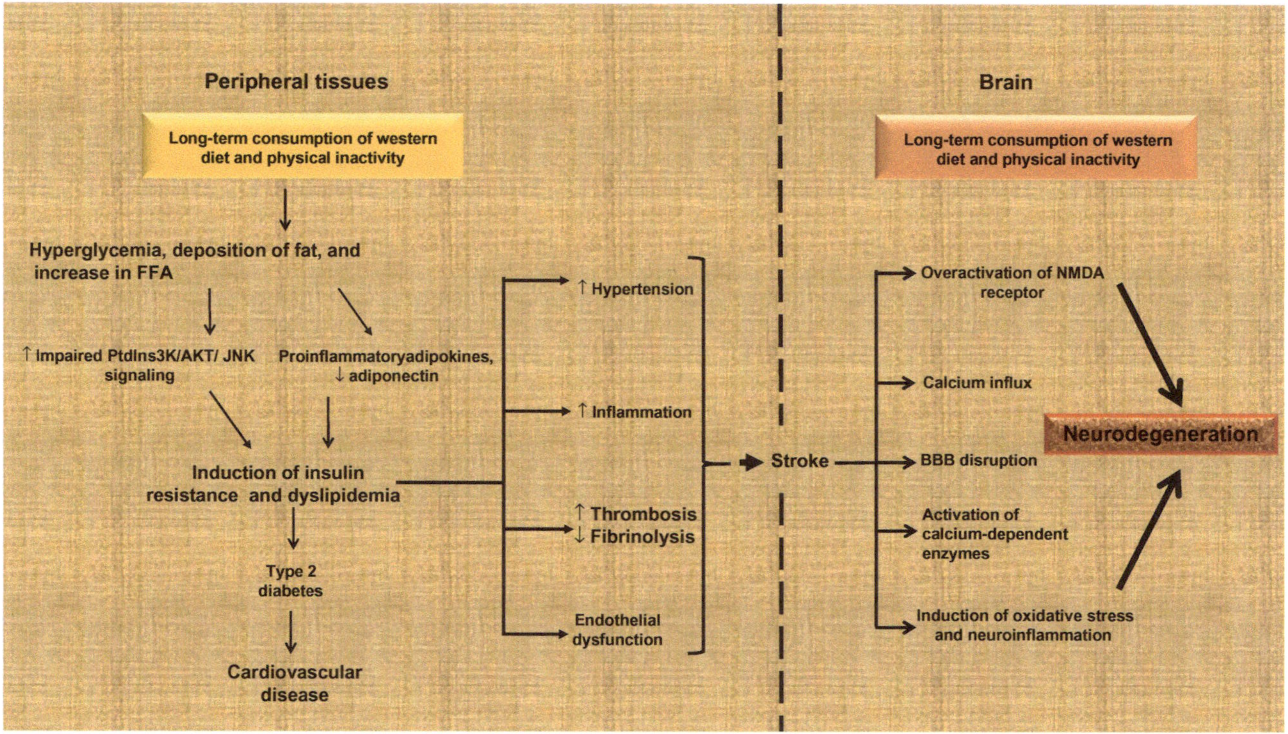

Figure 5.1 Biochemical changes in associated with stroke.
AKT, Protein kinase B; *BBB*, blood–brain barrier; *FFAs*, free fatty acids; *NMDA*, *N*-methyl-D-aspartate; *PtdIns 3K*, phosphatidylinositol 3-kinase; *tPA*, tissue-type plasminogen activator.

vasculature adipose tissue is critical for thyroid function, immune response, bone health maintenance, reproduction, and blood clotting. Furthermore, adipose tissue is an active endocrine organ secreting free fatty acids (FFAs), leptin, adiponectin, adipsin, complement factor 3, interleukin-6 (IL-6), tumor necrosis factor-α (TNF-α), angiotensinogen, and plasminogen activation inhibitor-1 (PAI-1), among others (Fig. 5.2). Abnormal signaling and the deficiency of abovementioned hormones result in deleterious effects (Kissebah et al., 1982; Xing et al., 1998). Thus, during obesity, adipose tissue becomes dysfunctional. It not only loses its thermogenic capacity, but also secretes proinflammatory adipokines, which induce endothelial dysfunction and infiltration of inflammatory cells, promoting the development of atherosclerosis. Insulin resistance can also be induced by dietary habits. Thus vitamin D deficiency (i.e., hypovitaminosis D) promotes insulin resistance by impairs insulin secretion. This results in poorly controlled glucose homeostasis (Leung, 2016).

Long-term consumption of high-fat diet (HFD) not only contributes to obesity, but is also associated with reduction in the levels of neuronal

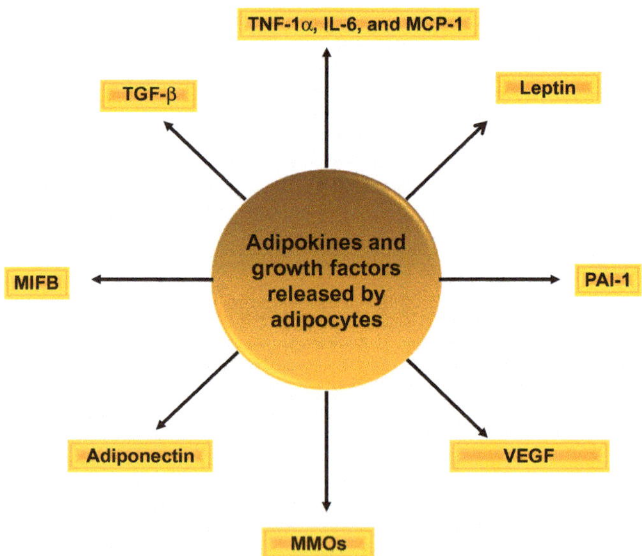

Figure 5.2 Role of adipocytes in metabolism.
IL-6, Interleukin-6; *MCP-1*, monocyte chemoattractant protein-1; *MIFB*, macrophage migration inhibitory factor; *PAI-1*, plasminogen activation inhibitor-1; *TGF-β*, transforming growth factor beta; *TNF-α*, tumor necrosis factor-alpha; *VEGF*, vascular endothelial growth factor.

plasticity-related proteins, specifically brain-derived neurotrophic factor (BDNF) and synaptophysin (SYN), in the hippocampus. The molecular mechanisms contributing to the above processes are not fully understood. However, it is well known that ERS plays a key role in regulating gene expression and protein production by affecting stress signaling pathways and ER functions of protein folding and posttranslational modification in peripheral tissues of obese rodent models (Cai et al., 2016). As mentioned earlier, long-term consumption of HFD can also contribute to hyperglycemia. The induction of hyperglycemia not only impair insulin signaling, but also promote cognitive dysfunction health in HFD mice. Regular aerobic exercise is known to reduce hyperglycemia, decreases central inflammation, and elevates hippocampal BDNF and SYN levels in obese rats. The molecular mechanism associated with this process is not known. However, it is suggested that aerobic exercise produces its effect by activating the Nrf2–HO-1 pathway. This pathway not only relieves ERS, but also elevates BDNF and SYN production in obese rats (Cai et al., 2016).

It is well known that human brain uses glucose as a primary fuel; insulin secreted by the pancreas cross the blood–brain barrier (BBB) and reaches neurons and glial cells. BBB is a physical barrier composed of endothelial, mural and glia cells, astrocytes, and macrophages. Among these cells, endothelial cells, which are located in blood vessels, play a major role in BBB proper functioning (Daneman and Prat, 2015). BBB permits the passage of water, some gases and lipophilic molecules by passive diffusion and the selective transport of certain molecules (e.g., glucose) and protects against external toxins and pathogens. Oxidative stress affects the integrity of BBB. Thus increase in ROS production contributes to endothelium dysfunction and increased permeability of BBB (Enciu et al., 2013). These alterations are mainly attributed to the redistribution and/or altered expression of critical tight junction proteins such as claudin-5 and occludin (Lochhead et al., 2010).

Insulin exerts a region-specific effect on glucose metabolism in human brain. Glucose homeostasis is critical for energy generation, neuronal maintenance, neurogenesis, neurotransmitter regulation, cell survival, and synaptic plasticity. It also plays a key role in cognitive function. In an insulin resistance, there is a reduction in sensitivity to insulin resulting in hyperinsulinemia; this condition persists for several years before becoming full-blown diabetes. Collective evidence suggests that insulin resistance not only contributes to the defects in insulin receptor function, abnormalities in insulin signaling, alterations in glucose metabolism, induction of hyperinsulinemia, hyperglycemia, and inflammation, but also increases blood pressure (Wang and Jin, 2009). Insulin resistance is observed in approximately 90% of patients with type 2 diabetes and in 66% of individuals with impaired glucose tolerance. Insulin resistance together with β-cell dysfunction and apoptosis are the two fundamental mechanisms for the development of type 2 diabetes (Urbanavičius et al., 2013; Butler et al., 2003). As mentioned in Chapter 2, Insulin resistance, diabetes, and metabolic syndrome, insulin resistance per se doubles the risk for cardiovascular disease, which is the ultimate cause of death in about 80% of patients with type 2 diabetes. This observation suggests that induction of hyperglycemia and proatherogenic effects may be important risk factor not only in the pathogenesis of cardiovascular diseases such as coronary artery disease (CAD) and peripheral arterial disease (Hanley et al., 2002; Meigs et al., 2007; Ergul et al., 2012; Schaper et al., 2000), but also contribute to neurological disorders such as stroke and Alzheimer's disease (Farooqui, 2013). Acute stroke not only induces a local neuroinflammatory reaction, but also alters peripheral immune homeostasis. On the basis of several studies, it is suggested that insulin resistance and hyperglycemia in type 2 diabetes may decrease in cerebral blood flow (CBF) and size of lesions leading to hypoperfusion of brain tissue, a process, which may lead neuroglial energy crisis, synaptic dysfunction, decrease in LTP, and brain atrophy (Fig. 5.3). The development of these parameters may markedly increase the risk of stroke (Laing et al., 2003). Type 2 diabetes patients make up the vast majority of diabetic stroke and have more than fourfold higher rates of stroke at all ages of the disease according to pathway shown in Fig. 5.3 (Janghorbani et al., 2007; Hardigan et al., 2016; Ergul et al., 2012).

Hypertension and pathogenesis of stroke

Hyperglycemia is the technical term used for high blood glucose (>200 mg glucose/dL). High blood glucose happens when the body has too little insulin or when the body can't use insulin properly. Two diagnostic categories of hyperglycemia have been described: (1) hospital-related hyperglycemia is defined as when fasting glucose concentration reaches >126 mg/dL or random glucose >200 mg/dL without evidence of previous diabetes in blood and urine (American Diabetes Association) and (2) preexisting diabetes with deterioration of preillness glycemic

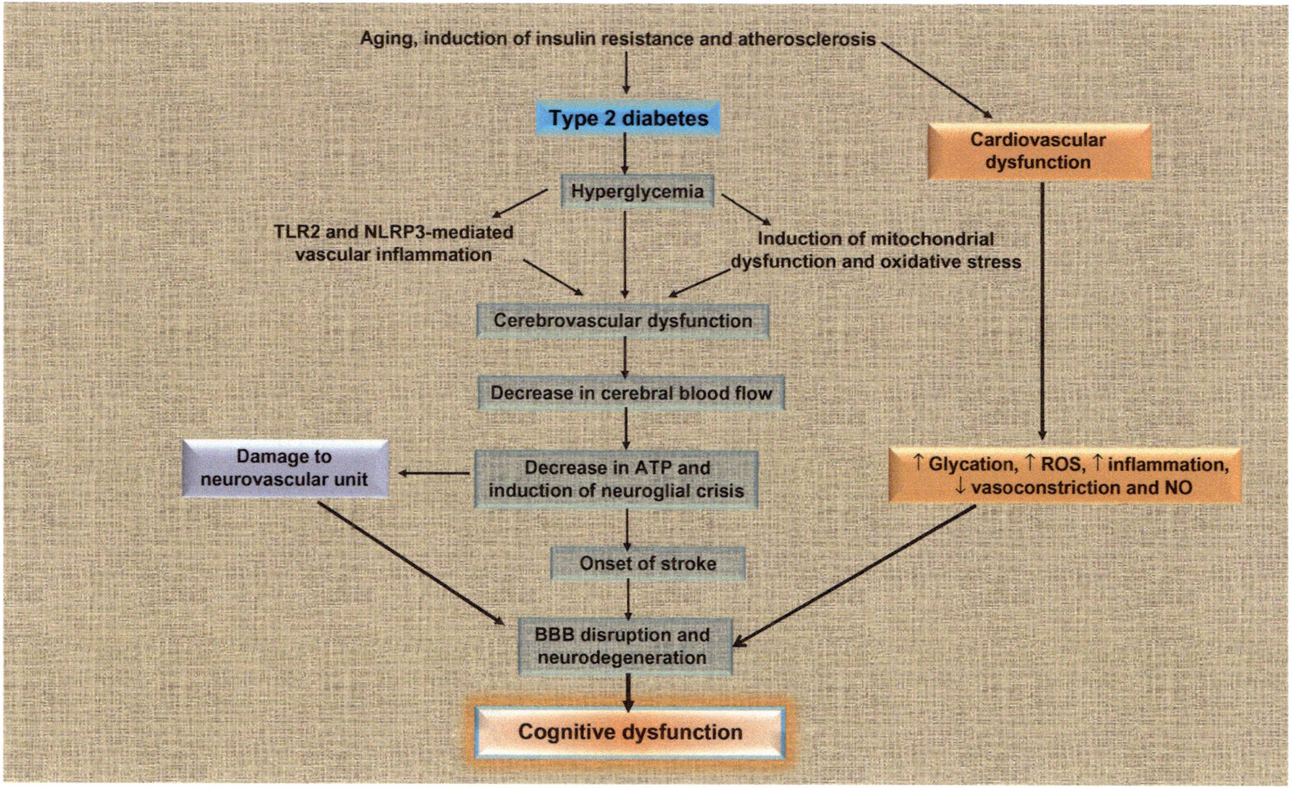

Figure 5.3 Pathways contributing to stroke in insulin resistance diabetic patients. *NLRP3*, NACHT, LRR, and PYD domains-containing protein 3; *TLR2*, Toll-like receptor 2.

control (Dungan et al., 2009). Symptoms of hyperglycemia include: (1) high glucose levels ($>$200 mg glucose/dL) in blood and urine, (2) frequent urination, and (3) increased thirst. Hyperglycemia at the time of acute ischemic stroke increases chances of death in stroke patients. Acute hyperglycemia not only increases the risk of severe ischemia and intracerebral hemorrhage after reperfusion, but also decreases the likelihood of neurologic recovery. Molecular mechanisms by which hyperglycemia contribute to worsening of functional outcome following acute ischemic stroke are not fully understood. However, on the basis of animal model studies, it is suggested that, in focal brain ischemia, hyperglycemia consistently increase infarct size by several mechanisms including blood coagulation and in fibrinolytic pathways (Vaidyula et al., 2006; Gentile et al., 2007), the decrease in reperfusion of the damaged brain area caused by the disturbances in metabolism of nitric oxide (NO), and increase in reperfusion injury which is the result of the detrimental effects of oxidative stress and inflammation (Bémeur et al., 2007). The effects of the abovementioned mechanisms alter the recovery of the ischemic penumbra a part of the ischemic area, which may still recover if proper reperfusion is restored within hours after the stroke onset. In contrast, other studies suggest that the association between hyperglycemia and poor outcome after stroke is stronger in patients with large-vessel thromboembolic stroke than in those with lacunar stroke and this is understandable considering that hyperglycemia primarily exerts its detrimental effects at the level of the ischemic penumbra which is usually not present in lacunar subtype (Uyttenboogaart et al., 2007; Bruno et al., 1999). In addition, hyperglycemia-mediated metabolic derangements may also involve lactic acid–mediated brain acidosis, cytotoxic brain edema, hemorrhagic transformation of infarcts, impaired thrombolysis, BBB disruption, slow recovery from calcium influx, mitochondrial dysfunction, and induction of oxidative stress and neuroinflammation. All these processes may be associated with neurodegeneration and death in diabetes linked stroke (Fig. 5.4). Another potential mechanism of hyperglycemia-mediated brain damage may involve detrimental changes in cerebral endothelial cells within hours after stroke (Martini and Kent, 2007). Acute hyperglycemia can impair cerebrovascular autoregulation, resulting in deleterious reperfusion predisposing to hemorrhagic transformation of infarcts in stroke patients. Hyperglycemia can also promote and support inflammation through the activation of nuclear factor kappaB (NF-κB), leading to increase in expression of proinflammatory cytokines and chemokine,

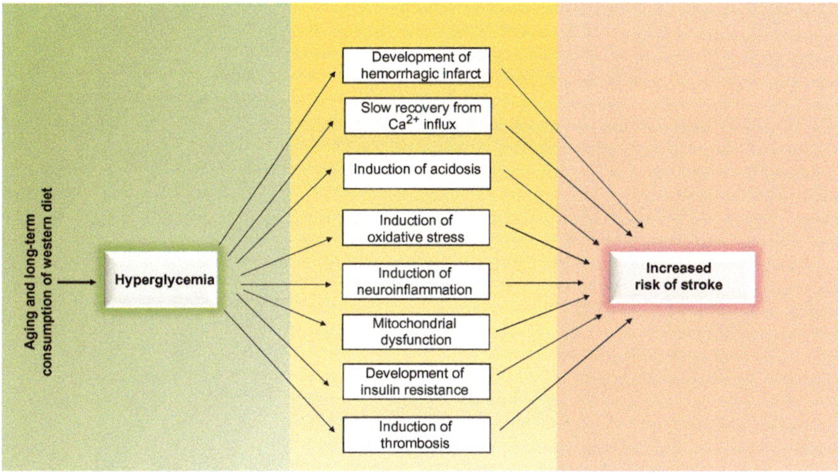

Figure 5.4 Effect of hyperglycemia on the brain.

intracellular adhesion molecule 1, vascular cellular adhesion molecule 1, and E-selectin (Martini and Kent, 2007). The adhesion molecules promote leukocyte adhesion to postcapillary venule walls, obstructing blood flow (Zoppo and Mabuchi, 2003). In addition, hyperglycemia in diabetic patients also increases coagulation by stimulating not only thrombin production, but also the tissue factor pathway (Gentile et al., 2007). Hyperglycemia also reduces the fibrinolytic activity of tissue-type plasminogen activator (tPA) by increasing the production of PAI-1 (Pandolfi et al., 2001). Additionally, hyperglycemia during stroke may not only exacerbate or accelerate some of the pathologic processes involved in ischemic brain injury (Bruno et al., 2010), but also increase the risk of cerebral hemorrhage in acute stroke patients after treatment with intravenous tPA (Bruno et al., 2002; Putaala et al., 2011).

Contribution of diet, microbiota, and insulin resistance in the pathogenesis of stroke

It is becoming increasingly evident that there exist a bidirectional communication and interaction between the gut and brain through autonomic, neuroendocrine, enteric, and immune system pathways (Fig. 5.5) (Zhao et al., 2018). Gut microbiota communicate with the brain by producing a number of neuroactive and immunocompetent substances (hormones, neurotransmitters, neuropeptides, and short-chain fatty acids) (Fig. 5.5).

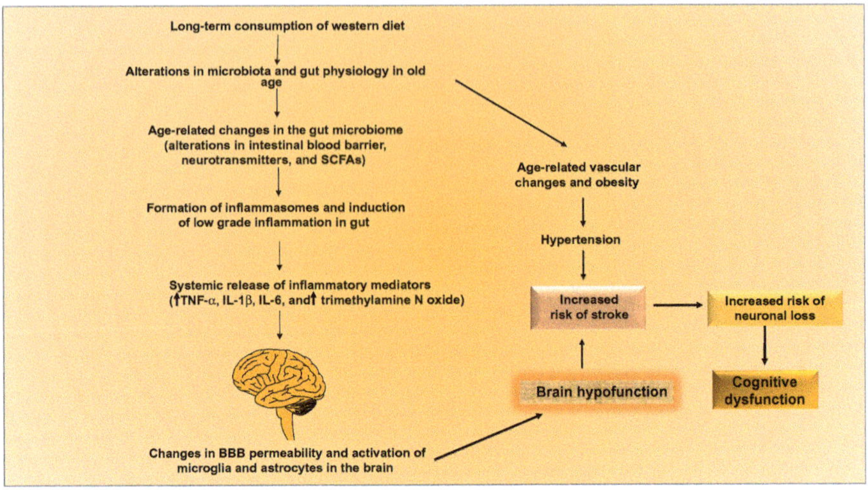

Figure 5.5 Alterations in microbiota and risk of stroke.
IL-1β, Interleukin-1β; *SCFAs*, short-chain fatty acids.

These substances are transported from gut to the brain through lymphatic subsystem, which allows for direct or indirect transport (Winek et al., 2016a,b). This specialized lymphatic system drains from the cerebrospinal fluid in the adjacent subarachnoid space and the interstitial fluid (ISF) to the deep cervical lymph nodes and may be visualized by molecular imaging (Aspelund et al., 2015; Louveau et al., 2015; Sharon et al., 2016). Gut microbiota also modulate afferent sensory nerves, for example, through the inhibition of calcium-dependent potassium channels (Kunze et al., 2009), and regulate the mucosal immune function (McDermott and Huffnagle, 2014). Furthermore, microbiota influence the specific phenotypes and physiological functions of T and B cells in the gut mucosa layer (Honda and Littman, 2016). These cells play a pivotal role in immune homeostasis by defending against foreign antigens and by maintaining the integrity of the gut mucosal barrier. A balanced gut–microbiota stimulates resident macrophages to release large amounts of IL-10 and transforming growth factor beta (Schieber et al., 2015), thus promoting the induction of regulatory T cells (Tregs) and preventing an increase in the number of proinflammatory T helper 17 (Th17) cells in the gut (Rivollier et al., 2012). In fact, a peripheral tolerance is maintained by the correct balance between gut–bacteria population and responses by the host. Conversely, the brain can in turn alter microbial composition, gut microenvironment, and behavior via the autonomic nervous system. Changes in this microenvironment can lead to a broad spectrum of physiological and behavioral effects including hypothalamic–pituitary–adrenal axis activation, and altered activity of neurotransmitter systems (γ-Amino butyric acid (GABA), serotonin, acetylcholine, histamine, and melatonin) and immune function. Collective evidence suggests that microbiota–gut–brain axis is a complex multidirectional crosstalk system between the gut microbiota, the enteric nervous system (ENS), and the brain. It has been suggested that interactions between brain and gut may shape the structure and function of brain regions involved in the control of emotions, cognition, and physical activity. While an appropriate, coordinated physiological response (immune or stress response) is necessary for survival of neural cells, a dysfunctional physiological response can be detrimental to the host contributing to the development of a number of CNS disorders. After stroke, up to 50% of patients experience gastrointestinal complications, including gut dysmotility, gut microbiota dysbiosis, "leaky" gut, gut hemorrhage, and even gut-origin sepsis (Wen and Wong, 2017). Stroke patients associated with gastrointestinal complications often have poor outcomes, with increased mortality rates and deteriorating neurologic function (Camara-Lemarroy et al., 2014). Gut dysbiosis or gut-origin sepsis are both reported to occur after stroke in aged animals (Singh et al., 2016; Ritzel et al., 2018). The underlying mechanisms of stroke-associated gastrointestinal complications, as well as the poor stroke outcome, remain understudied. The communication between the brain and the gut takes place through several complex signaling pathways involving vagus nerves to the ENS, the neuronal–glial–endothelial interactions, as well as, damage-associated molecular patterns- and cytokines-induced activation of gut inflammatory and immune cells (Collins et al., 2012; Matteoli and Boeckxstaens, 2013). The afferent fiber of the vagus nerve expresses receptor to sense microbiota metabolite, gut peptides like ghrelin, and leptin to transfer gut information to the brain (Bonaz et al., 2018; de Lartigue et al., 2011).

As mentioned in Chapter 1, Insulin resistance and obesity, gut microbiota generate trimethylamine (TMA), which can be further oxidized as trimethylamine N oxide (TMAO) in the host liver by flavin monooxygenases (Wang et al., 2011; Koeth et al., 2013). TMAO is known to show the proatherogenic properties. Circulating high TMAO levels are associated not only with prevalence of cardiovascular disease and myocardial infarction, but also with stroke (Wang et al., 2011; Tang et al., 2013). It is suggested that TMAO promote atherosclerosis and thrombosis not only by activating endothelial cell MAPK and NF-κB signaling, promoting expression of inflammatory genes, and inducing NLRP3 inflammasome formation (Sun et al., 2016; Boini et al., 2017; Chen et al., 2017), but also by increasing endoplasmic recticulum calcium release in platelet cell, consequently leading to platelet aggregation and thrombosis (Zhu et al., 2016). Decreasing the consumption of TMA precursors in diet and inhibition of TMAO production has been used to protect from harmful effects of microbiota. In addition, gut microbiota also play an important role in regulation of BBB. Germ-free animals show reduced expression of tight junction proteins, which results in increased permeability of the BBB for various molecules (Winek et al., 2016a,c). Complete depletion of gut microbiota with a broad-spectrum antibiotic cocktail results in increased mortality rates and development of severe colitis in a mouse model of stroke (Winek et al., 2016c) supporting the view that gut microbiota may contribute to the stability of BBB and disruption of BBB is associated with the pathogenesis of stroke. It is reported that antibiotic-mediated microbiota dysbiosis is accompanied by an increase in regulatory T cells, suppression of IL-17-positive γδ T cells and thus reducing the trafficking of effector T cells from the gut to the leptomeninges after stroke, resulting in smaller infarcts and improved behavioral outcome (Benakis et al., 2016). Converging evidence suggests that multiple mechanisms, including endocrine and neurocrine pathways, may be involved in gut microbiota-to-brain signaling. Targeting the microbiota in therapy for stroke is a promising approach (Zinnhardt et al., 2018). Various microbiota-improving methods including fecal microbiota transplantation, probiotics, prebiotics, a healthy diet, and healthy lifestyle have shown the capability to promote the function of the gut–brain, microbiota–gut–brain axis, and brain (Kostic et al., 2014; Zinnhardt et al., 2018). These methods may partially restore bacterial species diversity and improve stroke outcome via a lymphocyte-dependent mechanism (Singh et al., 2016). More studies are needed on possibility of treating metabolic diseases (type 2 diabetes, obesity, and inflammatory bowel disease) and neurological disorders (stroke and Alzheimer's disease) by improving gut microbiota composition through diet.

Stroke-mediated changes in the brain

Stroke is caused by severe reduction or blockade in CBF due to the formation of a clot. This blockade not only decreases delivery of oxygen and glucose to the brain tissue, but also results in the breakdown of BBB and buildup of potentially toxic products in brain (Farooqui, 2018). Deprivation of oxygen to the brain for just 30 minutes during ischemic stroke produces devastating consequences. It results in the death of ~1.9 million neurons and ~14 million synapses every minute (Saver, 2006). Neurons and their synapses perish because without sufficient O_2, mitochondria are unable to reduce O_2 to H_2O to support ATP synthesis (Bailey et al., 2009). Two types of strokes are known to occur in human population namely ischemic stroke and hemorrhagic stroke is caused by the bursting of the blood vessels. Stroke is confirmed by brain CT and/or MRI in baseline conditions and represents the second most common cause of mortality and the third most common cause of disability in developed countries. In the Western world, most strokes are ischemic (88%), whereas in Asian countries majorities of strokes are of hemorrhagic (12%) in origin. The most ischemic stroke is due to the middle cerebral artery occlusion, resulting in the brain tissue damage in the affected territory, which is followed by inflammatory and immune response. An ischemic injury produces two distinguishable areas: the infarcted core, which is the region supplied by the occluded vessel; and the penumbra, which is the area between the lethally damaged core and normally perfused territory, which receives some collateral blood flow from unaffected vessels (Astrup et al., 1981; Hossmann, 1994). Neural cells within the ischemic core are often irreversibly damaged even if blood flow is reestablished. The ischemic penumbra, however, can be defined by a moderate reduction in CBF where collateral blood vessels provide neural cells with limited metabolic nutrients to temporarily maintain homeostasis during the initial stages of ischemia, but it is nonfunctional (Heiss et al., 2004). Mechanisms of neurodegeneration in ischemic core and penumbra are different (Smith, 2004). Necrosis and apoptosis, two major modes of cell death, are implicated in ischemia. Necrosis is predominant in core tissue, whereas both necrosis and apoptosis are dominant in the penumbra (Smith, 2004). Collective evidence suggests that stroke-mediated injury often leads to gut dysmotility, gut microbiota dysbiosis, "leaky" gut, gut hemorrhage, and even gut-origin sepsis, which is often associated with poor prognosis. Emerging evidence suggests that gut inflammatory and immune response play a key role in the pathophysiology of stroke. Ischemic brain injury produces damage to molecular patterns to initiate innate and adaptive immune response both locally and systemically through the specialized pattern-recognition receptors (e.g., toll-like receptors) (Farooqui, 2018).

After stroke, innate immune cells including neutrophils, microglia or macrophages, mast cells, innate lymphocytes (IL-17 secreting $\gamma\delta$ T-cell), and natural killer T-cell respond within hours, followed by the adaptive immune response through activation of T and B lymphocytes. Subpopulations of T-cells can help or worsen ischemic brain injury (Farooqui, 2018). Proinflammatory Th1, Th17, and $\gamma\delta$ T-cells are often associated with increased inflammatory damage, whereas regulatory T-cells are known to suppress postischemic inflammation by increasing the secretion of antiinflammatory cytokine IL-10. Although known to play a key role, research in the gut inflammatory and immune response after stroke is still in its initial stage (Farooqui, 2018).

Location of stroke on the left hemisphere promote disturbance in language and comprehension, which reduce the ability to communicate (Pirmoradi et al., 2016). In contrast, when stroke affects the right hemisphere, the intuitive thinking, reasoning, solving problems as well as the perception, judgment and the visualspatial functions may be impaired (Cumming et al., 2013; Sun et al., 2014; Tiozzo et al., 2015; Save-Pédebos et al., 2016). Two types of risk factors (modifiable and nonmodifiable) are known to modulate stroke-mediated injury. Nonmodifiable factors include family history of cerebrovascular diseases, older age, male sex, and Hispanic or Black race along with lifestyle (Allen and Bayraktutan, 2008). Modifiable risk factors include hypertension, diabetes, heart disease, hypercholesterolemia (atherosclerosis), atrial fibrillation (most common sustained cardiac arrhythmia), high alcohol consumption, cigarette smoke, and oral contraceptive (Allen and Bayraktutan, 2008).

Molecular mechanisms contributing to stroke-mediated brain damage

Ischemic brain injury not only involves excitotoxicity, oxidative stress, apoptosis, and neuroinflammation, but also activation of glial cells (astrocytes and microglia) and infiltration of leukocytes (Fig. 5.6) (Moskowitz et al., 2010; Farooqui, 2018). These processes result in abnormal signal transduction pathways, which contribute to necrosis in the ischemic core of the infarct within minutes (Moskowitz et al., 2010). In contrast, in penumbra, which remains partially perfused—usually through collateral vessels—and neurons die predominantly through apoptotic cell death (Hara and Snyder, 2007). Importantly, many researchers suggest that cell death in the penumbra can be prevented with suitable neuroprotective drugs. At the molecular level, stroke is accompanied by the release of

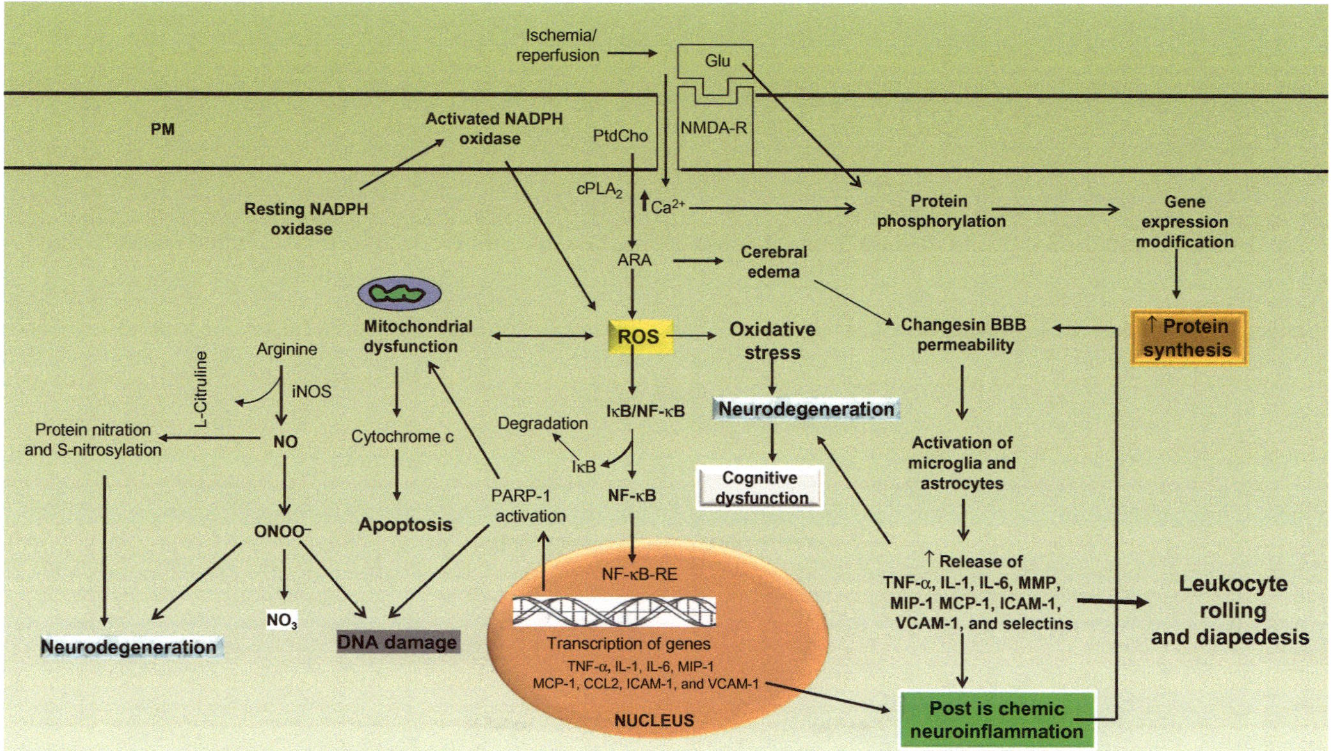

Figure 5.6 Processes contributing to stroke-mediated brain injury.

ARA, Arachidonic acid; *COX-2*, cyclooxygenase-2; *cPLA*$_2$, cytosolic phospholipase A$_2$; *Glu*, glutamate; *ICAMs*, intercellular adhesion molecules; *iNOS*, inducible nitric oxide synthase; *MIP1*, macrophage inflammatory protein 1; *MCP-1*, monocyte chemoattractant protein-1; *NF-κB*, nuclear factor-κB; *NO*, nitric oxide; *NMDA-R*, *N*-methyl-D-aspartate receptor; $ONOO^-$, peroxynitrite; *PM*, plasma membrane; *PtdCho*, phosphatidylcholine; *ROS*, reactive oxygen species; *VCAM*, vascular cell adhesion protein 1.

glutamate, overactivation of NMDA receptors, influx of Ca^{2+}, increase in levels of FFAs, generation, and activation of nicotinamide adenine dinucleotide phosphate (NADPH) oxidase. Influx of Ca^{2+} is associated with activation of Ca^{2+}-dependent enzymes including phospholipases A_2 (PLA_2), cyclooxygenases (COXs), lipoxygenases (LOXs), protein kinases, nitric oxide synthases (NOS), endonucleases, and matrix metalloproteinases (MMPs) (Moskowitz et al., 2010; Farooqui, 2018). The massive influx of Ca^{2+} not only activates the neuronal NOS (nNOS) and calpain I, but also results in depolarization of mitochondrial membrane and the opening of the mitochondrial permeability transition pore, causing mitochondrial permeability transition along with increase in ROS, and neuronal cell death (Farooqui, 2018). PLA_2 and COX-1 and COX-2, and LOXs contribute to the breakdown of neural membrane phospholipid (Fig. 5.7). The degradation of neural membrane phospholipids results in the release of arachidonic acid (ARA) and synthesis of lysophospholipids. Enzymic oxidation of ARA by COX and LOX results in generation of prostaglandins, leukotrienes, and thromboxanes. These metabolites are collectively known as eicosanoids. They produce wide ranges of biological effects including neuroinflammation, vasodilation, vasoconstriction, apoptosis, and immune responses. These processes are accompanied by the activation of microglia, which develop macrophage-like capabilities including phagocytosis, cytokine production, antigen presentation, and the release of MMPs. As mentioned above, activation of MMPs weakens the BBB. This process contributes to transmigration of leukocytes across the blood vessels and into the brain contributing to more inflammation with increase in ROS production. The nonenzymic oxidation of ARA not only produces 4-hydroxynonenal, isoprostanes, acrolein, and malondialdehyde, but also generate a variety of reactive oxygen species (ROS) including superoxide radical ($O_2^{\bullet -}$) (Farooqui, 2014). The initial product, $O_2^{\bullet -}$, results from the addition of a single electron to molecular oxygen. $O_2^{\bullet -}$ is rapidly dismutated by superoxide dismutase (SOD), yielding H_2O_2 and O_2, which can be reused to generate superoxide radical. In the presence of reduced transition metals (iron and copper), H_2O_2, although less reactive than $O^{\bullet -}{}_2$ and highly diffusible can be converted into the highly reactive $^{\bullet}OH$. ROS are not only produced by uncontrolled ARA cascade, but also by mitochondrial respiratory chain, and activation of NADPH oxidases (Farooqui, 2014). Under physiological conditions, low levels of ROS are involved in signal transduction processes, which are needed for growth, adaptation responses, neuronal development, and

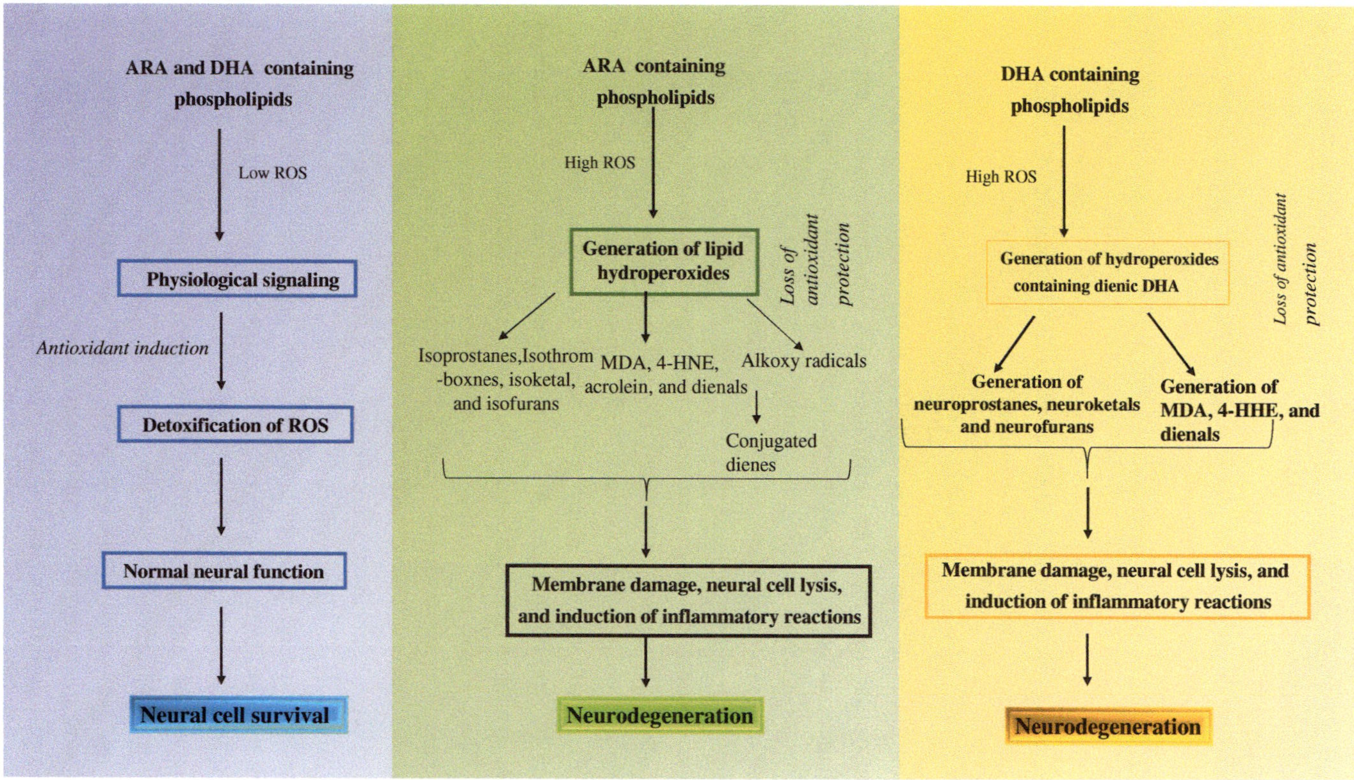

Figure 5.7 Effect of various levels of oxidative stress on the brain.
DHA, Docosahexaenoic acid; *4-HHE*, 4-hydroxyhexanal; *4-HNE*, 4-hydroxynonenal; *MDA*, malondialdehyde.

normal functioning of immune system. In contrast, under ischemic conditions, high levels of ROS may lead to the endothelial dysfunction by reducing the bioavailability of endothelium-derived NO and formation of advanced glycosylation end products (AGEs). The binding of AGE with receptor for AGE (RAGE) accelerates the atherosclerotic process by promoting low-density lipoprotein uptake and their oxidation. These processes may lead to foam cell formation and the diversion of glucose into the aldose reductase pathway, which ultimately results in the activation of one or more isozymes of protein kinase C (Schmidt et al., 1999; Farooqui, 2013). Furthermore, high levels of ROS facilitate the migration of NF-κB from cytoplasm to the nucleus, where it interacts with NF-κB response element to facilitate the expression of proinflammatory cytokines (TNF-α, IL-1β, IL-6, and IL-12), chemokines (MIP-1α and MCPP1), and adhesion molecules (ICAM and VCAM) leading to induction of more neuroinflammation. The restoration of CBF (postischemic reperfusion) enhances tissue oxygenation exacerbating ROS production, inflammatory responses and further oxidative damage to neural membrane phospholipids, proteins, and nucleic acids (Farooqui, 2014) along with inactivation of enzymes, release of Ca^{2+} from intracellular store, and damage to the cytoskeletal structure and chemotaxis. In addition, excessive levels ROS also produce diminish LTP and induce synaptic dysfunction and decrease neuroplasticity (Salim, 2017). The detoxification of ROS is a major prerequisite for the survival of neural cells. The detoxification is performed by several enzymic and nonenzymic antioxidant mechanisms, which are available to the neural cells in different cellular compartments (Droge, 2002; Farooqui, 2010). Enzymic mechanisms involve enzymes like superoxide dismutases (Mn-SODs), which are located in the mitochondria and cytosol (Cu-Zn-SOD). These enzymes convert $O_2^{\bullet-}$ into diatomic oxygen and hydrogen peroxide. Glutathione peroxidases and catalase convert H_2O_2 into water.

Neural membrane phospholipids contain polyunsaturated fatty acids, which are susceptible to lipid peroxidation. Superoxide and hydroxyl radicals attack long-chain fatty acids [ARA and docosahexaenoic acid (DHA)] and generate several metabolites. Nonenzymic metabolites of ARA metabolism include endoperoxides, isoprostanes, isoketals, and isofurans. In contrast, nonenzymic oxidation of DHA generates endoperoxides, neuroprostanes, neuroketals, and neurofurans. These metabolites are known to produce many pharmacological and toxicological effects in the brain. These effects are known to contribute to neurodegeneration in the brain (Fig. 5.7) (Farooqui and Horrocks, 2007). Free radical attack on fatty acid chains of neural membrane phospholipids disturbs the assembly of cell membranes, which inevitably has impact not only on membrane fluidity, lipid–lipid, and lipid–protein interaction dynamics, membrane permeability, and physicochemical properties, but also ion, and nutrient transport, membrane-initiated signaling pathways, and metabolic processes. Alterations in these processes may lead to neuronal cell death (Farooqui and Horrocks, 2007; Adibhatla and Hatcher, 2010). Intensive research work performed over the last decades has also revealed that under physiological conditions, low levels of lipid peroxidation products may be involved in cellular signaling associated with variety of cellular functions as well as normal aging (Pamplona, 2008). However, under ischemic condition, high levels of ARA- and DHA-derived metabolites are associated with abnormal neural function and neurodegeneration. In addition to fatty acids, human brain neural membranes contain large amounts of cholesterol. This lipid is essential for normal brain function. It not only plays an essential role in determining the fluidity and electrical and permeability, but it also regulates activities of membrane bound enzymes, receptors, and ion channels (Simons and Ikonen, 2000). Nonenzymic oxidation of cholesterol by ROS results in formation of oxysterols. The chemical structures of oxysterols vary depending upon the number and position of oxygenated functional groups present on steroid ring. Oxysterols may include keto-, hydroxyperoxy-, and epoxy-steroid (Simons and Ehehalt, 2002). Generation of these metabolites may contribute to neurodegeneration under ischemic conditions.

Proteins are the major target for free radical attack at multiple sites due to the presence of amino acids (side-chain and polypeptide backbone), which are susceptible to oxidation (Davies, 2005, 2016). Thus oxidative damage to amino acid side chains in the presence of metal ions and hydrogen peroxide produces semialdehyde amino acids, with the majority of these reactions occurring with lysine, arginine, and proline residues (Stadtman and Berlett, 1991). In addition, modification of protein may result not only in increased side-chain hydrophilicity, side-chain and backbone fragmentation, but also aggregation of proteins via covalent cross-linking or hydrophobic interactions, protein unfolding and altered conformation, altered interactions with biological partners and modified turnover (Davies, 2005, 2016). High levels of peroxyl and hydroxyl radicals contribute to protein peroxidation; the latter account for up to 70% of the initial oxidant flux. It is stated that reduction of one electron results in additional radicals and chain reactions with alcohols and carbonyls as major products; the latter are commonly used markers of protein damage. In addition, direct oxidation of cysteine (and less commonly) methionine residues is a major reaction; this is typically faster than with H_2O_2, and results in altered protein activity and function (Davies, 2005, 2016). The oxidative damage to proteins not only has deleterious effects on their structure, but also influences biological activities. This effects neural cell viability. Most oxidative damage to protein is nonrepairable. Thus oxidant-mediated damage to proteins results in various modifications such as glycation, carbonylation, and conjugation with products of lipid peroxidation. These modifications may lead to nonspecific interactions among proteins, which may promote the formation of high-molecular-weight protein aggregates (Höhn et al., 2013). Formation of protein aggregates provides a basis for many senescence-associated changes, which ultimately may contribute to a range of neurological pathologies. For the removal of protein aggregates and maintenance of it functions, neural cells use two systems (the proteasomal system, the autophagy-lysosomal system) to removal of oxidized and modified proteins.

Attack of hydroxyl radical permanently damage the DNA by inflicting injuries to purines, pyrimidines, and deoxyribose, but most notably, it is mitochondrial DNA (mtDNA) that is prone to oxidative damage since mitochondria are the main site of ROS production and mtDNA is in direct contact with ROS (Grimm and Eckert, 2017). Detailed investigations have shown that the attack of ROS on DNA and RNA produces many oxidized products of nucleic acid such as such as 8-oxo-G, 2,6-diamino-4-hydroxy-5-formamidopyrimidine (FaPyG) and 7,8-dihydro-8-oxoadenine (8-oxo-A) (Fig. 5.8). The accumulation of DNA lesions, including oxidative base modifications and strand breaks, triggers cell death in neurons and other vulnerable cell populations in the ischemic brain. DNA repair systems, particularly base excision repair, are endogenous defense mechanisms that combat oxidative DNA damage. The capacity for DNA repair may affect the susceptibility of neurons to ischemic stress and influence the pathological outcome of stroke (Li et al., 2011). In mtDNA, ROS attack also contribute to strand breaks. This process produces a reduction in the efficiency of the electron transport chain complex leading to decrease in ATP production. ROS attack on mtDNA also induces mtDNA mutations (Hsieh et al., 1994; Lu and Liu, 2010; Damsma and Cramer, 2009). These mutations occur at sites of known mtDNA transcription and replication regulatory elements and may

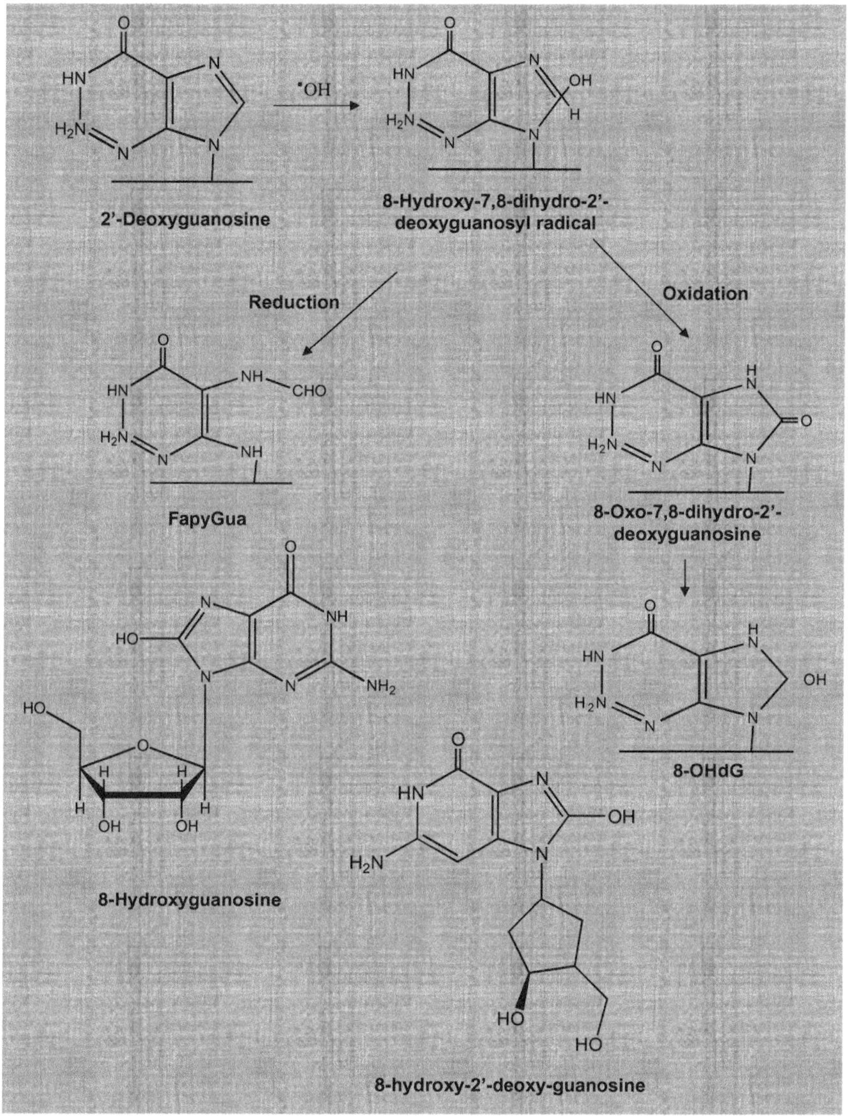

Figure 5.8 Effects of oxidative stress on nonenzymic oxidation of nucleic acids and chemical structures of nucleic acid–derived metabolites.
FaPyG, 8-Oxo-G, 2,6-diamino-4-hydroxy-5-formamidopyrimidine; *8-oxo-A*, 7,8-dihydro-8-oxoadenine.

involve a reduction in transcript levels of and deleterious functional consequences for mitochondrial homeostasis. Many studies have also indicated that mtDNA is associated with longevity (Brand et al., 1992).

ROS-mediated changes in mtDNA structure may decrease longevity and contribute to neural cell death. However, other investigators indicate that associations between mtDNA and longevity are weak (Castri et al., 2009). Mitochondria have been reported to interact with the nucleus (Ryan and Hoogenraad, 2007). While the precise details of these interactions remain elusive and controversial, there is ample evidence for crosstalk between the mitochondria and nucleus in humans. For example, many defects in mtDNA maintenance are caused by mutations in nuclear genes of the replisome (Kaukonen et al., 2000; Spelbrink et al., 2001), and the mitochondrial dysfunction may result in mtDNA depletion or damage activates responses in a large number of nuclear genes (Hansson et al., 2004). DNA repair systems, particularly base excision repair, are endogenous defense mechanisms that combat oxidative DNA damage. The capacity for DNA repair may affect the susceptibility of neurons to ischemic injury and may influence the pathological outcome of stroke (Li et al., 2011). Oxidative damage to RNA occurs more frequently than DNA, because RNA molecules are mostly single stranded and its bases are less protected by hydrogen bonding. Moreover, most of the mRNAs are not associated with chromatin and are distributed in the cytoplasm, closer to the site, where ROS generation occurs (Radak and Boldogh, 2010). There is evidence that oxidized mRNA causes errors in translation, eventually leading to the production of abnormal proteins (Tanaka et al., 2007), such as Aβ in AD, α-synuclein in PD, and mutated huntingtin in HD (Nunomura et al., 2009).

Metabolic links between insulin resistance and stroke

The most consistent links between insulin resistance and stroke are the induction of hyperglycemia, hypercholesterolemia, hypertension, decrease in CBF, endothelial, and vascular dysfunctions (Fig. 5.9). These parameters often develop due to aging and long-term consumption of western diet (Farooqui, 2015). As mentioned above, during the onset of stroke, the decrease in oxygen and glucose delivery to the brain, the breakdown of BBB, and buildup of potentially toxic products in the brain results in transmigration of numerous immune system cells (macrophages, lymphocytes, neutrophils, and dendritic cells) producing hyperpermeability promoted by enhanced transcytosis and gap formation between endothelial cells (Bousser, 2012; Gelderblom et al., 2009). The decrease in ATP generation, loss of ion homeostasis, mitochondrial dysfunction, production of ROS, such as superoxide, and hydroxyl anion and reactive nitrogen

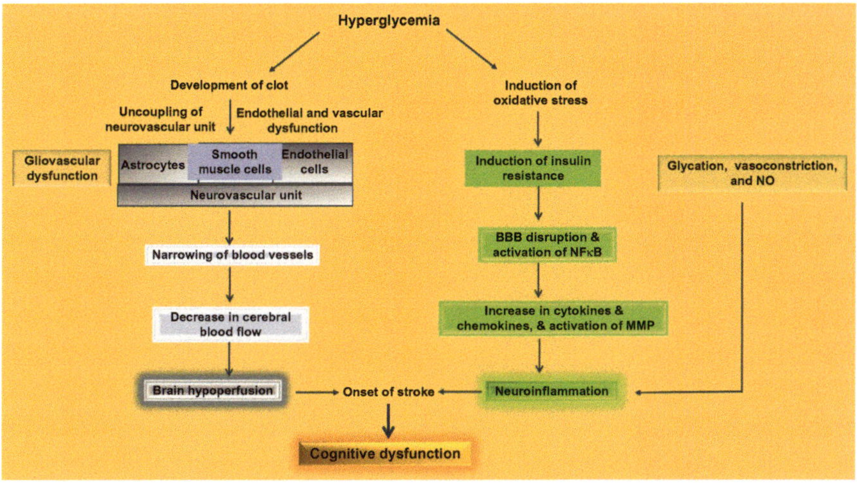

Figure 5.9 Effect of hyperglycemia on neurovascular unit and expression of proinflammatory cytokines.

species, such as NO and $ONOO^-$, and changes in the redox status of neural cells. These processes also contribute to cerebral edema, which is the primary cause of patient mortality after stroke. High glucose levels in the blood cause oxidative stress, which activates NF-κB/I-κB complex. This activation promotes the dissociation of NF-κB/I-κB complex and free NF-κB migrates to the nucleus, where it interacts with NF-κB response element and upregulates the expression of multiple inflammatory cytokines, including TNF-α, interleukin-1β (IL-1β), and IL-6. The expression of these cytokines is not only linked with neuroinflammation, a key component in the pathophysiology of cerebral ischemia (Iadecola and Anrather, 2011) as well as insulin resistance (Wang et al., 2012). During the induction of neuroinflammation, the first cells to react to the ischemic injury are microglia cells (i.e., the resident immune cells of the brain). In the pathogenesis of stroke, neuroinflammation is also supported by the disruption of BBB (Fernández-López et al., 2012; Uchida et al., 2017). Neuroinflammation may also play an essential role in cleaning up dead brain cells and their repair (Iadecola and Anrather, 2011; Sochocka et al., 2017).

As mentioned above, neurovascular unit is composed of neurons, glial cells, smooth muscle cells, and endothelial cells. Neurovascular unit not only controls CBF and BBB permeability, but also maintains the chemical composition of the neuronal "milieu," which modulates functioning of neuronal circuits. In addition, neurovascular unit also controls neuronal activity by modulating glucose uptake. A decreased cerebral glucose metabolism is an early event in the pathogenesis of stroke and may precede the neuropathological changes associated with the pathogenesis stroke. Moderate to severe reduction in CBF not only reduces ATP synthesis and diminishes (Na^+, K^+) ATPase activity, but also reduces synaptic plasticity by decreasing protein synthesis. These alterations inhibit the ability of neurons to generate action potentials producing alterations in ion homeostasis (Iadecola, 2004; Kalaria, 2010). In addition, reduction in CBF lowers the pH not only by altering electrolyte balances and water gradients, but also by promoting the development of cerebral edema, white matter lesions, and the accumulation of glutamate and proteinaceous toxins (amyloid-β and hyperphopshorylated tau). These processes contribute to excitotoxicity and oxidative stress in the brain. Furthermore, decrease in CBF also impairs the clearance of neurotoxic molecules that accumulate and/or are deposited in the ISF, nonneuronal cells and neurons. These processes promote cognitive impairments following stroke. Stroke patients show reduction in cerebral autoregulation, which modulates and maintains CBF across a wide range of arterial perfusion pressures (Kastrup et al., 1986). Collective evidence suggests that in normal brain, CBF is regulated by three major regulatory mechanisms: (1) cerebral autoregulation, (2) endothelium-dependent vasomotor function, and (3) neurovascular coupling response. Interactions among these mechanisms provide moment-to-moment adjustment of CBF, preventing both cerebral hypo- and hyperperfusions. This process ensures adequate delivery of oxygen and nutrients to the brain supporting the view that optimal CBF is necessary for the maintenance of healthy brain structure and function. Formation of clot in stroke patients disturbs CBF causing brain damage and subsequent cognitive impairment. Thus optimal CBF is very important for proper cerebral perfusion and neuronal function. Collective evidence suggests that vascular health is not only necessary for CBF, but also for the stability of the BBB. Therefore micro- and macrovascular pathological conditions of the brain may have profound effects on neurologic function in stroke patients especially when a secondary injury is superimposed on this existing pathology (Ergul et al., 2012).

Molecular link among advanced glycated end products, insulin resistance, and stroke

AGEs are a heterogeneous group of molecules, formed in vivo through nonenzymic reaction between reducing sugar and free amino groups of proteins, lipids, or nucleic acids. AGEs mediate their effect by binding to their receptors (RAGEs) (Brownlee, 2001, 2005; Giacco and Brownlee, 2010; Farooqui, 2013). RAGE is a pattern-recognition receptor present in diverse cell types (Lipscomb et al., 2001), and it has been shown to bind as well to non-AGE-related ligands, such as S100/calgranulins (Hofmann et al., 1999) and HGMB1 (Lin et al., 2009). The binding of AGEs with their receptors (RAGE) induce a variety of microvascular and macrovascular complications of diabetes by forming of cross-links between molecules in the basement membrane of the extracellular matrix. At the molecular level, the binding of AGEs with RAGE elicits more oxidative stress, which subsequently evokes inflammatory responses in various types of cells including neural cells, endothelial cells, hepatocytes, and hepatic stellate cells (Farooqui, 2013). Furthermore, the effects of AGEs-mediated oxidative stress in individuals with type 2 diabetes and insulin resistance are compounded by the inactivation of two critical antiatherosclerotic enzymes: endothelial NOS and prostacyclin synthase (Fig. 5.10). Normal aging is accompanied by the formation and accumulation of AGEs at a slow rate. The accumulation of AGEs occurs at an extremely accelerated rate in a variety of pathological conditions such as type 2 diabetes, heart disease, neurotraumatic diseases (stroke), neurodegenerative diseases (Alzheimer's disease), and various types of cancers (Farooqui, 2013; Clarke et al., 2016).

AGEs damage neural and peripheral cells by three general mechanisms including (1) modification of intracellular proteins by AGEs results in alterations in function, such as resistance glycated proteins to lysosomal enzyme degradation and poor recognition of glycated hemoglobin by lipoprotein receptors and scavenger receptors (Zimmermann et al., 2001), (2) modification of extracellular matrix components by AGE precursors produces abnormalities in between other matrix components and matrix integrins receptors, which are located on the surface of cells, and (3) modification of plasma proteins by AGEs. AGEs interact with AGE receptors (RAGE), which are found on macrophages, vascular endothelial cells, vascular smooth muscle cells, neurons, astrocytes, and microglial cells to induce the production of ROS, which in turn activates the pleiotropic transcription factor NF-κB, triggering multiple pathological changes in gene expression involved in proinflammatory events (Lue et al., 2001; Sasaki et al., 2001; Goldin et al., 2006). Furthermore, the AGE–RAGE interaction can activate the inducible NOS (iNOS). The iNOS resides mainly in inflammatory cells, is regulated by inflammatory cytokines, and

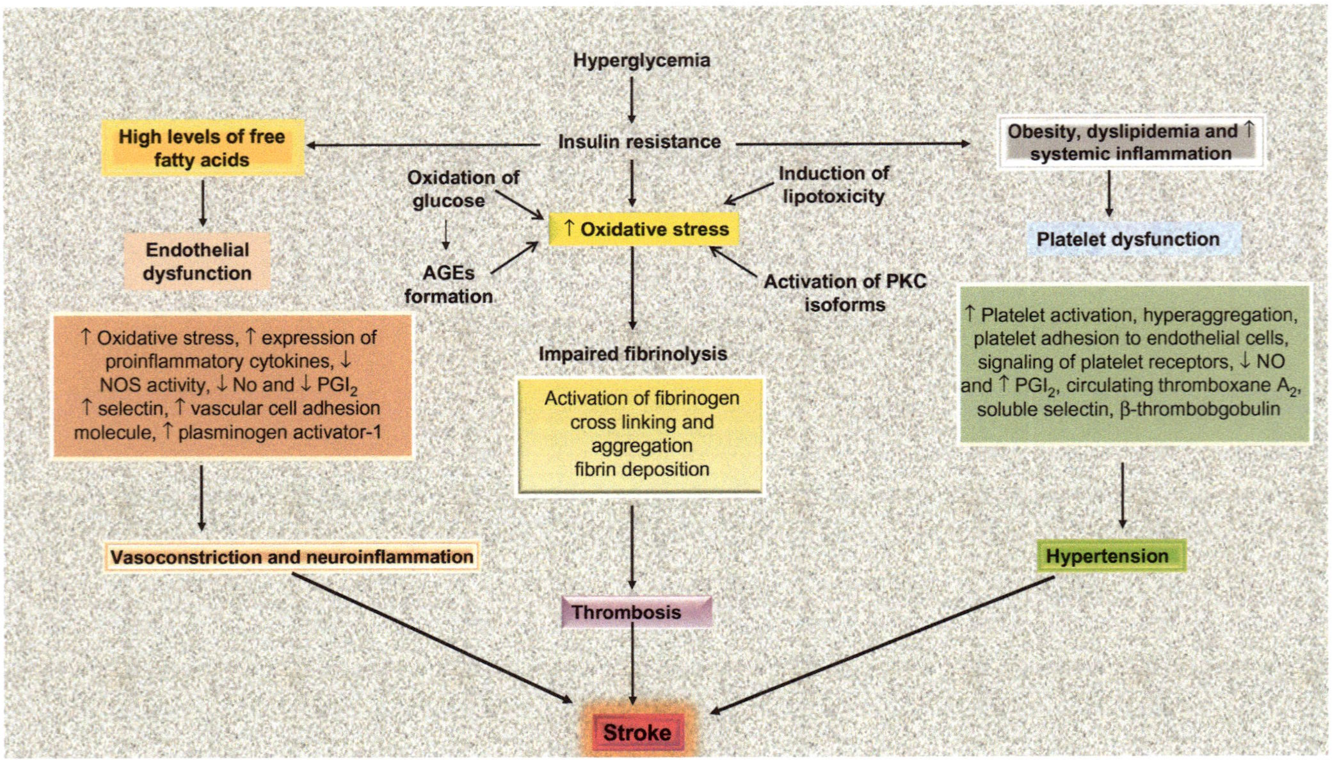

Figure 5.10 Molecular mechanisms contributing to the pathogenesis of stroke.

can be stimulated by oxidative stress in an NFκB-dependent manner, resulting in toxic concentrations of NO (Griscavage et al., 1996).

In stroke and diabetes, hyperglycemia promotes the generation of AGEs, which by interacting with RAGE produce oxidative stress and neuroinflammation. These processes contribute to the complications of diabetes (Brownlee, 2001; Giacco and Brownlee, 2010; Farooqui, 2013). Compelling epidemiological evidence suggests that elevated AGEs may be a significant risk factor for stroke, but also for beta cell injury (Zhao et al., 2009) and for peripheral insulin resistance (Cai et al., 2007, 2008). It has been proposed that hyperglycemia-mediated insulin resistance is a key component of various health problems caused by the prolonged consumption of high-carbohydrate diet resulting in positive energy imbalance (overeating) (Hoehn et al., 2009). These processes are supported by the development of oxidative stress and low-grade inflammation. In addition, increase in AGEs production causes a cascade of cellular metabolic alterations such as hypertension, obesity, and downstream effects may lead to blood-flow abnormalities, increased vascular permeability, angiogenesis, capillary occlusion, and proinflammatory gene expression (Rask-Madsen and King, 2013; Kolluru et al., 2012). These alterations play a major role in the pathogenesis of stroke-mediated neuronal injury.

Adipokines, insulin resistance, and stroke

As mentioned above, adipose tissue is a highly specialized organ that stores excess energy and releases it when needed by other tissues (Trayhurn and Beattie, 2001). Obesity develops when the intake of calories exceeds energy expenditure (Trayhurn and Beattie, 2001). Adipokines are hormones secreted by adipose tissue. Examples of adipokines are leptin, adipolneptin, resistin, and visfatin (Fig. 5.11). Adipokines have structural homology to cytokines that actively participate in regulating many biological processes such as immune regulation and energy balance, lipid and glucose metabolism, blood pressure, insulin sensitivity, and angiogenesis (Trayhurn and Beattie, 2001). In addition to above processes, adipokines have also been found to regulate an expanding array of physiological functions, including hemostasis, blood pressure regulation, insulin sensitivity, and angiogenesis (Guri and Bassaganya-Riera, 2010; Trayhurn and Wood, 2005). Under pathological conditions, adipokines not only play an important role in the pathogenesis of a low-grade inflammation in obesity and insulin resistance, but also contribute to the pathogenesis of neurological

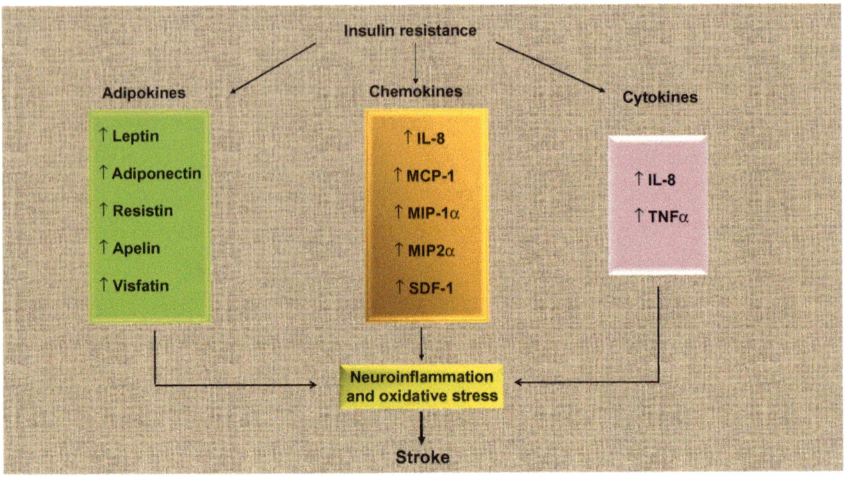

Figure 5.11 Involvement of adipokines, chemokines, and cytokines in pathogenesis of stroke.
adipsin, Acylation-stimulating protein; *IL-8*, interleukin-8; *MCP-1*, monocyte chemoattractant protein-1; *MIP-1α*, macrophage inflammatory protein-1α; *MIP-2α*, macrophage inflammatory protein-2α; *SDF-1*, stromal cell-derived factor-1.

conditions such as stroke, Alzheimer's disease, and Parkinson's disease (Farooqui, 2013; Rocha et al 2014; Gustafson, 2014; Kantorová et al., 2015; Opatrilova et al., 2018). Among adipokines, leptin and adiponectin are linked to the development of coronary heart disease and may be involved in the underlying biological mechanism of ischemic stroke. Resistin, a proinflammatory cytokine, is predictive of atherosclerosis and poor clinical outcomes in patients with CAD and ischemic stroke. The changes in serum levels of novel adipokines apelin and visfatin are also associated with acute ischemic stroke. These adipokines have been proposed as potential prognostic biomarkers of cardiovascular mortality/morbidity and therapeutic targets in patients with cardiometabolic diseases.

Molecular link between hypertension and brain damage

It is well known that hypertension quietly damages our body for years before its symptoms develop. If left uncontrolled, one may end up with a fatal heart attack or stroke. Cardiovascular and cerebrovascular changes, which occur in human tissues due to changes heart rate, insulin resistance, and hypertension in middle and old age may cause silent brain injuries such as white matter lesions or white matter atrophy (including the white

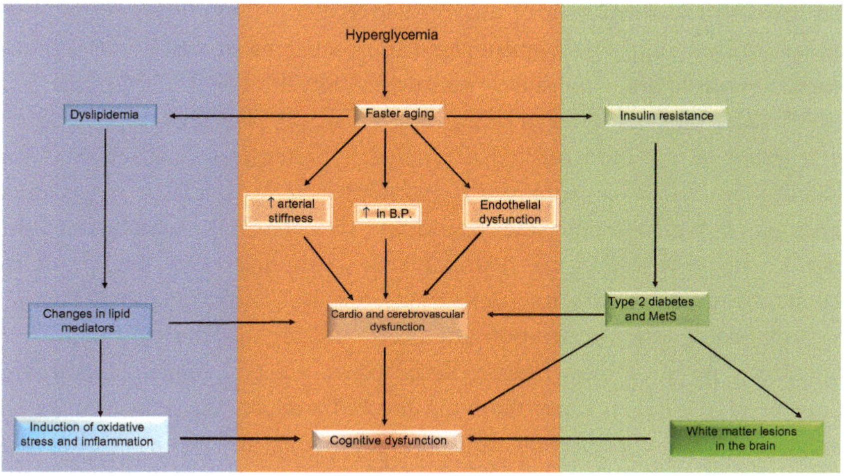

Figure 5.12 Hypertension-mediated changes during brain damage.

matter hyperintensities evident on magnetic resonance imaging scans) (Fig. 5.12) (Iadecola and Davisson, 2008; Maniega et al., 2015; Verhaaren et al., 2013; Capone et al., 2011, 2012). These changes can be quantified on diffusion tensor imaging and FLAIR before the development of white matter lesions suggesting that these changes develop gradually, and that visually appreciable white matter lesions are only the tip of the iceberg of white matter pathology (de Groot et al., 2013). It is also reported that white matter microstructural integrity may be an important neural correlate of executive function even in cognitively intact cardioarterial disease patients. These studies support the view that white matter damage may be relevant to subtle cognitive decline in a population that may have early risk for cognitive decline and progression to dementia (Santiago et al., 2015). The mechanisms contributing to white matter lesions are not fully understood. However, it is becoming increasingly evident that during brain injury process, hypertension not only damages endothelial cells in cerebral blood vessels, but also disrupts neurovascular unit and mechanism, which links energy needs neuronal and glial cells with CBF in the white matter area of the brain (Iadecola and Davisson, 2008; Unverzagt et al., 2011; Jennings et al., 2005; Iadecola, 2010) suggesting that midlife hypertension-mediated brain damage is not only associated with subcortical white matter lesions, but may also increase the risk of stroke, vascular cognitive impairment, and Alzheimer's disease (Vermeer et al., 2003; van Dijk et al., 2004). At the molecular level, development of midlife hypertension may involve the formation of angiotensin-II (Ang-II)-induced ROS along the subfornical organ-paraventricular (SFO) nucleus of the hypothalamus-rostral ventrolateral medulla pathway (SFOPVN–RVLM pathway) (Fig. 5.13) (Braga et al., 2011; Capone et al., 2011). This hypothesis has been tested in a mouse model of gradual hypertension. It is reported that chronic administration (subpressor doses) of Ang-II results in the overexpression of CuZn-superoxide dismutase (CuZnSOD), which not only retards the alteration in neurovascular coupling and endothelium-dependent responses in somatosensory cortex, but also induce hypertension by suppressing the generation of free radicals in the SFO nucleus. Alterations in SFO may induce dysfunction via two signaling pathways (Capone et al., 2011). One involving the SFO-dependent activation of the paraventricular hypothalamic nucleus, elevations in plasma vasopressin, upregulation of endothelin-1 in cerebral resistance arterioles, and activation of endothelin type A receptors. The other pathway is associated with the activation of cerebrovascular AngII type 1 (AT1) receptors by AngII. Both pathways mediate vasomotor dysfunction by inducing vascular oxidative stress. The findings implicate that SFO and its efferent hypothalamic pathways in the cerebrovascular alterations induced by AngII, and identify vasopressin and endothelin-1 as potential therapeutic targets to counteract the devastating effects of hypertension on the brain (Capone et al., 2011).

Collective evidence suggests that brain tissue critically depends on a continuous and well-regulated blood supply to support its dynamic needs for oxygen and glucose and to remove metabolic by-products of brain activity (Iadecola and Nedergaard, 2007; Moskowitz et al., 2010). Complex regulatory mechanisms ensure that the brain receives sufficient CBF to maintain the homeostasis of the cerebral microenvironment. Thus active neurons evoke powerful increases in CBF to match substrate delivery with the metabolic requirements of activation (functional hyperemia), whereas endothelial cells release potent vasoactive substances that regulate the distribution of microvascular flow. Midelife hypertension disrupts these control mechanisms and increases the susceptibility of the brain to vascular insufficiency (Iadecola and Davisson, 2008).

Effect of microbiota composition on stroke outcome

As mentioned in Chapter 1, Insulin resistance and obesity, there is a bidirectional communication between gut microbiota and brain through the brain–gut–microbiome axis (Mayer et al., 2015; Zhao et al., 2018).

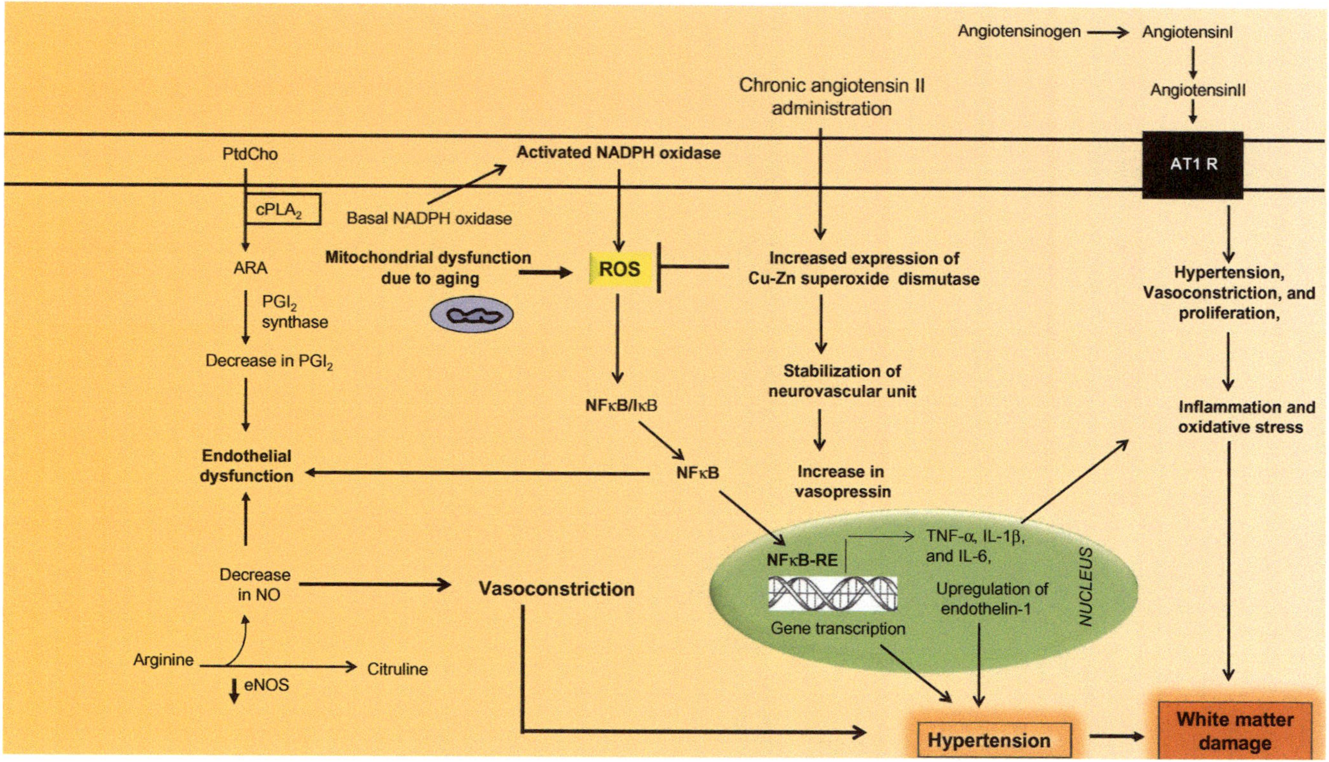

Figure 5.13 Hypothetical diagram showing molecular mechanism of hypertension-mediated white matter damage. *AT1 receptor*, Angiotensin II type-1 receptor; *eNOS*, endothelial nitric oxide synthase; *IκB*, inhibitory subunit of NF-κB; *LOX*, lipoxygenase; *NADPH oxidase*, nicotinamide adenine dinucleotide phosphate-oxidase; *NF-κB-RE*, nuclear factor kappaB response element.

The structure and function of the brain can be modulated by the gut microbiota and conversely, the brain regulates the gut microenvironment and microbiota composition. Recent evidence suggests a key role of the gut microbiota not only in the outcome of neurological and neuropsychiatric disorders such as stroke, Parkinson's disease (de Vos and de Vos, 2012), autism spectrum disorders (Cryan and Dinan, 2012; Mayer et al., 2014), disorders of mood and affect (Cryan and Dinan, 2012; Park et al., 2013), and chronic pain but also in the development of autoimmune diseases. Studies on two distinct models of stroke have indicated that large stroke lesions cause gut microbiota dysbiosis, which in turn affects stroke outcome via immune-mediated mechanisms (Singh et al., 2016). Reduced species diversity and bacterial overgrowth of bacteroidetes have been identified as hallmarks of poststroke dysbiosis (Singh et al., 2016), which contribute to intestinal barrier dysfunction and reduction in intestinal motility. Recolonizing germ-free mice with dysbiotic poststroke microbiota exacerbates lesion volume and functional deficits after experimental stroke compared with the recolonization with a normal control microbiota (Singh et al., 2016). In addition, recolonization of mice with a dysbiotic microbiome induces a proinflammatory T-cell polarization in the intestinal immune compartment and in the ischemic brain. In vivo cell-tracking studies have demonstrated that the migration of intestinal lymphocytes to the ischemic brain. Furthermore, therapeutic transplantation of fecal microbiota normalizes brain lesion-induced dysbiosis and improves stroke outcome. These results support a novel mechanism in which the gut microbiome is not only a target of stroke-induced systemic alterations, but also an effector, which also has substantial impact on stroke outcome (Singh et al., 2016).

Conclusion

Insulin resistance is a metabolic disorder, which is characterized by diminished tissue sensitivity to insulin. It originates from environmental factors such as a sedentary lifestyle, central obesity, and genetic predisposition. Insulin resistance is a pivotal pathophysiologic contributor to the increased risk of stroke. Stroke with insulin resistance is the leading cause of disability and death in adults and seniors. Acute cerebral ischemia with insulin resistance accounts for 75%–80% of mortality in stroke patients. The pathogenesis of insulin resistance and ischemic stroke is very complex. It involves effects of hyperglycemia not only on vascular tissues and coagulation, and aberrations in blood pressure regulation, but also on changes in lipid metabolism, endothelial function, vascular inflammation, lipid metabolism, smooth muscle cell proliferation, and fibrinolysis. Insulin resistance enhances platelet adhesion, activation, and aggregation which are conducive to the occurrence of ischemic stroke. A persistently high level of insulin over an extended period causes a decrease in the body's sensitivity to insulin, which results in insulin resistance. When cells are not able to absorb glucose, their primary source of energy, it causes an increased risk of many health problems. Insulin resistance also induces hemodynamic disturbances, which contributes to the onset of ischemic stroke. In addition, insulin resistance may augment the role of the modifiable risk factors in ischemic stroke and induce the occurrence of ischemic stroke. Preclinical and clinical studies have supported that improving insulin resistance may be an effective measure to prevent or delay ischemic stroke. In contrast, ischemic stroke triggers a complex and highly interconnected cascade of cellular and molecular events.

Pathogenesis of stroke involves not only by decrease in ATP production, loss of neural cell homeostasis, excitotoxicity, and activation of PLA_2, CaMKs, MAPKs, NOS, calpains, calcinurin, and endonucleases but also disruption of the BBB with infiltration of leukocytes. In addition, ischemia/reperfusion injury promotes the activation of neuronal and glial cells, release of high levels of proinflammatory cytokines and chemokines, mitochondrial dysfunction, high levels of formation of superoxides, hydroxyl radicals, and peroxynitrite. These processes contribute to neurodegeneration in stroke patients. The most effective strategies to prevent stroke among people with diabetes include blood pressure control, antiplatelet therapy, and statin therapy. Tight glycemic control is recommended to prevent microvascular disease, but the effect on macrovascular disease, including stroke, has not been proven.

References

Adibhatla, R.M., Hatcher, J.F., 2010. Lipid oxidation and peroxidation in CNS health and disease: from molecular mechanisms to therapeutic opportunities. Antioxid. Redox. Signal. 12, 125–169.

Allen, C.L., Bayraktutan, U., 2008. Risk factors for ischaemic stroke. Int. J. Stroke. 3, 105–116.

Aspelund, A., Antila, S., Proulx, S.T., Karlsen, T.V., Karaman, S., Detmar, M., 2015. A dural lymphatic vascular system that drains brain interstitial fluid and macromolecules. J. Exp. Med. 212.

Astrup, J., Siesjö, B.K., Symon, L., 1981. Thresholds in cerebral ischemia—the ischemic penumbra. Stroke 12, 723–725.
Bailey, D.M., Bärtsch, P., Knauth, M., Baumgartner, R.W., 2009. Emerging concepts in acute mountain sickness and high-altitude cerebral edema: from the molecular to the morphological. Cell. Mol. Life Sci. 66, 3583–3594.
Bémeur, C., Ste-Marie, L., Montgomery, J., 2007. Increased oxidative stress during hyperglycemic cerebral ischemia. Neurochem. Int. 50, 890–904.
Benakis, C., Brea, D., Caballero, S., Faraco, G., Moore, J., Murphy, M., et al., 2016. Commensal microbiota affects ischemic stroke outcome by regulating intestinal γδ T cells. Nat. Med. 22, 516–523.
Boini, K.M., Hussain, T., Li, P.L., Koka, S., 2017. Trimethylamine-N-oxide instigates NLRP3 inflammasome activation and endothelial dysfunction. Cell. Physiol. Biochem. 44, 152–162.
Bonaz, B., Bazin, T., Pellissier, S., 2018. The vagus nerve at the interface of the microbiota-gut-brain axis. Front. Neurosci. 12, 49.
Bousser, M.G., 2012. Stroke prevention: an update. Front. Med. 6, 22–34.
Braga, V.A., Medeiros, I.A., Ribeiro, T.P., França-Silva, M.S., Botelho-Ono, M.S., Guimarães, D.D., 2011. Angiotensin-II-induced reactive oxygen species along the SFO-PVN-RVLM pathway: implications in neurogenic hypertension. Braz. J. Med. Biol. Res. 44, 871–876.
Brancati, F.L., Wang, N.Y., Mead, L.A., Liang, K.Y., Klag, M.J., 1999. Body weight patterns from 20 to 49 years of age and subsequent risk for diabetes mellitus: the Johns Hopkins precursors study. Arch. Intern. Med. 159, 957–963.
Brand, F.N., Kiely, D.K., Kannel, W.B., Myers, R.H., 1992. Family patterns of coronary heart disease mortality: the Framingham Longevity Study. J. Clin. Epidemiol. 45, 169–174.
Brownlee, M., 2001. Biochemistry and molecular cell biology of diabetic complications. Nature 414, 813–820.
Brownlee, M., 2005. The pathobiology of diabetic complications: a unifying mechanism. Diabetes 54, 1615–1625.
Brüning, J.C., Winnay, J., Bonner-Weir, S., Taylor, S.I., Accili, D., Kahn, C.R., 1997. Development of a novel polygenic model of NIDDM in mice heterozygous for IR and IRS-1 null alleles. Cell. 88, 561–572.
Bruno, A., Biller, J., Adams Jr, H.P., Clarke, W.R., Woolson, R.F., et al., 1999. Acute blood glucose level and outcome from ischemic stroke. Trial of ORG 10172 in Acute Stroke Treatment (TOAST) Investigators. Neurology 52, 280–284.
Bruno, A., Levine, S.R., Frankel, M.R., Brott, T.G., Lin, Y., Tilley, B.C., et al., 2002. Admission glucose level and clinical outcomes in the NINDS rt-PA stroke trial. Neurology 59, 669–674.
Bruno, A., Liebeskind, D., Hao, Q., Raychev, R., 2010. Investigators US Diabetes mellitus, acute hyperglycemia, and ischemic stroke. Curr. Treat. Options Neurol. 12, 492–503.
Butler, A.E., Janson, J., Bonner-Weir, S., Ritzel, R., Rizza, R.A., Butler, P.C., 2003. Beta-cell deficit and increased beta-cell apoptosis in humans with type 2 diabetes. Diabetes 52, 102–110.
Cai, W., He, J.C., Zhu, L., Chen, X., Wallenstein, S., Striker, G.E., et al., 2007. Reduced oxidant stress and extended lifespan in mice exposed to a low glycotoxin diet: association with increased AGER1 expression. Am. J. Pathol. 170, 1893–1902.
Cai, W., He, J.C., Zhu, L., Chen, X., Zheng, F., Striker, G.E., et al., 2008. Oral glycotoxins determine the effects of calorie restriction on oxidant stress, age-related diseases, and lifespan. Am. J. Pathol. 173, 327–336.
Cai, M., Wang, H., Li, J.J., Zhang, Y.L., Xin, L., Li, F., et al., 2016. The signaling mechanisms of hippocampal endoplasmic reticulum stress affecting neuronal plasticity-related

protein levels in high fat diet-induced obese rats and the regulation of aerobic exercise. Brain Behav. Immun. 57, 347–359.
Camara-Lemarroy, C.R., Ibarra-Yruegas, B.E., Gongora-Rivera, F., 2014. Gastrointestinal complications after ischemic stroke. J. Neurol. Sci. 346, 20–205.
Capone, C., Faraco, G., Park, L., Cao, X., Davisson, R.L., Iadecola, C., 2011. The cerebrovascular dysfunction induced by slow pressor doses of angiotensin II precedes the development of hypertension. Am. J. Physiol. 300, H397–H407.
Capone, C., Faraco, G., Peterson, J.R., Coleman, C., Anrather, J., Milner, T.A., et al., 2012. Central cardiovascular circuits contribute to the neurovascular dysfunction in angiotensin-II hypertension. J. Neurosci. 32, 4878–4886.
Castri, L., Melendez-Obando, M., Villegas-Palma, R., Barrantes, R., Raventos, H., et al., 2009. Mitochondrial polymorphisms are associated both with increased and decreased longevity. Hum. Hered. 67, 147–153.
Chang, L., Karin, M., 2001. Mammalian MAP kinase signalling cascades. Nature 410, 37–40.
Chen, M.L., Zhu, X.H., Ran, L., Lang, H.D., Yi, L., Mi, M.T., 2017. Trimethylamine-N-oxide induces vascular inflammation by activating the NLRP3 inflammasome through the SIRT3-SOD2-mtROS signaling pathway. J. Am. Heart Assoc. 6, e006347.
Cheng, Z., Tseng, Y., White, M.F., 2010. Insulin signaling meets mitochondria in metabolism. Trends Endocrinol. Metab. 21, 589–598.
Clarke, R.E., Dordevic, A.L., Tan, S.M., Ryan, L., Coughlan, M.T., 2016. Dietary advanced glycation end products and risk factors for chronic disease: a systematic review of randomised controlled trials. Nutrients 8, 125.
Collins, S.M., Surette, M., Bercik, P., 2012. The interplay between the intestinal microbiota and the brain. Nat. Rev. Microbiol. 10, 735–742.
Cryan, J.F., Dinan, T.G., 2012. Mind-altering microorganisms: the impact of the gut microbiota on brain and behaviour. Nat. Rev. Neurosci. 13, 701–712.
Cumming, T.B., Marshall, R.S., Lazar, R.M., 2013. Stroke, cognitive deficits and rehabilitation: still an in complete picture. Int. J.Stroke 8, 38–45.
Damsma, G.E., Cramer, P., 2009. Molecular basis of transcriptional mutagenesis at 8-oxoguanine. J. Biol. Chem. 284, 31658–31663.
Daneman, R., Prat, A., 2015. The blood-brain barrier. Cold Spring Harb. Perspect. Biol. 2015 (7), a020412.
Davies, M.J., 2005. The oxidative environment and protein damage. Biochim. Biophys. Acta 1703, 93–109.
Davies, M.J., 2016. Protein oxidation and peroxidation. Biochem. J. 473, 805–825.
de Groot, M., Verhaaren, B.F., de Boer, R., Klein, S., Hofman, A., van der Lugt, A., et al., 2013. Changes in normal-appearing white matter precede development of white matter lesions. Stroke 44, 1037–1042.
de Lartigue, G., de La Serre, C.B., Raybould, H.E., 2011. Vagal afferent neurons in high fat diet-induced obesity; intestinal microflora, gut inflammation and cholecystokinin. Physiol. Behav. 105, 100–105.
de Vos, W.M., de Vos, E.A., 2012. Role of the intestinal microbiome in health and disease: from correlation to causation. Nutr. Rev. 70 (Suppl. 1), S45–S56.
Droge, W., 2002. Free radicals in the physiological control of cell function. Physiol. Rev. 82, 47–95.
Dungan, K.M., Braithwaite, S.S., Praser, J.C., 2009. Stress hyperglycemia. Lancet 373, 1798–1807.
Enciu, A.M., Gherghiceanu, M., Popescu, B.O., 2013. Triggers and effectors of oxidative stress at blood-brain barrier level: relevance for brain ageing and neurodegeneration. Oxid. Med. Cell Longev. 2013, 297512.

Ergul, A., Kelly-Cobbs, A., Abdalla, M., Fagan, S.C., 2012. Cerebrovascular complications of diabetes: focus on stroke. Endocr. Metab. Immune Disord. Drug Targets 12, 148–158.
Farooqui, A.A., 2010. Neurochemical Aspects of Neurotraumatic and Neurodegenerative Diseases. Springer, New York.
Farooqui, A.A., 2013. Metabolic Syndrome. Springer Science + Bussiness Media, New York.
Farooqui, A.A., 2014. Inflammation and Oxidative Stress in Neurological Disorders. Springer International Publishing, Switzerland.
Farooqui, A.A., 2015. High Calorie Diet and the Human Brain. Springer International Publishing, Switzerland.
Farooqui, A.A., 2018. Ischemic and Traumatic Brain and Spinal Cord Injuries. Academic Press, San Diego, CA.
Farooqui, A.A., Horrocks, L.A., 2007. Glycerophospholipids in Brain. Springer, New York.
Fernández-López, D., Faustino, J., Daneman, R., Zhou, L., Lee, S.Y., Derugin, N., et al., 2012. Blood-brain barrier permeability is increased after acute adult stroke but not neonatal stroke in the rat. J. Neurosci. 32, 9588–9600.
Gelderblom, M., Leypoldt, F., Steinbach, K., Behrens, D., Choe, C.-U., Siler, D.A., et al., 2009. Temporal and spatial dynamics of cerebral immune cell accumulation in stroke. Stroke 40, 1849–1857.
Gentile, N.T., Vaidyula, V.R., Kanamalla, U., DeAngelis, M., Gaughan, J., Rao, A.K., 2007. Factor VIIa and tissue factor procoagulant activity in diabetes mellitus after acute ischemic stroke: impact of hyperglycemia. Thromb. Haemost. 98, 1007–1013.
Giacco, F., Brownlee, M., 2010. Oxidative stress and diabetic complications. Circ. Res 107, 1058–1070.
Goldin, A., Beckman, J.A., Schmidt, A.M., Creager, M.A., 2006. Advanced glycation end products:sparking the development of diabetic vascular injury. Circulation 114, 597–605.
Grimm, A., Eckert, A., 2017. Brain aging and neurodegeneration: from a mitochondrial point of view. J. Neurochem. 143, 418–431.
Griscavage, J.M., Wilk, S., Ignarro, L.J., 1996. Inhibitors of the proteasome pathway interfere with induction of nitric oxide synthase in macrophages by blocking activation of transcription factor NF-kappa B. Proc. Natl. Acad. Sci. U.S.A. 93, 3308–3312.
Guri, A.J., Bassaganya-Riera, J., 2010. Systemic effects of white adipose tissue dysregulation and obesity-related inflammation. Obesity 19, 689–700.
Gustafson, D.R., 2014. Adiposity hormones and dementia. J. Neurol. Sci. 299, 30–34.
Hara, M.R., Snyder, S.H., 2007. Cell signaling and neuronal death. Annu. Rev. Pharmacol. Toxicol. 47, 117–141.
Hanley, A.J., Williams, K., Stern, M.P., Haffner, S.M., 2002. Homeostasis model assessment of insulin resistance in relation to the incidence of cardiovascular disease: the San Antonio Heart Study. Diabetes Care 25, 1177–1184.
Hansson, A., Hance, N., Dufour, E., Rantanen, A., Hultenby, K., Clayton, D.A., et al., 2004. A switch in metabolism precedes increased mitochondrial biogenesis in respiratory chain-deficient mouse hearts. Proc. Natl. Acad. Sci. U.S.A. 101, 3136–3141.
Hardigan, T., Ward, R., Ergul, A., 2016. Cerebrovascular complications of diabetes: focus on cognitive dysfunction. Clin. Sci. (Lond.) 130, 1807–1822.
Heiss, W.D., Sobesky, J., Hesselmann, V., 2004. Identifying thresholds for penumbra and irreversible tissue damage. Stroke 35, 2671–2674.
Hoehn, K.L., Salmon, A.B., Hohnen-Behrens, C., Turner, N., Hoy, A.J., Maghzal, G.J., et al., 2009. Insulin resistance is a cellular antioxidant defense mechanism. Proc. Natl. Acad. Sci. USA 106, 17787–17792.

Hofmann, M.A., Drury, S., Fu, C., et al., 1999. RAGE mediates a novel proinflammatory axis: a central cell surface receptor for S100/calgranulin polypeptides. Cell. 97, 889–901.
Höhn, A., König, J., Grune, T., 2013. Protein oxidation in aging and the removal of oxidized proteins. J. Proteomics 92, 132–159.
Honda, K., Littman, D.R., 2016. The microbiota in adaptive immune homeostasis and disease. Nature 535, 75–84.
Hossmann, K.A., 1994. Viability thresholds and the penumbra of focal ischemia. Ann. Neurol. 36, 557–565.
Hotamisligil, G.S., 2008. Inflammation and endoplasmic reticulum stress in obesity and diabetes. Int. J. Obes. (Lond.) 32 (Suppl. 7), S52–S54.
Hsieh, R.H., Hou, J.H., Hsu, H.S., Wei, Y.H., 1994. Age-dependent respiratory function decline and DNA deletions in human muscle mitochondria. Biochem. Mol. Biol. Int. 32, 1009–1022.
Iadecola, C., 2004. Neurovascular regulation in the normal brain and in Alzheimer's disease. Nat. Rev. Neurosci. 5, 347–360.
Iadecola, C., 2010. The overlap between neurodegenerative and vascular factors in the pathogenesis of dementia. Acta Neuropathol. 120, 287–296.
Iadecola, C., Anrather, J., 2011. The immunology of stroke: from mechanisms to translation. Nat. Med. 17, 796.
Iadecola, C., Davisson, R.L., 2008. Hypertension and cerebrovascular dysfunction. Cell. Metab. 7, 476–484.
Iadecola, C., Nedergaard, M., 2007. Glial regulation of the cerebral microvasculature. Nat. Neurosci. 10, 1369–1376.
Janghorbani, M., Hu, F.B., Willett, W.C., Li, T.Y., Manson, J.E., Logroscino, G., et al., 2007. Prospective study of type 1 and type 2 diabetes and risk of stroke subtypes: the Nurses' Health Study. Diabetes Care 30, 1730–1735.
Jennings, J.R., Muldoon, M.F., Ryan, C., Price, J.C., Greer, P., Sutton-Tyrrell, K., et al., 2005. Reduced cerebral blood flow response and compensation among patients with untreated hypertension. Neurology 64, 1358–1365.
Kalaria, R.N., 2010. Vascular basis for brain degeneration: faltering controls and risk factors for dementia. Nutr. Rev. 68 (Suppl. 2), S74–S87.
Kantorová, E., Jesenská, Ľ., Čierny, D., Zeleňák, K., Sivák, Š., Stančík, M., et al., 2015. The intricate network of adipokines and stroke. Int. J. Endocrinol. 2015, 967698.
Kastrup, J., Rørsgaard, S., Parving, H.H., Lassen, N.A., 1986. Impaired autoregulation of cerebral blood flow in long-term type I (insulin-dependent) diabetic patients with nephropathy and retinopathy. Clin. Physiol. 6, 549–559.
Kaukonen, J., Juselius, J.K., Tiranti, V., Kyttälä, A., Zeviani, M., et al., 2000. Role of adenine nucleotide translocator 1 in mtDNA maintenance. Science 289, 782–785.
Kissebah, A.H., Vydelingum, N., Murray, R., Evans, D.J., Hartz, A.J., et al., 1982. Relation of body fat distribution to metabolic complications of obesity. J. Clin. Endocrinol. Metab. 54, 254–260.
Koeth, R.A., Wang, Z., Levison, B.S., Buffa, J.A., Org, E., Sheehy, B.T., et al., 2013. Intestinal microbiota metabolism of l-carnitine, a nutrient in red meat, promotes atherosclerosis. Nat. Med. 19, 576–585.
Kolluru, G.K., Bir, S.C., Kevil, C.G., 2012. Endothelial dysfunction and diabetes: effects on angiogenesis, vascular remodeling, and wound healing. Int. J. Vasc. Med. 2012, 918267.
Kostic, A.D., Xavier, R.J., Gevers, D., 2014. The microbiome in inflammatory bowel disease: current status and the future ahead. Gastroenterology 146, 1489–1499.
Kunze, W.A., Mao, Y.K., Wang, B., Huizinga, J.D., Ma, X., Forsythe, P., et al., 2009. Lactobacillus reuteri enhances excitability of colonic AH neurons by inhibiting calcium-dependent potassium channel opening. J. Cell. Mol. Med. 13, 2261–2270.

Laing, S.P., Swerdlow, A.J., Carpenter, L.M., Slater, S.D., Burden, A.C., Botha, J.L., et al., 2003. Mortality from cerebrovascular disease in a cohort of 23 000 patients with insulin-treated diabetes. Stroke 34, 418–421.
Leung, P.S., 2016. The potential protective action of vitamin D in hepatic insulin resistance and pancreatic islet dysfunction in type 2 diabetes mellitus. Nutrients 8, pii: E147.
Li, P., Hu, X., Gan, Y., Gao, Y., Liang, W., Chen, J., 2011. Mechanistic insight into DNA damage and repair in ischemic stroke: exploiting the base excision repair pathway as a model of neuroprotection. Antioxid. Redox. Signal. 14, 1905–1918.
Lin, L., Park, S., Lakatta, E.G., 2009. RAGE signaling in inflammation and arterial aging. Front. Biosci. 14, 1403–1413.
Lipscomb, E.A., Sarmiere, P.D., Freeman, R.S., 2001. SM-20 is a novel mitochondrial protein that causes caspase-dependent cell death in nerve growth factor-dependent neurons. J. Biol. Chem. 276, 5085–5092.
Lochhead, J.J., McCaffrey, G., Quigley, C.E., Finch, J., DeMarco, K.M., Nametz, N., et al., 2010. Oxidative stress increases blood-brain barrier permeability and induces alterations in occludin during hypoxia-reoxygenation. J. Cereb. Blood Flow Metab. 30, 1625–1636.
Louveau, A., Smirnov, I., Keyes, T.J., Eccles, J.D., Rouhani, S.J., Peske, J.D., et al., 2015. Structural and functional features of central nervous system lymphatic vessels. Nature 523, 337–341.
Lu, J., Liu, Y., 2010. Deletion of Ogg1 DNA glycosylase results in telomere base damage and length alteration in yeast. EMBO J. 29, 398–409.
Lue, L.F., Walker, D.G., Brachova, L., Beach, T.G., Roger, J., Schmidt, A.M., et al., 2001. Involvement of microglial receptor for advanced glycation endproducts (RAGE) in Alzheimer's disease: identification of a cellular activation mechanism. Exp. Neurol. 171, 29–45.
Martini, S.R., Kent, T.A., 2007. Hyperglycemia in acute ischemic stroke: a vascular perspective. J. Cereb. Blood Flow Metab. 27, 435–451.
Mayer, E.A., Padua, D., Tillisch, K., 2014. Altered brain-gut axis in autism: comorbidity or causative mechanisms? Bioessays 36, 933–939.
Mayer, E.A., Tillisch, K., Gupta, A., 2015. Gut/brain axis and the microbiota. J. Clin. Invest. 125, 926–938.
McAuley, K., Mann, J., 2006. Thematic review series: patient-oriented research. nutritional determinants of insulin resistance. J. Lipid Res. 47, 1668–1676.
McDermott, A.J., Huffnagle, G.B., 2014. The microbiome and regulation of mucosal immunity. Immunology 142, 24–31.
Maggio, C.A., Pi-Sunyer, F.X., 1997. The prevention and treatment of obesity: application to type 2 diabetes. Diabetes Care 20, 1744–1766.
Maniega, S.M., Valdes Hernandez, M.C., Clayden, J.D., Royle, N.A., Murray, C., Morris, Z., et al., 2015. White matter hyperintensities and normal-appearing white matter integrity in the aging brain. Neurobiol. Aging 36, 909–918.
Matteoli, G., Boeckxstaens, G.E., 2013. The vagal innervation of the gut and immune homeostasis. Gut 62, 1214–1222.
Meigs, J.B., Rutter, M.K., Sullivan, L.M., Fox, C.S., D'Agostino Sr, R.B., Wilson, B.W., 2007. Impact of insulin resistance on risk of type 2 diabetes and cardiovascular disease in people with metabolic syndrome. Diabetes Care 30, 1219–1225.
Montgomery, M.K., Turner, N., 2015. Mitochondrial dysfunction and insulin resistance: an update. Endocr. Connect. 4, R1–R15.
Morino, K., Petersen, K.F., Shulman, G.I., 2006. Molecular mechanisms of insulin resistance in humans and their potential links with mitochondrial dysfunction. Diabetes 55 (Suppl. 2), S9–S15.

Moskowitz, M.A., Lo, E.H., Iadecola, C., 2010. The science of stroke: mechanisms in search of treatments. Neuron 67, 181–198.

Newburgh, L., 1972. Control of the hyperglycemia of obese "diabetics" by weight reduction. Ann. Intern. Med. 17, 935–942.

Nunomura, A., Hofer, T., Moreira, P.T., Castellani, R.J., Smith, M.A., Perry, G., 2009. RNA oxidation in Alzheimer disease and related neurodegenerative disorders. Acta Neuropathol. 118, 151–166.

Opatrilova, R., Caprnda, M., Kubatka, P., Valentova, V., Uramova, S., Nosal, V., et al., 2018. Adipokines in neurovascular diseases. Biomed. Pharmacother. 98, 424–432.

Pamplona, R., 2008. Membrane phospholipids, lipoxidative damage and molecular integrity: a causal role in aging and longevity. Biochim. Biophys. Acta 1777, 1249–1262.

Pandolfi, A., Giaccari, A., Cilli, C., Alberta, M.M., Morviducci, L., De Filippis, E.A., et al., 2001. Acute hyperglycemia and acute hyperinsulinemia decrease plasma fibrinolytic activity and increase plasminogen activator inhibitor type 1 in the rat. Acta Diabetol. 38, 71–76.

Park, A.J., Collins, J., Blennerhassett, P.A., Ghia, J.E., Verdu, E.F., Bercik, P., et al., 2013. Altered colonic function and microbiota profile in a mouse model of chronic depression. Neurogastroenterol. Motil. 25 (733), e575.

Pirmoradi, M., Jemel, B., Gallagher, A., Tremblay, J., D'Hondt, F., et al., 2016. Verbal memory and verbal fluency tasks used for language localization and lateralization during magnetoencephalography. Epilepsy Res. 119, 1–9.

Putaala, J., Liebkind, R., Gordin, D., Thorn, L.M., Haapaniemi, E., Forsblom, C., et al., 2011. Diabetes mellitus and ischemic stroke in the young: clinical features and long-term prognosis. Neurology 76, 1831–1837.

Radak, Z., Boldogh, I., 2010. 8-Oxo-7,8-dihydroguanine: links to gene expression, aging, and defense against oxidative stress. Free Rad. Biol. Med. 49, 587–596.

Rask-Madsen, C., King, G.L., 2013. Vascular complications of diabetes: mechanisms of injury and protective factors. Cell. Metab. 17, 20–33.

Redmon, J.B., Reck, K.P., Raatz, S.K., et al., 2005. Two-year outcome of a combination of weight loss therapies for type 2 diabetes. Diabetes Care 28, 1311–1315.

Ritzel, R.M., Lai, Y.J., Crapser, J.D., Patel, A.R., Schrecengost, A., Grenier, J.M., et al., 2018. Aging alters the immunological response to ischemic stroke. Acta Neuropathol. 136, 89–110.

Rivollier, A., He, J., Kole, A., Valatas, V., Kelsall, B.L., 2012. Inflammation switches the differentiation program of Ly6Chi monocytes from antiinflammatory macrophages to inflammatory dendritic cells in the colon. J. Exp. Med. 209, 139–155.

Rocha, B.N., Scalzo, P.L., Barbosa, I.G., et al., 2014. Circulating levels of adipokines in Parkinson's disease. J. Neurol. Sci. 339, 64–68.

Ryan, M.T., Hoogenraad, N.J., 2007. Mitochondrial-nuclear communications. Annu. Rev. Biochem. 76, 701–722.

Salim, S., 2017. Oxidative stress and the central nervous system. J. Pharmacol. Exp. Ther. 360, 201–205.

Santiago, C., Herrmann, N., Swardfager, W., Saleem, M., Oh, P.I., Black, S.E., et al., 2015. White matter microstructural integrity is associated with executive function and processing speed in older adults with coronary artery disease. Am. J. Geriatr. Psychiatry 23, 754–763.

Sasaki, N., Toki, S., Chowei, H., Saito, T., Nakano, N., Hayashi, Y., et al., 2001. Immunohistochemical distribution of the receptor for advanced glycation end products in neurons and astrocytes in Alzheimer's disease. Brain Res. 888, 256–262.

Saver, J.L., 2006. Time is brain—quantified. Stroke 37, 263–266.

Save-Pédebos, J., Pinabiaux, C., Dorfmuller, G., Sorbets, S.F., Delalande, O., et al., 2016. The development of pragmatic skills in children after hemispherotomy:contribution from left and right hemispheres. Epilepsy Behav. 55, 139–145.
Schaper, N.C., Nabuurs-Franssen, M.H., Huijberts, M.S., 2000. Peripheral vascular disease and type 2 diabetes mellitus. Diabetes Metab. Res. Rev. 16 (Suppl. 1), S11–S15.
Schenk, S., Saberi, M., Olefsky, J.M., 2008. Insulin sensitivity: modulation by nutrients and inflammation. J. Clin. Invest. 118, 2992–3002.
Schieber, A.M., Lee, Y.M., Chang, M.W., Leblanc, M., Collins, B., Downes, M., et al., 2015. Disease tolerance mediated by microbiome *E. coli* involves inflammasome and IGF-1 signaling. Science 350, 558–563.
Schmidt, A.M., Yan, S.D., Wautier, J.L., Stern, D., 1999. Activation of receptor for advanced glycation end products: a mechanism for chronic vascular dysfunction in diabetic vasculopathy and atherosclerosis. Circ. Res. 84, 489–497.
Sharon, G., Sampson, T.R., Geschwind, D.H., Mazmanian, S.K., 2016. The central nervous system and the gut microbiome. Cell. 167, 915–932.
Shoelson, S.E., Lee, J., Goldfine, A.B., 2006. Inflammation and insulin resistance. J. Clin. Invest. 116, 1793–1801.
Simons, K., Ikonen, E., 2000. How cells handle cholesterol. Science 290, 1721–1726.
Simons, K., Ehehalt, R., 2002. Cholesterol, lipid rafts, and disease. J. Clin. Invest. 110, 597–603.
Smith, W.S., 2004. Pathophysiology of focal cerebral ischemia: a therapeutic perspective. J. Vasc. Interv. Radiol. 15, S3–S12.
Singh, V., Roth, S., Llovera, G., Sadler, R., Garzetti, D., Stecher, B., et al., 2016. Microbiota dysbiosis controls the neuroinflammatory response after stroke. J. Neurosci. 36, 7428–7440.
Sochocka, M., Diniz, B.S., Leszek, J., 2017. Inflammatory response in the CNS: friend or foe? Mol. Neurobiol. 54, 8071–8089.
Solinas, G., Becattini, B., 2016. NK at the crossroad of obesity, insulin resistance, and cell stress response. Mol. Metab. 6, 174–184.
Spelbrink, J., Li, F.Y., Tiranti, V., Nikali, K., Yuan, Q.P., et al., 2001. Human mitochondrial DNA deletions associated with mutations in the gene for twinkle, a phage t7 gene 4-like protein localized to mitochondrial nucleoids. Nat. Genet. 28, 223–231.
Stadtman, E.R., Berlett, B.S., 1991. Metal-catalyzed oxidation of proteins. Physiological consequences. J. Biol. Chem. 266, 17201–17211.
Sun, J.-H., Tan, L., Yu, J.-T., 2014. Post-stroke cognitive impairment: epidemiology, mechanisms and management. Ann. Transl. Med. 2, 80.
Sun, X., Jiao, X., Ma, Y., Liu, Y., Zhang, L., He, Y., et al., 2016. Trimethylamine N-oxide induces inflammation and endothelial dysfunction in human umbilical vein endothelial cells via activating ROS-TXNIP-NLRP3 inflammasome. Biochem. Biophys. Res. Commun. 481, 63–70.
Tanaka, M., Chock, P.B., Stadtman, E.R., 2007. Oxidized messenger RNA induces translation errors. Proc. Natl. Acad. Sci. U.S.A 104, 66–71.
Tang, W.H., Wang, Z., Levison, B.S., Koeth, R.A., Britt, E.B., Fu, X., et al., 2013. Intestinal microbial metabolism of phosphatidylcholine and cardiovascular risk. N. Engl. J. Med. 368, 1575–1584.
Thaler, J.P., Schwartz, M.W., 2010. Minireview: Inflammation and obesity pathogenesis: the hypothalamus heats up. Endocrinology 151, 4109–4115.
Thaler, J.P., Yi, C.X., Schur, E.A., Guyenet, S.J., Hwang, B.H., Dietrich, M.O., et al., 2012. Obesity is associated with hypothalamic injury in rodents and humans. J. Clin. Invest. 122, 153–162.
Tiozzo, E., Youbi, M., Dave, K., Perez-Pinzon, M., Rundek, T., et al., 2015. Aerobic, resistance and cognitive exercise training post stroke. Stroke 46, 2012–2016.

Trayhurn, P., Beattie, J.H., 2001. Physiological role of adipose tissue: white adipose tissue as an endocrine and secretory organ. Proc. Nutr. Soc. 60, 329–339.
Trayhurn, P., Wood, I.S., 2005. Signalling role of adipose tissue: adipokines and inflammation in obesity. Biochem. Soc. Trans. 33, 1078–1081.
Uchida, T., Mori, M., Uzawa, A., Masuda, H., Muto, M., Ohtani, R., et al., 2017. Increased cerebrospinal fluid metalloproteinase-2 and interleukin-6 are associated with albumin quotient in neuromyelitis optica: their possible role on blood-brain barrier disruption. Mult. Scler. 23, 1072–1084.
Unverzagt, F.W., McClure, L.A., Wadley, V.G., Jenny, N.S., Go, R.C., Cushman, M., et al., 2011. Vascular risk factors and cognitive impairment in a stroke-free cohort. Neurology 77, 1729–1736.
Urbanavičius, V., Abalikšta, T., Brimas, G., Abraitienė, A., Gogelienė, L., Strupas, K., 2013. Comparison of changes in blood glucose, insulin resistance indices, and adipokine levels in diabetic and non-diabetic subjects with morbid obesity after laparoscopic adjustable gastric banding. Medicina (Kaunas) 49, 9–14.
Uyttenboogaart, M., Koch, M.W., Stewart, R.E., Vroomen, P.C., Luijckx, G.J., et al., 2007. Moderate hyperglycaemia is associated with favourable outcome in acute lacunar stroke. Brain 130, 1626–1630.
Vaidyula, V.R., Rao, A.K., Mozzoli, M., Homko, C., Cheung, P., et al., 2006. Effects of hyperglycemia and hyperinsulinemia on circulating tissue factor procoagulant activity and platelet CD40 ligand. Diabetes 55, 202–208.
van Dijk, E.J., Breteler, M.M., Schmidt, R., Berger, K., Nilsson, L.G., Oudkerk, M., et al., 2004. The association between blood pressure, hypertension, and cerebral white matter lesions: cardiovascular determinants of dementia study. Hypertension 44, 625–630.
Verhaaren, B.F., Vernooij, M.W., de Boer, R., Hofman, A., Niessen, W.J., van der Lugt, A., et al., 2013. High blood pressure and cerebral white matter lesion progression in the general population. Hypertension 61, 1354–1359.
Vermeer, S.E., Hollander, M., van Dijk, E.J., Hofman, A., Koudstaal, P.J., Breteler, M. M., 2003. Silent brain infarcts and white matter lesions increase stroke risk in the general population: the Rotterdam Scan Study. Stroke 34, 1126–1129.
Wang, Q., Jin, T., 2009. The role of insulin signaling in the development of beta-cell dysfunction and diabetes. Islets 1, 95–101.
Wang, Z., Klipfell, E., Bennett, B.J., Koeth, R., Levison, B.S., Dugar, B., et al., 2011. Gut flora metabolism of phosphatidylcholine promotes cardiovascular disease. Nature 472, 57–63.
Wang, J., Li, G., Wang, Z., Zhang, X., Yao, L., Wang, F., et al., 2012. High glucose-induced expression of inflammatory cytokines and reactive oxygen species in cultured astrocytes. Neuroscience 202, 58–68.
Weickert, M.O., 2012. Nutritional modulation of insulin resistance. Scientifica (Cairo) 2012, 424780.
Wen, S.W., Wong, C.H., 2017. An unexplored brain-gut microbiota axis in stroke. Gut Microb. 8, 601–606.
Winek, K., Dirnagl, U., Meisel, A., 2016a. The gut microbiome as therapeutic target in central nervous system diseases: implications for stroke. Neurotherapeutics 13, 762–774.
Winek, K., Meisel, A., Dirnagl, U., 2016b. Gut microbiota impact on stroke outcome: Fad or fact? J. Cereb. Blood Flow Metab. 36, 891–898.
Winek, K., Engel, O., Koduah, P., Heimesaat, M.M., Fischer, A., Bereswill, S., et al., 2016c. Depletion of cultivatable gut microbiota by broad-spectrum antibiotic pretreatment worsens outcome after murine stroke. Stroke 47, 1354–1363.
Xing, Z., Gauldie, J., Cox, G., Baumann, H., Jordana, M., et al., 1998. IL-6 is an antiinflammatory cytokine required for controlling local or systemic acute inflammatory responses. J. Clin. Invest. 101, 311–320.

Zhao, Z., Zhao, C., Zhang, X.H., Zheng, F., Cai, W., Vlassara, H., et al., 2009. Advanced glycation end products inhibit glucose-stimulated insulin secretion through nitric oxide-dependent inhibition of cytochrome c oxidase and adenosine triphosphate synthesis. Endocrinology 150, 2569–2576.

Zhao, L., Xiong, Q., Stary, C.M., Mahgoub, O.K., Ye, Y., Gu, L., et al., 2018. Bidirectional gut-brain-microbiota axis as a potential link between inflammatory bowel disease and ischemic stroke. J. Neuroinflamm. 15, 339.

Zhu, W., Gregory, J.C., Org, E., Buffa, J.A., Gupta, N., Wang, Z., et al., 2016. Gut microbial metabolite TMAO enhances platelet hyperreactivity and thrombosis risk. Cell. 165, 111–124.

Zimmermann, R., Panzenböck, U., Wintersperger, A., Levak-Frank, S., Graier, W., Glatter, O., 2001. Lipoprotein lipase mediates the uptake of glycated LDL in fibroblasts, endothelial cells, and macrophages. Diabetes 50, 1643–1653.

Zinnhardt, B., Wiesmann, M., Honold, L., Barca, C., Schäfers, M., Kiliaan, A.J., et al., 2018. In vivo imaging biomarkers of neuroinflammation in the development and assessment of stroke therapies—towards clinical translation. Theranostics 8, 2603–2620.

Zoppo, G.J., Mabuchi, T., 2003. Cerebral microvessel responses to focal ischemia. J. Cereb. Blood Flow Metab. 23, 879–894.

Further reading

Calleja, A.I., Garcia-Bermejo, P., Cortijo, E., Bustamante, R., Rojo Martinez, E., Gonzalez Sarmiento, E., et al., 2011. Insulin resistance is associated with a poor response to intravenous thrombolysis in acute ischemic stroke. Diabetes Care 34, 2413–2417.

Cosentino, F., Battista, R., Scuteri, A., De Sensi, F., De Siati, L., Di Russo, C., et al., 2009. Impact of fasting glycemia and regional cerebral perfusion in diabetic subjects: a study with technetium-99m-ethyl cysteinate dimer single photon emission computed tomography. Stroke 40, 306–308.

Evans, J.L., Goldfine, I.D., Maddux, B.A., Grodsky, G.M., 2002. Endocr. Rev. 23, 599–622.

Hacke, W., Kaste, M., Bluhmki, E., Brozman, M., Davalos, A., Guidetti, D., et al., 2008. Thrombolysis with alteplase 3 to 4.5 hours after acute ischemic stroke. N. Engl. J. Med. 359, 1317–1329.

Keymeulen, B., Jacobs, A., de Metz, K., de Sadeleer, C., Bossuyt, A., Somers, G., 1995. Regional cerebral hypoperfusion in long-term type 1 (insulin-dependent) diabetic patients: relation to hypoglycaemic events. Nucl. Med. Commun. 16, 10–16.

Mocan Hognogi, L.D., Goidescu, C.M., Farcas, A.D., 2018. Usefulness of the adipokines as biomarkers of ischemic cardiac dysfunction. Dis. Markers 2018, 3406028.

Molina, C.A., 2011. Reperfusion therapies for acute ischemic stroke current pharmacological and mechanical approaches. Stroke 42, S16–S19.

CHAPTER 6

Insulin resistance and Alzheimer's disease

Introduction

Insulin resistance represents a physiological state in which the action of insulin in target tissues is impaired. As a consequence, the body is not effectively able to adapt to its metabolic or energy demands, also described as metabolic inflexibility (Stinkens et al., 2015). Insulin resistance is characterized by defects in uptake and oxidation of glucose, a decrease in glycogen synthesis, and, to a lesser extent, the ability to suppress lipid oxidation (Diehl et al., 2017). Insulin resistance develops simultaneously in multiple organs and the severity of insulin resistance may vary between different organs and is associated with numerous metabolic and neurological disorders such as obesity, type 2 diabetes, metabolic syndrome, aging, stroke, and various types of dementia (Farooqui, 2013; Diehl et al., 2017). These disorders are characterized by altered glucose homeostasis, hyperinsulinemia, and hypertriglyceridemia. In peripheral metabolic disorders (obesity, type 2 diabetes, and metabolic syndrome), prolonged metabolic stress and proinflammatory signaling contribute to altered insulin signaling and decrease in cellular responsiveness to insulin (Gregor and Hotamisligil, 2011). This pathological state is called as insulin resistance. It not only impairs glucose homeostasis and mitochondrial function leading to reduction in production of energy, but also promoting pancreatic β cells to secrete more insulin, in a process known as compensatory hyperinsulinemia (Leney and Tavare, 2009). Chronic peripheral hyperinsulinemia leads to the downregulation of insulin transporters at the blood–brain barrier (BBB), which in turn decreases the amount of insulin that may enter brain (Banks et al., 2012).

Insulin resistance is plaguing our society for years. As stated in Chapter 1, Insulin resistance and obesity, insulin is involved in the regulation of energy and lipid metabolism via the activation of an intracellular signaling cascade involving the insulin receptor, insulin receptor substrate (IRS) proteins, phosphoinositol 3-kinase (PtdIns 3K), and protein kinase

Insulin Resistance as a Risk Factor in Visceral and Neurological Disorders
DOI: https://doi.org/10.1016/B978-0-12-819603-8.00006-7

B (Akt), which in turn affect master regulatory switches of cell metabolism, cell survival, growth and differentiation, such as the mammalian target of rapamycin (mTOR), and glycogen synthase kinase 3 (GSK3) (Tzatsos, 2009; Zhang and Liu, 2014). In peripheral tissues, insulin resistance and activation of the TNF-α/JNK pathway are linked to major inflammatory/stress signaling networks, including endoplasmic reticulum (ER) stress and the stress kinases I-κBα kinase (IKK) and double-stranded RNA-dependent protein kinase (PKR) (Nakamura et al., 2010). It should also be noted that vascular function is tightly coupled to insulin signaling. Central to this relationship is endothelial dysfunction, which manifests through deficient vasodilation and improper vasoconstriction throughout the body in the setting of insulin resistance (Cersosimo and DeFronzo, 2006). The vasodilator effects of insulin are mediated by the PtdIns 3K signaling pathway, which leads not only to nitric oxide (NO) production in endothelial cells, but also increases levels of cyclic guanosine 3′,5′-monophosphate (cGMP) in vascular smooth muscle. Vasoconstrictor effects of insulin are mediated through endothelin-1 (Muniyappa and Sowers, 2013). Thus insulin signaling increases NO production in dose-dependent manner (Zeng and Quon, 1996), whereas impaired PtdIns 3K signaling decreases NO and cGMP, leading to decrease in vasodilation (Francis et al., 2010). In addition, NO also inhibits platelet aggregation, monocyte adhesion, and thrombosis, all of which damage the vessel wall (Celermajer, 1997). Microvascular disruption leads to superoxide production, which, among other events, leads to a rise in advanced glycation end products. Pathological activation of the receptor for these advanced glycation end products (RAGE) increases oxidative stress, exacerbating vascular inflammation, thrombosis, and vascular damage (Kook et al., 2012). Impaired endothelial cell–mediated vasodilation may also be caused by excess free fatty acids (FFAs), which flow with the blood stream (Steinberg et al., 1997). FFAs, which are often elevated in diabetic patients. They inhibit NF-κB kinase subunit beta (IKKβ). This subunit not only modulates NF-κB but also inhibits the production of NO resulting in decreased vasodilation, deterioration of cardiovascular function, and exacerbation of insulin resistant state (Kim et al., 2005).

Insulin signaling in the brain

As mentioned in Chapter 1, Insulin resistance and obesity, insulin exerts its actions by interacting with the extracellular α subunit of the insulin receptor. These interactions lead to a conformational change that autophosphorylates the intracellular β subunit of the receptor via tyrosine kinase activation (Draznin, 2006). The activation of tyrosine kinase promotes the recruitment and phosphorylation of the IRS1/2, which represent the first node in the insulin signaling cascade and exerts downstream effects on several key regulatory proteins of cell metabolism, cell survival, growth, and differentiation, including the mTOR, PKB, and GSK3 (Fig. 6.1) (Tzatsos, 2009; Sarbassov et al., 2005). As mentioned in Chapter 1, Insulin resistance and obesity, insulin and insulin-like growth factors regulate many neurochemical processes such as axonal growth, protein synthesis, cell growth, gene expression, proliferation, differentiation, and development. Among these, the energy metabolism and synaptic plasticity are the major transduction processes regulated by insulin, which are the core objectives for learning and memory. It the molecular level, insulin modifies neuronal activity promoting synaptic plasticity (Schmitz et al., 2018) and improving memory function in mammalian brain (Benedict et al., 2008). Importance of glial insulin signaling in the brain is gaining lots of attention in recent years. Thus insulin modulates proinflammatory cytokine secretion in microglia and astrocytes in vitro (Spielman et al., 2015; Kurochkin et al., 2018). It has been proposed that hyperinsulinemia induced insulin resistance results in injury to the central nervous system by the activation of GSK3β which is the key ailment in the cognitive decline. Hence, the endogenous brain-specific insulin impairments and signaling account for the majority of Alzheimer's abnormalities.

Insulin signaling pathway largely converge downstream onto Akt activation, primarily through PtdIns 3K-mediated activation of 3-phosphoinositide-dependent protein kinase (PDK) (Zhang and Liu, 2014; Pessin and Saltiel, 2000; Lopez-Lopez et al., 2004). Mammalian tissues contain highly conserved PtdIns 3K/Akt/mTOR pathways, which play a pivotal role in nutrition uptake and storage (Saltiel and Kahn, 2001; Engelman et al., 2006). Three classes of lipid kinases (classes I, II, and III) are found in mammalian tissues. Ligands for PtdIns 3K/Akt/mTOR pathway include growth factors, cytokines, hormones, and G-protein-coupled receptors (GPCRs). GPCRs directly interact with PtdIns 3Ks through Gα or Gβγ subunits. Activated Class I lipid kinase (PtdIns 3K) phosphorylates the substrate phosphatidylinositol 4,5-biphosphate (PtdIns 4,5-P_2) to form phosphatidylinositol 3,4,5-triphosphate (PtdIns 3,4,5-P_3) on intracellular membranes, subsequently recruiting signaling proteins, including Akt (PKB) (Vanhaesebroeck et al., 2010; Abeyrathna and Su, 2015; Manning and Toker, 2017). Akt is

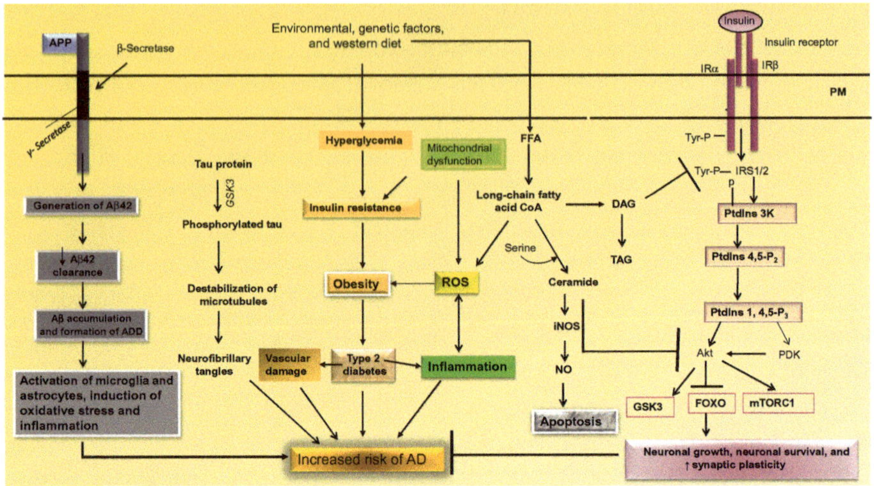

Figure 6.1 Hypothetical diagram showing signal transduction pathways associated with the pathogenesis of Alzheimer's disease.
Aβ, Beta-amyloid; *Akt*, serine/threonine protein kinase or protein kinase B; *APP*, amyloid precursor protein; *DAG*, diacylglycerol; *FFAs*, free fatty acids; *FOXO*, forkhead box O; *GSK3*, glycogen synthase kinase 3; *mTOR*, mammalian target of rapamycin; *PtdIns 3K*, phosphoinositide 3-kinase; *PtdIns 4,5-P_2*, phosphatidylinositol 4,5-bisphosphate; *PtdIns 3,4,5-P_3*, phosphatidylinositol 3,4,5-triphosphate; *ROS*, reactive oxygen species; *TAG*, triacylglyceroland.

activated through two pivotal phosphorylation processes. First, phosphorylation of the threonine 308 (Akt1) in the kinase domain by phosphoinositide-dependent protein kinase 1 (PDK1) initiates the activation process (Alessi et al., 1997), subsequent phosphorylation at serine 473 (Akt1) in the carboxy-terminal regulatory domain through mTOR complex 2 (mTORC2) (Sarbassov et al., 2005; Zhang et al., 2015), which is activated by a PtdIns 3K-dependent mechanism, completely activates Akt (Liu et al., 2015). Similar phosphorylation events are observed at corresponding residues in Akt2 (T309 and S474) and Akt3 (T305 and S472) (Manning and Toker, 2017). Phosphorylation of both residues is necessary for maximum activation of Akt. Protein phosphatase 2A (PP2A) (Andjelkovic et al., 1996) and PH domain leucine-rich repeat protein phosphatases (PHLPP1 and PHLPP2) (Gao et al., 2005) dephosphorylate Akt T308 and S473, respectively, inducing the inactivation of Akt. Recently, endomembranes that contain PtdIns 3,4,5-P_3 and PtdIns 4,5-P_2 have also been shown to directly contribute to Akt activation (Jethwa et al., 2015; Braccini et al., 2015).

In response to dietary nutrient, the upstream signaling molecule PtdIns 3K is activated. This leads to the activation of Akt (Cantley, 2002). The Akt kinase has two major metabolic target molecules, FOXO and mTOR (Huang and Tindall, 2007; Hay, 2005). Phosphorylation by Akt inactivates FOXO, a key transcription factor involved in catabolism, while activates the mTOR complex and its downstream targets. The increase in glucose and insulin levels activates mTOR within metabolic tissues to control whole body metabolic homeostasis. mTOR, a conserved serine/threonine kinase that plays a key role in controlling a balance between protein synthesis and degradation (Loewith et al., 2002; Wullschleger et al., 2006). In conjugation with other protein it forms two complexes: mTOR complex 1 (mTORC1) and mTOR complex 2 (mTORC2). Both complexes are phosphorylated by Akt dependent PtdIns 3K, in which localization of both is solely cytoplasmic, but the field of operation is completely different. mTORC1 having raptor as its integral component promotes protein synthesis, ribosome biogenesis, proliferation, migration, and differentiation, by stimulating p70 S6 kinase (S6K1) and inhibiting 4E-binding protein (4EBP1) and elF4E (an RNA helicase 2). While mTORC2 having rictor as its integral component promotes cell survival, cell cycle progression, and actin remodeling by its actions through PKC and SGK1(Ser 422) (Fig. 6.2) (Loewith et al., 2002; Wullschleger et al., 2006). mTORC2 also regulates glucose homeostasis through Akt. This enzyme promotes glucose uptake by increasing glucose transporter type 4 (GLUT4) translocation to the membrane in adipocytes (Kohn et al., 1996). In addition, Akt phosphorylates and deactivates GSK3, which decreases the rate of phosphorylation of glycogen synthase. This leads to an increase in glycogen synthase activity, thus elevating the accumulation of glycogen, which is especially important in the muscles and liver (Manning and Cantley, 2007). Akt also controls glucose homeostasis by phosphorylating and inhibiting FOXO1, a transcription factor that regulates gluconeogenesis (Accili and Arden, 2004). By suppressing glucose production in the liver and stimulating glucose uptake in muscle and fat, insulin reduces blood glucose levels to maintain glucose homeostasis in humans and animals. Insulin also regulates many important anabolic processes including protein and glycogen synthesis in muscle and liver, lipid synthesis and storage in liver and adipocytes. Insulin also inhibits fatty acid oxidation, glycogenolysis, and gluconeogenesis. Similarly, in brain, the intrinsic communication of mTOR complexes with the metabolic control of glycogenesis and lipogenesis is essential to maintain central homeostasis

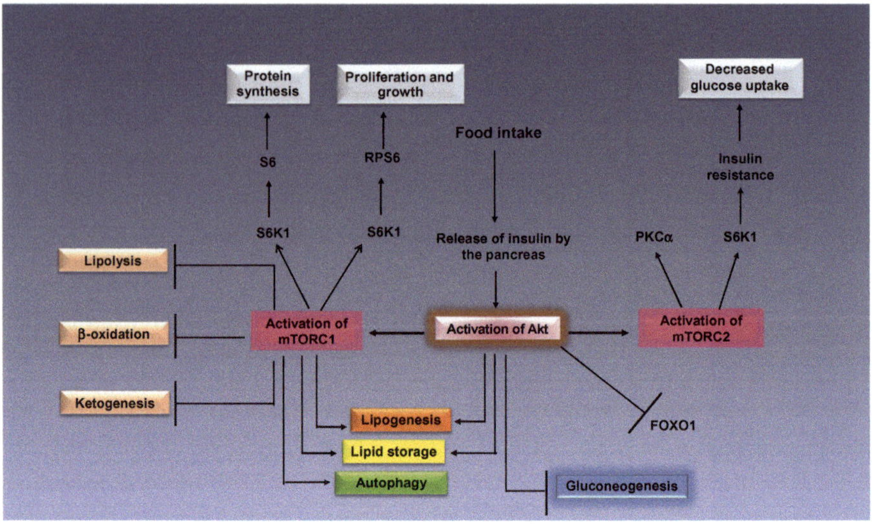

Figure 6.2 Signal transduction pathways associated with consumption of food.

(Zoncu et al., 2011; Cornu et al., 2013), since neural cells are highly dependent on the continued supply of glucose and other energy substrates (e.g., ketone bodies) to maintain the ATP/AMP ratio. This dynamic allows for the correct regulation of autophagy systems, essential for the clearance of malfunctioning organelles and misfolded proteins, which have been found to be dysregulated in central diseases such as AD (Perluigi et al., 2015).

Insulin resistance develops through complex interactions of genotype and lifestyle (lack of exercise and over-nutrition) (Muniyappa et al., 2007; Romao and Roth, 2008; Stolerman and Florez, 2009). As mentioned in earlier chapters, insulin sensitivity in target tissues is regulated physiologically by several circulating factors such as plasma lipids, circulating hormones and adipokines, and their respective signaling pathways (Ahima and Lazar, 2008; Zac-Varghese et al., 2010). Cross-talk among these various signaling pathways with the insulin signaling pathways constantly tunes insulin sensitivity. Adipose tissue, along with the brain and the gut, constitute a neuroendocrine axis that regulates metabolism in large measure by regulating insulin sensitivity in target tissues (Ahima and Lazar, 2008; Zac-Varghese et al., 2010). Insulin resistance is also regulated by alterations in systemic lipid metabolism including: (1) high levels of plasma triacylglycerols (TAG), (2) low levels of high-density lipoprotein, and (3) the appearance of small dense low-density lipoproteins. These three conditions along with endothelial dysfunction induces aberrant insulin signaling and atherosclerotic plaque formation in blood vessels.

Pathogenesis of Alzheimer's disease

Alzheimer's disease (AD) is a neurodegenerative disease, which is characterized by the accumulation of extracellular β-amyloid (Aβ) plaques (senile plaques) and intracellular neurofibrillary tangles (NFTs) composed of Tau (Hardy and Selkoe, 2002; Hardy, 2009; Hardy, 2006; Selkoe, 2008). AD is also characterized by a loss or decline in memory and other intellectual functions that can lead to impairments in everyday performance. Two forms of AD (familial and sporadic) have been reported to occur in human population. Majority of AD cases (>90%–95%) are of sporadic (late-onset form). These patients are older than 65 years. Only 5%–7% cases are primarily genetic (early-onset familial form) involving apolipoprotein E (APOE), amyloid precursor protein (APP), presenilin 1 (PS 1), and presenilin 2 (PS 2) genes (van der Flier and Scheltens, 2005; Duthey, 2013). The accumulation of senile plaques leads to the loss of synapses and loss of neurons in cortical and subcortical areas and hippocampus, not only producing impairments in memory and other cognitive functions, but also promoting changes in personality and behavior affecting daily activities (Reddy and Beal, 2008). How these factors ultimately contribute to memory impairments and cognitive deficits in AD remain elusive (Hardy and Selkoe, 2002; Hardy, 2006, 2009; Selkoe, 2008). Senile plaques and tangles are composed of Aβ and hyperphosphorylated tau protein, respectively. Aβ is a 38–43 amino acids long hydrophobic fragment derived from proteolytic cleavage of the APP, a transmembrane protein with a single membrane-spanning domain, which may have a trophic function by α- and β-secretases (Aguzzi and Haass, 2003; O'Brien and Wong, 2011; Weyer et al., 2011). Excessive generation of Aβ42 peptide not only produces cellular toxicity directly through altered Aβ42 peptide–plasma membrane interactions and channel-mediated destabilization of ionic homeostasis, but also through direct binding of Aβ with cell adhesion molecules such as neuroligins and neurexins located in the postsynaptic cleft (Connelly et al., 2012; Jang et al., 2014; Sindi and Dodd, 2015). Aβ is degraded by many enzymes includes, but not limited to, insulin-degrading enzyme (IDE; a 110 kDa zinc metalloendopeptidase) and neprilysin (a 90–110 kDa neutral zinc metalloendopeptidase). Activities of these enzymes are reduced in AD (Dong et al., 2012). The imbalance between

Aβ production and clearance causes Aβ to accumulate in the extracellular space. Aβ1−42 readily forms insoluble clumps and initiates a cascade of events leading to apoptosis and neuronal dysfunction or death.

Another consequence of amyloid deposition in the brain is the activation of immune responses. Thus Aβ deposits in the brain activate the microglia, the brain innate immune cells, leading to an inflammatory response in the CNS (neuroinflammation). An acute, self-limiting neuroinflammatory response leads to Aβ clearance, what is beneficial to neuronal protection (Penke et al., 2017; Fuster-Matanzo et al., 2013). However, during aging process, onset of certain changes in the brain result in occurrence of immune system (immunosenescence) and microglia-related immune response (Pawelec, 2017). Persistent microglia activation (reactive microglia) is associated with chronic inflammatory response that supports the neurotoxic processes, which cause brain injury and neuronal death (Sochocka et al., 2017; Heneka et al., 2015). Moreover, reactive microglial cells activate astrocytes, another kind of glial cell that supports neuron functioning. Presence of reactive astrocytes, as well as reactive microglia are associated with neuroinflammatory burden and BBB dysfunction. BBB, an important CNS protection system, is responsible for the strict control of the molecules transported in and out of the brain. In differentiated SK-N-BE neuroblastoma cells, Aβ monomers inhibit apoptosis and allow autophagy with intracellular accumulation of autophagosomes and elevation of levels of Beclin-1 (BECN1; Bcl-2 interacting protein) and microtubule-associated protein 1 light chain 3 (LC3-II), resulting in an inhibition of substrate degradation due to an inhibitory action on lysosomal activity (Guglielmotto et al., 2014). Aβ oligomers induce the formation of the Bcl-2-BECN1 complex promoting apoptosis. In addition, Aβ oligomers cause a less profound increase in BECN1 and LC3-II levels than monomers without affecting the autophagic flux supporting the view that there is a link between autophagy and apoptosis with monomers and oligomers, respectively (Guglielmotto et al., 2014).

The imbalance in Aβ homeostasis, that is, production versus clearance of Aβ peptides not only leads to their accumulation, aggregation (in either fibrillary, oligomeric, or β-sheet conformations), but also in deposition within the extracellular space (Sadigh-Eteghad et al., 2015). The extracellular Aβ deposits trigger active "gliosis" that primarily consists in hyperactive astrocytes and microglia among the surrounding parenchyma (Kurz and Perneczky, 2011; Wildsmith et al., 2013). In addition, Aβ aggregates into Aβ-derived diffusible ligands (ADDLs), which acts as an initiator of AD pathogenesis not only by interacting with a number of specialized synaptic proteins (scaffolding proteins and ion channels) inducing synaptic loss and progressive cognitive decline, but also by mediating the development of Tau pathology and synaptic dysfunction (Fig. 6.2) (Tu et al., 2014; Viola and Klein, 2015; Selkoe, 2008; Bloom, 2014). ADDLs also mediate their neurotoxic effects by enhancing the generation of ROS through indirect mechanisms (Fang et al., 2010; Borger et al., 2013). These processes cause alterations in neurotransmission and memory formation before plaques build up. Based on these results, it is proposed that ADDLs may be directly contributed to the pathogenesis of AD (Haass and Selkoe, 2007).

Tau is a soluble microtubule-associated protein that is expressed in mature neurons and is localized within axons. It is associated with maintenance of the stability of neuronal microtubules. Hyperphosphorylation of Tau is catalyzed by protein kinases such as GSK3, cyclin-dependant kinase-5 (Cdk-5), CaM kinase II, casein kinase II, stress-activated kinase, c-Jun N-terminal kinase (SAPK/JNK), kinase p38, and Fyn kinase (Fig. 6.2) (Gong and Iqbal, 2008; Avila et al., 2010). Hyperphosphorylation not only makes Tau resistant to calcium activated proteases (calpains) and the ubiquitin-proteasome pathway, but also worsen the accumulation of insoluble fibrillar Tau (Fibrillar Tau). Hyperphosphorylation of Tau not only promotes in its dissociation from microtubules, but also result in aggregation in the form of NFTs inside of neuronal cell bodies and neurites. These processes trigger gliosis and are closely associated with the pathogenesis of AD (Kadavath et al., 2015). Synaptic and neuronal loss as a result of Tau pathology and Aβ deposition have been shown to best correlate with the cognitive impairment seen in AD (Spires-Jones and Hyman, 2014). These pathological features may appear early in subcortical nuclei such as the locus coeruleus, then affect the hippocampus and entorhinal cortex, and eventually spread to other cortical areas as the disease progresses (Holtzman et al., 2011). Hyperphosphorylated tau exerts its neurotoxic effects by increasing oxidative stress, promoting neuronal apoptosis, inducing mitochondrial dysfunction. These processes result in collapse of the microtubule-based cytoskeleton, inducing neuritic dystrophy, and subsequent neuronal demise (Mandelkow et al., 2003; Arnaud et al., 2006; Oddo et al., 2008). It is reported that senile plaques first appear in the frontal cortex, and then spread over the entire cortical region while hyperphosphorylated tau and insoluble tangles initially appear in the limbic system (entorhinal cortex, hippocampus, and dentate gyrus) and then progress to the cortical region. NFTs appear before the deposition of plaque in AD brains, and that tangle pathology is more closely associated with disease severity than the plaque load (Braak and Braak, 1998; Josephs et al., 2008). The amyloid hypothesis has been challenged in recent years as a result of several findings. One is the failure of human trials using therapies targeting the β-amyloid as neurotoxic agent. Another evidence against β-amyloid hypothesis comes from positron emission tomography neuroimaging. It has been demonstrated that increased β-amyloid accumulation in normal human subjects and an absence of correlation between β-amyloid toxicity and severity of disease in AD patients and in cognitively normal individuals (Morris et al., 2014; Chetelat et al., 2013; Farooqui, 2017).

In addition to above changes, AD is also characterized by the decrease in cerebral blood flow (CBF); disruption of BBB (Zlokovic, 2011); alterations in lipid and glucose metabolism; onset of diabetes and metabolic syndrome; activation of microglial and astroglial cells; induction of oxidative stress and neuroinflammation; and deterioration of synapse (Mrak and Griffin, 2005; Prokop et al., 2013; Farooqui, 2013; Farooqui, 2014). Finally, there is a growing consensus that iron dyshomeostasis may play a pivotal pathological role in AD and increased levels of iron have been proposed as the primary driver of neurodegeneration in AD (Belaidi and Bush, 2016). Among abovementioned possibilities, decrease in CBF, disruption of BBB, induction of oxidative stress, and neuroinflammation at early stages seems to be common processes in the pathogenesis of AD.

The metabolic pathway contributing to the pathogenesis of sporadic AD is based on the fact that insulin signaling is impaired in the brains of AD patients (Craft, 2007; Steen et al., 2005; Bomfim et al., 2012; Talbot et al., 2012). Neuronal insulin resistance can be induced in primary cultures of hippocampal neurons by treatment with Aβ and in vivo by intracerebroventricular injection of Aβ oligomer in mice and monkeys (Ferreira et al., 2018). It is suggested that Aβ oligomer induces it effect through TNF-α activation and IRS inhibition. These processes have major impact on synaptic function, synaptic plasticity, and synapse loss (Townsend et al., 2007; De Felice et al., 2009; Bomfim et al., 2012; Batista et al., 2018). Furthermore, injection of Aβ oligomers (icv) also induce peripheral glucose intolerance and classic hallmarks of peripheral insulin resistance, a process also observed in transgenic AD mice models (Clarke et al., 2015) and that may underlie increased risk for diabetes in AD (Janson et al., 2004). Additionally, antidiabetic drugs exert beneficial effects on cognition, synapse protection, insulin signaling deficits, and other AD-related pathological mechanisms, such as ER stress and chronic inflammation (Batista et al., 2018; Tai et al., 2018). Studies performed by de la Monte's group have demonstrated significant decreases in insulin and insulin growth factor (IGF1) receptor levels in frontal cortex, hippocampus, and hypothalamus of sporadic AD patients (Steen et al., 2005; de la Monte and Wands, 2008; Chami et al., 2016).

In addition, de la Monte's group has also demonstrated the decrease in gene expression and protein levels of insulin, IGF1 receptors, increased GSK-3β level, and other downstream molecules, with impaired acetylcholine production and cognitive performance in brains of sporadic AD patients (de la Monte and Wands, 2008). These changes are inversely correlate with neuropathological features of AD, which indicates the relationship between impaired insulin signaling and progression of the disease (Steen et al., 2005). Other studies have also indicated that there are significant alterations in mRNA expression profiles of genes related to insulin signaling in the cortex and hippocampus, a brain region associated with the cognitive processing (Hokama et al., 2014) suggesting alterations in insulin signaling may contribute to the pathogenesis of sporadic AD (Kullmann et al., 2016). Moreover, the activation of mTOR also enhances Aβ generation and deposition by modulating APP metabolism and upregulating β- and γ-secretases. mTOR, an inhibitor of autophagy, decreases Aβ clearance by scissoring autophagy function. mTOR regulates Aβ generation or Aβ clearance by regulating several key signaling pathways, including phosphoinositide 3-kinase (PtdIns 3K)/Akt, GSK-3, AMP-activated protein kinase (AMPK), and IGF-1 (Ma et al., 2010; Cai et al., 2012; Pozueta et al., 2013; Lafay-Chebassier et al., 2005). The activation of mTOR is also a contributor to aberrant hyperphosphorylated tau (Mueed et al., 2019). Rapamycin, the inhibitor of mTOR, may mitigate cognitive impairment and inhibit the pathologies associated with amyloid plaques and NFTs by promoting autophagy. Collective evidence suggests that mTOR signaling also contribute to the accumulation of two hallmarks of AD (Aβ plaques and NFTs) and cognitive impairment in clinical presentation, respectively (Ma et al., 2010; Cai et al., 2012; Pozueta et al., 2013; Lafay-Chebassier et al., 2005). It is also proposed that the induction of abovementioned processes produce multiple toxic effects on neurons and glia leading to oxidative damage, mitochondrial dysfunction, calcium dysregulation, inflammation and ER stress, which together may set in programmed death of neurons (Dong et al., 2012).

Risk factors for AD include the presence of apolipoprotein E4 (*APOE4*) and triggering receptor expressed on myeloid cells 2 (*TREM2*) genes. *APOE4* increases AD risk both by Aβ-dependent and Aβ-independent effects (Liu et al., 2013). APOE4 protein not only regulates Aβ aggregation and clearance, but also an essential regulator of brain cholesterol metabolism. APOE4 plays an important role in cerebral energy metabolism, modulation of chronic inflammation, neurovascular function, neurogenesis, and synaptic plasticity (Kim et al., 2009; Kim et al., 2014). The *TREM2* gene is located on human chromosome 6p21 (Allcock et al., 2003). TREM2 protein (~25.4 kDa cell surface Type-1 transmembrane protein) is an innate immune receptor expressed on the cell surface of microglia, macrophages, osteoclasts, and immature dendritic cells. It triggers the activation of immune responses. The expression of TREM2 is increased in AD patients (Lue et al., 2015; Celarain et al., 2016; Perez et al., 2017). The expression of TREM2 is associated with the recruitment of microglia to amyloid plaques (Perez et al., 2017). Interestingly, aging is also associated with increased expression of TREM2 (Raha et al., 2017). Furthermore, TREM2 expression is also increased in chronic inflammation observed in AD.

As mentioned in Chapter 5, In insulin resistance, gut microbiota communicate with the brain by producing a number of neuroactive and immunocompetent substances [hormones, γ-aminobutyric acid (GABA), catecholamines, histamine, and acetylcholine], neuropeptides, serotonin, kynurenine, melatonin, and short-chain fatty acids (Barrett et al., 2012; Lyte, 2011). These substances are transported from gut to the brain through lymphatic subsystem, which allows for direct or indirect transport (Winek et al., 2016a,b). The composition of the intestinal microbiota is modulated by the circadian rhythm. Disruption of circadian rhythm may influence intestinal microbiota. The imbalance between the microbiota and host organism leads to dysbiosis (changes in the microbiome) and immune imbalance. Dysbiosis also contributes to the dormant blood microbiome (atopobiosis) and directly promotes systemic inflammation via the amyloidogenic formation and shedding of potent inflammagens such as lipopolysaccharides.

Thus alterations in the gut microbiota may also be associated with elevation in the permeability of the gut barrier and immune activation leading to systemic inflammation. This process may in turn impair the BBB and promote neuroinflammation and oxidative stress. These processes ultimately result in neurodegeneration (Fig. 6.3). Furthermore, it is reported that

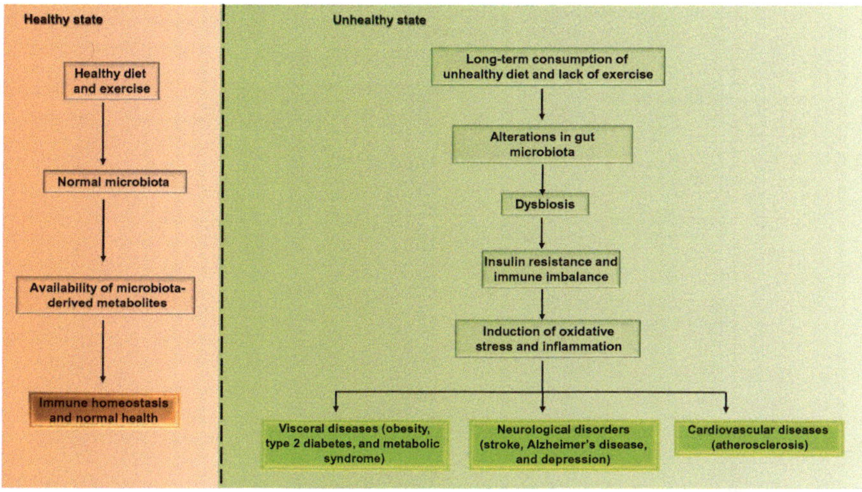

Figure 6.3 Effects of healthy and unhealthy diet on gut microbiota composition and development of dysbiosis on insulin resistance-related diseases.

Gram-negative bacteria–derived lipopolysaccharides can also induce metabolic endotoxaemia, inflammation, impaired glucose metabolism, insulin resistance, obesity, and contribute to the development of metabolic syndrome, type 2 diabetes, inflammarory bowel diseases, and autoimmunity. Abnormalities in the composition of the microbiota, and translocation of bacterial antigens into the systemic circulation and the brain, have also become areas of intense research in the pathogenesis of AD (Vogt et al., 2017; Zhao et al., 2017; Pretorius et al., 2018). On the basis of these studies, it is suggested that gut microbiota can secrete large amounts of amyloids and lipopolysaccharides, which may contribute to the modulation of signaling pathways and the production of proinflammatory cytokines associated with the pathogenesis of AD (Fig. 6.4).

Recent studies have also indicated that Aβ is an antimicrobial peptide, which participates in the innate immune responses. However, in the dysregulated state, Aβ production may induce harmful effects. Importantly, bacterial amyloids through molecular mimicry may elicit cross-seeding of misfolding and induce microglial priming (Kowalski and Mulak, 2019). The Aβ seeding and propagation may occur at different levels of the brain–gut–microbiota axis. The potential mechanisms of amyloid spreading include neuron-to-neuron or distal neuron spreading, direct BBB crossing or via other cells as astrocytes, fibroblasts, microglia, and immune system cells. A growing body of experimental and clinical data have

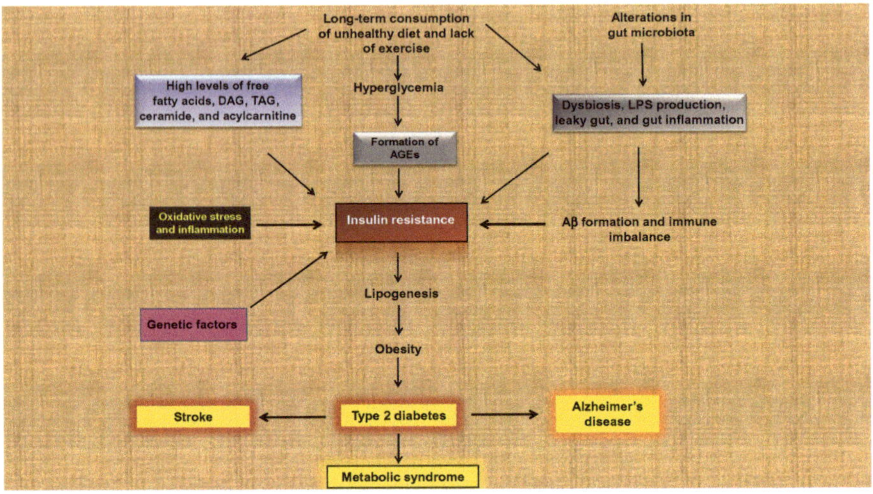

Figure 6.4 Effects of unhealthy diet and lack of exercise on insulin resistance. *DAG*, Diacylglycerol; *TAG*, triacylglycerol.

indicated that gut microbiota–host interactions and gut dysbiosis play an important role in neurodegeneration. It is proposed that induction of gut-mediated inflammatory responses and poor diet in the elderly (seniors) may contribute to the pathogenesis of AD (Jiang et al., 2017; Kowalski and Mulak, 2019).

Dysregulation of kynurenine pathway in tryptophan metabolism may also be associated with the pathogenesis of AD (Willette et al., 2017). Thus depressive symptoms in AD may be due to the involvement of metabolites of kynurenine pathway (Fig. 6.5). Tryptophan is an essential amino acid, which is required for synthesis of proteins. Tryptophan can cross BBB and can be converted into the key neurotransmitters (serotonin and tryptamine). In mammals >95% of tryptophan is metabolized through the kynurenine pathway leading to the synthesis of nicotinamide adenosine dinucleotide. Other metabolites of tryptophan metabolism such as 3-hydroxykynurenine and quinolinic acid produce neurotoxic effects, whereas other metabolites, such as kynurenic acid promote neuroprotective effects. Neurotoxic effects of 3-hydroxykynurenine are due to its antagonistic activity against the strychnine–insensitive glycine-binding site of the *N*-methyl-D-aspartate (NMDA) receptor (Kessler et al., 1989). This metabolite not only produce a weak antagonistic activity against α-amino-3-hydroxy-5-methyl-4-isoxazolepropionate (AMPA) and kainate receptors (Birch et al., 1988), but also inhibits α_7 nicotinic receptor

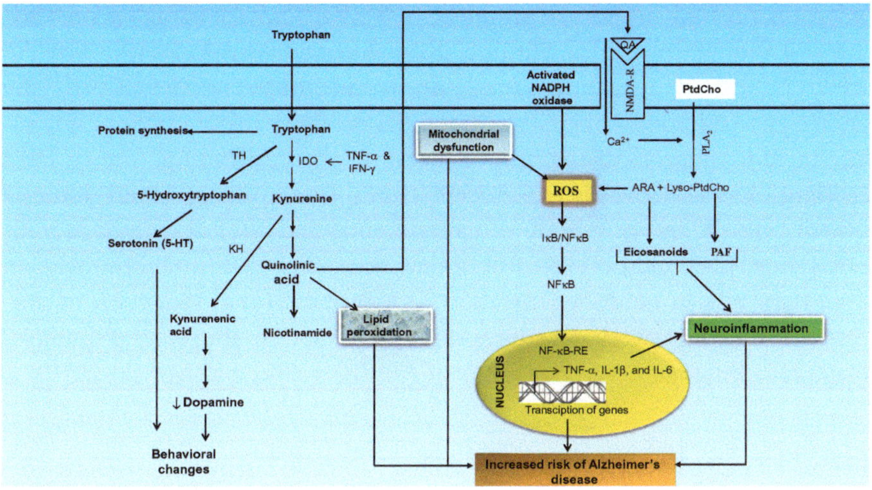

Figure 6.5 Metabolism of tryptophan in brain and effect of quinolinic acid–mediated excitotoxicity in the brain.
ARA, Arachidonic acid; *cPLA₂*, cytosolic phospholipase A_2; *IDO*, indolamine 2,3-dioxygenase; *I-κBα*, I-κB kinaseα; *IL-1β*, interleukin-1beta; *INF-γ*, interferon-gama; *KMO*, kynurenine 3-monooxygenase; *NF-κB*, nuclear factor-kappaB; *NMDA-R*, *N*-methyl-D-aspartate receptor; *PAF*, platelet activating factor; *PtdCho*, phosphatidylcholine; *TH*, tryptophan hydroxylase; *TNF-α*, tumor necrosis factor-α; Quinolenic acid stimulates NMDA receptor and promote the generation of eicosanoids and platelet activating factor. These products induce oxidative stress and facilitate neuroinflammation.

(Hilmas et al., 2001; Albuquerque and Schwarcz, 2013), which contribute to the regulation of presynaptic glutamate release. Similarly, neurotoxic effects of quinolinic acid are due to interactions with NMDA receptors (Schwarcz et al., 1983). These effects modulate the release or reuptake inhibition of glutamate (Tavares et al., 2002). Glutamate not only mediates excitotoxicity-mediated neuronal death, but also promote lipid peroxidation induced neuronal injury in AD (Santamaria et al., 2001). Generation of 3-hydroxykynurenine also leads apoptotic cell death in neuronal cell cultures (Santamaria et al., 2001).

Alterations in the gut microbiota are also associated with elevation in the permeability of the gut barrier and immune activation leading to systemic inflammation. This process in turn impairs the BBB and promotes neuroinflammation, neural injury, and ultimately neurodegeneration. Recent studies have indicated that Aβ is an antimicrobial peptide, which participates in the innate immune responses. However, in the dysregulated state, Aβ production may induce harmful effects. Importantly, bacterial amyloids through molecular mimicry may elicit cross-seeding of misfolding and induce microglial priming (Kowalski and Mulak, 2019). The Aβ seeding and propagation may occur at different levels of the brain–gut–microbiota axis. The potential mechanisms of amyloid spreading include neuron-to-neuron or distal neuron spreading, direct BBB crossing or via other cells as astrocytes, fibroblasts, microglia, and immune system cells. A growing body of experimental and clinical data have indicated that gut microbiota–host interactions and gut dysbiosis play an important role in neurodegeneration. It is proposed that induction of gut-mediated inflammatory responses and poor diet in the elderly (seniors) may contribute to the pathogenesis of AD (Kowalski and Mulak, 2019).

Insulin receptor, insulin signaling, and insulin resistance in the brain

As mentioned in Chapter 1, Insulin resistance and obesity, insulin plays an important role in carbohydrate and lipid metabolism. It modulates glucose uptake and its storage as glycogen in the liver, muscles, and fat cells (Duckworth et al., 1997). Most of the available insulin in the brain is derived from circulating blood levels, and there is very little de novo synthesis of insulin in the brain (Bilotta et al., 2017). However, brain is an insulin sensitive organ and insulin receptor signaling in the brain is important for the growth of neural cells, their survival, metabolism, gene expression, protein synthesis, cytoskeletal assembly, synapse formation, neurotransmitter function, and plasticity (de la Monte and Wands, 2005; D'Ercole and Ye, 2008). In addition, insulin and IGF signaling play critical roles in maintaining cognitive function. In brain, insulin also inhibits the firing of neurons in the hippocampus and hypothalamus; blocks the reuptake of norepinephrine; modulates catecholamine turnover in the hypothalamus; and regulates norepinephrine and dopamine transporter mRNA levels in neurons (Craft, 2007). Insulin also contributes to the modulation of NMDA, AMPA, and GABA receptors signaling. Signal transduction processes associated with these receptors support synaptic plasticity (Fernandez and Torres-Aleman, 2012). Emerging evidence suggests that insulin signaling in the brain plays important role in neuronal development (dendritic outgrowth), neuronal survival, circuit development, synaptic plasticity, postsynaptic neurotransmitter receptor trafficking, glucoregulation, feeding behavior, body weight, and cognitive processes (Blazquez et al., 2014).

Neurochemical links among insulin resistance, type 2 diabetes, and Alzheimer's disease

The central function of insulin receptors is not only to regulate of the energy metabolism in the hypothalamus (Benedict et al., 2011; Thienel et al., 2017), but also to modulate learning and memory in the hippocampus (McNay et al., 2010; Grillo et al., 2015). The onset of insulin resistance in normal aged brain and brains of diabetes and AD patients occurs several years before the patients start to experience the symptoms and are diagnosed (Dankner et al., 2009). Induction of insulin resistance in the brain increases risk of AD risk by at least twofold (Sims-Robinson et al., 2010), and this deleterious effect is not only due to changes in brain vasculature (Biessels and Reijmer, 2014; Frosch et al., 2017), but also due to direct effects on Aβ aggregation or tau phosphorylation. Abnormalities in insulin signaling in the brain not only impair cognitive functions, but markedly affect eating behavior, a process closely associated with the pathogenesis of obesity, insulin resistance, and type 2 diabetes. Among 25% of prediabetic patients (65 years and above) have higher risk of dementia (Luchsinger, 2010). These patients also have a two- to threefold increased risk for AD compared to normal subjects without diabetes (Accardi et al., 2012). Type 2 diabetes and AD are linked with each other through several metabolic pathways such as enzymic degradation of Aβ, forkhead box protein O1 (FOXO1) signaling, and insulin signaling. Aβ monomers augment β-secretase gene transcription activation not only through the MAPK8/JNK1-MAPK9/JNK2 signaling pathway, but also by interfering with its lysosomal degradation leading to an amyloid vicious cycle (Guglielmotto et al., 2011, 2014). Abnormal insulin signaling is the hallmark of type 2 diabetes as well as AD. Furthermore, type 2 diabetes and AD are accompanied by the reduction in insulin sensitivity, decrease in insulin receptors density in the brain, and increased accumulation of Aβ aggregates and hyperphosphorylation of Tau protein (Fig. 6.6) (Craft et al., 2013; Correia et al., 2012, 2011; Cholerton et al., 2013). Accumulation of Aβ aggregates induces neuronal insulin resistance in type 2 diabetic and AD brain not only by inhibiting the insulin network by targeting the insulin/Akt pathway (Townsend et al., 2007), but also through the dissociation of insulin receptors from the dendrites of synaptic sites (Zhao et al., 2008). Furthermore, both conditions are accompanied by hyperphosphorylation of Tau protein. In this process, impairment in insulin signaling does not efficiently block GSK3β and therefore, the

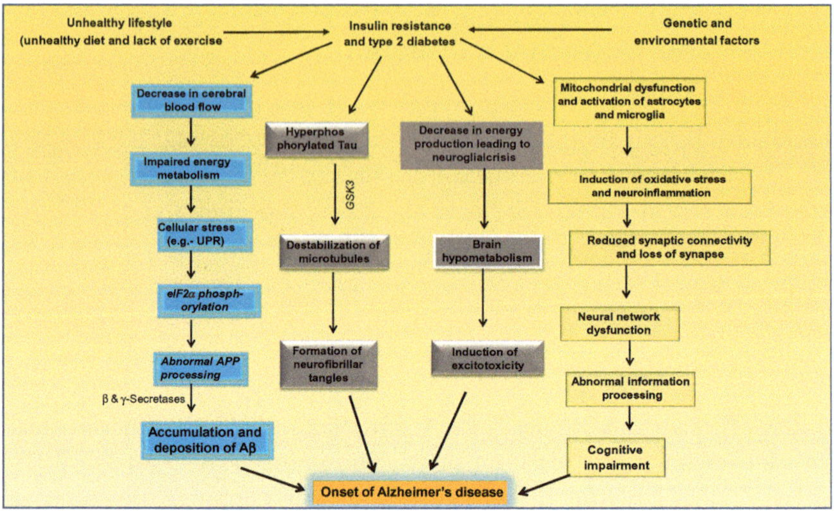

Figure 6.6 Signal transduction processes contributing to insulin resistance and type 2 diabetes–mediated increased risk of Alzheimer's disease.
UPR, Unfolded-protein response.

activated GSK3β increases Tau phosphorylation (Hooper et al., 2008). This results in simultaneous induction of two major pathological processes (accumulation of Aβ and formation of NFTs) in type 2 diabetes and AD. These investigations are supported by postmortem studies on brains of type 2 diabetes and AD patients (also in animal models of AD) (Steen et al., 2005; Moloney et al., 2010). In both pathological conditions are also accompanied by atrophy in hippocampus and amygdala (den Heijer et al., 2003). These alterations are accompanied by increased APP mRNA expression (Steen et al., 2005). Recent studies on animal models of diabetes and AD transgenic mice (5XFAD model) have indicated that insulin deficiency not only alters APP processing by increasing the expression of β-site APP-cleaving enzyme 1 (BACE-1 or β-secretase), but also induces increase in upregulation of APP through the PERK-eIF2α phosphorylation pathway (Devi et al., 2012). In addition, type 2 diabetes and AD have some common risk factors such as increasing age, higher cholesterol (Ristow, 2004), abnormal glucose and lipid metabolism (Doble and Woodgett, 2003), deposition of aggregated Aβ (Ristow, 2004), cardiovascular disease, development of oxidative stress and inflammation (Haan, 2006), ApoE4 (Qiu and Folstein, 2006), induction of apoptosis, etc. Furthermore, amylin deposition in the pancreas is more common in AD than in normal aging, and although there is no significant elevation in cerebral Aβ deposition in type 2 diabetes, the extent of Aβ accumulation in AD correlates with type 2 diabetes duration.

Type 2 diabetes and AD patients show induction of hyperglycemia a process, which is caused by high levels of glucose in the blood. Hyperglycemia not only accelerates Aβ aggregation, but also induces oxidative stress (a process, which is caused by an imbalance between the production of oxidant species and the antioxidant defenses). Oxidative stress affects cellular redox homeostasis leading to molecular alterations and thus resulting in brain damage due to the formation of acrolein, malondialdehyde, and 4-hydroxy-2-nonenal (Farooqui, 2014). In addition to the induction of above processes, hyperglycemia also results in formation of advanced glycation products (AGEs), which are generated by a nonenzymic reaction of glucose, free amino groups, lipids, and nucleic acids (Fig. 6.7) (Sims-Robinson et al., 2010). Glycation reactions may lead to the production of more reactive oxygen species (ROS) leading to high oxidative stress. The nonenzymatic glycation of plasma proteins such as albumin, fibrinogen, and globulins may produce various deleterious effects including platelet activation, generation of oxygen free radicals, and impairment in immune system regulation (Negre-Salvayre et al., 2009). Receptors for AGEs (RAGE) are highly expressed in both microglia and neurons (Fig. 6.7) and are responsible for the pathological consequences

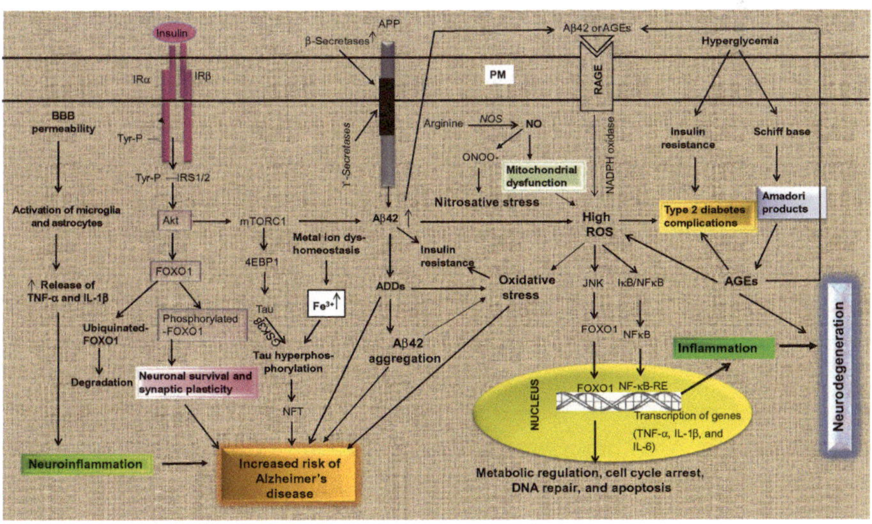

Figure 6.7 Contribution of decrease in cerebral blood flow on the pathogenesis of AD.

(Lue et al., 2001). Aβ binds to RAGE receptor and this binding exaggerates neuronal stress, leading to more accumulation of Aβ, impairment of learning and memory, and induction of neuroinflammation (Chen et al., 2007). Additionally, AD patients with type 2 diabetes have higher levels of AGEs than nondiabetic AD individuals (Valente et al., 2010). Increased AGE levels in hyperglycemia may also contribute to RAGE-mediated disruption of BBB integrity and further promoting brain pathology leading to AD. Proinflammatory cytokines reaching the brain through compromised BBB form a toxic environment for neurons, leading to neuronal insulin resistance, and synaptic dysfunction (Bomfim et al., 2012; Kiernan et al., 2016; Vieira et al., 2017). Collective evidence suggests that the pathogenesis of type 2 diabetes and AD is mediated by Aβ-mediated oxidative stress (Arancio et al., 2004; Moreira, 2012). On the basis of these observations, there is a growing interest in identifying free radical scavenging molecules that can prevent cell death following oxidative stress-induced damage of neural cell membranes. In this perspective, over the last few years, the endocannabinoid system has attracted considerable attention because recent studies have indicated the existence of cross-talk between endocannabinoids and oxidative stress generating lipid mediators (Fig. 6.8) (Gallelli et al., 2018; Lipina and Hundal, 2016). Therefore it is proposed that knowledge of pathways involved in the metabolism of endocannabinoids and generation of lipid-related mediators may contribute to the modulation of oxidative stress and lipid peroxidation. Modulation of these processes may represent a significant research area that may yield novel pharmaceutical strategies for the treatment of diseases associated with regulation of redox imbalance (Gallelli et al., 2018; Lipina and Hundal, 2016). Another study has indicated that insulin resistance may alter APP processing through the activation of autophagy (Son et al., 2012). In addition, insulin signaling provides a physiological defense mechanism against Aβ oligomer-induced synapse loss, a process, which involves the downregulation of Aβ oligomer binding sites in neurons (De Felice et al., 2009). Insulin also produces multiple antiamyloidogenic effects on human neuronal cells, including preventing the translocation of the APP intracellular domain fragment into the nucleus, increasing the transcription of antiamyloidogenic proteins, and increasing the α-secretase-dependent APP-processing pathway (Pandini et al., 2013). On other hand, Aβ oligomers can inhibit insulin signaling via the JNK/TNFα pathway (Bomfim et al., 2012) suggesting a positive feed-forward mechanism.

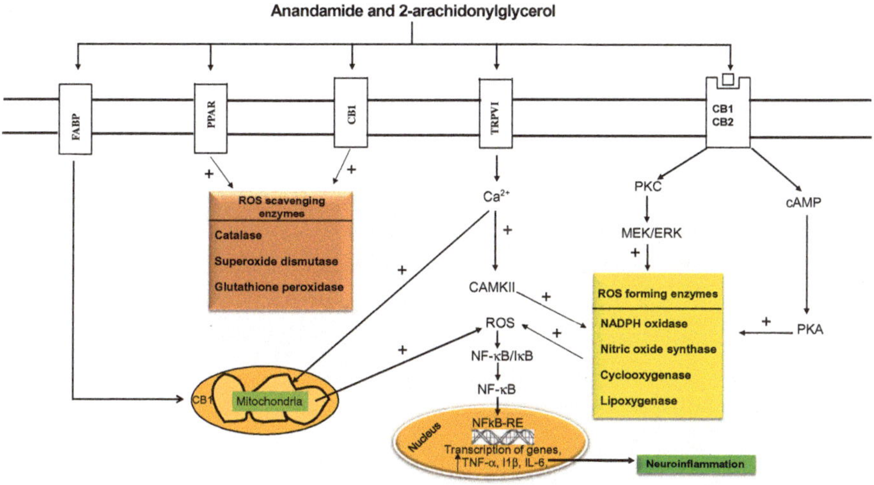

Figure 6.8 Cross-talk between endocannabinoid system and oxidative stress system. *AEA*, Anandamide (*N*-arachidonoylethanolamine); *2-AG*, 2-arachidonoylglycerol; *CB1*, cannabinoid receptors type 1; *CB2*, cannabinoid receptors type 2; *ROS*, reactive oxygen species; *cAMP*, adenosine 3′,5′-cyclic monophosphate; *CAMKII*, Ca^{2+}/calmodulin-dependent protein kinase; *PKC*, protein kinase C; *PKA*, protein kinase A; *MAPK*, mitogen-activated protein kinase, *MEK/ERK*, extracellular signal-regulated kinase; *LOX*, lipoxygenase; *COX*, cyclooxygenase.

Accumulating evidence suggests the importance of Aβ-induced cerebrovascular dysfunction in diabetes and AD.

Brain pathological and neurochemical abnormalities in diabetes and AD are also linked with insulin resistance and neuroinflammation through downregulation of peroxisome proliferator activator receptors (PPARs) (Govindarajulu et al., 2018). These receptors are ligand-activated nuclear transcription factors that regulate peripheral lipid and glucose metabolism. Three subtypes of PPAR receptors (α, γ, β/δ) are found in mammalian tissues. These receptors control the expression of a large number of regulatory genes in lipid metabolism, insulin sensitization, inflammation, and cell proliferation (Dubuquoy et al., 2006) and can inhibit the activation of NFκB, mitogen-activated protein kinase (MAPK), and cyclooxygenase 2 (COX-2) pathways leading to reduction of proinflammatory mediators (cytokines and prostaglandins). The synthetic ligands for PPARα (fibrates) and PPARγ (thiazolidinediones) are currently used for the management of dyslipidemia and type 2 diabetes. Antidiabetic drugs can also be used for the treatment of AD not only in animal models but also in AD patients (Cheng et al., 2015; Govindarajulu et al., 2018). Detailed investigations have indicated that antidiabetic drugs significantly reduce Aβ and Tau pathology measured in samples from AD patients. The anti-inflammatory properties of these drugs improve glucose metabolism and cognitive function not only in AD patients but also in transgenic animal models of type 2 diabetes and AD (Masciopinto et al., 2012; Jiang et al., 2012). The use of these drugs for restoration of cognitive deficits in diabetes and AD is limited due to their poor bioavailability in the brain and off-target effects (Kermani and Garg, 2003; Landreth et al., 2008). Therefore there is a critical need to develop PPARγ targeted agents that display improved tolerability. To understand the significance of these chemical and pharmacological standpoints, the molecular structure and how PPARγ modulates different cellular targets need to be more thoroughly evaluated. Given abovementioned similarities and correlations between type 2 diabetes and AD, it has been proposed that there may be common underlying mechanism(s) that predispose to both type 2 diabetes patient to AD (Asih et al., 2017). Collective evidence suggests that in addition to vascular complications, insulin resistance may represent a molecular link between type 2 diabetes and AD.

Brain insulin resistance in obesity, type 2 diabetes, and AD is accompanied by the brain atrophy (Kullmann et al., 2015; Farooqui, 2013, 2017; Femminella et al., 2018), which can be global as well as region specific. Changes in brain atrophy may involve the loss of neurons, axodendritic pruning, and reduction in synaptic plasticity. These changes are not only observed in anatomical and neurochemical studies, but have been confirmed by advanced magnetic resonance imaging studies in normal aging, obesity, type 2 diabetes, and AD type of dementia (Savva et al., 2009; Farooqui, 2013). These patients show a reduction in gray matter volume and cortical thickness as well as a loss of white matter integrity (Bischof and Park, 2015; Kullmann et al., 2015). A longitudinal study over 6 years in older adults identified body mass index as the strongest predictor of declining gray matter volume, particularly in the frontal lobe and subcortical regions such as the hippocampus (Bischof and Park, 2015; Raji et al., 2010). The medial temporal lobe, including the hippocampus, seems to be particularly affected by type 2 diabetes and AD. Hippocampal atrophy, a marker of neurodegeneration, has been identified in individuals with impaired glucose tolerance and insulin resistance (Convit et al., 2003; Ursache et al., 2012).

Cerebral blood flow in type 2 diabetes and Alzheimer's disease

It is well known that a healthy cardiovascular system is required for brain development, function, and survival. To meet metabolic demands efficiently, the blood flow to the brain needs to be approximately 750 mL/min or about 50/100 mL of brain tissue per minute (Ito et al., 2005). Global CBF declines with age (Chen et al., 2011; Liu et al., 2016; Tarumi et al., 2014), which may be due to regional atrophy (although reduced CBF has been shown independent of regional atrophy) (Chen et al., 2011), lower cerebral metabolism, increased cerebrovascular resistance, or vascular dysfunction (Tarumi et al., 2014). In older adults, there is a positive association between self-reported volumes of physical activity and CBF in gray matter regions (Zimmerman et al., 2014). Similarly, studies investigating blood flow parameters in the middle cerebral artery (MCA) also report an association between cardiorespiratory fitness and MCA blood velocity (Ainslie et al., 2008). CBF declines with age, and greater global CBF or MCA blood velocity may be protective in reducing the negative impact of aging on brain health. This hypothesis is based on the idea that: (1) greater levels of CBF may prevent or slow the accumulation of neuropathology, in particular amyloid-β (Mattsson et al., 2014; Cselenyi and Farde, 2015); and (2) greater levels of CBF may allow for higher cognitive reserve for a given level of neuropathological burden (Stern, 2012; Davenport et al., 2012). CBF has been reported to be decreased in patients with diabetes and AD (30%–40%) compared to age-matched control subjects (Erol, 2008) causing cerebrovascular hypoperfusion and hypometabolism (Qiu et al., 2003; Roher et al., 2012; Ngwa et al., 2018). Sustained cerebral hypoperfusion has been suggested to be the cause of white matter attenuation, a key feature common to both AD and dementia associated with cerebral small vessel disease. Midlife white matter changes increase the risk for stroke, dementia, and disability. At the molecular level, decrease in CBF may cause brain hypometabolism leading not only to an energy crisis in neurons, but also affecting the function of ion pumps such as Na^+/K^+ ATPases that maintain the resting potential in neurons. These processes produce depolarization, excessive calcium entry into neurons, and induction of abnormalities in signal transduction pathways related to excitotoxicity in the aged brain. Reduction in CBF due to cerebrovascular diseases may also promote cognitive problems induced by the accumulation of endogenous toxic products such as ammonia, lactic acid, proinflammatory eicosanoids, cytokines, and chemokines. Increased expression of cytokines and chemokines is associated with increased tau phosphorylation and decreased synaptophysin levels, indicating the role of cytokines and chemokines in cytoskeletal and synaptic changes in AD (Quintanilla et al., 2004). To this end, targeting the increased circulating levels of IL-1β with a neutralizing antibody has been shown to reduce the activity of several tau kinases and levels of phosphorylated tau, and also to reduce the load of oligomeric and fibrillar Aβ in brains of triple-transgenic AD mice (3xTg) (Kitazawa et al., 2011). Interestingly, treatment with a drug targeting the p38 MAPK pathway (Munoz et al., 2007) has been not only shown to normalize levels of proinflammatory cytokines, but also attenuate synaptic protein loss and impaired synaptic plasticity in AD mouse models (Bachstetter et al., 2012). Increased expression and accumulation of cytokines and chemokines may also contribute to the pathogenesis of diabetic neuropathy, ischemic stroke, and AD due to mitochondrial dysfunction (Fig. 6.4). Increasing evidence support the view that vascular and AD neuropathology often coexist with metabolic dysfunction (insulin resistance, neuroinflammation, and oxidative stress) along with neurovascular dysfunction. These factors play a critical role in the development or progression of AD (Iadecola, 2013; Farooqui, 2013, 2015). These events may be one of the very early events in the pathogenesis of AD (Iadecola, 2004, 2013). This suggestion is supported by the observation that β and γ-secretases are regulated in response to stress caused by energy deprivation (O'Connor et al., 2008). An insufficient supply of glucose induces phosphorylation of the translation initiation factor eIF2alpha (eIF2α), which consequently increases β and γ-secretases, resulting in the overproduction of Aβ through amyloidogenic pathway (Fig. 6.4). Furthermore, the lack of oxygen may also cause the impairment of brain clearance of Aβ through stimulation of serum response factor (SRF) and myocardin (MYOCD) expressions (Bell and Zlokovic, 2009), which are much more active in the blood vessels of AD patients than in normal subjects (Chow et al., 2007). Overexpression of SRF and MYOCD in cerebrovascular smooth muscle cells negatively regulates the expression of low-density lipoprotein receptor-related protein-1 (LRP1), which is the major Aβ clearance receptor in the BBB. The stimulation of LRP1 along with the transactivation properties of the sterol regulatory element binding transcription factor 2, ultimately leads to toxic Aβ accumulation. On the basis of these studies, it is suggested that the components of this signaling pathway may be involved in reduced blood flow. Improving blood flow by exercise, healthy eating, or using dietary supplements may effectively decrease the prevalence of AD. Decrease in CBF may negatively affect the synthesis of proteins required for learning and memory and eventually promote neuritic injury and neuronal death leading to cognitive decline and loss of memory. Furthermore, impaired clearance of Aβ due to decrease in CBF may lead to hypoxia, breakdown of the BBB, and brain atrophy. These processes interfere with functioning of the neurovascular unit and promote the deposition of Aβ in cerebral blood vessels causing cerebral amyloid angiopathy, a pathological condition associated with cognitive decline in AD (Bell and Zlokovic, 2009; Kelleher and Soiza, 2013; Sagare et al., 2013). Furthermore, hyperglycemia increases the production of free radicals such as ROS. This overproduction of free radicals can exhaust a cell's antioxidant capacity and lead to the induction of oxidative stress, which is a hallmark of both type 2 diabetes and AD. Oxidative stress-mediated damage protein and nucleic acid not only results in neuronal damage, but can also cause cell death in patients with type 2 diabetes or AD compared with healthy controls (Pratico and Sung, 2004; Farooqui, 2014). In type 2 diabetes and AD, the induction of neuroinflammation is supported by the activation of inflammatory pathways (Farooqui, 2014) and the release of inflammatory mediators such as C-reactive protein and interleukin 6 into the circulation (Farooqui, 2014). Local neuroinflammation initiated by activation of microglial cells and astrocytes, which surround extracellular Aβ plaques. This can lead to activation of the complement cascade and neuronal cell damage (Akiyama et al., 2000). Under normal conditions, this transcription factor NF-κB is present in cytoplasm in an inactivated form. Induction of oxidative stress activates NF-κB, which migrates to the nucleus, where it binds to the NF-κB-RE and promote the expression of inflammatory cytokines and chemokines, including interleukins, tumor necrosis factor, and macrophage inflammatory protein 1α in type diabetes and AD (Farooqui, 2014). Collective evidence suggests that there is considerable overlap in the pathogenesis of type 2 diabetes and AD.

Ceramode-mediated insulin resistance in type 2 diabetes and Alzheimer's disease

Ceramides are essential constituents of cell membranes and important member of the sphingolipid family. It is not only an essential precursor for complex sphingolipids, but also plays a key role in signal transductions of crucial physiological processes such as growth, differentiation, proliferation, migration, apoptosis, adhesion, senescence, vasoconstriction, and cell death (Hannun and Obeid, 2008; Cogolludo et al., 2009; Moreno et al., 2014). In neural and nonneural tissues, ceramides are biosynthesized by the de novo and salvage pathways (Fig. 6.9). The de novo synthesis of ceramide involves the condensation of palmitoylCoA with serine to form ceramide. The second pathway produces ceramide either through the reacylation of sphingosine (salvage pathway) or through the N-acylation of a sphingoid base, a reaction, which is catalyzed by ceramide synthase. Ceramide induces insulin resistance by inhibiting Akt/PKB activity. Two independent mechanisms contribute to the ceramide-mediated inhibition of Akt/PKB: (1) the first mechanism involves the stimulation of Akt dephosphorylation via protein phosphatase 2A (PP2A) and (2) the second mechanism involves the blocking of translocation of Akt via PKCζ (Stratford et al., 2004). Ceramide-induced activation of PP2A also inhibits the action of Akt/PKB by impairing Akt serine phosphorylation. The process decreases the translocation of GLUT4 to the plasma membrane and hence decreases the uptake of glucose. An important point, which should be mentioned here is that these studies have been performed on

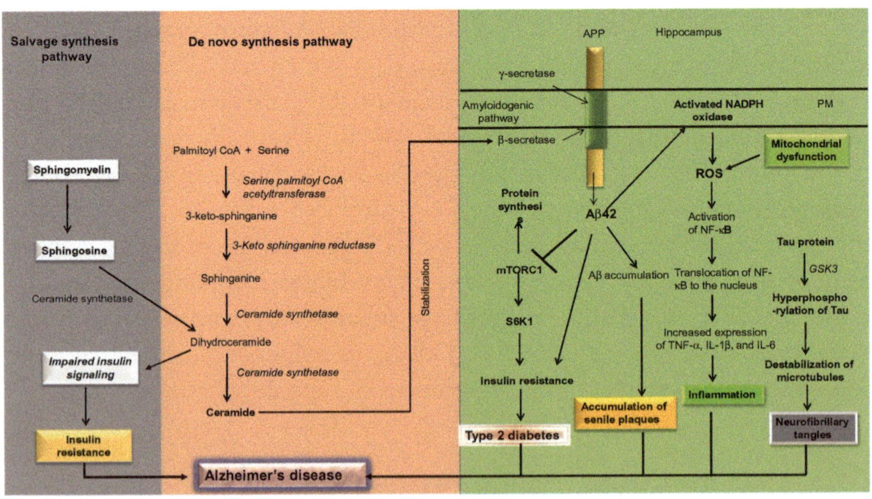

Figure 6.9 Synthesis of ceramide by de novo and salvage pathways and effect of ceramide on the pathogenesis of Alzheimer's disease.
Aβ, Beta-amyloid; *APP*, amyloid precursor protein; *mTORC1*, mammalian target of rapamycin complex 1; *PM*, plasma membrane; *S6K1*, ribosomal protein S6 kinase beta-1.

cell-permeable short-chain ceramides with unphysiologic concentrations (Chavez and Summers, 2012). Alternatively, PKCζ activation of the fatty acid transporter CD36 has been proposed as a mechanism for ceramide-induced hepatic steatosis (Xia et al., 2015). Such a mechanism has the advantage of linking hepatic ceramides to other lipid mediators (DAG and TAG), which are known to induce hepatic insulin resistance. During the pathogenesis of AD, ceramide can enter brain from blood circulation by crossing the BBB. Once in the brain, ceramide can promote oxidative stress, neuroinflammation, mediate insulin resistance, and facilitate the pathogenesis of AD by following mechanisms (Tong and de la Monte, 2009). Ceramide is known to enhance formation of Aβ through post-translational stabilization of β-secretase (BACE1), which stimulates proteolytic modifications of APP (de La Monte, 2012; Jazvinšćak Jembrek et al., 2015; Ahmad et al., 2017). As mentioned above, the deposition of Aβ peptide plays a key role in the pathogenesis of AD. Nevertheless, increase in Aβ level has also been reported in patients with insulin resistance (de la Monte and Wands, 2008; Willette et al., 2015). It is also proposed that ceramide may also act by inducing upregulating NADPH oxidase. Activation of this enzyme not only results in generation of a large amount of superoxide anions ($O_2^{\bullet -}$), depletion of GSH, and mitochondrial abnormalities, but also oxidation of lipids, proteins, and nucleic acids (de La Monte, 2012; Jazvinšćak Jembrek et al., 2015; Ahmad et al., 2017). It is also reported that there is a link between ceramides and inflammation, possibly through direct activation of the NLRP3 inflammasome formation. This link may provide another mechanism for ceramide-induced insulin resistance (Henao-Mejia et al., 2012; Vandanmagsar et al., 2011). In addition, ceramide initiates inflammatory signaling pathways, leading to the activation of both c-jun NH_2-terminal kinase (JNK) and NF-κB/inducer of κ kinase (Ruvolo, 2003). All these processes are closely associated with the development of insulin resistance (Cai et al., 2005, 2015; Chung et al., 2008; Ikonen and Vainio, 2005).

Contribution of adipokines (leptin and adiponectin) in insulin resistance and Alzheimer's disease

As mentioned in Chapter 1, Insulin resistance and obesity, adipokines exert their pleiotropic effects on different tissues and regulate numerous important physiological functions such as appetite, energy expenditure, insulin sensitivity and secretion, fat distribution, lipid and glucose metabolism, endothelial function, blood pressure, homeostasis, neuroendocrine, and immune functions (Blüher and Mantzoros, 2015). In hypothalamus, leptin plays an important role in energy homeostasis; in hippocampus, midbrain and hindbrain leptin contributes to neurogenesis, axonal growth, and synaptogenesis (Fig. 6.10) (Bouret, 2010). In hippocampus, leptin modulates synaptic plasticity. In rodents (*db/db* mice or *fa/fa* rats), mutation in leptin receptor is associated with impairments in hippocampal LTP and long-term depression as well as deficits in hippocampal-specific memory tasks (Li et al., 2002). Direct administration of leptin into the hippocampus improves learning and memory performance (Farr et al., 2006; Oomura et al., 2006) and facilitates hippocampal LTP (Wayner et al., 2004). Experimental studies in cell lines and AD transgenic mice have indicated that in brain leptin modulates APP-Aβ trafficking, Aβ accumulation and clearance, β-secretase expression and phosphorylation of tau protein (Chakrabarti et al., 2015). Low levels of leptin in brains of AD patient contribute to cognitive decline (Baranowska-Bik et al., 2015). Collective evidence suggests that peripheral leptin administration in mice may reduce the brain Aβ levels and therefore decrease the development of sporadic AD (Fewlass et al., 2004). In contrast, administration of exogenous adiponectin

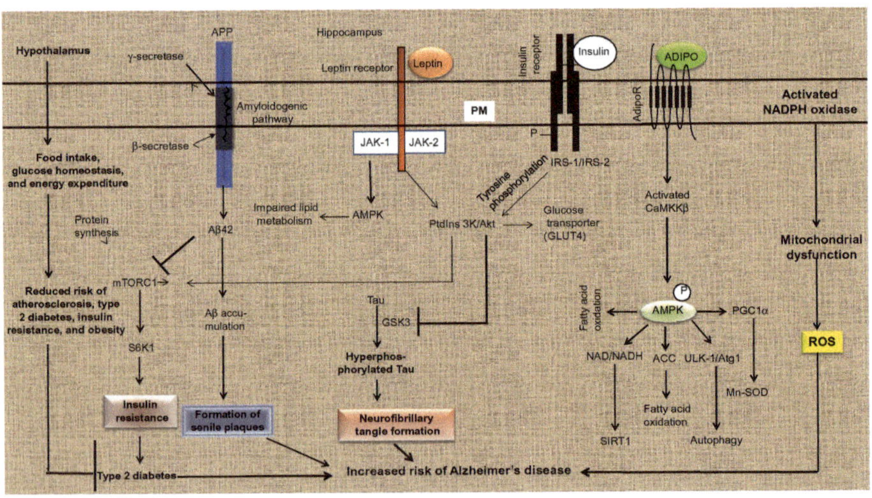

Figure 6.10 Interactions between leptin and insulin signaling and their involvement in the development of Alzheimer disease.

ACC, Acetyl-CoA carboxylase; *ADIPO*, adiponectin; *AdipoR*, adiponectin receptor; *AMPK*, AMP-activated protein kinase; *Atg1*, yeast homolog of ULK1; *GLUT4*, glucose transporter4; *IRS-1*, insulin receptor substrate 1; *JAK*, Janus kinase; *PGC1α*, peroxisome proliferator-activated response-gamma coactivator-1alpha; *SIRT1*, sirtuin1; *ULK1*, serine/threonine kinase.

in the brain produces neuroprotective effects against Aβ-induced neurotoxicity in a cell culture model (Chan et al., 2012). In the brain, adiponectin also modulates brain metabolism and sensitivity of insulin regulating memory and cognitive dysfunction (Song and Lee, 2013). Both adiponectin and leptin may contribute to the AD by connecting peripheral and central pathogenic mechanisms. In rodents, the deletion of adiponectin gene leads to insulin resistance (Kubota et al., 2002). In humans, a reduced serum concentration of adiponectin incurs obesity, insulin resistance, and type 2 diabetes (Hotta et al., 2000). Impaired proximal signaling of insulin receptor also mediates insulin resistance. Decreased IRS protein levels contribute to insulin resistance in rodents and humans (Craft et al., 1999). The IRS protein levels are also decreased in streptozotocin-induced dementia in rat model which have been used to study AD compared with sham group (normal control group) in the hippocampus and in the cortex (Song and Lee, 2013). In sporadic AD, insulin system dysfunction incurs severe pathology such as cognitive decline suggesting that adiponectin may have an important target for sporadic AD.

Conclusion

Sporadic AD and type 2 diabetes are multifactorial disorders, which are characterized by insulin resistance. It is reported that insulin signaling in the brain stimulates numerous molecular cascades, such as cholesterol metabolism, energy expenditure, glucose homeostasis, feeding behavior, synaptogenesis, neurotrophy, neurotransmitters, cognition, memory, inflammation, and apoptosis. In addition, insulin regulates the metabolism of peripheral Aβ and hyperphosphorylated tau protein. Both sporadic AD and type 2 diabetes are accompanied by the accumulation of Aβ plaques and aggregation of hyperphosphorylated tau protein in NFTs. These processes are supported by activation of microglial cells and astrocytes, release of proinflammatory cytokines, abnormalities in CBF, BBB permeability, and loss of neurons and synapses. Mechanisms, which contribute to neurodegeneration, include energy and metabolism deficits, oxidative and ER stress, mitochondrial dysfunction, accumulation of AGEs with increased induction of neuroinflammation, and onset of proapoptosis cascade. Impairment in insulin receptor function and increased expression and activation of IDE have also been implicated. These processes not only compromise neuronal and glial function, but also alter neurotransmitter homeostasis. Insulin resistance facilitate the accumulation of ADD and oligomeric fibrils or insoluble larger aggregated fibrils in the form of plaques that are neurotoxic. Additionally, there is production and accumulation of hyperphosphorylated tau which can exacerbate cytoskeletal collapse and promote synaptic disconnection. Furthermore, vascular defects in AD and type 2 diabetes may reduce CBF and disrupt BBB leading to brain damage.

References

Abeyrathna, P., Su, Y., 2015. The critical role of Akt in cardiovascular function. Vasc. Pharmacol. 74, 38–48.

Accardi, G., Caruso, C., Colonna-Romano, G., Camarda, C., Monastero, R., Candore, G., 2012. Can Alzheimer disease be a form of type 3 diabetes? Rejuvenation Res. 15, 217–221.

Accili, D., Arden, K.C., 2004. Foxos at the crossroads of cellular metabolism, differentiation, and transformation. Cell 117, 421–426.

Aguzzi, A., Haass, C., 2003. Games played by rogue proteins in prion disorders and Alzheimer's disease. Science 302, 814–818.

Ahima, R.S., Lazar, M.A., 2008. Adipokines and the peripheral and neural control of energy balance. Mol. Endocrinol. 22, 1023–1031.

Ahmad, W., Ijaz, B., Shabbiri, K., Ahmed, F., Rehman, S., 2017. Oxidative toxicity in diabetes and Alzheimer's disease: mechanisms behind ROS/RNS generation. J. Biomed. Sci. 24, 76.

Ainslie, P.N., Cotter, J.D., George, K.P., Lucas, S., Murrell, C., Shave, R., et al., 2008. Elevation in cerebral blood flow velocity with aerobic fitness throughout healthy human ageing. J. Physiol. 586, 4005–4010.

Akiyama, H., Barger, S., Barnum, S., Bradt, B., Bauer, J., Cole, G.M., et al., 2000. Inflammation and Alzheimer's disease. Neurobiol. Aging 21, 383–421.

Albuquerque, E.X., Schwarcz, R., 2013. Kynurenic acid as an antagonist of α7 nicotinic acetylcholine receptors in the brain: facts and challenges. Biochem. Pharmacol. 85, 1027–1032.

Alessi, D.R., Deak, M., Casamayor, A., Caudwell, F.B., Morrice, N., Norman, D.G., et al., 1997. 3-Phosphoinositide-dependent protein kinase-1 (PDK1): structural and functional homology with the Drosophila DSTPK61 kinase. Curr Biol. 7, 776–789.

Allcock, R.J., Barrow, A.D., Forbes, S., Beck, S., Trowsdale, J., 2003. The human TREM gene cluster at 6p21.1 encodes both activating and inhibitory single IgV domain receptors and includes NKp44. Eur. J. Immunol. 33, 567–577.

Andjelkovic, M., Jakubowicz, T., Cron, P., Ming, X.F., Han, J.W., Hemmings, B.A., 1996. Activation and phosphorylation of a pleckstrin homology domain containing protein kinase (RAC-PK/PKB) promoted by serum and protein phosphatase inhibitors. Proc. Natl. Acad. Sci. U. S. A. 93, 5699–5704.

Arancio, O., Zhang, H.P., Chen, X., Lin, C., Trinchese, F., Puzzo, D., et al., 2004. RAGE potentiates Aβ-induced perturbation of neuronal function in transgenic mice. EMBO J. 23, 4096–4105.

Arnaud, L., Robakis, N.K., Figueiredo-Pereira, M.E., 2006. It may take inflammation, phosphorylation and ubiquitination to 'tangle' in Alzheimer's disease. Neurodegener. Dis. 3, 313–319.

Asih, P.R., Tegg, M.L., Sohrabi, H., Carruthers, M., Gandy, S.E., Saad, F., et al., 2017. Multiple mechanisms linking type 2 diabetes and Alzheimer's disease: testosterone as a modifier. J. Alzheimer Dis. 59, 445–466.

Avila, J., Gomez de Barreda, E., Engel, T., Lucas, J.J., Hernandez, F., 2010. Tau phosphorylation in hippocampus results in toxic gain-of-function. Biochem. Soc. Trans. 38, 977–980.

Bachstetter, A.D., Norris, C.M., Sompol, P., Wilcock, D.M., Goulding, D., Neltner, J.H., et al., 2012. Early stage drug treatment that normalizes proinflammatory cytokine production attenuates synaptic dysfunction in a mouse model that exhibits age-dependent progression of Alzheimer's disease-related pathology. J. Neurosci. 32, 10201–10210.

Banks, W.A., Owen, J.B., Erickson, M.A., 2012. Insulin in the brain: there and back again. Pharmacol. Ther. 136, 82–93.

Baranowska-Bik, A., Bik, W., Styczynska, M., Chodakowska-Zebrowska, M., Barcikowska, M., Wolinska-Witort, E., et al., 2015. Plasma leptin levels and free leptin index in women with Alzheimer's disease. Neuropeptides 52, 73–78.

Barrett, E., Ross, R.P., O'Toole, P.W., et al., 2012. γ-Aminobutyric acid production by culturable bacteria from the human intestine. J. Appl. Microbiol. 113, 411–417.

Batista, A.F., Forny-Germano, L., Clarke, J.R., Lyra e Silva, N.M., Brito-Moreira, J., Boehnke, S.E., et al., 2018. The diabetes drug liraglutide reverses cognitive impairment in mice and attenuates insulin receptor and synaptic pathology in a non-human primate model of Alzheimer's disease. J. Pathol. 245, 85–100.

Belaidi, A.A., Bush, A.I., 2016. Iron neurochemistry in Alzheimer's disease and Parkinson's disease: targets for therapeutics. J. Neurochem. 139 (Suppl. 1), 179–197.

Bell, R.D., Zlokovic, B.V., 2009. Neurovascular mechanisms and blood-brain barrier disorder in Alzheimer's disease. Acta Neuropathol. 118, 103–113.

Benedict, C., Kern, W., Schultes, B., Born, J., Hallschmid, M., 2008. Differential sensitivity of men and women to anorexigenic and memory-improving effects of intranasal insulin. J. Clin. Endocrinol. Metab. 9, 1339–1344.

Benedict, C., Brede, S., Schiöth, H.B., Lehnert, H., Schultes, B., Born, J., et al., 2011. Intranasal insulin enhances postprandial thermogenesis and lowers postprandial serum insulin levels in healthy men. Diabetes 60, 114–118.

Biessels, G.J., Reijmer, Y.D., 2014. Brain changes underlying cognitive dysfunction in diabetes: what can we learn from MRI? Diabetes 63, 2244–2252.

Bilotta, F., Lauretta, M.P., Tewari, A., Haque, M., Hara, N., Uchino, H., et al., 2017. Insulin and the brain: a sweet relationship with intensive care. J. Intensive Care Med. 32, 48–58.

Birch, P.J., Grossman, C.J., Hayes, A.G., 1988. Kynurenate and FG9041 have both competitive and non-competitive antagonist actions at excitatory amino acid receptors. Eur. J. Pharmacol. 151, 313–315.

Bischof, G.N., Park, D.C., 2015. Obesity and aging: consequences for cognition, brain structure, and brain function. Psychosom. Med. 77, 697–709.

Blazquez, E., Velazquez, E., Hurtado-Carneiro, V., Ruiz-Albusac, J.M., 2014. Insulin in the brain: its pathophysiological implications for states related with central insulin resistance, type 2 diabetes and Alzheimer's disease. Front. Endocrinol (Lausanne) 5, 161.

Bloom, G.S., 2014. Amyloid-β and tau: the trigger and bullet in Alzheimer disease pathogenesis. JAMA Neurol. 71, 505–508.

Blüher, M., Mantzoros, C.S., 2015. From leptin to other adipokines in health and disease: facts and expectations at the beginning of the 21st century. Metabolism. 64, 131–145.

Bomfim, T.R., Forny-Germano, L., Sathler, L.B., Brito-Moreira, J., Houzel, J.C., Decker, H., et al., 2012. An anti-diabetes agent protects the mouse brain from defective insulin signaling caused by Alzheimer's disease-associated Aβ oligomers. J. Clin. Invest. 122, 1339–1353.

Borger, E., Aitken, L., Du, H., Zhang, W., Gunn-Moore, F.J., Yan, S.S., 2013. Is amyloid binding alcohol dehydrogenase a drug target for treating Alzheimer's disease? Curr. Alzheimer Res. 10, 21–29.

Bouret, S.G., 2010. Neurodevelopmental actions of leptin. Brain Res. 1350, 2–9.

Braak, H., Braak, E., 1998. Evolution of neuronal changes in the course of Alzheimer's disease. J. Neural Transm. Suppl. 53, 127–140.

Braccini, L., Ciraolo, E., Campa, C.C., Perino, A., Longo, D.L., Tibolla, G., et al., 2015. PI3K-C2gamma is a Rab5 effector selectively controlling endosomal Akt2 activation downstream of insulin signalling. Nat. Commun. 6, 7400.

Cai, D., Yuan, M., Frantz, D.F., Melendez, P.A., Hansen, L., Lee, J., et al., 2005. Local and systemic insulin resistance resulting from hepatic activation of IKK-beta and NF-kappaB. Nat. Med. 11, 183–190.

Cai, Z., Zhao, B., Li, K., Zhang, L., Li, C., Quazi, S.H., et al., 2012. Mammalian target of rapamycin: a valid therapeutic target through the autophagy pathway for Alzheimer's disease? J. Neurosci. Res. 90, 1105–1118.

Cai, Z., Xiao, M., Chang, L., Yan, L.J., 2015. Role of insulin resistance in Alzheimer's disease. Metab. Brain Dis. 30, 839–851.

Cantley, L.C., 2002. The phosphoinositide 3-kinase pathway. Science 296, 1655–1657.

Celarain, N., Sanchez-Ruiz de Gordoa, J., Zelaya, M.V., Roldan, M., Larumbe, R., et al., 2016. TREM2 upregulation correlates with 5-hydroxymethycytosine enrichment in Alzheimer's disease hippocampus. Clin. Epigenetics. 8, 37.

Celermajer, D.S., 1997. Endothelial dysfunction: does it matter? Is it reversible? J. Am. Coll. Cardiol. 30, 325–333.

Cersosimo, E., DeFronzo, R.A., 2006. Insulin resistance and endothelial dysfunction: the road map to cardiovascular diseases. Diabetes Metab. Res. Rev. 22, 423–436.

Chakrabarti, S., Khemka, V.K., Banerjee, A., Chatterjee, G., Ganguly, A., Biswas, A., 2015. Metabolic risk factors of sporadic Alzheimer's disease: implications in the pathology, pathogenesis and treatment. Aging Dis. 6, 282–299.

Chami, B., Steel, A.J., De La Monte, S.M., Sutherland, G.T., 2016. The rise and fall of insulin signaling in Alzheimer's disease. Metab. Brain Dis. 31, 497–515.

Chan, K.H., Lam, K.S., Cheng, O.Y., Kwan, J.S., Ho, P.W., Cheng, K.K., et al., 2012. Adiponectin is protective against oxidative stress induced cytotoxicity in amyloid-beta neurotoxicity. PLoS One 7, e52354.

Chavez, J.A., Summers, S.A., 2012. A ceramide-centric view of insulin resistance. Cell Metab. 15, 585–594.

Chen, X., Walker, D.G., Schmidt, A.M., Arancio, O., Lue, L.F., Yan, S.D., 2007. RAGE: a potential target for Aβ-mediated cellular perturbation in Alzheimer's disease. Curr. Mol. Med. 7, 735–742.

Chen, J.J., Rosas, H.D., Salat, D.H., 2011. Age-associated reductions in cerebral blood flow are independent from regional atrophy. Neuroimage 55, 468–478.

Cheng, H., Shang, Y., Jiang, L., Shi, T.L., Wang, L., 2015. The peroxisome proliferators activated receptor-gamma agonists as therapeutics for the treatment of Alzheimer's disease and mild-to-moderate Alzheimer's disease: a meta-analysis. Int. J. Neurosci. 126, 299–307.

Chetelat, G., La Joie, R., Villain, N., Perrotin, A., de La Sayette, V., Eustache, F., et al., 2013. Amyloid imaging in cognitively normal individuals, at-risk populations and preclinical Alzheimer's disease. Neuroimage Clin. 2, 356–365.

Cholerton, B., Baker, L.D., Craft, S., 2013. Insulin, cognition, and dementia. Eur. J. Pharmacol. 719, 170–179.

Chow, N., Bell, R.D., Deane, R., Streb, J.W., Chen, J., Brooks, A., et al., 2007. Serum response factor and myocardin mediate arterial hypercontractility and cerebral blood flow dysregulation in Alzheimer's phenotype. Proc. Natl. Acad. Sci. U.S.A. 104, 823–828.

Chung, J., Nguyen, A.K., Henstridge, D.C., Holmes, A.G., Chan, M.H., Mesa, J.L., et al., 2008. HSP72 protects against obesity-induced insulin resistance. Proc. Natl. Acad. Sci. U.S.A. 105, 1739–1744.

Clarke, J.R., Lyra e Silva, N.M., Figueiredo, C.P., Frozza, R.L., Ledo, J.H., Beckman, D., et al., 2015. Alzheimer-associated Aβ oligomers impact the central nervous system to induce peripheral metabolic deregulation. EMBO Mol. Med. 7, 190–210.

Cogolludo, A., Moreno, L., Frazziano, G., Moral-Sanz, J., Menendez, C., Castaneda, J., et al., 2009. Activation of neutral sphingomyelinase is involved in acute hypoxic pulmonary vasoconstriction. Cardiovasc. Res. 82, 296–302.

Connelly, L., Jang, H., Arce, F.T., Capone, R., Kotler, S.A., Ramachandran, S., et al., 2012. Atomic force microscopy and MD simulations reveal pore-like structures of all-D-enantiomer of Alzheimer's β-amyloid peptide: relevance to the ion channel mechanism of AD pathology. J. Phys. Chem. B 116, 1728–1735.

Convit, A., Wolf, O.T., Tarshish, C., de Leon, M.J., 2003. Reduced glucose tolerance is associated with poor memory performance and hippocampal atrophy among normal elderly. Proc. Natl. Acad. Sci. U.S.A. 100, 2019–2022.

Cornu, M., Albert, V., Hall, M.N., 2013. mTOR in aging, metabolism, and cancer. Curr. Opin. Genet. Dev. 23, 53–62.

Correia, S.C., Santos, R.X., Perry, G., Zhu, X., Moreira, P.I., Smith, M.A., 2011. Insulin-resistant brain state: the culprit in sporadic Alzheimer's disease? Ageing Res. Rev. 10, 264–273.

Correia, S.C., Santos, R.X., Carvalho, C., Cardoso, S., Candeias, E., Santos, M.S., et al., 2012. Insulin signaling, glucose metabolism and mitochondria: major players in Alzheimer's disease and diabetes interrelation. Brain Res. 1441, 64–78.

Craft, S., 2007. Insulin resistance and Alzheimer's disease pathogenesis: potential mechanisms and implications for treatment. Curr. Alzheimer Res. 4, 147–152.

Craft, S., Asthana, S., Newcomer, J.W., Wilkinson, C.W., Matos, I.T., Baker, L.D., et al., 1999. Enhancement of memory in Alzheimer disease with insulin and somatostatin, but not glucose. Arch. Gen. Psychiatry 56, 1135–1140.

Craft, S., Cholerton, B., Baker, L.D., 2013. Insulin and Alzheimer's disease: untangling the web. J. Alzheimer Dis. 33 (Suppl 1), S263–S275.

Cselenyi, Z., Farde, L., 2015. Quantification of blood flow-dependent component in estimates of beta-amyloid load obtained using quasi-steady-state standardized uptake value ratio. J. Cereb. Blood Flow Metab. 35, 1485–1493.

Dankner, R., Chetrit, A., Shanik, M.H., Raz, I., Roth, J., 2009. Basal-state hyperinsulinemia in healthy normoglycemic adults is predictive of type 2 diabetes over a 24-year follow-up: a preliminary report. Diabetes Care 32, 1464–1466.

Davenport, M.H., Hogan, D.B., Eskes, G.A., Longman, R.S., Poulin, M.J., 2012. Cerebrovascular reserve: the link between fitness and cognitive function? Exerc. Sport Sci. Rev. 40, 153–158.

De Felice, F.G., Vieira, M.N., Bomfim, T.R., Decker, H., Velasco, P.T., Lambert, M.P., et al., 2009. Protection of synapses against Alzheimer's-linked toxins: insulin signaling prevents the pathogenic binding of Aβ oligomers. Proc. Natl. Acad. Sci. U.S.A. 106, 1971–1976.

D'Ercole, A.J., Ye, P., 2008. Expanding the mind: insulin-like growth factor I and brain development. Endocrinology 149, 5958–5962.

de La Monte, S.M., 2012. Brain insulin resistance and deficiency as therapeutic targets in Alzheimers disease. Curr. Alzheimer Res. 9, 35–66.

de la Monte, S.M., Wands, J.R., 2005. Review of insulin and insulin-like growth factor expression, signaling, and malfunction in the central nervous system: relevance to Alzheimer's disease. J. Alzheimer Dis. 7, 45–61.
de la Monte, S.M., Wands, J.R., 2008. Alzheimer's disease is type 3 diabetes-evidence reviewed. J. Diabetes Sci. Technol. 2, 1101–1113.
den Heijer, T., Vermeer, S.E., van Dijk, E.J., Prins, N.D., Koudstaal, P.J., Hofman, A., et al., 2003. Type 2 diabetes and atrophy of medial temporal lobe structures on brain MRI. Diabetologia 46, 1604–1610.
Devi, L., Alldred, M.J., Ginsberg, S.D., Ohno, M., 2012. Mechanisms underlying insulin deficiency-induced acceleration of β-amyloidosis in a mouse model of Alzheimer's disease. PLoS One 7, e32792.
Diehl, T., Mullins, R., Kapogiannis, D., 2017. Insulin resistance in Alzheimer's disease. Transl. Res. 183, 26–40.
Doble, B.W., Woodgett, J.R., 2003. GSK-3: tricks of the trade for a multi-tasking kinase. J. Cell Sci. 116, 1175–1186.
Dong, S., Duan, Y., Hu, Y., Zhao, Z., 2012. Advances in the pathogenesis of Alzheimer's disease: a re-evaluation of amyloid cascade hypothesis. Transl. Neurodegener. 1, 1–18.
Draznin, B., 2006. Molecular mechanisms of insulin resistance: serine phosphorylation of insulin receptor substrate-1 and increased expression of p85alpha: the two sides of a coin. Diabetes 55, 2392–2397.
Dubuquoy, L., Rousseaux, C., Thuru, X., Peyrin-Biroulet, L., Romano, O., Chavatte, P., et al., 2006. PPARgamma as a new therapeutic target in inflammatory bowel diseases. Gut 55, 1341–1349.
Duckworth, W.C., Bennett, R.G., Hamel, F.G., 1997. The significance of intracellular insulin to insulin action. J. Investig. Med. 45, 20–27.
Duthey, B., 2013. Background paper 6.11: Alzheimer disease other dementias. Public Health Approach Innovat. 5, 4–74.
Engelman, J.A., Luo, J., Cantley, L.C., 2006. The evolution of phosphatidylinositol 3-kinases as regulators of growth and metabolism. Nat. Rev. Genet. 7, 606–619.
Erol, A., 2008. An integrated and unifying hypothesis for the metabolic basis of sporadic Alzheimer's disease. J. Alzheimer Dis. 13, 241–253.
Fang, C.L., Wu, W.H., Liu, Q., Sun, X., Ma, Y., Zhao, Y.F., et al., 2010. Dual functions of beta-amyloid oligomer and fibril in Cu(II)-induced H_2O_2 production. Regul. Pept. 163, 1–6.
Farooqui, A.A., 2013. Metabolic Syndrome: An Important Risk Factor for Stroke, Alzheimer, and Depression. Spinger Science-Business, New York.
Farooqui, A.A., 2014. Inflammation and Oxidative Stress in Neurological Disorders. Springer International Publishing, Switzerland.
Farooqui, A.A., 2015. High Calorie Diet and the Human Brain. Springer International Publishing, Switzerland.
Farooqui, A.A., 2017. Neurochemical Aspects of Alzheimer's Disease. Academic Press, an imprint of Elsevier, San Diego, CA.
Farr, S.A., Banks, W.A., Morley, J.E., 2006. Effects of leptin on memory processing. Peptides 27, 1420–1425.
Femminella, G.D., Thayanandan, T., Calsolaro, V., Komici, K., Rengo, G., Corbi, G., 2018. Imaging and molecular mechanisms of Alzheimer's disease: a review. Int. J. Mol. Sci. 19. pii: E3702.
Fernandez, A.M., Torres-Aleman, I., 2012. The many faces of insulin-like peptide signalling in the brain. Nat. Rev. Neurosci. 13, 225–239.
Ferreira, L.S.S., Fernandes, C.S., Vieira, M.N.N., De Felice, F.G., 2018. Insulin resistance in Alzheimer's disease. Front. Neurosci. 12, 830.

Fewlass, D.C., Noboa, K., Pi-Sunyer, F.X., Johnston, J.M., Yan, S.D., Tezapsidis, N., 2004. Obesity-related leptin regulates Alzheimer's Abeta. FASEB J. 18, 1870–1878.

Francis, S.H., Busch, J.L., Corbin, J.D., Sibley, D., 2010. cGMP-dependent protein kinases and cGMP phosphodiesterases in nitric oxide and cGMP action. Pharmacol. Rev. 62, 525–563.

Frosch, O.H., Yau, P.L., Osorio, R.S., Rusinek, H., Storey, P., Convit, A., 2017. Insulin resistance among obese middle-aged is associated with decreased cerebrovascular reactivity. Neurology 89, 249–255.

Fuster-Matanzo, A., Llorens-Martín, M., Hernández, F., et al., 2013. Role of neuroinflammation in adult neurogenesis and Alzheimer disease: therapeutic approaches. Mediat. Inflamm. 260925.

Gallelli, C.A., Calcagnini, S., Romano, A., Koczwara, J.B., de Ceglia, M., Dante, D., et al., 2018. Modulation of the oxidative stress and lipid peroxidation by endocannabinoids and their lipid analogues, Antioxidants (Basel), 7. pii: E93.

Gao, T., Furnari, F., Newton, A.C., 2005. PHLPP: a phosphatase that directly dephosphorylates Akt, promotes apoptosis, and suppresses tumor growth. Mol. Cell 18, 13–24.

Gong, C.X., Iqbal, K., 2008. Hyperphosphorylation of microtubule-associated protein tau: a promising therapeutic target for Alzheimer disease. Curr. Med. Chem. 15, 2321–2328.

Govindarajulu, M., Pinky, P.D., Bloemer, J., Ghanei, N., Suppiramaniam, V., Amin, R., 2018. Signaling mechanisms of selective PPARγ modulators in Alzheimer's disease. PPAR Res. 2018:2010675.

Gregor, M.F., Hotamisligil, G.S., 2011. Inflammatory mechanisms in obesity. Annu. Rev Immunol. 29, 415–445.

Grillo, C.A., Piroli, G.G., Lawrence, R.C., Wrighten, S.A., Green, A.J., Wilson, S.P., et al., 2015. Hippocampal insulin resistance impairs spatial learning and synaptic plasticity. Diabetes 64, 3927–3936.

Guglielmotto, M., Monteleone, D., Giliberto, L., Fornaro, M., Borghi, R., Tamagno, E., et al., 2011. Amyloid-β42 activates the expression of BACE1 through the JNK pathway. J. Alzheimer Dis. 27, 871–883.

Guglielmotto, M., Monteleone, D., Piras, A., Valsecchi, V., Tropiano, M., Ariano, S., et al., 2014. Aβ1–42 monomers or oligomers have different effects on autophagy and apoptosis. Autophagy 10, 1827–1843.

Haan, M.N., 2006. Therapy insight: type 2 diabetes mellitus and the risk of late-onset Alzheimer's disease. Nat. Clin. Pract. Neurol. 2, 159–166.

Haass, C., Selkoe, D.J., 2007. Soluble protein oligomers in neurodegeneration: lessons from the Alzheimer's amyloid beta-peptide. Nat. Rev. Mol. Cell Biol. 8, 101–112.

Hannun, Y.A., Obeid, L.M., 2008. Principles of bioactive lipid signalling: lessons from sphingolipids. Nat. Rev. Mol. Cell Biol. 9, 139–150.

Hardy, J., 2006. Alzheimer's disease: the amyloid cascade hypothesis: an update and reappraisal. J. Alzheimer Dis. 9 (Suppl. 3), 151–153.

Hardy, J., 2009. The amyloid hypothesis for Alzheimer's disease: a critical reappraisal. J. Neurochem. 110, 1129–1134.

Hardy, J., Selkoe, D.J., 2002. The amyloid hypothesis of Alzheimer's disease: progress and problems on the road to therapeutics. Science 297, 353–356.

Hay, N., 2005. The Akt-mTOR tango and its relevance to cancer. Cancer Cell 8, 179–183.

Henao-Mejia, J., Elinav, E., Jin, C., Hao, L., Mehal, W.Z., Strowig, T., et al., 2012. Inflammasome-mediated dysbiosis regulates progression of NAFLD and obesity. Nature 482, 179–185.

Heneka, M.T., Carson, M.J., Khoury, J.E., Landreth, G.E., Brosseron, F., Feinstein, D. L., et al., 2015. Neuroinflammation in Alzheimer's disease. Lancet Neurol. 14, 388–405.

Hilmas, C., Pereira, E.F., Alkondon, M., Rassoulpour, A., Schwarcz, R., Albuquerque, E. X., 2001. The brain metabolite kynurenic acid inhibits alpha7 nicotinic receptor activity and increases non-alpha7 nicotinic receptor expression: physiopathological implications. J. Neurosci. 21, 7463–7473.

Hokama, M., Oka, S., Leon, J., Ninomiya, T., Honda, H., Sasaki, K., et al., 2014. Altered expression of diabetes-related genes in Alzheimer's disease brains: the Hisayama study. Cereb. Cortex 24, 2476–2488.

Holtzman, D.M., Morris, J.C., Goate, A.M., 2011. Alzheimer's disease: the challenge of the second century. Sci. Transl. Med. 3, 77sr1.

Hooper, C., Killick, R., Lovestone, S., 2008. The GSK3 hypothesis of Alzheimer's disease. J Neurochem. 104, 1433–1439.

Hotta, K., Funahashi, T., Arita, Y., Takahashi, M., Matsuda, M., Okamoto, Y., et al., 2000. Plasma concentrations of a novel, adipose-specific protein, adiponectin, in type 2 diabetic patients. Arterioscler. Thromb. Vasc. Biol. 20, 1595–1599.

Huang, H., Tindall, D.J., 2007. Dynamic FoxO transcription factors. J. Cell. Sci. 120, 2479–2487.

Iadecola, C., 2004. Neurovascular regulation in the normal brain and in Alzheimer's disease. Nat. Rev. Neurosci. 5, 347–360.

Iadecola, C., 2013. The pathobiology of vascular dementia. Neuron. 80, 844–866.

Ikonen, E., Vainio, S., 2005. Lipid microdomains and insulin resistance: is there a connection? Sci. STKE pe3.

Ito, H., Kanno, I., Fukuda, H., 2005. Human cerebral circulation: positron emission tomography studies. Ann. Nucl. Med. 19, 65–74.

Jang, H., Arce, F.T., Ramachandran, S., Kagan, B.L., Lal, R., Nussinov, R., 2014. Disordered amyloidogenic peptides may insert into the membrane and assemble into common cyclic structural motifs. Chem. Soc. Rev. 43, 6750–6764.

Janson, J., Laedtke, T., Parisi, J.E., O'Brien, P., Petersen, R.C., Butler, P.C., 2004. Increased risk of type 2 diabetes in Alzheimer disease. Diabetes 53, 474–481.

Jazvinšćak Jembrek, M., Hof, P.R., Šimić, G., 2015. Ceramides in Alzheimer's disease: key mediators of neuronal apoptosis induced by oxidative stress and Aβ accumulation. Oxid. Med. Cell. Longev. 2015, 346783.

Jethwa, N., Chung, G.H.C., Lete, M.G., Alonso, A., Byrne, R.D., Calleja, V., et al., 2015. Endomembrane PtdIns(3,4,5)P-3 activates the PI3K-Akt pathway. J. Cell Sci. 128, 3456–3465.

Jiang, L.-Y., Tang, S.-S., Wang, X.-Y., Liu, L.P., Long, Y., Hu, M., et al., 2012. PPARγ agonist pioglitazone reverses memory impairment and biochemical changes in a mouse model of type 2 diabetes mellitus. CNS Neurosci. Therap. 18, 659–666.

Jiang, C., Li, G., Huang, P., Liu, Z., Zhao, B., 2017. The gut microbiota and Alzheimer's disease. J. Alzheimer Dis. 58, 1–15.

Josephs, K.A., Whitwell, J.L., Ahmed, Z., Shiung, M.M., Weigand, S.D., Knopman, D. S., et al., 2008. Beta-amyloid burden is not associated with rates of brain atrophy. Ann. Neurol. 63, 204–212.

Kadavath, H., Hofele, R.V., Biernat, J., Kumar, S., Tepper, K., Urlaub, H., et al., 2015. Tau stabilizes microtubules by binding at the interface between tubulin heterodimers. Proc. Natl. Acad. Sci. U. S. A. 112, 7501–7506.

Kelleher, R.J., Soiza, R.L., 2013. Evidence of endothelial dysfunction in the development of Alzheimer's disease: is Alzheimer's a vascular disorder? Am. J. Cardiovasc. Dis. 3, 197–226.

Kermani, A., Garg, A., 2003. Thiazolidinedione-associated congestive heart failure and pulmonary edema. Mayo Clin. Proc. 78, 1088–1091.

Kessler, M., Terramani, T., Lynch, G., Baudry, M., 1989. A glycine site associated with N-methyl-D-aspartic acid receptors: characterization and identification of a new class of antagonists. J. Neurochem. 52, 1319–1328.

Kiernan, E.A., Smith, S.M.C., Mitchell, G.S., Watters, J.J., 2016. Mechanisms of microglial activation in models of inflammation and hypoxia: implications for chronic intermittent hypoxia. J. Physiol. 594, 1563–1577.

Kim, F., Tysseling, K.A., Rice, J., Pham, M., Haji, L., Gallis, B.M., et al., 2005. Free fatty acid impairment of nitric oxide production in endothelial cells is mediated by IKKβ. Arterioscler. Thromb. Vasc. Biol. 25, 989–999.

Kim, J., Basak, J.M., Holtzman, D.M., 2009. The role of apolipoprotein E in Alzheimer's disease. Neuron 63, 287–303.

Kim, J., Yoon, H., Basak, J., Kim, J., 2014. Apolipoprotein E in synaptic plasticity and Alzheimer's disease: potential cellular and molecular mechanisms. Mol. Cell 37, 767–776.

Kitazawa, M., Cheng, D., Tsukamoto, M.R., Koike, M.A., Wes, P.D., Vasilevko, V., et al., 2011. Blocking IL-1 signaling rescues cognition, attenuates tau pathology, and restores neuronal beta-catenin pathway function in an Alzheimer's disease model. J. Immunol. 187, 6539–6549.

Kohn, A.D., Summers, S.A., Birnbaum, M.J., Roth, R.A., 1996. Expression of a constitutively active Akt Ser/Thr kinase in 3T3-L1 adipocytes stimulates glucose uptake and glucose transporter 4 translocation. J. Biol. Chem. 271, 31372–31378.

Kook, S.Y., Hong, H.S., Moon, M., Ha, C.M., Chang, S., Mook-Jung, I., 2012. $A\beta_{1-42}$-RAGE interaction disrupts tight junctions of the blood-brain barrier via Ca^{2+}-calcineurin signaling. J. Neurosci. 32, 8845–8854.

Kowalski, K., Mulak, A., 2019. Brain-gut-microbiota axis in Alzheimer's disease. J. Neurogastroenterol. Motil. 25, 48–60.

Kubota, N., Terauchi, Y., Yamauchi, T., Kubota, T., Moroi, M., Matsui, J., et al., 2002. Disruption of adiponectin causes insulin resistance and neointimal formation. J. Biol. Chem. 277, 25863–25866.

Kullmann, S., Schweizer, F., Veit, R., Fritsche, A., Preissl, H., 2015. Compromised white matter integrity in obesity. Obes. Rev. 16, 273–281.

Kullmann, S., Heni, M., Hallschmid, M., Fritsche, A., Preissl, H., Häring, H.U., 2016. Brain insulin resistance at the crossroads of metabolic and cognitive disorders in humans. Physiol. Rev. 96, 1169–1209.

Kurochkin, I.V., Guarnera, E., Berezovsky, I.N., 2018. Insulin-degrading enzyme in the fight against Alzheimer's disease. Trends Pharmacol. Sci. 39, 49–58.

Kurz, A., Perneczky, R., 2011. Amyloid clearance as a treatment target against Alzheimer's disease. J. Alzheimers. Dis. 24 (Suppl. 2), 61–73.

Lafay-Chebassier, C., Paccalin, M., Page, G., Barc-Pain, S., Perault-Pochat, M.C., Gil, R., et al., 2005. mTOR/p70S6k signalling alteration by Abeta exposure as well as in APP-PS1 transgenic models and in patients with Alzheimer's disease. J. Neurochem. 94, 215–225.

Landreth, G., Jiang, Q., Mandrekar, S., Heneka, M., 2008. PPARgamma agonists as therapeutics for the treatment of Alzheimer's disease. Neurotherapeutics 5, 481–489.

Leney, S.E., Tavare, J.M., 2009. The molecular basis of insulin-stimulated glucose uptake: signalling, trafficking and potential drug targets. J. Endocrinol. 203, 1–18.

Li, X.L., Aou, S., Oomura, Y., Hori, N., Fukunaga, K., Hori, T., 2002. Impairment of long-term potentiation and spatial memory in leptin receptor-deficient rodents. Neuroscience 113, 607–615.

Lipina, C., Hundal, H.S., 2016. Modulation of cellular redox homeostasis by the endocannabinoid system. Open Biol. 6, 150276.

Liu, C.C., Kanekiyo, T., Xu, H., Bu, G., 2013. Apolipoprotein E and Alzheimer disease: risk, mechanisms and therapy. Nat. Rev. Neurol. 9, 106–118.

Liu, P., Gan, W., Chin, Y.R., Ogura, K., Guo, J., Zhang, J., et al., 2015. PtdIns(3,4,5)P3-dependent activation of the mTORC2 kinase complex. Cancer Discov. 5, 1194–1209.

Liu, W., Lou, X., Ma, L., 2016. Use of 3D pseudo-continuous arterial spin labeling to characterize sex and age differences in cerebral blood flow. Neuroradiology 58, 943–948.

Loewith, R., Jacinto, E., Wullschleger, S., Lorberg, A., Crespo, J.L., Bonenfant, D., et al., 2002. Two TOR complexes, only one of which is rapamycin sensitive, have distinct roles in cell growth control. Mol. Cell 10, 457–468.

Lopez-Lopez, C., LeRoith, D., Torres-Aleman, I., 2004. Insulin-like growth factor I is required for vessel remodeling in the adult brain. Proc. Natl. Acad. Sci. U.S.A. 101, 9833–9838.

Luchsinger, J.A., 2010. Type 2 diabetes, related conditions, in relation and dementia: an opportunity for prevention? J. Alzheimer Dis. 20, 723–736.

Lue, L.F., Walker, D.G., Brachova, L., Beach, T.G., Rogers, J., Schmidt, A.M., et al., 2001. Involvement of microglial receptor for advanced glycation endproducts (RAGE) in Alzheimer's disease: identification of a cellular activation mechanism. Exp. Neurol. 171, 29–45.

Lue, L.F., Schmitz, C.T., Serrano, G., Sue, L.I., Beach, T.G., Walker, D.G., 2015. TREM2 protein expression changes correlate with Alzheimer's disease neurodegenerative pathologies in post-mortem temporal cortices. Brain Pathol. 25, 469–480.

Lyte, M., 2011. Probiotics function mechanistically as delivery vehicles for neuroactive compounds: microbial endocrinology in the design and use of probiotics. Bioessays 33, 574–581.

Ma, T., Hoeffer, C.A., Capetillo-Zarate, E., Yu, F., Wong, H., Lin, M.T., et al., 2010. Dysregulation of the mTOR pathway mediates impairment of synaptic plasticity in a mouse model of Alzheimer's disease. PLoS One 5, e12845.

Mandelkow, E.M., Stamer, K., Vogel, R., Thies, E., Mandelkow, E., 2003. Clogging of axons by tau, inhibition of axonal traffic and starvation of synapses. Neurobiol. Aging 24, 1079–1085.

Manning, B.D., Cantley, L.C., 2007. AKT/PKB signaling: navigating downstream. Cell 129, 1261–1274.

Manning, B.D., Toker, A., 2017. AKT/PKB Signaling: Navigating the Network. Cell 169, 381–405.

Masciopinto, F., Di Pietro, N., Corona, C., Bomba, M., Pipino, C., Curcio, M., et al., 2012. Effects of long-term treatment with pioglitazone on cognition and glucose metabolism of PS1-KI, 3xTg-AD and wild-type mice. Cell Death Dis. 3, e448–e448.

Mattsson, N., Tosun, D., Insel, P.S., Simonson, A., Jack Jr, C.R., Beckett, L.A., et al., 2014. Association of brain amyloid-beta with cerebral perfusion and structure in Alzheimer's disease and mild cognitive impairment. Brain 137, 1550–1561.

McNay, E.C., Ong, C.T., McCrimmon, R.J., Cresswell, J., Bogan, J.S., Sherwin, R.S., 2010. Hippocampal memory processes are modulated by insulin and high-fat-induced insulin resistance. Neurobiol. Learn. Mem. 93, 546–553.

Moloney, A.M., Griffin, R.J., Timmons, S., O'Connor, R., Ravid, R., O'Neill, C., 2010. Defects in IGF-1 receptor, insulin receptor and IRS-1/2 in Alzheimer's disease indicate possible resistance to IGF-1 and insulin signalling. Neurobiol. Aging 31, 224–243.

Moreira, P.I., 2012. Alzheimer's disease and diabetes: an integrative view of the role of mitochondria, oxidative stress, and insulin. J. Alzheimer Dis. 30, S199–S215.
Moreno, L., Moral-Sanz, J., Morales-Cano, D., Barreira, B., Moreno, E., Ferrarini, A., et al., 2014. Ceramide mediates acute oxygen sensing in vascular tissues. Antioxid. Redox Signal. 20, 1–14.
Morris, G.P., Clark, I.A., Vissel, B., 2014. Inconsistencies and controversies surrounding the amyloid hypothesis of Alzheimer's disease. Acta Neuropathol. Commun. 2, 135.
Mrak, R.E., Griffin, W.S., 2005. Potential inflammatory biomarkers in Alzheimer's disease. J. Alzheimer Dis. 8, 369–375.
Mueed, Z., Tandon, P., Maurya, S.K., Deval, R., Kamal, M.A., Poddar, N.K., 2019. Tau and mTOR: the hotspots for multifarious diseases in Alzheimer's development. Front. Neurosci. 12, 1017.
Muniyappa, R., Sowers, J.R., 2013. Role of insulin resistance in endothelial dysfunction. Rev. Endocr. Metab. Disord. 14, 5–12.
Muniyappa, R., Montagnani, M., Koh, K.K., Quon, M.J., 2007. Cardiovascular actions of insulin. Endocr. Rev. 28, 463–491.
Munoz, L., Ralay, R.H., Roy, S.M., Hu, W., Craft, J.M., et al., 2007. A novel p38 alpha MAPK inhibitor suppresses brain proinflammatory cytokine up-regulation and attenuates synaptic dysfunction and behavioral deficits in an Alzheimer's disease mouse model. J. Neuroinflamm. 4, 21.
Nakamura, T., Furuhashi, M.P., Li, P., Cao, H., Tuncman, G., Sonenberg, N., et al., 2010. Double-stranded RNA-dependent protein kinase links pathogen sensing with stress and metabolic homeostasis. Cell 140, 338–348.
Negre-Salvayre, A., Salvayre, R., Augé, N., Pamplona, R., Portero-Otín, M., 2009. Hyperglycemia and glycation in diabetic complications. Antioxid. Redox Signal. 11, 3071–3109.
Ngwa, J.S., Fungwe, T.V., Ntekim, O., Allard, J.S., Johnson, S.M., 2018. Alzheimer's disease neuroimaging initiative. associations of pulse and blood pressure with hippocampal volume by APOE and cognitive phenotype: the Alzheimer's disease neuroimaging initiative (ADNI). Dement. Geriatr. Cogn. Disord. 45, 66–78.
O'Brien, R.J., Wong, P.C., 2011. Amyloid precursor protein processing and Alzheimer's disease. Annu. Rev. Neurosci. 34, 185–204.
O'Connor, T., Sadleir, K.R., Maus, E., Velliquette, R.A., Zhao, J., Cole, S.L., et al., 2008. Phosphorylation of the translation initiation factor eIF2α increases BACE1 levels and promotes amyloidogenesis. Neuron 60, 988–1009.
Oddo, S., Caccamo, A., Tseng, B., Cheng, D., Vasilevko, V., Cribbs, D.H., et al., 2008. Blocking Abeta42 accumulation delays the onset and progression of tau pathology via the C terminus of heat shock protein70-interacting protein: a mechanistic link between Abeta and tau pathology. J. Neurosci. 28, 12163-1275.
Oomura, Y., Hori, N., Shiraishi, T., Fukunaga, K., Takeda, H., Tsuji, M., et al., 2006. Leptin facilitates learning and memory performance and enhances hippocampal CA1 long term potentiation and CaMK II phosphorylation in rats. Peptides 27, 2738–2749.
Pandini, G., Pace, V., Copani, A., Squatrito, S., Milardi, D., Vigneri, R., 2013. Insulin has multiple antiamyloidogenic effects on human neuronal cells. Endocrinology 154, 375–387.
Pawelec, G., 2017. Age and immunity: what is 'immunosenescence'? Exp. Gerontol. 105, 4–9.
Penke, B., Bogár, F., Fülöp, L., 2017. β-Amyloid and the pathomechanisms of Alzheimer's disease: a comprehensive view. Molecules (Basel Switz) 22, 1692.

Perez, S.E., Nadeem, M., He, B., Miguel, J.C., Malek-Ahmadi, M.H., Chen, K., et al., 2017. Neocortical and hippocampal TREM2 protein levels during the progression of Alzheimer's disease. Neurobiol. Aging 54, 133–143.
Perluigi, M., Di Domenico, F., Butterfield, D.A., 2015. mTOR signaling in aging and neurodegeneration: at the crossroad between metabolism dysfunction and impairment of autophagy. Neurobiol. Dis. 84, 39–49.
Pessin, J.E., Saltiel, A.R., 2000. Signaling pathways in insulin action: molecular targets of insulin resistance. J. Clin. Invest. 106, 165–169.
Pozueta, J., Lefort, R., Shelanski, M.L., 2013. Synaptic changes in Alzheimer's disease and its models. Neuroscience 251, 51–65.
Pratico, D., Sung, S., 2004. Lipid peroxidation and oxidative imbalance: early functional events in Alzheimer's disease. J. Alzheimer Dis. 6, 171–175.
Pretorius, L., Kell, D.B., Pretorius, E., 2018. Iron dysregulation and dormant microbes as causative agents for impaired blood rheology and pathological clotting in Alzheimer's type dementia. Front. Neurosci. 12, 851.
Prokop, S., Miller, K.R., Heppner, F.L., 2013. Microglia actions in Alzheimer's disease. Acta Neuropathol. 126, 461–477.
Qiu, W.Q., Folstein, M.F., 2006. Insulin, insulin-degrading enzyme and amyloid-beta peptide in Alzheimer's disease: review and hypothesis. Neurobiol. Aging 27, 190–198.
Qiu, C., Winblad, B., Viitanen, M., Fratiglioni, L., 2003. Pulse pressure and risk of Alzheimer disease in persons aged 75 years and older: a community-based, longitudinal study. Stroke 34, 594–599.
Quintanilla, R.A., Orellana, D.I., Gonzalez-Billault, C., Maccioni, R.B., 2004. Interleukin-6 induces Alzheimer-type phosphorylation of tau protein by deregulating the cdk5/p35 pathway. Exp. Cell Res. 295, 245–257.
Raha, A.A., Henderson, J.W., Stott, S.R., Vuono, R., Foscarin, S., Friedland, R.P., et al., 2017. Neuroprotective effect of TREM-2 in aging and Alzheimer's disease model. J. Alzheimer Dis. 55, 199–217.
Raji, C.A., Ho, A.J., Parikshak, N.N., Becker, J.T., Lopez, O.L., Kuller, L.H., et al., 2010. Brain structure and obesity. Hum. Brain Mapp. 31, 353–364.
Reddy, P.H., Beal, M.F., 2008. Amyloid beta, mitochondrial dysfunction and synaptic damage: Implications for cognitive decline in aging and Alzheimer's disease. Trends Mol. Med. 14, 45–53.
Ristow, M., 2004. Neurodegenerative disorders associated with diabetes mellitus. J. Mol. Med. 82, 510–529.
Roher, A.E., Debbins, J.P., Malek-Ahmadi, M., Chen, K., Pipe, J.G., Maze, S., et al., 2012. Cerebral blood flow in Alzheimer's disease. Vasc. Health Risk Manag. 8, 599–611.
Romao, I., Roth, J., 2008. Genetic and environmental interactions in obesity and type 2 diabetes. J. Am. Diet Assoc. 108, S24–S28.
Ruvolo, P.P., 2003. Intracellular signal transduction pathways activated by ceramide and its metabolites. Pharmacol. Res. 47, 383–392.
Sadigh-Eteghad, S., Sabermarouf, B., Majdi, A., Talebi, M., Farhoudi, M., Mahmoudi, J., 2015. Amyloid-beta: a crucial factor in Alzheimer's disease. Med. Princ. Pract. 24, 1–10.
Sagare, A.P., Bell, R.D., Zlokovic, B.V., 2013. Neurovascular defects and faulty amyloid-beta vascular clearance in Alzheimer's disease. J. Alzheimer Dis. 33 (Suppl. 1), S87–S100.
Saltiel, A.R., Kahn, C.R., 2001. Insulin signalling and the regulation of glucose and lipid metabolism. Nature 414, 799–806.

Santamaria, A., Galvan-Arzate, S., Lisy, V., Ali, S.F., Duhart, H.M., Osorio-Rico, L., et al., 2001. Quinolinic acid induces oxidative stress in rat brain synaptosomes. Neuroreport 12, 871–874.

Sarbassov, D.D., Guertin, D.A., Ali, S.M., Sabatini, D.M., 2005. Phosphorylation and regulation of Akt/PKB by the rictor-mTOR complex. Science 307, 1098–1101.

Savva, G.M., Wharton, S.B., Ince, P.G., Forster, G., Matthews, F.E., Brayne, C., 2009. Medical research council cognitive function and ageing study. Age, neuropathology, and dementia. N. Engl. J. Med. 360, 2302–2309.

Schmitz, L., Kuglin, R., Bae-Gartz, I., Janoschek, R., Appel, S., Mesaros, A., et al., 2018. Hippocampal insulin resistance links maternal obesity with impaired neuronal plasticity in adult offspring. Psychoneuroendocrinology 89, 46–52.

Schwarcz, R., Whetsell Jr, W.O., Mangano, R.M., 1983. Quinolinic acid: an endogenous metabolite that produces axon-sparing lesions in rat brain. Science 219, 316–318.

Selkoe, D.J., 2008. Soluble oligomers of the amyloid β-protein impair synaptic plasticity and behavior. Behav. Brain Res. 192, 106–113.

Sims-Robinson, C., Kim, B., Rosko, A., Feldman, E.L., 2010. How does diabetes accelerate Alzheimer disease pathology? Nat. Rev. Neurol. 6, 551–559.

Sindi, I.A., Dodd, P.R., 2015. New insights into Alzheimer's disease pathogenesis: the involvement of neuroligins in synaptic malfunction. Neurodegener. Dis. Manag. 5, 137–145.

Sochocka, M., Diniz, B.S., Leszek, J., 2017. Inflammatory response in the CNS: friend or foe? Mol. Neurobiol. 54, 8071–8089.

Son, S.M., Song, H., Byun, J., Park, K.S., Jang, H.C., Park, Y.J., et al., 2012. Accumulation of autophagosomes contributes to enhanced amyloidogenic APP processing under insulin-resistant conditions. Autophagy 8, 1842–1844.

Song, J., Lee, J.E., 2013. Adiponectin as a new paradigm for approaching Alzheimer's disease. Anat. Cell Biol. 46, 229–234.

Spielman, L.J., Bahniwal, M., Little, J.P., Walker, D.G., Klegeris, A., 2015. Insulin modulates in vitro secretion of cytokines and cytotoxins by human glial cells. Curr. Alzheimer Res. 12, 684–693.

Spires-Jones, T.L., Hyman, B.T., 2014. The intersection of amyloid beta and tau at synapses in Alzheimer's disease. Neuron 82, 756–771.

Steen, E., Terry, B.M., Rivera, E.J., Cannon, J.L., Neely, T.R., Tavares, R., et al., 2005. Impaired insulin and insulin-like growth factor expression and signaling mechanisms in Alzheimer's disease—is this type 3 diabetes? J. Alzheimer Dis. 7, 63–80.

Steinberg, H.O., Tarshoby, M., Monestel, R., Hook, G., Cronin, J., Johnson, A., et al., 1997. Elevated circulating free fatty acid levels impair endothelium-dependent vasodilation. J. Clin. Invest. 100, 1230–1239.

Stern, Y., 2012. Cognitive reserve in ageing and Alzheimer's disease. Lancet Neurol. 11, 1006–1012.

Stinkens, R., Goossens, G.H., Jocken, J.W., Blaak, E.E., 2015. Targeting fatty acid metabolism to improve glucose metabolism. Obes. Rev. 16, 715–757.

Stolerman, E.S., Florez, J.C., 2009. Genomics of type 2 diabetes mellitus: implications for the clinician. Nat. Rev. Endocrinol. 5, 429–936.

Stratford, B.A.P.K., Hoehn, S., Liu, K.L.F., Summers, S.A., 2004. Regulation of insulin action by ceramide. J. Biol. Chem. 279, 36608–36615.

Tai, J., Liu, W., Li, Y., Li, L., Hölscher, C., 2018. Neuroprotective effects of a triple GLP-1/GIP/glucagon receptor agonist in the APP/PS1 transgenic mouse model of Alzheimer's disease. Brain Res. 1678, 64–74.

Talbot, K., Wang, H., Kazi, H., Han, L., Bakshi, K.P., Stucky, A., et al., 2012. Demonstrated brain insulin resistance in Alzheimer' s disease patients is associated with

IGF-1 resistance, IRS-1 dysregulation, and cognitive decline. J. Clin. Invest. 122, 1316–1338.

Tarumi, T., Ayaz Khan, M., Liu, J., Tseng, B.Y., Parker, R., Riley, J., et al., 2014. Cerebral hemodynamics in normal aging: central artery stiffness, wave reflection, and pressure pulsatility. J. Cereb. Blood Flow Metab. 34, 971–978.

Tavares, R.G., Tasca, C.I., Santos, C.E., Alves, L.B., Porciúncula, L.O., Emanuelli, T., et al., 2002. Quinolinic acid stimulates synaptosomal glutamate release and inhibits glutamate uptake into astrocytes. Neurochem. Int. 40, 621–627.

Thienel, M., Wilhelm, I., Benedict, C., Born, J., Hallschmid, M., 2017. Intranasal insulin decreases circulating cortisol concentrations during early sleep in elderly humans. Neurobiol. Aging 54, 170–174.

Tong, M., de la Monte, S.M., 2009. Mechanisms of ceramide-mediated neurodegeneration. J. Alzheimer Dis. 16, 705–714.

Townsend, M., Mehta, T., Selkoe, D.J., 2007. Soluble Abeta inhibits specific signal transduction cascades common to the insulin receptor pathway. J. Biol. Chem. 282, 33305–33312.

Tu, S., Okamoto, S., Lipton, S.A., Xu, H., 2014. Oligomeric Aβ-induced synaptic dysfunction in Alzheimer's disease. Mol. Neurodegener. 9, 48.

Tzatsos, A., 2009. Raptor binds the SAIN (Shc and IRS-1 NPXY binding) domain of insulin receptor substrate-1 (IRS-1) and regulates the phosphorylation of IRS-1 at Ser-636/639 by mTOR. J. Biol. Chem. 284, 22525–22534.

Ursache, A., Wedin, W., Tirsi, A., Convit, A., 2012. Preliminary evidence for obesity and elevations in fasting insulin mediating associations between cortisol awakening response and hippocampal volumes and frontal atrophy. Psychoneuroendocrinology 37, 1270–1276.

Valente, T., Gella, A., Fernàndez-Busquets, X., Unzeta, M., Durany, N., 2010. Immunohistochemical analysis of human brain suggests pathological synergism of Alzheimer's disease and diabetes mellitus. Neurobiol. Dis. 37, 67–76.

Vandanmagsar, B., Youm, Y.H., Ravussin, A., Galgani, J.E., Stadler, K., Mynatt, R.L., et al., 2011. The NLRP3 inflammasome instigates obesity-induced inflammation and insulin resistance. Nat. Med. 17, 179–188.

van der Flier, W.M., Scheltens, P., 2005. Epidemiology and risk factors of dementia. J. Neurol. Neurosurg. Psychiatry 76, 2–7.

Vanhaesebroeck, B., Guillermet-Guibert, J., Graupera, M., Bilanges, B., 2010. The emerging mechanisms of isoform-specific PI3K signalling. Nat. Rev. Mol. Cell Biol. 11, 329–341.

Vieira, M.N.N., Lima-Filho, R.A.S., De Felice, F.G., 2017. Connecting Alzheimer's disease to diabetes: underlying mechanisms and potential therapeutic targets. Neuropharmacology 136, 160–171.

Viola, K.L., Klein, W.L., 2015. Amyloid β oligomers in Alzheimer's disease pathogenesis, treatment, and diagnosis. Acta Neuropathol. 129, 183–206.

Vogt, N.M., Kerby, R.L., Dill-McFarland, K.A., Harding, S.J., Merluzzi, A.P., Johnson, S. C., et al., 2017. Gut microbiome alterations in Alzheimer's disease. Sci. Rep. 7, 13537.

Wayner, M.J., Armstrong, D.L., Phelix, C.F., Oomura, Y., 2004. Orexin-A (Hypocretin-1) and leptin enhance LTP in the dentate gyrus of rats in vivo. Peptides 25, 991–996.

Weyer, S.W., Klevanski, M., Delekate, A., Voikar, V., Aydin, D., Hick, M., et al., 2011. APP and APLP2 are essential at PNS and CNS synapses for transmission, spatial learning and LTP. EMBO J. 30, 2266–2280.

Wildsmith, K.R., Holley, M., Savage, J.C., Skerrett, R., Landreth, G.E., 2013. Evidence for impaired amyloid β clearance in Alzheimer's disease. Alzheimer Res. Ther. 5, 33.

Willette, A.A., Johnson, S.C., Birdsill, A.C., Sager, M.A., Christian, B., Baker, L.D., et al., 2015. Insulin resistance predicts brain amyloid deposition in late middle-aged adults. Alzheimer Dement. 11, 504–510.e1.

Willette, A.A., Webb, J.L., Dantzer, R., 2017. Neuroinflammation, depression, and Alzheimer's disease: insight from the kynurenine pathway. Alzheimer's Dement 13 (Suppl.), P181–P182.

Winek, K., Dirnagl, U., Meisel, A., 2016a. The gut microbiome as therapeutic target in central nervous system diseases: implications for stroke. Neurotherapeutics 13, 762–774.

Winek, K., Meisel, A., Dirnagl, U., 2016b. Gut microbiota impact on stroke outcome: Fad or fact? J. Cereb. Blood Flow Metab. 36, 891–898.

Wullschleger, S., Loewith, R., Hall, M.N., 2006. TOR signaling in growth and metabolism. Cell 124, 471–484.

Xia, J.Y., Holland, W.L., Kusminski, C.M., Sun, K., Sharma, A.X., Pearson, M.J., et al., 2015. Targeted induction of ceramide degradation leads to improved systemic metabolism and reduced hepatic steatosis. Cell Metab. 22, 266–278.

Zac-Varghese, S., Tan, T., Bloom, S.R., 2010. Hormonal interactions between gut and brain. Discov. Med. 10, 543–552.

Zeng, G., Quon, M.J., 1996. Insulin-stimulated production of nitric oxide is inhibited by wortmannin. Direct measurement in vascular endothelial cells. J. Clin. Invest. 98, 894–898.

Zhang, J., Liu, F., 2014. Tissue-specific insulin signaling in the regulation of metabolism and aging. IUBMB Life 66, 485–495.

Zhang, P., Liang, X., Shan, T., Jiang, Q., Deng, C., Zheng, R., et al., 2015. mTOR is necessary for proper satellite cell activity and skeletal muscle regeneration. Biochem. Biophys. Res. Commun. 463, 102–108.

Zhao, W.Q., De Felice, F.G., Fernandez, S., Chen, H., Lambert, M.P., et al., 2008. Amyloid beta oligomers induce impairment of neuronal insulin receptors. FASEB J. 22, 246–260.

Zhao, Y., Jaber, V., Lukiw, W.J., 2017. Secretory products of the human GI tract microbiome and their potential impact on Alzheimer's Disease (AD): detection of lipopolysaccharide (LPS) in AD hippocampus. Front Cell Infect. Microbiol. 7, 318.

Zimmerman, B., Sutton, B.P., Low, K.A., Fletcher, M.A., Tan, C.H., Schneider-Garces, N., et al., 2014. Cardiorespiratory fitness mediates the effects of aging on cerebral blood flow. Front. Aging Neurosci. 6, 59.

Zlokovic, B.V., 2011. Neurovascular pathways to neurodegeneration in Alzheimer's disease and other disorders. Nat. Rev. Neurosci. 12, 723–738.

Zoncu, R., Efeyan, A., Sabatini, D.M., 2011. mTOR: from growth signal integration to cancer, diabetes and ageing. Nat. Rev. Mol. Cell Biol. 12, 21–35.

Further reading

Baglietto-Vargas, D., Shi, J., Yaeger, D.M., Ager, R., LaFerla, F.M., 2016. Diabetes and Alzheimer's disease crosstalk. Neurosci. Biobehav. Rev. 64, 272–287.

Escribano, L., Simón, A.M., Gimeno, E., Cuadrado-Tejedor, M., López de Maturana, R., García-Osta, A., et al., 2010. Rosiglitazone rescues memory impairment in Alzheimer's transgenic mice: mechanisms involving a reduced amyloid and tau pathology. Neuropsychopharmacology 35, 1593–1604.

Górski, J., 2012. Ceramide and insulin resistance: how should the issue be approached? Diabetes 61, 3081–3308.

Jay, T.R., von Saucken, V.E., Landreth, G.E., 2017. TREM2 in neurodegenerative diseases. Mol. Neurodegener. 12, 56.

Li, Y., Liu, L., Barger, S.W., Griffin, W.S., 2003. Interleukin-1 mediates pathological effects of microglia on tau phosphorylation and on synaptophysin synthesis in cortical neurons through a p38-MAPK pathway. J. Neurosci. 23, 1605–1611.

Sharma, M.D., Garber, A.J., Farmer, J.A., 2008. Role of insulin signaling in maintaining energy homeostasis. Endocr. Pract. 14, 373–380.

Statovci, D., Aguilera, M., MacSharry, J., Melgar, S., 2017. The impact of western diet and nutrients on the microbiota and immune response at mucosal interfaces. Front. Immunol. 8, 838.

Xu, H., Moore, E., Meiri, N., Quon, M.J., Alkon, D.L., 1999. Brain insulin receptors and spatial memory. J. Biol. Chem. 274, 34839–34842.

Yeh, F.L., Hansen, D.V., Sheng, M., 2017. TREM2, microglia, and neurodegenerative diseases. Trends Mol. Med. 23, 512–533.

CHAPTER 7

Insulin resistance and Parkinson's disease

Introduction

Parkinson's disease (PD) is a multifactorial and progressive neurological disorder with unknown etiology, in which genetic, environmental factors, and lifestyle factors play important roles. PD is characterized by the selective loss of dopaminergic neurons in substantia nigra pars compacta as well as the formation of intracellular inclusion bodies called as Lewy bodies (LBs). Before the induction of prominent motor signs, PD presents a range of nonmotor symptoms (NMS) that precede the clinical motor phase by many years. Some are well known, such as olfactory and gastrointestinal (GI) dysfunction, sleep disorders, circadian changes, and cognitive impairment (Lima et al., 2012; Lima, 2013; Videnovic and Golombek, 2013). Moreover, neuropathological studies support the association of these early-phase disturbances based on the identification of LBs in nondopaminergic nuclei in early Braak stages, prior to significant substantia nigra pars compacta degeneration and motor signs (Del Tredici and Braak, 2016). Recent epidemiological studies have indicated that NMS can appear up to 25 years before the onset of clinical PD (Tolosa and Pont-Sunyer, 2011), and it is well established that patients report sleep disruption at least a decade before the first motor symptoms. Aging is the main risk factor for developing PD. Its incidence in human population ranges from approximately 1% in people over 65 years old to 4% in people over 86 years old. In the United States, more than one million people suffer from PD (Goldenberg, 2008; Tysnes and Storstein, 2017). PD is more common in men (about 1.5 times) than in women (Davie, 2008), and higher incidences of PD have been reported in developed countries (Bove et al., 2005), due to an increase in the aged population (Cannon and Greenamyre, 2011; Stuendl et al., 2016). Eighty percent of PD patients develop dementia as PD progresses (Hely et al., 2008). Lewy body dementia (LBD), PD, and Parkinson's disease dementia (PDD) have been grouped under the umbrella term of LBD spectrum due to the

Insulin Resistance as a Risk Factor in Visceral and Neurological Disorders
DOI: https://doi.org/10.1016/B978-0-12-819603-8.00007-9

overlap in symptom profile, similar treatment response, and common underlying neuropathology (Francis, 2009). Converging evidence suggests that LBD, PD, and PDD, patients share the presence of α-synuclein, a presynaptic amyloidogenic protein that regulates the release of neurotransmitters from synaptic vesicles in the brain. α-Synuclein has also been implicated in response to cellular stress. α-Synuclein aggregates, including LBs, are features of both sporadic and familial forms of PD. These aggregates undergo several key stages of fibrillation, oligomerization, and aggregation (Merdes et al., 2003; Recchia et al., 2004; Fujishiro et al., 2008; Ince, 2011). Recent studies have indicated that α-synuclein is also found in the nucleus of neurons (Maroteaux et al., 1988; Surguchov, 2015), suggesting that this protein plays additional cellular roles beyond the synapse. On the basis of immunolocalization studies, it is proposed that in the nucleus, α-synuclein may be involved in either direct or indirect interactions with DNA, including modulating histone modification state (Yu et al., 2007) or direct DNA binding (Vasudevaraju et al., 2012). Some forms of aggregated α-synuclein have been shown to have DNA endonuclease activity (Vasquez et al., 2017). In overexpression models, which increase specific aggregated forms of α-synuclein potentially relevant to disease, neuronal toxicity is increased, possibly due to downregulated transcription of DNA repair genes (Paiva et al., 2019), or increases in prooxidant species that result in DNA damage (Milanese et al., 2018). Nuclear α-synuclein has also been shown to influence neuronal cell death; in *Drosophila*, nuclear forms of the protein promote neurotoxicity (Rousseaux et al., 2016), while in human cells higher molecular weight α-synuclein species contribute to reduction in neurotoxicity (Pinho et al., 2019). α-Synuclein's interaction with DNA has also been argued to regulate normal cell function by influencing transcription (Siddiqui et al., 2012; Kim et al., 2014). α-Synuclein is also involved in the pathogenesis several lysosomal storage diseases, including Gaucher's disease and Krabbe's disease (KD) or globoid cell leukodystrophy (Moore et al., 2005; Smith et al., 2014). In addition, accumulation of β-amyloid (Aβ) deposition also frequently occurs in PD and LBD (Merdes et al., 2003). In addition, PD is accompanied by additional atypical features defining the Parkinson-plus syndromes, such as multiple system atrophy, progressive supranuclear palsy, and corticobasal degeneration.

The molecular mechanisms contributing to the pathogenesis of PD remain unknown. However, based on earlier studies, it is suggested that the neurodegeneration of dopaminergic neurons in PD result in the depletion

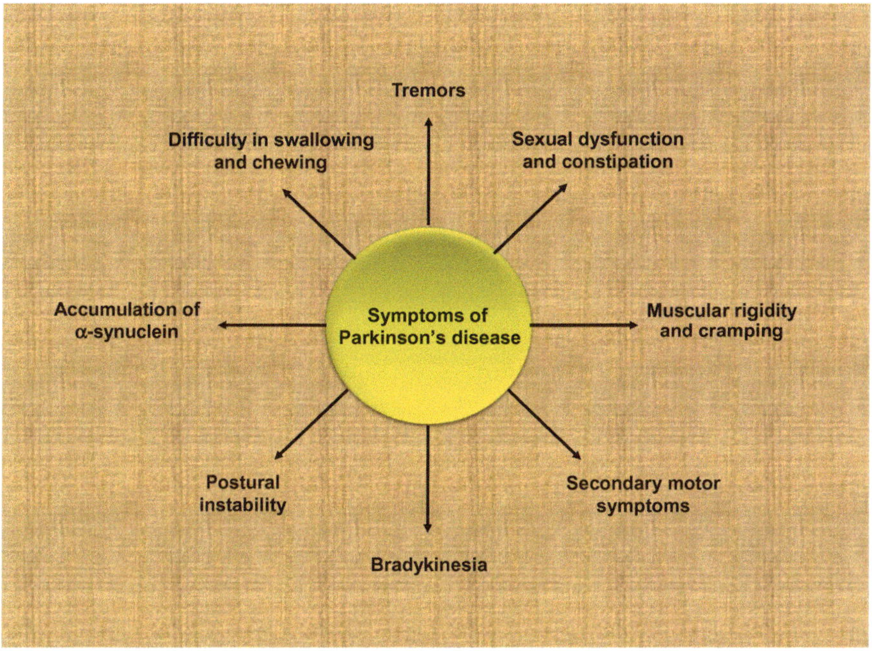

Figure 7.1 Symptoms of Parkinson's disease.

of dopamine leading to abnormal dopaminergic neurotransmission in the basal ganglia motor circuits producing resting tremor, muscular rigidity, akinesia, bradykinesia, and posture instability (Fig. 7.1) (Jankovic, 2008; Hornykiewicz, 2008; Maiti et al., 2017). Symptoms of PD also include cognitive impairment, hallucinations, depression, intermittent confusion, postural instability, and fatigue. Among these symptoms, fatigue is a very common complain. Fatigue includes both mental and physical aspects. Up to 70% of individuals with PD experience fatigue everyday (Friedman et al., 2007). Fatigue dramatically impairs quality of life (Friedman et al., 2016). Fatigue is a complex syndrome emerging from dysfunction in the nervous, endocrine, and immune system (Klimas et al., 2012). From a clinical point of view, fatigue is frequently associated with other nonmotor syndromes, such as sleepiness, apathy, depression, and autonomic dysfunction (Friedman et al., 2007; Chou et al., 2017). The underlying mechanisms associated with fatigue are not fully understood. However, it is proposed that the activation of the inflammatory cytokine network may contribute to fatigue (Klimas et al., 2012; Bower, 2007). Thus high serum levels of inflammatory markers [interleukin-6 (IL-6), IL1-Ra, sIL-2R, and VCAM-1] can be used as potential biomarkers of fatigue (Herlofson et al., 2018; Pereira et al., 2016). In PD patients, high levels of neuroinflammatory marker-mediated processes may promote glutamate dysregulation and further influence neuronal activity and neuroplasticity, and impact neuronal circuits mediating distress and motivation in PD (Eyre and Baune, 2012; Miller et al., 2013; Lindqvist et al., 2012).

GI tract symptoms also occur from the earliest stage of PD, and constipation is known to precede the motor symptoms of the disease (Cersosimo et al., 2013). In addition, the extent of GI dysfunction corresponds with widespread enteric nervous system (ENS) synucleinopathy, supporting the view that the abnormal deposition of α-synuclein in the ENS may be the main cause of GI dysfunction in PD (Cersosimo et al., 2013). It has been reported that GI tract dysfunction in PD occurs prior to manifestation of classic motor deficits (Pfeiffer, 2003), and LB pathology in the myenteric plexus may precede a clinical diagnosis of PD by many years (Wakabayashi et al., 1988). The GI tract is innervated by parasympathetic cholinergic neurons that may be particularly sensitive to the pathologic changes responsible for PD dopaminergic neuronal dysfunction. The lacrimal gland is innervated by cholinergic and dopaminergic neurons (Dartt, 2004). Stimulation of adrenergic and dopaminergic pathways is associated with the secretion of proteins into tears. It is proposed that PD-dependent alterations in cholinergic stimulation and/or in the trophic effects of cholinergic stimulation on lacrimal gland acinar cells function may lead to a characteristic tear protein profile in PD patients that can be measured cost-effectively and noninvasively (Dartt, 2009; Hamm-Alvarez et al., 2019).

At the molecular level, PD is accompanied by the accumulation of misfolded proteins (α-synuclein), ubiquitin-proteasome system dysfunction, increase in iron level, formation and accumulation of LBs, and deposition of Aβ and hyperphosphorylation of tau along with elevation in oxidative stress and neuroinflammation (Gerlach et al., 2006; Jankovic, 2008; Hornykiewicz, 2008; Maiti et al., 2017; Davie, 2008; Michel et al., 2016; Si et al., 2017). These parameters are prognostic markers for cognitive deficits in PD (Sengupta et al., 2015).

Interestingly, besides the "clinical overlapping," α-synuclein has been shown to physically interact with tau (Sengupta et al., 2015; Andersen et al., 2017) or Aβ (Parnetti et al., 2013; Andersen et al., 2017) and to induce the formation of hybrid oligomers ("heteroaggregates") in PD patients' brains. In this context, it has been demonstrated that α-syn forms heteroaggregates with Aβ or tau in red blood cells (RBCs) of healthy subjects, too (Daniele et al., 2017) in the same study, a novel home-made immunoenzymic assay has been employed to quantify such oligomers, highlighting their accumulation with increasing age and decreasing anti-oxidant capability (Daniele et al., 2017). The molecular mechanisms of dopaminergic neuronal loss are not fully understood. However, it is suggested that the degeneration of dopaminergic neurons in the substantia nigra pars compacta may be due to monoamine oxidase–mediated abnormal dopamine metabolism and generation of hydrogen peroxide. These processes lead to induction of oxidative stress. Majority of PD patients are sporadic, which may be caused by complex interactions among genetic factors, environmental exposures to toxins (paraquat, rotenone, herbicide, and insecticide), and aging of genetic variants with environmental risk factors (Lesage and Brice, 2009). Some cases of PD (familial PD) are caused by mutation in seven genes. These genes include α-synuclein, Parkin, PTEN-induced putative kinase 1 (*PINK1*), Protein DJ-1 (*DJ1*), Leucine rich repeat kinase 2 (*LRRK2*), ubiquitin carboxyl-terminal hydrolase isozyme 1 (*UCHL1*), *VPS35*, and *GBA1* (Fig. 7.2) (Maiti et al., 2017;

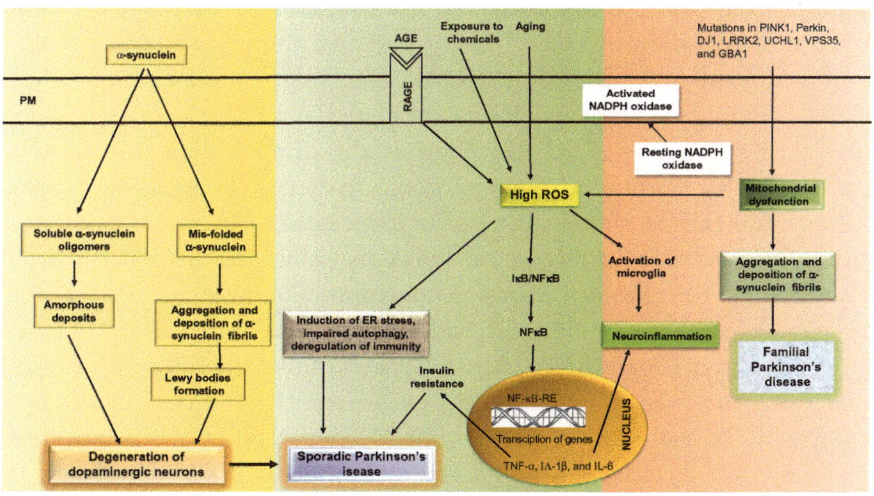

Figure 7.2 Hypothetical diagram showing the pathogenesis of sporadic and familial Parkinson's disease.

AGEs, Advanced glycated end products; *DJ-1*, protein DJ-1; *ER*, endoplasmic reticulum; *GBA1*, glucocerebrosidase; *IL-1β*, interleukin-1β; *IL-6*, interleukin-6; *NF-κB*, nuclear factor-κB; *NF-κB-RE*, nuclear factor-κB response element; *PINK1*, PTEN-induced putative kinase 1; *PM*, plasma membrane; *ROS*, reactive oxygen species; *TNF-α*, tumor necrosis factor-α; *UCHL1*, ubiquitin carboxyl-terminal hydrolase isozyme 1; *VPS35*, vacuolar protein sorting-associated protein 35.

Zeng et al., 2018; Videira and Castro-Caldas, 2018). Both types of PD are characterized by the accumulation of misfolded and aggregated α-synuclein.

Familial Parkinson's disease

Mutations in α-synuclein are not only closely associated with the development of PD, but also with potential mechanisms for loss of dopaminergic neurons (dopamine metabolism, endoplasmic reticulum stress, impaired autophagy, and deregulation of immunity). In addition, mutations in other genes such as *DJ1* also contribute to maintenance of energy balance and glucose homeostasis by regulating brown adipose tissue activity, mitochondrial dysfunction, and oxidative stress. DJ-1-deficient mice show reduction in body mass, increase in energy expenditure and improvement in insulin sensitivity. Furthermore, DJ-1 deletion also results in retardation of high-fat-diet–mediated obesity and insulin resistance (Wu et al., 2017). At the molecular level, DJ-1–mediated activation of PtdIns 3K/Akt activation is associated with neuronal protection and tumorigenesis (Yang et al., 2005). PTEN (phosphatase and tensin homologue deleted on chromosome 10), a putative phosphatase, has a dominant inhibitory role in PtdIns 3K/Akt signaling (Di Cristofano et al., 1998). These processes are closely associated with the pathogenesis of familial PD (Maiti et al., 2017; Zeng et al., 2018; Videira and Castro-Caldas, 2018). Parkin is a ubiquitin ligase largely implicated in PD. Recent studies have shown that parkin loss exacerbates inflammation and promotes survival of activated microglia by inhibiting necroptosis, thus contributing to chronic neuroinflammation (Dionísio et al., 2019), whereas in astrocytes it induces endoplasmic reticulum stress. Whether such effects can negatively impact on synaptic function and mediate synapse loss, however, remains to be elucidated. In humans LRRK2 is encoded by the *PARK8/LRRK2* gene (Paisán-Ruíz et al., 2004). This gene not only encodes a large multidomain protein kinase containing an ankyrin-like– and a leucine-rich–repeat region, but also a Rab-like GTPase domain, a tyrosine-kinase-like domain and a WD-like domain in LRRK2. This enzyme regulates a variety of cellular processes critical for homeostasis and cell survival. LRRK2 not only controls the neurite morphology (Civiero et al., 2015; MacLeod et al., 2006) and regulates synaptic vesicle recycling/endocytosis (Arranz et al., 2015), but also plays an important role in dopamine receptor trafficking (Rassu et al., 2017).

Furthermore, LRRK2 is also linked with the intertwined pathways regulating inflammation, protein degradation, mitochondrial- and autophagy/lysosomal functions (Manzoni and Lewis, 2017). The deficiency of LRRK2 selectively modulates insulin-dependent intracellular signaling. After stimulation with insulin, the rapid intracellular translocation of GLUT4 to the cell surface fails in LRRK2-deficient cells from 6-month-old animals. This malfunction is accompanied by slight elevation of protein kinase B (PKB, Akt) phosphorylation. Furthermore, this defect is restored during aging by prolonged insulin-dependent activation of Akt and Akt substrate of 160 kDa (AS160/TBC1D4), and is compensated by elevated basal levels of GLUT4 on the plasma membrane. Dysregulation of LRRK2 in fibroblasts from PD patients results in dynamic insulin-triggered changes associated with phosphorylation of Rab10 supporting the view that in PD, LRRK2 plays a crucial role in insulin-driven translocation and/or fusion of GLUT4-vesicles to the plasma membrane through the phosphorylation of Rab10 (Funk et al., 2019).

Sporadic Parkinson's disease

The pathogenesis of sporadic PD involves oxidative stress, excitotoxicity, mitochondrial dysfunction, energy failure, neuroinflammation, misfolding and aggregation of α-synuclein, impairment of protein clearance pathways, cell-autonomous mechanisms, and deficits in proteasomal function or autophagy-lysosomal degradation of defective proteins (e.g., α-synuclein) (Fig. 7.3) (Alexander, 2004; Davie, 2008; Michel et al., 2016; Si et al., 2017; Maiti et al., 2017; Franco-Iborra et al., 2016; Truban et al., 2017; Moors et al., 2016). α-Synuclein is encoded by the α-synuclein gene, located on the long arm of chromosome 4 in humans (4q21.3−q22) (Chen et al., 1995). Full-length α-synuclein is a 140 amino acid protein, however, alternative splicing in exons 3 and 5 can result in 126, 112, or 98 amino acid isoforms (Beyer et al., 2006). α-Synuclein is primarily localized at the presynaptic terminals of neurons (Iwai et al., 1995). This protein lacks both cysteine and tryptophan residues. α-Synuclein is present in high concentration at presynaptic terminals and is found in both soluble- and membrane-associated fractions of the brain (Lee et al., 2002). It has two closely related homologs, β-synuclein and γ-synuclein. The N-terminus of α-synuclein is characterized by the presence of 11 amino acid repeats containing a KTKEGV consensus sequence (Jakes et al., 1994). This sequence is highly conserved between species

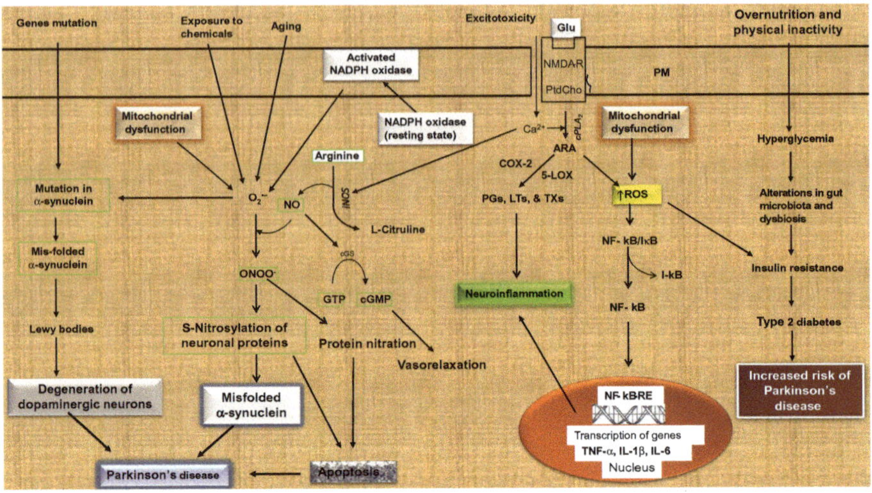

Figure 7.3 Hypothetical diagram showing molecular mechanisms contributing to the pathogenesis of sporadic PD.
ARA, Arachidonic acid; *COX-2*, cyclooxygenase-2; *cPLA*$_2$, cytosolic phospholipase A_2; *Glu*, glutamate; *iNOS*, inducible nitric oxide synthase; *LOX*, lipoxygenase; *LTs*, leukotriens; *NMDA-R*, *N*-methyl-D-aspartate receptor; *NO*·, nitric oxide; $O_2 \cdot^-$, superoxide; $ONOO^-$, peroxinitrite; *PGs*, prostaglandin; *PtdCho*, phosphatidylcholine; *TXs*, thromboxanes.

and within the synuclein family itself (Surguchov, 2008) and predicts an α helix secondary structure (Bussell and Eliezer, 2003). The central portion of α-synuclein is highly hydrophobic and is thought to underlie the aggregate prone nature of the protein. Under physiological conditions, α-synuclein functions in its native conformation as a soluble monomer. However, in PD, the accumulation of α-synuclein leads to it oligomerization. The oligomers of α-synuclein ultimately turn into insoluble α-synuclein aggregates (Uversky and Eliezer, 2009). The abnormal deposition of aggregated α-synuclein initially occurs in the olfactory bulb and dorsal motor nucleus of the vagus nerve in the medulla; thereafter, it spreads through less vulnerable nuclear gray and cortical areas in an ascending course (Braak et al., 2003). Investigators have also found α-synuclein aggregation in the ENS, occurring at the earliest stage of PD, or even preceding the onset of PD (Braak et al., 2006). The high correlation between α-synuclein burden and PD has led to the hypothesis that α-synuclein aggregation produces toxicity through a gain-of-function mechanism. However, α-synuclein has been implicated to function in a diverse range of essential cellular processes such as the regulation of neurotransmission and response to cellular stress. As such, an alternative hypothesis with equal explanatory power is that the aggregation of α-synuclein may result in toxicity because of a toxic loss of necessary α-synuclein function, following sequestration of functional forms α-synuclein into insoluble protein aggregates. Oligomers and protofibrils of α-synuclein have been identified to be the most toxic species, with their accumulation at presynaptic terminals affecting several steps of neurotransmitter release (Bridi and Hirth, 2018). In solution, α-synuclein is intrinsically disordered, but in the presence of lipid surfaces α-synuclein adopts a highly helical structure that is believed to mediate its normal function(s) (Fig. 7.4). The molecular basis of PD is tightly coupled to the aggregation of α-synuclein and the factors that affect its conformation. The aggregation of α-synuclein is a two-step process. Step 1 is initiated by a rate limiting nucleation phase in which soluble monomers associate into transient intermediate oligomers, which are built upon during the exponential elongation phase, producing primary filaments that are in turn integrated into fibrillary assembles (Morris et al., 2008). This process conforms to a generalized scheme of protein fibrillation established not only for α-synuclein (Fink, 2006), but also for

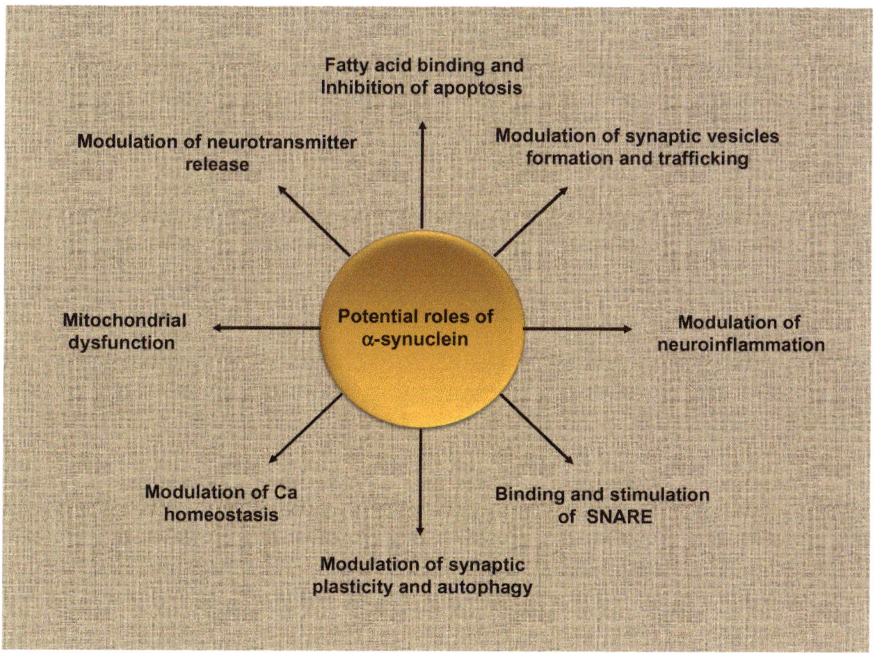

Figure 7.4 Functions of α-synuclein monomers in the brain.

other proteins such as tau (Kuret et al., 2005) or Aβ (Serpell, 2000). Step 2 involves the elongation. It requires small disordered oligomeric arrangements to adopt more stable ordered configuration, resistant to degradation and capable of promoting further fibrillation (Iljina et al., 2016). Recent studies have indicated that there is a homology between segments of α-synuclein and fatty acid-binding proteins, supporting the view that α-synuclein may act as a lipid-binding protein (Perrin et al., 2001). The binding of α-synuclein with polyunsaturated fatty acids (PUFAs), particularly with neuron-specific higher PUFAs (docosahexanoic acid) may contribute to its oligomerization and further aggregation (De et al., 2011), resulting in increased cytotoxicity. α-Synuclein aggregates may also interact with vesicular transport, and may embed themselves into lipid membranes forming pores, and affecting membrane permeability (Fecchio et al., 2013). Also, α-synuclein may also interact with neural membranes, fatty acids, and intracellular lipid droplets (Ruiperez et al., 2010). But rather than a fatty acid carrier, it is the binding to neural membranes that seems to be the intrinsic property of α-synuclein (Lucke et al., 2006). Interestingly, α-synuclein oligomers are formed upon interaction with peroxidation-prone PUFAs but not with monounsaturated fatty acids, while nonsaturated fatty acids actually reduce the level of oligomerization (Sharon et al., 2003). Based on above finding and the fact that binding of lipid peroxidation-promoting transition metals such as iron or copper exacerbates oligomerization of α-synuclein (Amer et al., 2006). It is suggested that α-synuclein may also have a specific affinity for peroxidized lipids. Thus monomeric α-synuclein may increase resistance to apoptosis, while α-synuclein oligomers have the opposite effect (Ruiperez et al., 2010). While lipid peroxidation features prominently in both apoptosis and α-synuclein-associated pathology, the detailed pro- and antiapoptotic properties of α-synuclein and its oligomeric forms, as well as their complex interactions with PUFAs in mitochondria membranes await more studies on this important topic (Ruiperez et al., 2010).

Misfolded α-synuclein not only undergoes phosphorylation, nitration, and truncatation, but also has abnormal solubility and has ability to prompt the production of oligomeric species, aggregates into fibrils, and is ubiquitinated (Hashimoto et al., 2004; Mukaetova-Ladinska and McKeith 2006). Other posttranslational modifications, such as ubiquitination (Rott et al., 2008), sumoylation (Dorval and Fraser, 2006; Krumova et al., 2011), glycation (Padmaraju et al., 2011), and glycosylation (Marotta et al., 2012), have also been reported. Collective evidence suggests that the etiology of both familial and sporadic PD cases is associated with the formation of insoluble α-synuclein aggregates, impaired mitochondrial function, increased oxidative stress, deficient antioxidant capacity, and induction of neuroinflammation (Perry et al., 2009; McCann et al., 2016). These findings are supported by postmortem studies of PD patients' brains. In addition, native α-synuclein plays an important role in the regulation of synaptic vesicle release and trafficking, maintenance of synaptic vesicle pools, fatty acid binding, neurotransmitter release, synaptic plasticity, and neuronal survival (Bridi and Hirth, 2018). As mentioned above, PD is accompanied by the accumulation and aggregation of α-synuclein, an amyloidogenic protein, which is the common feature of synucleinopathies.

Many studies have indicated that aggregated tau and α-synuclein co-occur in tauopathies and synucleinopathies, and in familial PDD (Badiola et al., 2011; Irwin et al., 2013; Rousseaux et al., 2016). Notably, LBs also occur in brains from AD patients (Moussaud et al., 2014). Furthermore, the presence of neurofibrillary tangles containing tau is observed in sporadic PD (Moussaud et al., 2014) and both tau and α-synuclein are enriched in synaptic fractions of brains affected by either tauopathy or synucleinopathy (Muntane et al., 2008). In vitro studies have shown that coincubation of tau and α-synuclein accelerates the fibrillization of both proteins (Giasson et al., 2003). Tau expression also enhances the toxicity and secretion of α-synuclein and promotes the formation of smaller α-synuclein inclusions in human neuroglioma (H4) cells and primary neuronal cultures (Badiola et al., 2011). These observations suggest that interactions between tau and α-synuclein interact may promote the formation of neuropathological lesions in the tauopathies and synucleinopathies (Zhang et al., 2018). Events that increase the interaction of tau with α-synuclein may also modulate the activity of protein kinases and other tau modifying enzymes; thereby further influencing tau pathology and disease progression (Moussaud et al., 2014; Wills et al., 2011; Zhang et al., 2018).

Gut microbiota, insulin resistance, and Parkinson's disease

As mentioned in Chapter 6, Insulin resistance and Alzheimer's disease, bidirectional communication occurs between the brain and the gut through the microbiome, which contains 10–100 trillion microbes (Bäckhed et al., 2005; Marsland and Gollwitzer, 2014). These microbes modulate brain function through immunological, neuroendocrine, and direct neural mechanisms. Dysregulation of the brain–gut–microbiota axis in PD may contribute to GI dysfunction, which is present in over 80% of PD subjects (Pfeiffer, 2011). These patients also show weight loss, dental deterioration, salivary excess, dysphagia, gastroparesis, and anorectal dysfunction (Pfeiffer, 2011; Liddle, 2018). Moreover, the dysregulation of brain–gut–microbiota axis and alterations in microbiota composition (dysbiosis) may also be associated with the pathogenesis of PD itself, supporting the view that the pathological process in PD spreads from the gut to the brain (Braak et al., 2006; Lebouvier et al., 2009). This suggestion is supported by following findings. Clinically, GI symptoms of PD often appear in patients before other neurological signs in the brain and aggregation of α-synuclein occurs in enteric nerves of PD patients (Liddle, 2018). Importantly, patients undergoing vagotomy have a reduced risk of developing PD. Furthermore, experimentally, aggregation of α-synuclein occurs in enteric nerves before it appears in the brain and injection of abnormal α-synuclein into the wall of the intestine spreads to the vagus nerve. Recent studies have indicated that ingestion of toxins and alterations in gut microbiota can promote the aggregation of α-synuclein aggregation in PD. However, it is not known how PD starts. Recent studies have indicated that sensory cells of the gut known as enteroendocrine cells (EECs) contain α-synuclein and synapse with enteric nerves, thus providing a connection from the gut to the brain. It is possible that abnormal α-synuclein first develops in EECs and spreads to the nervous system (Liddle, 2018). It is well known that Toll-like receptors (TLRs) are important pattern recognition receptors, which play a crucial role in innate immunity by recognizing conserved motifs primarily found in microorganisms and a dysregulation in their signaling may be implicated in PD (Caputi and Giron, 2018). Following TLR engagement by ligands of microbial origin, dendritic cells undergo maturation, which in turn activates the adaptive immune response. Pathogen-associated molecular patterns recognized by TLRs can come from bacteria, fungi, protozoans, insects, or viruses. Among TLRs, TLR2 and TLR4 recognizes lipopolysaccharides (LPS) (Poltorak et al., 1998). On the basis of several studies, it is suggested that an overstimulation of the innate immune system due to gut dysbiosis may provoke local and systemic inflammation and enteric neuroglial activation promote the development of α-synuclein pathology.

Involvement of TLR2 and TLR4 in the development of inflammatory processes in PD brains has been recently reported (Dzamko et al., 2017;

Hughes et al., 2019). The activation of TLR2 and TLR4 contributes to the recruitment of myeloid differentiation primary response gene 88 (MyD88) to its cytosolic toll/IL-1R (TIR) domain via the adaptor protein MyD88 adaptor-like (Mal). Activation of TLR2 and TLR4 also promotes the recruitment of a second-signaling adaptor protein, TIR domain-containing adaptor-inducing interferon-beta (TRIF), via the TRIF-related adaptor molecule (Fig. 7.5). It is now clear that oligomeric proteins such as α-synuclein are recognized by TLRs 2 and 4 (Dzamko et al., 2017; Codolo et al., 2013; Daniele et al., 2015; Kim et al., 2013; Rannikko et al., 2015). Studies on TLR2 expression in postmortem brain tissue from PD patients and matched controls have confirmed the involvement of TLR2 in the pathogenesis of PD patient's brain. It is also reported that increased levels of TLR2 correlate with the accumulation of α-synuclein. Expression of TLR2 occurs both in neurons as well as microglial cells. However, the increased expression of neuronal TLR2 rather than glial cells contribute to PD (Dzamko et al., 2017). TLR2 is strongly colocalized to α-synuclein positive LBs. In cell culture, activation of neuronal TLR2 is coupled with inflammatory responses, including the secretion of inflammatory cytokines [tumor necrosis factor alpha (TNF-α), interleukin 1β (IL-1β), and IL-6] and microglial-activating chemokines [monocyte chemoattractant protein-1 (MCP-1)], as well as the production of reactive oxygen species (ROS). Furthermore, activation of neuronal TLR2 is also linked with increased levels of endogenous α-synuclein protein, which in turn is associated with increased levels of the autophagy/lysosomal pathway marker p62. Finally, promoting autophagy with rapamycin or pharmacological inhibition of the TLR2 signaling pathway blocks the TLR2-mediated increase in α-synuclein in neuronal cell cultures. These results support the view that neuronal TLR2 expression may be involved in the pathogenesis of human PD (Dzamko et al., 2017). PD animal model studies have also indicated that microglial TLR2 is closely associated with inflammatory processes and progression of PD. It is proposed that aggregated α-synuclein, released by neurons, activates TLR2/TLR4 on microglia cells to produce proinflammatory mediators, such as IL-1β. IL-1β is one of the strongest proinflammatory cytokines. IL-1β is produced as an inactive mediator, and its maturation and activation require inflammasome formation (Codolo et al., 2013). This process not only activates microglia, but also contributes to increased ROS production and cathepsin B release into the cytosol. Thus excessive microglia activation may lead to synaptic loss and neuronal dysfunction. This process involves the stimulation of

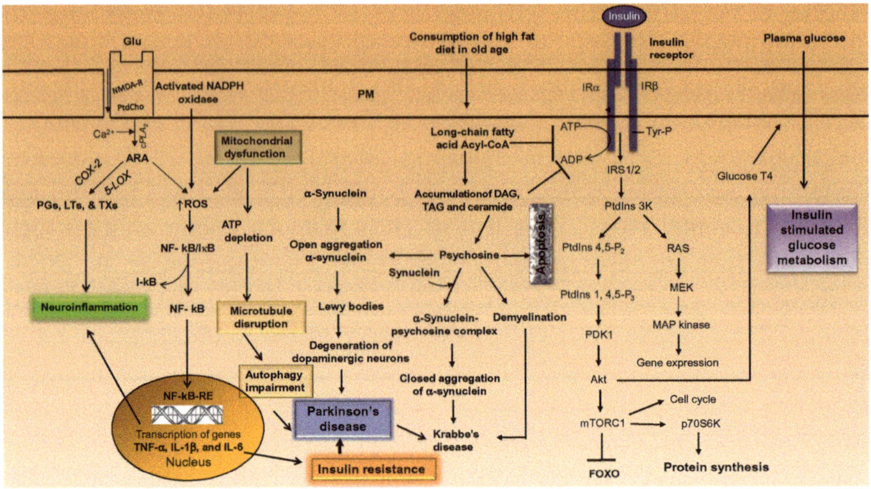

Figure 7.5 Hypothetical diagram showing the involvement of insulin signaling, phospholipid, sphingolopid-derived lipid mediators, and PtdIns 3K/AKT/mT0R signaling pathway in the pathogenesis of Parkinson's disease.
Akt, Protein kinase B; *FOXO,* Forkhead box O proteins; *GSK3,* glycogen synthase kinase 3; *IRS1/2,* insulin receptor substrate-1/2; *mTOR,* mammalian target of rapamycin; *PDK1,* phosphoinositide-dependent protein kinase 1; *PtdIns 3K,* phosphatidylinositiol 3-kinase; *PtdIns 4,5-P$_2$,* phosphatidylinositol-4,5-biphosphate; *PtdIns 3,4,5-P$_3$,* phosphatidylinositol-3,4,5-trisphosphate; *RAS,* GTPase.

TLR4 receptor. The stimulation of TLR4 receptor results in activation of Akt. This process precedes nuclear factor kappa B (NF-κB)-dependent transcription of proinflammatory genes in activated microglia (Saponaro et al., 2012). Remarkably, recent studies have shown that not only glial cells contribute to neuroinflammation, but stimulation of hippocampal neurons also promote and support neuroinflammation by releasing TNF-α and IL-1β via TLR4-mediated PtdIns 3K/AKT/NF-κB signaling (Zhao et al., 2014). Collectively, these finding support the notion that neuronal degeneration in PD may contribute to the release of fibrillar α-synuclein. This process likely acts as an endogenous trigger for inducing a strong inflammatory response in PD.

Pathologically, aggregated α-synuclein contributes to mitochondrial dysfunction by producing imbalance between mitochondria and α-synuclein. This process promotes the onset of PD by inducing neuronal impairment (Fujita et al., 2012; Faustini et al., 2017). Signal transduction processes contributing neuronal impairment are not clearly understood. However, Bose and Beal (2016) have proposed a new signaling pathways.

This pathway is called as retromer-trafficking pathway for the pathogenesis of PD. It not only involves the bioenergetic defects, mutations in mitochondrial DNA, nuclear DNA gene mutations, alterations in mitochondrial dynamics, and alterations in trafficking/transport, but also mitochondrial movement, impairment of transcription and the presence of mutated proteins associated with mitochondria (Bose and Beal, 2016). It is also reported that the aggregated α-synuclein may contribute to neurotoxicity through a gain-of-function mechanism (Benskey et al., 2016; Fujita et al., 2012). It is well known that α-synuclein is involved in a diverse range of essential cellular processes including the regulation of neurotransmission and response to cellular stress. Another alternative hypothesis is that the aggregation of α-synuclein results in toxicity because of a toxic loss of necessary α-synuclein functions leading to degeneration. The possibility that presynaptic aggregated α-synuclein interferes with the release of neurotransmitter is supported by the observation that neurochemical changes in C57/Bl6 mice brain slices are closely associated not only with the depletion of dopamine, but also with progressive impairments in neuronal excitability and connectivity. These changes lead to profound loss of dendritic spines (Day et al., 2006). The imbalance of dendritic spine changes in relation to the relative preservation of presynaptic terminals may be explained by the finding that the bidirectional synaptic plasticity is based on the morphological plasticity of the dendritic spines (Nagerl et al., 2004). This link between α-synuclein aggregation, synaptic pathology, and mitochondrial dysfunction paves the way toward explaining the clinical symptoms of PD. It also serves the basis for understanding the effect of L-DOPA therapy at the beginning of symptoms and its failure later in the disease process. In PD, the induction of oxidative stress not only produce nuclear membrane modifications, but also promote the translocation of α-synuclein from cytoplasm to the nucleus, where it can form complexes with histones leading to its oligomerization into insoluble fibrils. As mentioned above, aggregation and high levels of α-synuclein promote oxidant production or increase the level of oxidative stress. Within cells, α-synuclein normally adopts an α-helical conformation. However, during high levels of oxidative stress, α-synuclein undergoes a profound conformational transition to a β-sheet-rich structure that polymerizes to form toxic oligomers. Involvement of soluble oligomeric and protofibrillar forms of α-synuclein aggregates in the pathogenesis of PD is not only supported by the consistent detection of α-synuclein deposits in affected brain areas, but also by pathogenic mutations affecting the α-synuclein gene in familial PD and association of the α-synuclein locus with idiopathic PD in genome-wide association studies. Furthermore, in vitro studies on cell culture systems and animal models also support association of α-synuclein with PD (Simón-Sánchez et al., 2009). Recent studies on neurodegenerative potency of α-synuclein fibrils have indicated that toxicity of α-synuclein fibrils may be due to its ability to penetrate neural cell membranes (Volles et al., 2001; Pieri et al., 2012). Thus compounds that inhibit α-synuclein aggregation and fibrillization and stabilize it in a nontoxic state can therefore serve as therapeutic molecules for both prevention of accumulation of aggregated α-synuclein and maintenance of normal physiological concentrations of α-synuclein (Li et al., 2004).

Insulin resistance, type 2 diabetes, and Parkinson's disease

As mentioned in Chapter 1, Insulin resistance and obesity, the action of insulin in the brain includes food intake regulation, feeding behavior, body weight, and energy homeostasis (Blázquez et al., 2014; Gray et al., 2014). These effects may be mediated by two major components of the brain insulin transduction systems: phosphatidylinositol-3-kinase (PtdIns 3K)/Akt pathway (Fig. 7.6) and mitogen-activated protein kinases/Ras pathway (MAPKs/Ras) (Blázquez et al., 2014; Kleinridders et al., 2014). PtdIns 3K/Akt pathway phosphorylates many proteins such as glucose transporter 4 (GLUT4), glycogen synthase kinase 3 (GSK3), forkhead transcription factor (FOXO), and endothelial nitric oxide synthase (eNOS). Phosphorylation GLUT4 by Akt facilitates translocation of GLUT4 from cytoplasmic storage vesicles to the plasma membrane of adipocytes and skeletal muscle cells. Akt-mediated phosphorylation of GSK3 results in the activation of glycogen synthase and enhanced glycogen synthesis, while Akt-mediated phosphorylation of the FOXO 1 prevents translocation of the latter to the nucleus and inhibits expression of enzymes responsible for hepatocyte gluconeogenesis and glycogenolysis (Siddle, 2011). In endothelial cells, Akt phosphorylates and activates eNOS, leading to NOS (Wang et al., 2013). It is important to note that some of actions of insulin are specific for the brain. For example, insulin contributes to neuronal survival, participates in synaptic plasticity, and regulates the brain functioning including memory, cognition, learning, as well as attention through the regulation of neurotransmitters and modulation of long-term potentiation (LTP) and long-term depression (LTD)

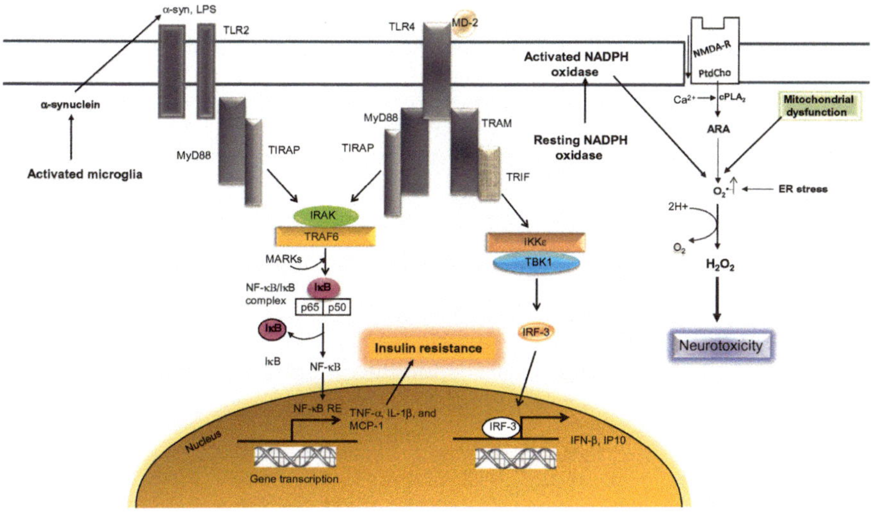

Figure 7.6 Hypothetical diagram showing the contribution of TLR receptors in the development of neuroinflammation in Parkinson's disease.
IFN-β, Interferon-beta; *I-κB*, inhibitory subunit of NF-κB; *IKK*, IκB kinase; *IP-10*, IFN inducible protein of 10 kDa; *IRF-3*, interferon regulatory transcription factor-3; *MCP1*, monocyte chemotactic protein-1; *MyD88*, adaptor protein; *NIK*, NF-κB-inducing kinase; *IRAK*, IL-1R-associated kinase; *TLR2 and TLR4*, Toll-like receptors 2 and 4; *TRAF6*, TNF receptor-associated factor adaptor protein 6; *TRIF*, TIR-domain-containing adapter-inducing interferon-β.

(Blázquez et al., 2014). Insulin resistance is defined as a condition in which insulin's target organs are resistant to its action, so that higher concentrations of this hormone are needed to obtain a normal biological effect. Thus hyperinsulinemia is an obvious consequence of insulin resistance. It contributes to the development of endothelial dysfunction, playing a key role in the establishment and progression of atherosclerosis (Dedoussis et al., 2007). Insulin resistance increases with age, and the organism maintains normal glucose levels as long as it can produce enough insulin. The molecular mechanism associated with insulin resistance is not fully understood. However, accumulation of lipids in the liver is considered to be one of the primary mechanisms involved in peripheral insulin resistance. In brain and visceral systems, insulin resistance is also accompanied by elevation in free fatty acids, triacylglycerol, diacylglycerol, acylcarnitines, and ceramide (Maciejczyk et al., 2019). Furthermore, the increase in proinflammatory cytokines (TNF-α and IL-6) and chemokines such as MCP-1 may also contribute to insulin resistance (Maciejczyk et al., 2019).

Hyperinsulinemia, in association with increased circulating levels of fatty acids promote peripheral insulin resistance, a process, which is a major factor contributing in progression of type 2 diabetes, a pathological condition, which is accompanied by chronic hyperglycemia, dyslipidemia, and lipotoxicity. Type 2 diabetes results in progressive deterioration of insulin secretion and insulin action (Taylor, 2012; Yazıcı and Sezer, 2017). Many epidemiological studies have indicated that there is a relationship between type 2 diabetes and an increased risk of developing PD (Schernhammer et al., 2011; Xu et al., 2011). However, other studies indicate that there is no relationship between type 2 diabetes and PD (Driver et al., 2008; Lu et al., 2014; Green et al., 2019). Despite these conflicting reports, it is becoming increasingly evident that type 2 diabetes and PD may have common neurochemical mechanisms underlying their onset (Santiago and Potashkin, 2013; Bosco et al., 2012; Driver et al., 2008). Thus a large proportion of PD and type 2 diabetes patients show impaired insulin signaling, insulin resistance, mitochondrial dysfunction, autophagy, and inflammation (Santiago and Potashkin, 2013; Bosco et al., 2012). In addition to sharing above biochemical processes, study of whole-genome transcriptome profiling of the substantia nigra of PD patients also provide evidence for genetic links between PD and type 2 diabetes (Moran and Graeber, 2008). Thus in both pathological conditions are accompanied by alterations not only in dysregulated priority genes, but also in various "hub" genes with multiple interactions with other genes, including those genes, which encode for GSK3β, insulin-like growth factor-1 (IGF-1), and IGF receptor (Yang et al., 2018). Insulin signaling in the brain involves GSK3β, an enzyme, which is downstream substrate of PtdIns 3K/Akt signaling following interactions between insulin and IGF-1. Furthermore, the genetic overexpression of GSK3β in cortex and hippocampus results in signs of neurodegeneration and spatial learning deficits in in vivo models of PD (Lucas et al., 2001) supporting the view that abnormalities in insulin singling may contribute to the pathogenesis of PD in some patients. It is interesting to note that induction of insulin resistance has also been observed in many PD patients even in the absence of type 2 diabetes (Bosco et al., 2012; Ashraghi et al., 2016). In the presence of type 2 diabetes in otherwise normal elderly individuals, motor features of PD include gait disturbance and rigidity, but not tremor or bradykinesia (Arvanitakis et al., 2004). Comorbid type 2 diabetes may contribute to motor impairments in PD. Thus a case—control study of PD subjects with and without antecedent type 2 diabetes has indicated that PD subjects with diabetes exhibited higher motor scores and received higher doses of dopaminergic medications (Cereda et al., 2012). These findings indicate that type 2 diabetes may preferentially exacerbate axial motor impairments. The more intensive dopamine replacement therapy in type 2 diabetic PD patients suggests that type 2 diabetes may be associated with greater nigrostriatal dopaminergic denervation. Axial motor dysfunctions, however, are generally less responsive to dopamine replacement and considerable data suggests that extranigral pathologies underlie axial motor dysfunctions (Devos et al., 2010). The molecular mechanisms associated with these processes remain unclear. However, it is becoming increasingly evident that type 2 diabetes increases the risk of developing PD in multiple ways. For instance, loss of Akt signaling, which is one of the main downstream targets of the insulin signaling pathway. This pathway is associated with the pathogenesis of several age-related neurodegenerative disorders, including AD and PD (Yang et al., 2018; Sánchez-Alegría et al., 2018). Postmortem analysis has also shown that the brains of PD patients have reduced levels of both total and phosphorylated Akt and single nucleotide polymorphism of the Akt gene. In addition, loss of Akt signaling also leads to apoptotic cell death of dopaminergic neurons in PD patients. Furthermore, impaired insulin signaling negatively regulates lysosomal systems that are responsible for the degradation of structurally/functionally abnormal cellular components, leading to increased aggregation of α-synuclein, a protein which is involved in the pathogenesis of PD (Yang et al., 2018). Under normal physiological condition, activation of insulin-Akt signaling is associated with the inhibition of GSK-3β, which in turn triggers autophagy and reduces the aggregation of α-synuclein. However, in the case of PD, significantly higher levels of GSK-3β and α-synuclein aggregates contribute to the pathogenesis of PD (Yang et al., 2018). Moreover, in PD, insulin resistance has been shown to alter not only mitochondrial protein levels, but also activity of mitochondrial complex I in the substantia nigra compacta. These processes result in excessive free radical generation, increased oxidative stress, and cell death. In 6-hydroxydopamine–induced rat model of PD, the treatment with insulin not only normalizes the production and functionality of dopamine, but also ameliorates motor impairments (Pang et al., 2016) indicating a link between insulin signaling and type 2 diabetes. The positive effect of insulin on cognition is mediated by PtdIns 3K activation and the increase in local glucose metabolism (McNay et al., 2010). Similarly, it is also reported that neuroprotective action of insulin may also involve the insulin/PtdIns 3K/Akt/mTOR pathway. This pathway modulates neuronal plasticity and cognitive functions in the brain. In fact, recently studies have indicated that Akt3 knockout mice exhibit microcephaly, cognitive defects, and reduction in protein synthesis in response to LTP via the inactivation of mTOR and reduction in protein synthesis associated with sustain neuroplasticity changes in the brain (Zhang et al., 2018). Different isoforms of PtdIns 3K have been implicated in synaptic plasticity and cognitive functions. Genetic deletion or overexpression of PtdIns 3Kγ disrupts LTD and reduces spatial learning tasks, while contextual fear memory is not affected (Kim et al., 2011; Choi et al., 2014). Similarly, activation of PtdIns 3K in the amygdala is associated with fear conditioning (Lin et al., 2001). Furthermore, dysregulation in PtdIns 3K/Akt signaling in neurons also contributes to several harmful effects in neurons due to elevation in ROS levels, membrane depolarization, fragmentation of mitochondria, and decrease in oxidative phosphorylation and ATP production (Liu et al., 2014; Kim et al., 2016). It is particularly interesting in relation to the involvement of amyloid peptide in the pathogenesis of AD. This peptide produces sustained activation of Akt, which in turn phosphorylates the mitochondrial fission protein Drp1. This mechanism has been proposed to be involved in the mitochondrial fragmentation observed in this neurodegenerative disease (Liu et al., 2014; Kim et al., 2016). Neuroprotective actions of insulin have also been reported in a variety of neurodegenerating insults, including glucose-oxygen deprivation (Sun et al., 2010), excitotoxicity (Kim and Han, 2005), and oxidative stress (Ribeiro et al., 2014).

In the PD brain, the activation of microglial cells is associated with immunological responses. Thus excessive microglia activation may lead to synaptic loss and neuronal dysfunction. This process involves the stimulation of TLR4 receptor. The stimulation of TLR4 receptor results in activation of Akt. This process precedes NF-κB–dependent transcription of proinflammatory genes in activated microglia (Saponaro et al., 2012). Remarkably, recent studies have shown that not only glial cells contribute to neuroinflammation, but hippocampal neurons can also contribute to neuroinflammation by releasing TNF-α and IL-1β via TLR4-mediated PtdIns 3K/AKT/NF-κB signaling (Zhao et al., 2014). Collective evidence suggests that the pathogenesis of type 2 diabetes and PD involves induction of oxidative stress and neuroinflammation. Oxidative stress is caused by a major increase in the amount of oxidized cellular components such as ROS, which include superoxide anions ($O_2\cdot^-$), hydroxyl ($^\bullet OH$), alkoxyl

($RO\cdot^{-}$), peroxyl radicals ($ROO\cdot$), and hydrogen peroxide (H_2O_2) (Sun et al., 2007; Farooqui, 2012, 2014). The major sources of ROS include mitochondrial respiratory chain dysfunction, activation of nicotinamide adenine dinucleotide phosphate (NADPH) oxidase in plasma membranes, and oxidation of arachidonic acid by cyclooxygenase and lipoxygenase in cytoplasm (Fig. 7.7) (Farooqui, 2014). $O_2\cdot^{-}$ are readily transformed by oxidoreduction reactions with transition metals or other redox cycling compounds into more aggressive radical species ($OH^{\bullet}$ and H_2O_2) (Hancock et al., 2001; Beal, 2005). ROS not only inactivate membrane proteins and DNA, but also promote peroxidation of neural membrane PUFAs associated with glycerophospholipids, enhance levels of ceramide and facilitate the formation of hydroxyl/ketocholesterol levels (Farooqui, 2012). Furthermore, in the brain, high ROS levels also retard induction of LTP, inhibit synaptic signaling, and decrease brain plasticity (Knapp and Klann, 2002; Forman et al., 2004) supporting the view that induction of high oxidative stress is hazardous to normal functioning of the brain. Induction of oxidative stress results in loss of cellular function. ROS modulate brain function by acting on several transcription factors including Forkhead homebox type O (FOXO) (Lu et al., 2012), nuclear factor E2-related factor 2 (Nrf2) (Ma, 2013), tumor suppressor 53 (p53) (He and Simon, 2013), NF-κB

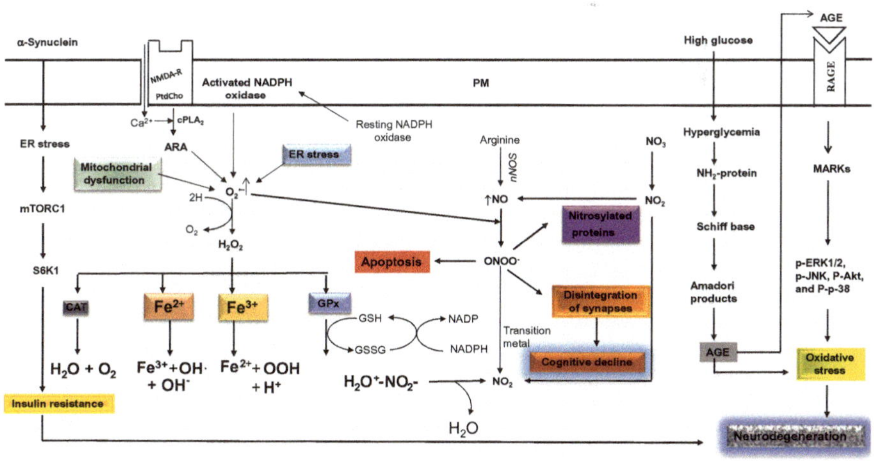

Figure 7.7 Hypothetical diagram showing the contribution of iron, and RAGE in the pathogenesis of sporadic Parkinson's disease.
AGEs, Advanced glycated products; *CAT*, catalase; *Fe^{3+}*, ferric ions; *Fe^{2+}*, ferrous; *GPx*, glutathione peroxidase; *GSH*, glutathione; *RAGEs*, receptor for advanced glycated products.

(Narayanan et al., 2014), activator protein-1 (Riera et al., 2015), and hypoxia inducible factor 1-a (Movafagh et al., 2015). FOXOs are viewed as molecular sensors for oxidative stress since their activity is regulated by H_2O_2. Once activated, FOXOs relay the stress signals to induce apoptosis, stress resistance, and senescence (Storz, 2011). Activation of NF-κB via oxidation-induced degradation of the inhibitory subunit IκB results in the activation of several antioxidant defence-related genes such as *Nrf2, HO-1* (Kostyuk et al., 2018). NF-κB also regulates the expression of genes regulating immune responses, such as IL-1β, IL-6, TNF-α, IL-8, and several adhesion molecules (Blaser et al., 2016). ROS-mediated lipid peroxidation disrupts the membrane lipid bilayer arrangement. This results in inactivation of membrane-bound receptors and enzymes leading to increased tissue permeability (Angelova et al., 2015). Oxidative modifications of the DNA include degradation of bases, breaks in the DNA strands, mutations, translocations, and abnormal cross-linking with proteins (Birben et al., 2012). Two neuroprotective mechanisms operate in the brain to tackle the threat posed by ROS: (1) the antioxidant enzyme systems such as catalase (CAT) and glutathione peroxidase and (2) the low-molecular-weight antioxidants such as vitamin E, melatonin (Kohen et al., 1999). The antioxidant enzyme system includes superoxide dismutase (SOD), glutathione reductase, glutathione peroxidase, and CAT (Griendling et al., 2000). The low-molecular-weight antioxidants include glutathione, uric acid, ascorbic acid, and melatonin, which offer neutralizing functions by causing chelation of transition metals (Halliwell, 2006).

Chronic inflammation in PD is characterized by marked increase in the generation of proinflammatory eicosanoids such as prostaglandins, leukotrienes, and thromboxanes and induction of long-standing chronic activation of microglia and astrocytes in the brain and macrophages in the visceral organs. This process results in sustained activation of NF-κB and increased expression of inflammatory cytokines and chemokines. The sustained expression and release of these inflammatory mediators causes an imbalance in the inflammatory cycle homeostasis by activating additional microglia in the brain and macrophages in the visceral tissues, promoting further release of inflammatory cytokines, such as TNF-α, IL-1β, and IL-6 (Farooqui, 2014). Induction of inflammation upregulates cerebrovascular pathology through the induction of proinflammatory cytokines and generation of nitric oxide (NO). Under physiological conditions in cardiovascular and cerebrovascular systems, NO is produced in a controlled manner and regulates physiological processes in the brain such as cerebral blood flow, consolidation of memory, modulation of LTP, and maintenance of sleep–wake cycles (Virarkar et al., 2013). NO is rapidly removed by diffusion through tissues into red blood cells, where it is rapidly converted to nitrate by reaction with oxyhemoglobin (Pacher et al., 2007). Under pathological conditions, such as PD and diabetes, NO reacts with superoxide anions to produce highly reactive peroxynitrite ($ONOO^-$), which at high concentration produces NO-mediated damage to biological molecules through the induction of nitrosative stress (Fig. 7.7) (Farooqui, 2014). Thus the direct reaction of $ONOO^-$ with transition metal centers is among the fastest reaction that occurs in vivo. $ONOO^-$ modifies proteins containing a heme prosthetic group, such as hemoglobin, myoglobin, and cytochrome *c*, oxidizing ferrous heme into the corresponding ferric forms (Pacher et al., 2007; Farooqui, 2014). Nitrosative stress-mediated neuroinflammation promotes long-term damage involving fatty acids, proteins, DNA, and mitochondria; these amplify and perpetuate several feedforward and feedback pathological loops. The latter includes dysfunctional energy metabolism (compromised mitochondrial ATP production), α-synuclein generation, endothelial dysfunction, and blood–brain barrier disruption.

These processes lead to decrease in cerebral blood flow and chronic cerebral hypoperfusion that may modulate metabolic dysfunction and induce neurodegeneration. In essence, hypoperfusion deprives the brain from its two paramount trophic substances, viz, oxygen and nutrients. Consequently, the brain suffers from synaptic dysfunction and neuronal degeneration/loss, leading to both gray and white matter atrophy, cognitive dysfunction, and PD (Farooqui, 2014). In type 2 diabetes and PD, the induction of oxidative stress and chronic inflammation in the brain promotes neurodegeneration through apoptotic cell death. In contrast, in visceral organs, induction of oxidative stress and chronic inflammation contributes to cell death through insulin resistance-mediated cell death (Farooqui, 2014). Collective evidence suggests that type 2 diabetes mellitus and PD are metabolic and neurodegenerative disorders, respectively. As mentioned above, these conditions are also supported by a progressive decrease in insulin action and gradual development in chronic hyperglycemia, and mitochondrial dysfunction.

Advanced glycation end products (AGEs) are formed when proteins or lipids become glycated after exposure to sugars. The formation of AGEs promotes the deposition of proteins due to the protease resistant cross-linking between the peptides and proteins. In peripheral tissues, hyperglycemia may result in nonenzymic glycation of lysine residues in serum proteins (albumin, hemoglobin, and LDL) and in the vessel wall proteins (collagen and fibronectin). In contrast, in the brain, hyperglycemia may lead to AGEs-mediated structural and functional changes due to insufficient endogenous scavengers and quality control systems. On the basis of many studies, it is proposed that AGEs impair neural cell signaling not only through direct covalent cross-linking of AGEs with various domains of its receptors, but also by interfering signal transduction processes modulated by AGE receptors (RAGEs). These receptors are found on macrophages, vascular endothelial cells, vascular smooth muscle cells, neurons, astrocytes, and microglial cells. RAGEs modulate many signal transductions pathways associated with generation of more oxidative stress and inflammatory events. In addition to AGE, RAGE is also activated by S100 and Aβ. The binding of these ligands with RAGE results in a sustained activation and migration of NF-κB to the nucleus, where its interaction with NF-κB response element promote the expression of proinflammatory cytokines (TNF-α, IL-1β, and IL-6), intercellular adhesion molecule-1, and vascular cell adhesion molecule-1 (Farooqui, 2010). RAGE expression has been found to be increased in incidental LB disease and is linked with oxidative stress, one of the main factors implicated in PD (Dalfó et al., 2005). During signal transduction process, the binding of AGE with RAGE stimulates activities of many enzymes including MAPKs, extracellular signal-regulated kinase 1/2 and p38, GTPases Cdc42 and Rac (Farooqui, 2010). In addition, the binding of AGEs with RAGE on endothelial cell surface also results in activation of NADPH oxidase leading to enhancement in the production of ROS (Sun et al., 2007). AGEs also interact with circulating proteins. These interactions vascular pathological changes in the vessel wall in visceral and brain tissues. Collective evidence suggests that AGEs generation is an irreversible process producing complications of PD and type 2 diabetes including nephropathy, neuropathy, retinopathy, and cardiovascular disease. Formation of AGEs in the brain accelerates the development of AD and PD (Farooqui, 2014).

Overlap between Parkinson's disease and Krabbe's disease

Aggregation of α-synuclein also occurs in KD or globoid cell leukodystrophy. KD is a neurological disorder, which is characterized by the deficiency of galactosylceramide-beta-galactosidase and accumulation of galactosylsphingosine (psychosine). Very little is known about factors, which influence the self-association of α-synuclein into toxic or nontoxic forms. However, mounting evidence suggests a central role for lysosomal function and glycosphingolipids metabolism in aggregation of α-synuclein (Moors et al., 2016). These patients have increased risk for PD due to mutation in lysosomal enzyme galactosyl-ceramidase (GALC). Mutations in the GALC gene result in the accumulation of psychosine in KD. Accumulation of psychosine in KD leads to destruction of oligodendroglia, the cells which produce myelin (Spassieva and Bieberich, 2016). Psychosine is also a potent inhibitor of neuronal functions in KD (Castelvetri et al., 2011; Cantuti-Castelvetri et al., 2012). It facilitates the formation of insoluble α-synuclein aggregates in vitro (Smith et al., 2014). Aggregates of α-synuclein are found in neurons in the cortex, hippocampus, midbrain, striatum, and brainstem in the twitcher (TWI) mouse model of KD, as well as in the brain of affected human patients (Smith et al., 2014). Interactions between psychosine and α-synuclein are based on cationic group of psychosine and the negatively charged C-terminus of α-synuclein. These interactions shed light on the potential mechanism of the formation of thio-S reactive aggregates in the brain. Although the initial binding is of low affinity, its positive cooperative nature allows the formation of a stable complex. The amino group and the sugar moiety of psychosine are involved in binding to α-synuclein and in hydrophilic clustering of the sphingolipid on the protein (Fig. 7.6). The acyl chain of the lipid does not contact α-synuclein and is free to engage in interactions with other proteins and lipids. This suggests that the protein can bind psychosine either in solution or in the context of the membrane bilayer. These studies suggest that there is a lot of overlap between mechanisms of PD and KD. Similarities between the two neurological disorders, including the pattern of α-synuclein aggregation in the brain of the twitcher mouse (the authentic murine model of KD), changes to lipid membrane dynamics, and possible dysfunction in synaptic function and macroautophagy, underscore a link between KD and late-onset synucleinopathies (PD). Silent GALC mutations may even constitute a risk factor for the development of Parkinson's in certain patients. More studies are required to identify definitively any link between the two diseases. This information will be invaluable for not only developing novel therapeutic targets for the treatment of PD and KD, but also identifying new biomarkers (Marshall and Bongarzone, 2016).

Overlap between Parkinson's disease and Gaucher's disease

An association between β-glucocerebrosidase (*GBA1*) gene mutations and PD was first observed in 2004. It is the most common genetic mutation that contributes to PD and up to 10% of PD patients in the United States carry it. Certain populations, including those of Ashkenazi Jews, are more apt to have *GBA1* mutations. Gaucher's disease is an autosomal-recessive lyososomal storage disorder caused by a deficiency of β-glucocerebrosidase. This enzyme catalyzes the hydrolysis of D-glucosyl-*N*-acylsphingosine (glucocerebroside) to D-glucose and *N*-acylsphingosine with the accumulation of glucosylceramide. The deficiency of glucocerebrosidase in Gaucher's disease has been linked to the development of synucleinopathies, such as PD (Chu et al., 2009; Dehay et al., 2010; Grabowski, 2008; Rocha et al., 2015; Blanz and Saftig, 2016; Mazzulli et al., 2016; Stojkovska et al., 2018). Increased risk of developing PD has been observed in both Gaucher's disease patients and carriers. It is estimated that glucocerebrosidase deficiency (GBA1 mutations) confer a 20- to 30-fold increased risk for the development of PD, and that at least 7%–10% of PD patients have a GBA1 mutation (Mazzulli et al., 2016). To date, mutations in the *GBA1* gene constitute numerically the most important risk factor for PD. The type of PD associated with GBA1 mutations (PD-GBA1) is almost identical to idiopathic PD, except for a slightly younger age of onset and a tendency to more cognitive impairment. Importantly, the pathology of PD-GBA1 is identical to idiopathic PD, with nigral dopamine cell loss, LBs, and neurites containing α-synuclein (Migdalska-Richards and Schapira, 2016). The mechanism by which GBA1 mutations increase the risk for PD is still unknown. However, given that clinical manifestation and pathological findings in PD-GBA1 patients are almost identical to those in idiopathic PD individuals, it is likely that, as in idiopathic PD, α-synuclein accumulation, mitochondrial dysfunction, autophagic impairment, oxidative, and endoplasmic reticulum stress may contribute to the development and progression of PD-GBA1. Other possibilities include both loss of function and toxic gain-of-function of abnormal β-glucocerebrosidase may be important for a close relationship between β-glucocerebrosidase and α-synuclein (Swan and Saunders-Pullman, 2013). Furthermore, recent studies have indicated that glucosylceramide stabilizes toxic oligomeric forms of α-synuclein, which not only effect the activity of β-glucocerebrosidase activity, but partially block the release of newly synthesized β-glucocerebrosidase from the endoplasmic reticulum amplifying the pathological effects of α-synuclein and ultimately resulting in neuronal cell death. It is also proposed that deficiency of lysosomal β-glucocerebrosidase may also contribute to the disruptions in lipid metabolism, protein trafficking and impaired protein quality control and these processes may be an important link between PD and Gaucher's disease (Blanz and Saftig, 2016; Mazzulli et al., 2016; Stojkovska et al., 2018).

Studies on the use a small-molecule modulator for the treatment of Gaucher's disease in pluripotent stem cell have indicated that this treatment not only increases the activity of lysosomal β-glucocerebrosidase in lysosomal compartments and reduces α-synuclein levels, but also ameliorates downstream toxicity. Furthermore, this treatment decreases β-glucocerebrosidase substrates and clears the α-synuclein accumulation regardless of the disease-causing mutations (Mazzulli et al., 2016). Importantly, the reduction of α-synuclein is sufficient to reverse downstream cellular pathologies induced by α-synuclein, including perturbations in hydrolase maturation and lysosomal dysfunction. These results indicate that enhancement of a single lysosomal hydrolase, β-glucocerebrosidase, can effectively reduce α-synuclein and provide therapeutic benefit in human midbrain neurons supporting the view that β-glucocerebrosidase activators may prove beneficial as treatments for PD and related synucleinopathies (Mazzulli et al., 2016).

Biomarkers for Parkinson's disease

The neuropathological diagnosis of PD is based on the detection and quantification of LBs, which are enriched in α-synuclein (Beach et al., 2008, 2009; Wakabayashi et al., 2007). The number of LBs in patients with mild to moderate loss of neurons in the substantia nigra pars compacta is higher than in patients with severe neuronal depletion indicating that LB-containing neurons are degenerating (Wakabayashi et al., 2007). It is well known that nigrostriatal dopamine levels and 3,4-dihydroxyphenylacetic acid (DOPAC) are depleted in CSF from PD patients. Furthermore, PD is accompanied by substantially increased Cys-DA/DOPAC in CSF. It is proposed that Cys-DA/DOPAC ratio in CSF can be used as a biomarker for PD (Goldstein et al., 2016). Levels of α-Synuclein in CSF can be used as biomarkers detected in CSF and blood (Gao et al., 2015). To this end, a specific enzyme-linked immunosorbent assay procedure has been developed using oligomeric α-synuclein in CSF and brain tissue (Tokuda et al., 2010; Paleologou et al., 2009). Some investigators have used the presence of α-synuclein in CSF and blood along with neuroimaging data [Positron emission tomography (PET), SPECT, and MRI] to diagnose PD, but these diagnostic tests are quite costly and are not always available in many hospitals (Walker et al., 2015).

Oxidative stress in Parkinson's disease

It is well known that induction of oxidative stress occurs when the concentration of prooxidants exceeds the level of antioxidants in cells resulting in redox imbalance between prooxidants and antioxidants in favor of the former ones, leading to oxidation of lipids, proteins, and nucleic acid.

In PD, a number of sources, mechanisms, and factors control the generation of ROS including the metabolism of dopamine itself, mitochondrial dysfunction, levels of iron, calcium, presence of neuroinflammatory cells, and aging. Among factors, the relationship between calcium and iron is quite important. Dysregulation of iron levels in the brain promotes calcium dyshomeostasis and abnormal calcium signaling, whereas increased calcium levels enhance redox-active iron levels. Neurodegenerative disorders including PD are accompanied by calcium/iron dysregulation (Pchitskaya et al., 2018). PD promoting genes (DJ-1, PINK1, parkin, α-synuclein, and LRRK2) and their products modulate mitochondrial function in complex ways leading to induction of oxidative stress (Dias et al., 2013).

NO· is another free radical, which is synthesized by the action of NOS on arginine. NO· reacts with $O_2\cdot^-$ to form the neurotoxic peroxynitrite ($ONOO^-$) (Fig. 7.3) (Bal-Price et al., 2002). ROS/RNS play an important role in cell signaling through redox signaling. To maintain proper cellular homeostasis and normal neural cell function, a balance must occur between ROS/RNS production and oxygen consumption and NO·. It has been hypothesized that $ONOO^-$ contribute to PINK1/Parkin-mediated mitophagy activation via triggering dynamin-related protein 1 (Drp1) recruitment leading to mitochondrial damage. Excessive ROS/RNS have to be either quenched by converting them into metabolically nondestructive molecules or be scavenged/neutralized right after their formation. This protective mechanism is called the antioxidant defence system preventing ROS/RNS mediated damage of cells leading to various diseases and aging (Yu, 1994; Winterbourn and Hampton, 2008). Another mechanism of redox signaling involves the glutathione thiol/disulfide redox couple (GSH/GSSG) is another predominant mechanism for maintaining the intracellular microenvironment in a highly reduced state that is essential for antioxidant/detoxification capacity, redox enzyme regulation, cell cycle progression, and transcription of antioxidant response elements (AREs) (Biswas et al., 2006; Fratelli et al., 2005).

$$2GSH + O_2 \rightarrow GSSG + 2H_2O_2$$

$$2GSH + 2H_2O_2 \rightarrow GSSG + 2H_2O \text{ (GSH peroxidase)}$$

$$2GSSG + NADH \rightarrow 2GSH + NADP \text{ (GSSG reductase)}$$

In response to oxidative and nitrosative stress, neural cells increase their antioxidant defenses through activation of nuclear factor erythroid 2—related factor (Nrf2), an important transcription factor (Maes et al., 2011). Nrf2 is a key component of this control system and recognizes the ARE found in the promoter regions of many genes that encode antioxidants and detoxification enzymes such as heme oxygenase 1 (HO-1), NAD(P)H dehydrogenase quinone 1, SOD1, glutathione peroxidase 1 (GPx1), and CAT (Itoh et al., 1997).

As mentioned above, deposition of misfolded α-synuclein, mitochondrial dysfunction and induction of oxidative stress are closely associated with the pathogenesis of PD (Blesa et al., 2015). Oxidative stress results in generation of higher levels of cholesterol hydroperoxide, MDA, 4-HNE, and OH8dG. One of the suggested cause of induction of oxidative stress in the substantia nigra pars compacta is the production of ROS during normal DA metabolism. In human substantia nigra pars compacta, the oxidation products of DA (mainly 6-hydroxydopamine) may polymerize to form neuromelanin, which may also be toxic by inducing apoptosis (Berman and Hastings, 1999). Furthermore, postmortem studies have indicated that decrease in GSH levels and increase in GSSG levels in the substantia nigra pars compacta. This can be a critical primary event that weakens or abrogates the natural antioxidant defence mechanisms of neural cells, thereby triggering degeneration of the nigral neurons and promoting the pathogenesis of PD (Gu et al., 2015). Since dysregulation of metal ion homeostasis is a potential catalyst to further production of ROS, the highly oxidative environment for DA interaction with α-synuclein, and the resulting oxidant-mediated toxicity and protein aggregation, is one of the most likely underlying mechanisms for PD. It is also proposed that neurodegeneration in PD may occur as a result of self-propagating reactions that involve not only DA, α-synuclein, mitochondrial dysfunction, but also involve the participation of redox-active metals (Carboni and Lingor, 2015). As mentioned above, PD inducing genes (DJ-1, PINK1, parkin, alpha-synuclein, and LRRK2) and their products also impact mitochondrial function leading to exacerbation of ROS generation and susceptibility to oxidative stress supporting the view that the interplay among various ROS producing mechanisms contributes to neurodegeneration in PD.

PDs are characterized by the abnormal accumulation or aggregation of proteins such as amyloid β, tau, α-synuclein, which become glycated and the extent of glycation is correlated with the pathologies of PD (Fig. 7.7) (Li et al., 2012). It is reported that AGE-mediated modification of glycated proteins triggers the sustain local oxidative stress and inflammatory response, eventually contributing to the pathological and clinical aspects of PD.

Neuroinflammation in Parkinson's disease

As mentioned above, neuroinflammation is closely associated with the pathogenesis of PD (Farooqui, 2014). Neuroinflammation is not only supported by the activation of microglia, astrocytes, neurons, but also by cells and humoral factors of the peripheral immune system, which are known to enter brain tissue (Phani et al., 2012; Farooqui, 2014). In PD, neuroinflammation is induced by not only increased synthesis of proinflammatory eicosanoids (PG, LT, and TX), but also by enhanced expression of TNF-α, IL-1β, IL-6, IL-8, IL-33, chemokine (C−C motif) ligand 2 (CCL2), chemokine (C−C motif) ligand 5 (CCL5), interferon-gamma (IFN-γ) in the midbrain, activation of matrix metalloproteinase, intracellular Ca^{2+} elevation, and activation of MAPKs and NF-kB (Farooqui, 2014; Kempuraj et al., 2016; Kempuraj et al., 2018). These data strongly suggest the involvement of immune components in PD pathogenesis. PET also indicate that there is pronounced activation of microglia in various regions of PD brain (Bartels et al., 2010). Moreover, activation of microglia in the substantia nigra compacta and striatum is profound in various types of PD animal models (Leal et al., 2013; Tansey et al., 2007). It is also proposed that in PD, the purpose of neuroinflammation is the removal of dead neurons to control the severity and progression of the diseases. Low intensity chronic neuroinflammation promotes neural cell homeostasis through the induction of resolution (Lucas et al., 2006). In contrast, induction of high intensity neuroinflammation contributes to neurodegeneration (Kielian, 2016). Furthermore, interplay between low levels of oxidative stress chronic neuroinflammation also promotes regeneration (Bollaerts et al., 2017) and in the presence of high oxidative stress, chronic neuroinflammation induces cytotoxic effects, which increase the severity of PD symptoms. Under physiological conditions, the quiescent state of microglia is maintained by a variety of immunomodulators, such as CX3CL1, CD200, CD22, CD47, CD95, and neural cell adhesion molecule, which are produced mainly by neuronal cells (Sheridan and Murphy, 2013). Interestingly, the receptors for these molecules are almost exclusively expressed by microglia in the CNS, indicating the critical role of neuron–microglia interactions in the regulation of neuroinflammation (Sheridan and Murphy, 2013). In addition to microglia, activation of astrocytes also contributes to neuroinflammation in PD. Like microglia, astrocytes respond to the inflammatory stimulations such as LPS, IL-1β, and TNF-α by producing proinflammatory cytokines both in vitro and in vivo (Tanaka et al., 2013). Reactive astrogliosis characterized by the increased expression levels of glial fibrillary acidic protein and hypertrophy of cell body and cell extensions have been reported in various PD animal models. Importantly, astrogliosis also exists in the affected brain regions of patients with PD, indicating the possible involvement of astrocytes in the immune processes in PD (Yamada et al., 1992). It is stated that astrocytic responses are relatively slower than microglial activation after stimulations. Microglia initiate the inflammatory responses after immune stimulations such as LPS treatment and α-synuclein aggregation. Astrocytes show increased expression of proinflammatory mediators and contribute to amplification of neuroinflammation (Fellner et al., 2013). Specific overexpression of mutant α-synuclein in astrocytes is associated with widespread astrogliosis, microglial activation, and degeneration of dopaminergic neurons and motor neurons in mice (Gu et al., 2010). Collective evidence suggests that in PD, the interactions between T and mast cell interaction with glial cells and neurons controls the intensity of neuroinflammation. In PD, the induction of chronic inflammatory processes is linked with the accumulation of misfolded α-synuclein, activation of microglia and astrocytes, neuronal loss, and cognitive dysfunction (Surendranathan et al., 2015; Streit and Xue, 2016; Farooqui, 2014; Kempuraj et al., 2016, 2018) along with a broad range of components of the innate and adaptive immune systems (Surendranathan et al., 2015). Evidence of neuroinflammation in PD is further supported by pathological and biomarker studies.

Furthermore, genetic and epidemiological studies also support a role for neuroinflammation in PD (Wang et al., 2015).

Depression and Parkinson's disease

PD patients develop depression compared to healthy individuals (Becker et al., 2011). Depressive symptoms affect 40%–50% of PD patients and significantly impact quality of life in PD (van Uem et al., 2016). In PD, depression is linked not only with inflammatory signaling (Politis et al., 2010), but also with a high availability of the serotonin transporter in the raphe nuclei and limbic regions of depressed PD patients (Boileau et al., 2008; Politis et al., 2010). Likewise, decrease in plasma levels of serotonin are found to be correlated with severity of depression (Tong et al., 2015). Depression correlates with a high serum level of IL-10 and IL-6 (Veselý et al., 2018). High levels of both sIL-2R and TNF-α in blood samples from PD patients were significantly associated with more severe depression and anxiety (Lindqvist et al., 2012, 2013). However, depression is not specific for PD. Chronic inflammation in physically ill patients is often associated with symptoms of depression and also occurs in normal aging (Chen et al., 2016; Brites and Fernandes, 2015). Moreover, PD in general is characterized by elevated levels of inflammatory cytokines, such as IL-6, TNF-α, IL-1β, IL-2, IL-10, C-reactive protein, and CCL5 (Qin et al., 2016).

Cognitive dysfunction in Parkinson's disease

Cognitive dysfunction is defined as the loss of intellectual functions such as thinking, remembering, and reasoning that interfere with daily activities. Human subjects with cognitive dysfunction loss their ability to learn, recall, concentrate, problem solving, and executive functions. Cognitive function is regulated by cerebrovascular aging, alterations in vascular tone (basal vessel diameter) and vascular reactivity (dynamic changes in vessel diameter), potential relationship between alterations in cerebral blood flow, and the deterioration of brain function, intricate synaptic changes, and neuronal and glial interactions (Morrison and Baxter, 2012). There is strong relationship between cellular metabolic capacity and regional cerebral blood flow leading to the conclusion that a clear understanding of age-related changes in the regulation of cerebral blood flow (including microvascular architecture, plasticity, and vessel reactivity) is essential for understanding the progressive decline in cellular metabolic activity and eventually tissue function with age (Sonntag et al., 2007). Furthermore, interactions and cross-talk between glial and neuronal metabolism also require an increase in cerebral blood flow, a process, which requires metabolic energy. It must be mentioned here that few studies have been performed on assessment of energy requirement of the dynamic processes, which modulate regional brain cerebral blood flow and cognitive function in aging human brain (Sonntag et al., 2007). Cognitive decline predisposes individuals to neurological and psychiatric disorders eventually affecting the quality of life. The intensity of cognitive decline is markedly increased in patients with neurological disorders such as stroke and neurodegenerative diseases (AD and PD) (Schuh et al., 2011; Farooqui, 2010). Cognitive deficits are common in PD patients. Cognitive function in PD deteriorates over time. Twenty-four percent of the PD patients have cognitive disturbances at onset of the disease and every second patient shows progressive cognitive decline in the first 3 years (Muslimovic et al., 2009). In the long-term follow-up studies, the cumulative prevalence rates of PD increase up to 80%. Therefore it seems that, regardless of the time of PD onset, the onset of PD occurs at around 70 years of age, and affects cognitive domains in a similar way (Reid et al., 2011). PD patients show cognitive deficits in attention, memory, visuospatial, and executive functions (Pagonabarraga and Kulisevsky, 2012). Neuroimaging studies have shown that in PD patients cognitive decline is linked with impaired corticostriatal connectivity. Thus impairment in corticostriatal connectivity contribute to decrease in integration among the striatum, mesolimbic cortex, and sensorimotor cortex. This process leads to mental "rigidity" in PD patients (Luo et al., 2014). Overall global cognitive performance in PD is shown to be associated with decreased functional connectivity in widespread regions including the paracentral lobe, superior parietal lobe, occipital regions, inferior frontal gyrus, and superior temporal gyrus (Olde et al., 2014). Decrease in functional connectivity in the frontoparietal network has been shown to be related to worsening executive function in PD with mild cognitive impairments (Amboni et al., 2014). Therefore it is speculated that decrease in connectivity not only occur in the frontostriatal region, but also in whole brain (Lin et al., 2018). This decrease in connectivity may contribute to cognitive deficits in PD. The cognitive decline in PD is not only accompanied by α-synuclein-mediated induction of oxidative stress and neuroinflammation, but also due to cortical thinning, hypometabolism, white matter changes, dopaminergic/cholinergic dysfunction, and increased α-synuclein burden (Jellinger, 2012; Hanganu et al., 2013). These alterations can not only cause destruction of essential neuronal networks, but also synapses rarefaction leading to cognitive dysfunction (Jellinger, 2012). These cognitive impairments are severe enough to impair their everyday functional abilities in PD. Furthermore, there is emerging evidence that healthy lifestyles along with optimal blood flow may decrease the rate of cognitive decline seen with aging and help delay the onset of cognitive symptoms in the setting of age-associated above diseases (Farooqui, 2012, 2018). Some investigators have implicated LBs in the neocortex, other investigators have pointed out that α-synuclein pathology in the hippocampus contribute to cognitive impairment and depression in PD (Yang and Yu, 2017).

In relation to insulin resistance and type 2 diabetes, PD patients also show microvascular changes leading to leukoariosis (white matter hyperintensities). A large recent study on cognitively asymptomatic elderly subjects has indicated that leukoariosis (white matter hyperintensities) is associated with executive function impairments, supporting the view that microvascular injury of subcortical white matter contributes to this aspect of cognitive impairment (Hedden et al., 2012). Similarly, an in vivo MRI study on leukoaraiosis do not differ between PD subjects with and without type 2 diabetes (Kotagal et al., 2013). Furthermore, more studies are needed to determine whether microstructural rather than macrostructural cerebral changes may play a significant role in the contribution of type 2 diabetes to the cognitive impairment syndrome in PD. Another recent MRI study in community-dwelling elderly has indicated not only a greater cerebral atrophy and reduced fractional anisotropy in the total white matter, but also greater mean diffusivity for the hippocampus and frontal cortex areas in type 2 diabetics subjects compared to nondiabetics (Falvey et al., 2013).

At the molecular level, vascular tone in cerebral vessels and cognitive decline are controlled by increase in cyclic GMP, generation of NO, immune response, mitochondrial dysfunction, decrease in growth factors (IGF), calcium homeostasis, induction of oxidative stress, and neuroinflammation (Sonntag et al., 2007; Loerch et al., 2008). However, how do abovementioned factors are linked together to modulate age-related alterations in cerebral blood flow, transport of nutrients from blood to brain, and cognitive dysfunction remain unknown.

Conclusion

PD is a heterogeneous neurodegenerative disease, which is typically diagnosed by its cardinal motor symptoms, including bradykinesia, hypokinesia, rigidity, resting tremor, and postural instability, which are subsumed under the syndrome of parkinsonism. The motor manifestations are attributable to the degeneration of dopaminergic neurons within the substantia nigra pars compacta, resulting in dopamine depletion and derangements of neuronal circuits in the basal ganglia target regions of these neurons. Another pathological hallmark of PD is the accumulation of misfolded and aggregated protein called α-synuclein-containing in neuronal perikaryal in the form of LBs and their processes (Lewy neurites). Nonmotor symptoms, such as autonomic dysfunction, sleep abnormalities, depression, and dementia, can contribute considerably to disability, as they usually are not responsive to dopamine replacement therapy. PD can also be triggered by genetic alterations, environmental/occupational exposures, and aging. However, the exact molecular mechanisms linking these PD risk factors to neurodegeneration and neuronal dysfunction are still unclear. However, it is becoming increasingly evident that induction of oxidative stress, excitotoxicity, mitochondrial dysfunction, energy failure, neuroinflammation, misfolding and aggregation of α-synuclein, impairment of protein clearance pathways, and deficits in proteasomal function or autophagy-lysosomal degradation of defective proteins (e.g., α-synuclein) are thought to be central components of neurodegeneration that contributes to the impairment of important homeostatic processes in dopaminergic cells. Detailed investigations have indicated that there is a homology between segments of α-synuclein and fatty acid–binding proteins, suggesting that α-synuclein may act as a lipid-binding protein. The binding of α-synuclein with docosahexanoic acid may contribute to its oligomerization and further aggregation, resulting in increased cytotoxicity. α-Synuclein aggregates may also interact with vesicular transport, and may embed themselves into lipid membranes forming pores, and affecting membrane permeability. Above mentioned moieties of α-synuclein are the most toxic and their accumulation at presynaptic terminals affects several steps in neurotransmitter release. α-Synuclein is expressed throughout the brain and enriched at presynaptic terminals. However, brain dopaminergic neurons are the most vulnerable structures in PD patients.

References

Alexander, G.E., 2004. Biology of Parkinson's disease: pathogenesis and pathophysiology of a multisystem neurodegenerative disorder. Dialogues Clin. Neurosci. 6, 259–280.

Amboni, M., Tessitore, A., Esposito, F., Santangelo, G., Picillo, M., Vitale, C., et al., 2014. Resting-state functional connectivity associated with mild cognitive impairment in Parkinson's disease. J. Neurol. 262, 425–434.

Amer, D.A., Irvine, G.B., El-Agnaf, O.M., 2006. Inhibitors of alpha-synuclein oligomerization and toxicity: a future therapeutic strategy for Parkinson's disease and related disorders. Exp. Brain Res. 173, 223–233.

Andersen, A.D., Binzer, M., Stenager, E., Gramsbergen, J.B., 2017. Cerebrospinal fluid biomarkers for Parkinson's disease—a systematic review. Acta Neurol. Scand. 135, 34–56.

Angelova, P.R., Horrocks, M.H., Klenerman, D., Gandhi, S., Abramov, A.Y., Shchepinov, M.S., 2015. Lipid peroxidation is essential for α-synuclein-induced cell death. J. Neurochem. 133, 582–589.

Arranz, A.M., Delbroek, L., Van Kolen, K., Guimarães, M.R., Mandemakers, W., Daneels, G., et al., 2015. LRRK2 functions in synaptic vesicle endocytosis through a kinase-dependent mechanism. J. Cell Sci. 128, 541–552.

Arvanitakis, Z., Wilson, R.S., Schneider, J.A., Bienias, J.L., Evans, D.A., Bennett, D.A., 2004. Diabetes mellitus and progression of rigidity and gait disturbance in older persons. Neurology 63, 996–1001.

Ashraghi, M.R., Pagano, G., Polychronis, S., Niccolini, F., Politis, M., 2016. Parkinson's disease, diabetes and cognitive impairment. Recent Pat. Endocr. Metab. Immune Drug Discov. 10, 11–21.

Bäckhed, F., Ley, R.E., Sonnenburg, J.L., Peterson, D.A., Gordon, J.I., 2005. Host-bacterial mutualism in the human intestine. Science 307, 1915–1920.

Badiola, N., de Oliveira, R.M., Herrera, F., Guardia-Laguarta, C., Gonçalves, S.A., Pera, M., et al., 2011. Tau enhances α-synuclein aggregation and toxicity in cellular models of synucleinopathy. PLoS One 6, e26609.

Bal-Price, A., Matthias, A., Brown, G.C., 2002. Stimulation of the NADPH oxidase in activated rat microglia removes nitric oxide but induces peroxynitrite production. J. Neurochem. 80, 73–80.

Bartels, A.L., Willemsen, A.T., Doorduin, J., de Vries, E.F., Dierckx, R.A., Leenders, K. L., 2010. [11C]-PK11195 PET: quantification of neuroinflammation and a monitor of anti-inflammatory treatment in Parkinson's disease? Parkinsonism Relat. Disord. 16, 57–59.

Beach, T.G., White, C.L., Hamilton, R.L., et al., 2008. Evaluation of alpha-synuclein immunohistochemical methods used by invited experts. Acta Neuropathol. 116, 277–288.

Beach, T.G., Adler, C.H., Lue, L., et al., 2009. Unified staging system for Lewy body disorders: correlation with nigrostriatal degeneration, cognitive impairment and motor dysfunction. Acta Neuropathol. 117, 613–634.

Beal, M.F., 2005. Mitochondria take center stage in aging and neurodegeneration. Ann. Neurol. 58, 495–505.

Becker, C., Brobert, G.P., Johansson, S., Jick, S.S., Meier, C.R., 2011. Risk of incident depression in patients with Parkinson disease in the UK. Eur. J. Neurol. 18, 448–453.

Benskey, M.J., Perez, R.G., Manfredsson, F.P., 2016. The contribution of alpha synuclein to neuronal survival and function—implications for Parkinson's disease. J. Neurochem. 137, 331–359.

Berman, S.B., Hastings, T.G., 1999. Dopamine oxidation alters mitochondrial respiration and induces permeability transition in brain mitochondria: implications for Parkinson's disease. J. Neurochem. 73, 1127–1137.

Beyer, K., Humbert, J., Ferrer, A., Lao, J.I., Carrato, C., et al., 2006. Low alpha-synuclein 126 mRNA levels in dementia with Lewy bodies and Alzheimer disease. Neuroreport 17, 1327–1330.

Birben, E., Sahiner, U.M., Sackesen, C., Erzurum, S., Kalayci, O., 2012. Oxidative stress and antioxidant defense. World Allergy Organ. J. 5, 9–19.

Biswas, S., Chida, A.S., Rahman, I., 2006. Redox modifications of protein-thiols: emerging roles in cell signaling. Biochem. Pharmacol. 71, 551–564.
Blanz, J., Saftig, P., 2016. Parkinson's disease: acid-glucocerebrosidase activity and alpha-synuclein clearance. J. Neurochem. 139 (Suppl. 1), 198–215.
Blaser, H., Dostert, C., Mak, T.W., Brenner, D., 2016. TNF and ROS crosstalk in inflammation. Trends Cell Biol. 26, 249–261.
Blázquez, E., Velázquez, E., Hurtado-Carneiro, V., Ruiz-Albusac, J.M., 2014. Insulin in the brain: its pathophysiological implications for states related with central insulin resistance, type 2 diabetes and alzheimer's disease. Front. Endocrinol. 5, 161.
Blesa, J., Trigo-Damas, I., Quiroga-Varela, A., Jackson-Lewis, V.R., 2015. Oxidative stress and Parkinson's disease. Front. Neuroanat. 9.
Boileau, I., Warsh, J.J., Guttman, M., Saint-Cyr, J.A., McCluskey, T., Rusjan, P., et al., 2008. Elevated serotonin transporter binding in depressed patients with Parkinson's disease: a preliminary PET study with [11C]DASB. Mov. Disord. 23, 1776–1780.
Bollaerts, I., Van Houcke, J., Andries, L., De Groef, L., Moons, L., 2017. Neuroinflammation as fuel for axonal regeneration in the injured vertebrate central nervous system. Mediators Inflamm. 9478542.
Bosco, D., Plastino, M., Cristiano, D., Colica, C., Ermio, C., et al., 2012. Dementia is associated with insulin resistance in patients with Parkinson's disease. J. Neurol. Sci. 315, 39–43.
Bose, A., Beal, M.F., 2016. Mitochondrial dysfunction in Parkinson's disease. J. Neurochem. 139 (Suppl. 1), 216–231.
Bove, J., Prou, D., Perier, C., Przedborski, S., 2005. Toxin-induced models of Parkinson's disease. NeuroRx. 2, 484–494.
Bower, J.E., 2007. Cancer-related fatigue: links with inflammation in cancer patients and survivors. Brain Behav. Immun. 21, 863–871.
Braak, H., Del Tredici, K., Rüb, U., de Vos, R.A., Jansen Steur, E.N., Braak, E., 2003. Staging of brain pathology related to sporadic parkinson's disease. Neurobiol. Aging 24, 197–211.
Braak, H., de Vos, R.A., Bohl, J., Del Tredici, K., 2006. Gastric alpha-synuclein immunoreactive inclusions in Meissner's and Auerbach's plexuses in cases staged for Parkinson's disease-related brain pathology. Neurosci. Lett. 396, 67–72.
Bridi, J.C., Hirth, F., 2018. Mechanisms of α-synuclein induced synaptopathy in Parkinson's disease. Front. Neurosci. 12, 80.
Brites, D., Fernandes, A., 2015. Neuroinflammation and depression: microglia activation, extracellular microvesicles and microRNA dysregulation. Front. Cell. Neurosci. 9, 476.
Bussell, R., Eliezer, D., 2003. A structural and functional role for 11-mer repeats in alpha-synuclein and other exchangeable lipid binding proteins. J. Mol. Biol. 329, 763–778.
Cannon, J.R., Greenamyre, J.T., 2011. The role of environmental exposures in neurodegeneration and neurodegenerative diseases. Toxicol. Sci. 124, 225–250.
Cantuti-Castelvetri, L., Zhu, H., Givogri, M.I., Chidavaenzi, R.L., Lopez-Rosas, A., Bongarzone, E.R., et al., 2012. Psychosine induces the dephosphorylation of neurofilaments by deregulation of PP1 and PP2A phosphatases. Neurobiol. Dis. 46, 325–335.
Caputi, V., Giron, M.C., 2018. Microbiome-gut-brain axis and toll-like receptors in Parkinson's disease. Int. J. Mol. Sci. 19 (6), pii: E1689.
Carboni, E., Lingor, P., 2015. Insights on the interaction of alpha-synuclein and metals in the pathophysiology of Parkinson's disease. Metallomics. 7, 395–404.
Castelvetri, L.C., Givogri, M.I., Zhu, H., Smith, B., Lopez-Rosas, A., Qiu, X., et al., 2011. Axonopathy is a compounding factor in the pathogenesis of Krabbe disease. Acta Neuropathol. 122, 35–48.

Cereda, E., Barichella, M., Cassani, E., Caccialanza, R., Pezzoli, G., 2012. Clinical features of Parkinson disease when onset of diabetes came first: a case-control study. Neurology 78, 1507–1511.
Cersosimo, M.G., Raina, G.B., Pecci, C., et al., 2013. Gastrointestinal manifestations in parkinson's disease: prevalence and occurrence before motor symptoms. J. Neurol. 260, 1332–1338.
Chen, X., de Silva, H.A., Pettenati, M.J., Rao, P.N., St George-Hyslop, P., et al., 1995. The human NACP/alpha-synuclein gene: chromosome assignment to 4q21.3–q22 and TaqI RFLP analysis. Genomics 26, 425–427.
Chen, W.W., Zhang, X., Huang, W.J., 2016. Role of neuroinflammation in neurodegenerative diseases (review). Mol. Med. Rep. 13, 3391–3396.
Choi, J.-H., Park, P., Baek, G.-C., Sim, S.-E., Kang, S.J., Lee, Y., et al., 2014. Effects of PI3Kβ overexpression in the hippocampus on synaptic plasticity and spatial learning. Mol. Brain 7, 78.
Chou, K.L., Gilman, S., Bohnen, N.I., 2017. Association between autonomic dysfunction and fatigue in Parkinson disease. J. Neurol. Sci. 377, 190–192.
Chu, Y., Dodiya, H., Aebischer, P., Olanow, C.W., Kordower, J.H., 2009. Alterations in lysosomal and proteasomal markers in Parkinson's disease: relationship to alpha-synuclein inclusions. Neurobiol. Dis. 35, 385–398.
Civiero, L., Cirnaru, M.D., Beilina, A., Rodella, U., Russo, I., Belluzzi, E., et al., 2015. Leucine-rich repeat kinase 2 interacts with p21-activated kinase 6 to control neurite complexity in mammalian brain. J. Neurochem. 135, 1242–1256.
Codolo, G., Plotegher, N., Pozzobon, T., Brucale, M., Tessari, I., Bubacco, L., et al., 2013. Triggering of inflammasome by aggregated α-synuclein, an inflammatory response in synucleinopathies. PLoS One 8, e55375.
Dalfó, E., Portero-Otín, M., Ayala, V., Martínez, A., Pamplona, R., Ferrer, I., 2005. Evidence of oxidative stress in the neocortex in incidental lewy body disease. J. Neuropathol. Exp. Neurol. 64, 816–830.
Daniele, S.G., Beraud, D., Davenport, C., Cheng, K., Yin, H., Maguire-Zeiss, K.A., 2015. Activation of MyD88-dependent TLR1/2 signaling by misfolded alpha-synuclein, a protein linked to neurodegenerative disorders. Sci. Signal. 8, ra45.
Daniele, S., Pietrobono, D., Fusi, J., Iofrida, C., Chico, L., Petrozzi, L., et al., 2017. α-synuclein aggregates with β-Amyloid or tau in human red blood cells: correlation with antioxidant capability and physical exercise in human healthy subjects. Mol. Neurobiol. 2017 (55), 2653–2675.
Dartt, D.A., 2004. Interaction of EGF family growth factors and neurotransmitters in regulating lacrimal gland secretion. Exp. Eye Res. 78, 337–345.
Dartt, D.A., 2009. Neural regulation of lacrimal gland secretory processes: relevance in dry eye diseases. Prog. Retin. Eye Res. 28, 155–177.
Davie, C.A., 2008. A review of Parkinson's disease. Br. Med. Bull. 86, 109–127.
Day, M., Wang, Z., Ding, J., et al., 2006. Selective elimination of glutamatergic synapses on striatopallidal neurons in Parkinson disease models. Nat. Neurosci. 9, 251–259.
De, F.G., Frare, E., Pivato, M., Relini, A., Penco, A., Greggio, E., et al., 2011. Structural and morphological characterization of aggregated species of alpha-synuclein induced by docosahexaenoic acid. J. Biol. Chem. 286, 22262–22274.
Dedoussis, G.V., Kaliora, A.C., Panagiotakos, D.B., 2007. Genes, diet and type 2 diabetes mellitus: a review. Rev. Diabet. Stud. 4, 13–24.
Dehay, B., Bové, J., Rodríguez-Muela, N., Perier, C., Recasens, A., Boya, P., et al., 2010. Pathogenic lysosomal depletion in Parkinson's disease. J. Neurosci. 30, 12535–12544.
Del Tredici, K., Braak, H., 2016. Review: sporadic Parkinson's disease: development and distribution of alpha-synuclein pathology. Neuropathol. Appl. Neurobiol. 42, 33–50.

Devos, D., Defebvre, L., Bordet, R., 2010. Dopaminergic and non-dopaminergic pharmacological hypotheses for gait disorders in Parkinson's disease. Fundam. Clin. Pharmacol. 24, 407–421.
Dias, V., Junn, E., Mouradian, M.M., 2013. The role of oxidative stress in Parkinson's disease. J. Parkinson Dis. 3, 461–491.
Di Cristofano, A., Pesce, B., Cordon-Cardo, C., Pandolfi, P.P., 1998. Pten is essential for embryonic development and tumour suppression. Nat. Genet. 19, 348–355.
Dionísio, P.E.A., Oliveira, S.R., Amaral, J., Rodrigues, C.M.P., 2019. Loss of microglial parkin inhibits necroptosis and contributes to neuroinflammation (2018) Mol. Neurobiol. (56), 2990–3004.
Dorval, V., Fraser, P.E., 2006. Small ubiquitin-like modifier (SUMO) modification of natively unfolded proteins tau and alpha-synuclein. J. Biol. Chem. 281, 9919–9924.
Driver, J.A., Smith, A., Buring, J.E., Gaziano, J.M., Kurth, T., Logroscino, G., 2008. Prospective cohort study of type 2 diabetes and the risk of Parkinson's disease. Diabetes Care 31, 2003–2005.
Dzamko, N., Gysbers, A., Perera, G., Bahar, A., Shankar, A., Gao, J., et al., 2017. Toll-like receptor 2 is increased in neurons in Parkinson's disease brain and may contribute to alpha-synuclein pathology. Acta Neuropathol. 133, 303–319.
Eyre, H., Baune, B.T., 2012. Neuroplastic changes in depression: a role for the immune system. Psychoneuroendocrinology 37, 1397–1416.
Falvey, C.M., Rosano, C., Simonsick, E.M., Harris, T., Strotmeyer, E.S., Satterfield, S., et al., 2013. Macro- and microstructural magnetic resonance imaging indices associated with diabetes among community-dwelling older adults. Diabetes Care 36, 677–682.
Farooqui, A.A., 2010. Neurochemical Aspects of Neurotraumatic and Neurodegenerative Diseases. Springer, New York.
Farooqui, A.A., 2012. Generation of reactive oxygen species in the brain: signaling for neural cell survival or suicide. In: Farooqui, T., Farooqui, A.A. (Eds.), Oxidative Stress in Vertebrate and Invertebrate: Molecular Aspects of Cell Signaling. Wiley-Blackwell, New York, pp. 3–15.
Farooqui, A.A., 2014. Inflammation and Oxidative Stress in Neurological Disorders. Springer International Publishing, Switzerland.
Farooqui, A.A., 2018. Neurochemical Aspects of Alzheimer's Disease. Academic Press, an imprint of Elsevier, San Diego, CA.
Faustini, G., Bono, F., Valerio, A., Pizzi, M., Spano, P., et al., 2017. Mitochondria and α-synuclein: friends or foes in the pathogenesis of Parkinson's disease? Genes (Basel), 8. pii: E377.
Fecchio, C., De Franceschi, G., Relini, A., Greggio, E., Dalla, S.M., Bubacco, L., et al., 2013. alpha-Synuclein oligomers induced by docosahexaenoic acid affect membrane integrity. PLoS One 8, e82732.
Fellner, L., Irschick, R., Schanda, K., Reindl, M., Klimaschewski, L., Poewe, W., et al., 2013. Toll-like receptor 4 is required for alpha-synuclein dependent activation of microglia and astroglia. Glia 61, 349–360.
Fink, A.L., 2006. The aggregation and fibrillation of alpha-synuclein. Acc. Chem. Res. 39, 628–634.
Forman, H.J., Fukuto, J.M., Torres, M., 2004. Redox signaling: thiol chemistry defines which reactive oxygen and nitrogen species can act as second messengers. Am. J. Physiol. Cell. Physiol. 287, C246–C256.
Francis, P.T., 2009. Biochemical and pathological correlates of cognitive and behavioural change in DLB/PDD. J. Neurol. 256, 280–285.
Franco-Iborra, S., Vila, M., Perier, C., 2016. The Parkinson disease mitochondrial hypothesis: where are we at? Neuroscientist 22, 266–277.

Fratelli, M., Goodwin, L.O., Ørom, U.A., et al., 2005. Gene expression profiling reveals a signaling role of glutathione in redox regulation. Proc. Natl. Acad. Sci. U.S.A. 102, 13998–14003.

Friedman, J.H., Beck, J.C., Chou, K.L., Clark, G., Fagundes, C.P., Goetz, C.G., et al., 2016. Fatigue in Parkinson's disease: report from a mutidisciplinary symposium. NPJ Parkinson Dis. 2, 15025.

Friedman, J.H., Brown, R.G., Comella, C., Garber, C.E., Krupp, L.B., Lou, J.S., et al., Working Group on Fatigue in Parkinson's Disease 2007. Fatigue in Parkinson's disease: a review. Mov. Disord. 22, 297–308.

Fujishiro, H., Ferman, T.J., Boeve, B.F., Smith, G.E., Graff-Radford, N.R., et al., 2008. Validation of the neuropathologic criteria of the third consortium for dementia with Lewy bodies for prospectively diagnosed cases. J. Neuropathol. Exp. Neurol. 67, 649–656.

Fujita, M., Sekigawa, A., Sekiyama, K., Takamatsu, Y., Hashimoto, M., 2012. Possible alterations in β-synuclein, the non-amyloidogenic homologue of α-synuclein, during progression of sporadic α-synucleinopathies. Int. J. Mol. Sci. 13, 11584–11592.

Funk, N., Munz, M., Ott, T., Brockmann, K., Wenninger-Weinzierl, A., Kühn, R., et al., 2019. The Parkinson's disease-linked Leucine-rich repeat kinase 2 (LRRK2) is required for insulin-stimulated translocation of GLUT4. Sci. Rep. 9, 4515.

Gao, L., Tang, H., Nie, K., Wang, L., Zhao, J., et al., 2015. Cerebrospinal fluid alpha-synuclein as a biomarker for Parkinson's disease diagnosis: a systematic review and meta-analysis. Int. J. Neurosci. 125, 645–654.

Gerlach, M., Double, K.L., Youdim, M.B., Riederer, P., 2006. Potential sources of increased iron in the substantia nigra of parkinsonian patients. J. Neural Transm. Suppl. 70, 133–142.

Giasson, B.I., Forman, M.S., Higuchi, M., Golbe, L.I., Graves, C.L., Kotzbauer, P.T., et al., 2003. Initiation and synergistic fibrillization of tau and alpha-synuclein. Science 300, 636–640.

Goldenberg, M.M., 2008. Medical management of Parkinson's disease. P T 33, 590–606.

Goldstein, D.S., Holmes, C., Sullivan, P., Jinsmaa, Y., Kopin, I.J., et al., 2016. Elevated cerebrospinal fluid ratios of cysteinyl-dopamine/3,4-dihydroxyphenylacetic acid in parkinsonian synucleinopathies. Parkinsonism Relat. Disord. 31, 79–86.

Grabowski, G.A., 2008. Phenotype, diagnosis, and treatment of Gaucher's disease. Lancet 372, 1263–1271.

Gray, S.M., Meijer, R.I., Barrett, E.J., 2014. Insulin regulates brain function, but how does it get there? Diabetes 63, 3992–3997.

Green, H., Tsitsi, P., Markaki, I., Aarsland, D., Svenningsson, P., 2019. Novel treatment opportunities against cognitive impairment in Parkinson's disease with an emphasis on diabetes-related pathways. CNS Drugs 33, 143–160.

Griendling, K.K., Sorescu, D., Lassègue, B., Ushio-Fukai, M., 2000. Modulation of protein kinase activity and gene expression by reactive oxygen species and their role in vascular physiology and pathophysiology. Arterioscler. Thromb. Vasc. Biol. 20, 2175–2183.

Gu, X.L., Long, C.X., Sun, L., Xie, C., Lin, X., Cai, H., 2010. Astrocytic expression of Parkinson's disease-related A53T alpha-synuclein causes neurodegeneration in mice. Mol. Brain 3, 12.

Gu, F., Chauhan, V., Chauhan, A., 2015. Glutathione redox imbalance in brain disorders. Curr. Opin. Clin. Nutr. Metab. Care. 18, 89–95.

Halliwell, B., 2006. Reactive species and antioxidants. Redox biology is a fundamental theme of aerobic life. Plant Physiol. 141, 312–322.

Hamm-Alvarez, S.F., Okamoto, C.T., Janga, S.R., Feigenbaum, D., Edman, M.C., Freire, D., et al., 2019. Oligomeric α-synuclein is increased in basal tears of Parkinson's patients. Biomark. Med. . Available from: https://doi.org/10.2217/bmm-2019-0167 [Epub ahead of print].

Hancock, J.T., Desikan, R., Neill, S.J., 2001. Role of reactive oxygen species in cell signalling pathways. Biochem. Soc. Trans. 29, 345–350.

Hanganu, A., Bedetti, C., Jubault, T., Gagnon, J.F., Mejia-Constain, B., et al., 2013. Mild cognitive impairment is linked with faster rate of cortical thinning in patients with Parkinson's disease longitudinally. Mov. Disord. 28, 1360–1369.

Hashimoto, M., Kawahara, K., Bar-On, P., Rockenstein, E., Crews, L., et al., 2004. The role of alpha-synuclein assembly and metabolism in the pathogenesis of Lewy body disease. J. Mol. Neurosci. 24, 343–352.

He, Z., Simon, H.U., 2013. A novel link between p53 and ROS. Cell Cycle 12, 201–202.

Hedden, T., Mormino, E.C., Amariglio, R.E., Younger, A.P., Schultz, A.P., Becker, J.A., et al., 2012. Cognitive profile of amyloid burden and white matter hyperintensities in cognitively normal older adults. J. Neurosci. 32, 16233–16242.

Hely, M.A., Reid, W.G.J., Adena, M.A., Halliday, G.M., Morris, J.G.L., 2008. The Sydney multicenter study of Parkinson's disease: the inevitability of dementia at 20 years. Mov. Disord. 23, 837–844.

Herlofson, K., Heijnen, C.J., Lange, J., Alves, G., Tysnes, O.B., Friedman, J.H., et al., 2018. Inflammation and fatigue in early, untreated Parkinson's disease. Acta Neurol. Scand. 138, 394–399.

Hornykiewicz, O., 2008. Basic research on dopamine in Parkinson's disease and the discovery of the nigrostriatal dopamine pathway: the view of an eyewitness. Neurodegener. Dis. 5, 114–117.

Hughes, C.D., Choi, M.L., Ryten, M., Hopkins, L., Drews, A., Botía, J.A., et al., 2019. Picomolar concentrations of oligomeric alpha-synuclein sensitizes TLR4 to play an initiating role in Parkinson's disease pathogenesis. Acta Neuropathol. 137, 103–120.

Iljina, M., Garcia, G.A., Horrocks, M.H., Tosatto, L., Choi, M.L., Ganzinger, K.A., et al., 2016. Kinetic model of the aggregation of alpha-synuclein provides insights into prion-like spreading. Proc. Natl. Acad. Sci. U.S.A. 113, E1206–E1215.

Ince, P.G., 2011. Dementia with Lewy bodies and Parkinson's disease with dementia. In: Dickson, D.W., Weller, R.O. (Eds.), Neurodegeneration, the Molecular Pathology of Dementia and Movement Disorders. Wiley-Blackwell, Oxford, pp. 224–237.

Irwin, D.J., Lee, V.M., Trojanowski, J.Q., 2013. Parkinson's disease dementia: convergence of α-synuclein, tau and amyloid-β pathologies. Nat. Rev. Neurosci. 14, 626–636.

Itoh, K., Chiba, T., Takahashi, S., Ishii, T., Igarashi, K., et al., 1997. An Nrf2/small maf heterodimer mediates the induction of phase II detoxifying enzyme genes through antioxidant response elements. Biochem. Biophys. Res. Commun. 236, 313–322.

Iwai, A., Masliah, E., Yoshimoto, M., Ge, N., Flanagan, L., et al., 1995. The precursor protein of non-A beta component of Alzheimer's disease amyloid is a presynaptic protein of the central nervous system. Neuron 14, 467–475.

Jakes, R., Spillantini, M.G., Goedert, M., 1994. Identification of two distinct synucleins from human brain. FEBS Lett. 345, 27–32.

Jankovic, J., 2008. Parkinson's disease: clinical features and diagnosis. J. Neurol. Neurosurg. Psychiatry 79, 368–376.

Jellinger, K.A., 2012. Neurobiology of cognitive impairment in Parkinson's disease. Expert. Rev. Neurother. 12, 1451–1466.

Kempuraj, D., Thangavel, R., Natteru, P.A., Selvakumar, G.P., Saeed, D., et al., 2016. Neuroinflammation induces neurodegeneration. J. Neurol Neurosurg. Spine 1, pii: 1003.

Kempuraj, D., Selvakumar, G.P., Zaheer, S., Thangavel, R., Ahmed, M.E., et al., 2018. Cross-talk between glia, neurons and mast cells in neuroinflammation associated with Parkinson's disease. J. Neuroimmune. Pharmacol. 13, 100–112.

Kielian, T., 2016. Multifaceted roles of neuroinflammation: the need to consider both sides of the coin. J. Neurochem. 136, 5–9.

Kim, S.J., Han, Y., 2005. Insulin inhibits AMPA-induced neuronal damage via stimulation of protein kinase B (Akt). J. Neural Transm. 112, 179–191.

Kim, J.-I., Lee, H.-R., Sim, S., Baek, J., Yu, N.-K., Choi, J.-H., et al., 2011. PI3Kγ is required for NMDA receptor–dependent long-term depression and behavioral flexibility. Nat. Neurosci. 14, 1447–1454.

Kim, C., Ho, D.-H., Suk, J.-E., You, S., Michael, S., Kang, J., et al., 2013. Neuron-released oligomeric α-synuclein is an endogenous agonist of TLR2 for paracrine activation of microglia. Nat. Commun. 4, 1562.

Kim, S., Park, J.M., Moon, J., Choi, H.J., 2014. Alpha-synuclein interferes with cAMP/PKA-dependent upregulation of dopamine beta-hydroxylase and is associated with abnormal adaptive responses to immobilization stress. Exp. Neurol. 252, 63–74.

Kim, D.I., Lee, K.H., Gabr, A.A., Choi, G.E., Kim, J.S., Ko, S.H., et al., 2016. Aβ-Induced Drp1 phosphorylation through Akt activation promotes excessive mitochondrial fission leading to neuronal apoptosis. Biochim. Biophys. Acta Mol. Cell Res. 1863, 2820–2834.

Kleinridders, A., Ferris, H.A., Cai, W., Kahn, C.R., 2014. Insulin action in brain regulates systemic metabolism and brain function. Diabetes 63, 2232–2243.

Klimas, N.G., Broderick, G., Fletcher, M.A., 2012. Biomarkers for chronic fatigue. Brain Behav. Immun. 26, 1202–1210.

Knapp, L.T., Klann, E., 2002. Role of reactive oxygen species in hippocampal long-term potentiation: contributory or inhibitory? J. Neurosci. Res. 70, 1–7.

Kohen, R., Beit-Yannai, E., Berry, E.M., Tirosh, O., 1999. Overall low molecular weight antioxidant activity of biological fluids and tissues by cyclic voltammetry. Methods Enzymol. 300, 285–296.

Kostyuk, S.V., Porokhovnik, L.N., Ershova, E.S., Malinovskaya, E.M., Konkova, M.S., Kameneva, L.V., et al., 2018. Changes of KEAP1/NRF2 and IKB/NF-κB expression levels induced by cell-free DNA in different cell types. Oxid. Med. Cell. Longevity. 2018, 1052413.

Kotagal, V., Albin, R.L., Muller, M.L., Koeppe, R.A., Frey, K.A., Bohnen, N.I., 2013. Diabetes is associated with postural instability and gait difficulty in Parkinson disease. Parkinsonism Relat. Disord. 19, 522–526.

Krumova, P., Meulmeester, E., Garrido, M., Tirard, M., Hsiao, H.H., Bossis, G., et al., 2011. Sumoylation inhibits alpha-synuclein aggregation and toxicity. J. Cell Biol. 194, 49–60.

Kuret, J., Chirita, C.N., Congdon, E.E., Kannanayakal, T., Li, G., Necula, M., et al., 2005. Pathways of tau fibrillization. Biochim. Biophys. Acta 1739, 167–178.

Leal, M.C., Casabona, J.C., Puntel, M., Pitossi, F.J., 2013. Interleukin-1beta and tumor necrosis factor-alpha: reliable targets for protective therapies in Parkinson's disease? Front. Cell. Neurosci. 7, 53.

Lebouvier, T., Chaumette, T., Paillusson, S., Duyckaerts, C., Bruley des Varannes, S., Neunlist, M., et al., 2009. The second brain and Parkinson's disease. Eur. J. Neurosci. 30, 735–741.

Lee, H.J., Choi, C., Lee, S.J., 2002. Membrane-bound α-synuclein has a high aggregation propensity and the ability to seed the aggregation of the cytosolic form. J. Biol. Chem. 277, 671–678.

Lesage, S., Brice, A., 2009. Parkinson's disease: from monogenic forms to genetic susceptibility factors. Hum. Mol. Genet. 18, R48–R59.
Li, J., Zhu, M., Rajamani, S., Uversky, V.N., Fink, A.L., 2004. Rifampicin inhibits alpha-synuclein fibrillation and disaggregates fibrils. Chem. Biol. 11, 1513–1521.
Li, J., Liu, D., Sun, L., Lu, Y., Zhang, Z., 2012. Advanced glycation end products and neurodegenerative diseases: mechanisms and perspective. J. Neurol. Sci. 317, 1–5.
Liddle, R.A., 2018. Parkinson's disease from the gut. Brain Res. 1693 (Pt B), 201–206.
Lima, M.M., 2013. Sleep disturbances in Parkinson's disease: the contribution of dopamine in REM sleep regulation. Sleep Med. Rev. 17, 367–375.
Lima, M.M., Martins, E.F., Delattre, A.M., Proenca, M.B., Mori, M.A., Carabelli, B., et al., 2012. Motor and non-motor features of Parkinson's disease—a review of clinical and experimental studies. CNS Neurol. Disord. Drug Targets 11, 439–449.
Lin, C.H., Yeh, S.H., Lin, C.H., Lu, K.T., Leu, T.H., Chang, W.C., et al., 2001. A role for the PI-3 kinase signaling pathway in fear conditioning and synaptic plasticity in the amygdala. Neuron 31, 841–851.
Lin, S.J., Baumeister, T.R., Garg, S., McKeown, M.J., 2018. Cognitive profiles and hub vulnerability in Parkinson's disease. Front. Neurol. 9, 482.
Lindqvist, D., Kaufman, E., Brundin, L., Hall, S., Surova, Y., Hansson, O., 2012. Non-motor symptoms in patients with Parkinson's disease—correlations with inflammatory cytokines in serum. PLoS One 7, e47387.
Lindqvist, D., Hall, S., Surova, Y., Nielsen, H.M., Janelidze, S., Brundin, L., et al., 2013. Cerebrospinal fluid inflammatory markers in Parkinson's disease: associations with depression, fatigue, and cognitive impairment. Brain Behav. Immun. 33, 183–189.
Liu, Q., Qiu, J., Liang, M., Golinski, J., van Leyen, K., Jung, J.E., et al., 2014. Akt and mTOR mediate programmed necrosis in neurons. Cell Death Dis. 5, e1084.
Loerch, P.M., Lu, T., Dakin, K.A., Vann, J.M., Isaacs, A., et al., 2008. Evolution of the aging brain transcriptome and synaptic regulation. PLoS One 3, e3329.
Lu, T., Chai, Q., Yu, L., d'Uscio, L.V., Katusic, Z.S., He, T., et al., 2012. Reactive oxygen species signaling facilitates FOXO-3a/FBXO-dependent vascular BK channel β1 subunit degradation in diabetic mice. Diabetes 61, 1860–1868.
Lu, L., Fu, D.-L., Li, H.-Q., Liu, A.-J., Li, J.-H., Zheng, G.-Q., 2014. Diabetes and risk of Parkinson's disease: an updated meta-analysis of case-control studies. PLoS One 9, e85781.
Lucas, J.J., Hernández, F., Gómez-Ramos, P., Morán, M.A., Hen, R., et al., 2001. Decreased nuclear beta-catenin, tau hyperphosphorylation and neurodegeneration in GSK-3beta conditional transgenic mice. EMBO J. 20, 27–39.
Lucas, S.M., Rothwell, N.J., Gibson, R.M., 2006. The role of inflammation in CNS injury and disease. Br. J. Pharmacol. 147, S232–S240.
Lucke, C., Gantz, D.L., Klimtchuk, E., Hamilton, J.A., 2006. Interactions between fatty acids and alpha-synuclein. J. Lipid Res. 47, 1714–1724.
Luo, C., Song, W., Chen, Q., Zheng, Z., Chen, K., Cao, B., et al., 2014. Reduced functional connectivity in early-stage drug-naive Parkinson's disease: a resting-state fMRI study. Neurobiol. Aging 35, 431–441.
Ma, Q., 2013. Role of nrf2 in oxidative stress and toxicity. Annu. Rev. Pharmacol. Toxicol. 53, 401–426.
Maciejczyk, M., Żebrowska, E., Chabowski, A., 2019. Insulin resistance and oxidative stress in the brain: what's new? Int. J. Mol. Sci. 20, pii: E874.
MacLeod, D., Dowman, J., Hammond, R., Leete, T., Inoue, K., Abeliovich, A., 2006. The familial Parkinsonism gene LRRK2 regulates neurite process morphology. Neuron 52, 587–593.
Maes, M., Galecki, P., Chang, Y.S., Berk, M., 2011. A review on the oxidative and nitrosative stress (O&NS) pathways in major depression and their possible contribution to

the (neuro)degenerative processes in that illness. Prog. Neuropsychopharmacol. Biol. Psychiatry 35, 676–692.
Maiti, P., Manna, J., Dunbar, G.L., 2017. Current understanding of the molecular mechanisms in Parkinson's disease: targets for potential treatments. Transl. Neurodegener. 6, 28.
Manzoni, C., Lewis, P.A., 2017. LRRK2 and autophagy. Adv. Neurobiol. 14, 89–105.
Maroteaux, L., Campanelli, J.T., Scheller, R.H., 1988. Synuclein: a neuron-specific protein localized to the nucleus and presynaptic nerve terminal. J. Neurosci. 8, 2804–2815.
Marotta, N.P., Cherwien, C.A., Abeywardana, T., Pratt, M.R., 2012. O-GlcNAc modification prevents peptide-dependent acceleration of alpha-synuclein aggregation. Chembiochem 13, 2665–2670.
Marshall, M.S., Bongarzone, E.R., 2016. Beyond Krabbe's disease: the potential contribution of galactosylceramidase deficiency to neuronal vulnerability in late-onset synucleinopathies. J. Neurosci. Res. 94, 1328–1332.
Marsland, B.J., Gollwitzer, E.S., 2014. Host-microorganism interactions in lung diseases. Nat. Rev. Immunol. 14, 827–835.
Mazzulli, J.R., Zunke, F., Tsunemi, T., Toker, N.J., Jeon, S., Burbulla, L.F., et al., 2016. Activation of β-glucocerebrosidase reduces pathological α-synuclein and restores lysosomal function in Parkinson's patient midbrain neurons. J. Neurosci. 36, 7693–7706.
McCann, H., Cartwright, H., Halliday, G.M., 2016. Neuropathology of alpha-synuclein propagation and braak hypothesis. Mov. Disord. 31, 152–160.
McNay, E.C., Ong, C.T., McCrimmon, R.J., Cresswell, J., Bogan, J.S., Sherwin, R.S., 2010. Hippocampal memory processes are modulated by insulin and high-fat-induced insulin resistance. Neurobiol. Learn. Mem. 93, 546–553.
Merdes, A.R., Hansen, L.A., Jeste, D.V., Galasko, D., Hofstetter, C.R., et al., 2003. Influence of Alzheimer pathology on clinical diagnostic accuracy in dementia with Lewy bodies. Neurology 60, 1586–1590.
Michel, P.P., Hirsch, E.C., Hunot, S., 2016. Understanding dopaminergic cell death pathways in Parkinson disease. Neuron 90, 675–691.
Migdalska-Richards, A., Schapira, A.H., 2016. The relationship between glucocerebrosidase mutations and Parkinson disease. J. Neurochem. 139 (Suppl. 1), 77–90.
Milanese, Cerri S., Ulusoy, A., Gornati, S.V., Plat, A., Gabriels, S., et al., 2018. Activation of the DNA damage response *in vivo* in synucleinopathy models of Parkinson's disease. Cell Death Dis. 9, 818.
Miller, A.H., Haroon, E., Raison, C.L., Felger, J.C., 2013. Cytokine targets in the brain: impact on neurotransmitters and neurocircuits. Depress. Anxiety 30, 297–306.
Moore, D.J., West, A.B., Dawson, V.L., Dawson, T.M., 2005. Molecular pathophysiology of Parkinson's disease. Annu. Rev. Neurosci. 28, 57–87.
Moors, T., Paciotti, S., Chiasserini, D., Calabresi, P., Parnetti, L., et al., 2016. Lysosomal dysfunction and alpha-synuclein aggregation in Parkinson's disease: diagnostic links. Mov. Disord. 31, 791–801.
Moran, L.B., Graeber, M.B., 2008. Towards a pathway definition of Parkinson's disease: a complex disorder with links to cancer, diabetes and inflammation. Neurogenetics 9, 1–13.
Morris, A.M., Watzky, M.A., Agar, J.N., Finke, R.G., 2008. Fitting neurological protein aggregation kinetic data via a 2-step, minimal/"Ockham's razor" model: the Finke-Watzky mechanism of nucleation followed by autocatalytic surface growth. Biochemistry 47, 2413–2427.
Morrison, J.H., Baxter, M.G., 2012. The aging cortical synapse: hallmarks and implications for cognitive decline. Nat. Rev. Neurosci. 13, 240–250.

Moussaud, S., Jones, D.R., Moussaud-Lamodiere, E.L., Delenclos, M., Ross, O.A., McLean, P.J., 2014. Alpha-synuclein and tau: teammates in neurodegeneration? Mol. Neurodegener. 9, 43.

Movafagh, S., Crook, S., Vo, K., 2015. Regulation of hypoxia-inducible factor-1a by reactive oxygen species: new developments in an old debate. J. Cell. Biochem. 116, 696–703.

Mukaetova-Ladinska, E.B., McKeith, I.G., 2006. Pathophysiology of synuclein aggregation in Lewy body disease. Mech. Ageing Dev. 127, 188–202.

Muntane, G., Dalfo, E., Martinez, A., Ferrer, I., 2008. Phosphorylation of tau and alpha-synuclein in synaptic-enriched fractions of the frontal cortex in Alzheimer's disease, and in Parkinson's disease and related alpha-synucleinopathies. Neuroscience 152, 913–923.

Muslimovic, D., Post, B., Speelman, J.D., de Haan, R.J., Schmand, B., 2009. Cognitive decline in Parkinson's disease: a prospective longitudinal study. J. Int. Neuropsychol. Soc. 15, 426–437.

Nagerl, U.V., Eberhorn, N., Cambridge, S.B., Bonhoeffer, T., 2004. Bidirectional activity-dependent morphological plasticity in hippocampal neurons. Neuron 44, 759–767.

Narayanan, A., Amaya, M., Voss, K., Chung, M., Benedict, A., Sampey, G., et al., 2014. Reactive oxygen species activate NFκB (p65) and p53 and induce apoptosis in RVFV infected liver cells. Virology 449, 270–286.

Olde, K.T.E., Deijen, J.B., Barkhof, F., 2014. Functional connectivity and cognitive decline over 3 years in Parkinson disease. Neurology 83, 2046–2053.

Pacher, P., Beckman, J.S., Liaudet, L., 2007. Nitric oxide and peroxynitrite in health and disease. Physiol. Rev. 87, 315–424.

Padmaraju, V., Bhaskar, J.J., Prasada Rao, U.J., Salimath, P.V., Rao, K.S., 2011. Role of advanced glycation on aggregation and DNA binding properties of alpha-synuclein. J. Alzheimer Dis. 24 (Suppl. 2), 211–221.

Pagonabarraga, J., Kulisevsky, J., 2012. Cognitive impairment and dementia in Parkinson's disease. Neurobiol. Dis. 46, 590–596.

Paisán-Ruíz, C., Jain, S., Evans, E.W., Gilks, W.P., Simón, J., van der Brug, M., et al., 2004. Cloning of the gene containing mutations that cause PARK8-linked Parkinson's disease. Neuron 44, 595–600.

Paiva, I., Pinho, R., Pavlou, M.A., Hennion, M., Wales, P., Schütz, A.L., et al., 2019. Sodium butyrate rescues dopaminergic cells from alpha-synuclein-induced transcriptional deregulation and DNA damage. Hum. Mol. Genet. 26, 2231–2246.

Paleologou, K.E., Kragh, C.L., Mann, D.M., Salem, S.A., Al-Shami, R., et al., 2009. Detection of elevated levels of soluble α-synuclein oligomers in post-mortem brain extracts from patients with dementia with Lewy bodies. Brain 132, 1093–1101.

Pang, Y., Lin, S., Wright, C., Shen, J., Carter, K., Bhatt, A., et al., 2016. Intranasal insulin protects against substantia nigra dopaminergic neuronal loss and alleviates motor deficits induced by 6-OHDA in rats. Neuroscience 318, 157–165.

Parnetti, L., Castrioto, A., Chiasserini, D., Persichetti, E., Tambasco, N., El-Agnaf, O., et al., 2013. Cerebrospinal fluid biomarkers in Parkinson disease. Nat. Rev. Neurol. 9, 131–140.

Pchitskaya, E., Popugaeva, E., Bezprozvanny, I., 2018. Calcium signaling and molecular mechanisms underlying neurodegenerative diseases. Cell Calcium 70, 87–94.

Pereira, J.R., Santos, L.V.D., Santos, R.M.S., Campos, A.L.F., Pimenta, A.L., de Oliveira, M.S., et al., 2016. IL-6 serum levels are elevated in Parkinson's disease patients with fatigue compared to patients without fatigue. J. Neurol. Sci. 370, 153–156.

Perrin, R.J., Woods, W.S., Clayton, D.F., George, J.M., 2001. Exposure to long chain polyunsaturated fatty acids triggers rapid multimerization of synucleins. J. Biol. Chem. 276, 41958–41962.

Perry, G., Moreira, P.I., Siedlak, S.L., Nunomura, A., Zhu, X., Smith, M.A., 2009. Natural oxidant balance in Parkinson disease. Arch. Neurol. 66, 1445.

Pfeiffer, R.F., 2003. Gastrointestinal dysfunction in Parkinson's disease. Lancet. Neurol. 2, 107–116.

Pfeiffer, R.F., 2011. Gastrointestinal dysfunction in Parkinson's disease. Parkinsonism Relat. Disord. 17, 10–15.

Phani, S., Loike, J.D., Przedborski, S., 2012. Neurodegeneration and inflammation in Parkinson's disease. Parkinsonism Relat. Disord. 18, 207–209.

Pieri, L., Madiona, K., Bousset, L., Melki, R., 2012. Fibrillar α-synuclein and huntingtin exon 1 assemblies are toxic to the cells. Biophys. J. 102, 2894–2905.

Pinho, R., Paiva, I., Jercic, K.G., Fonseca-Ornelas, L., Gerhardt, E., Fahlbusch, C., et al., 2019. Nuclear localization and phosphorylation modulate pathological effects of alpha-synuclein. Hum. Mol. Genet. 28, 31–50.

Politis, M., Wu, K., Loane, C., Turkheimer, F.E., Molloy, S., Brooks, D.J., et al., 2010. Depressive symptoms in PD correlate with higher 5-HTT binding in raphe and limbic structures. Neurology 75, 1920–1927.

Poltorak, A., He, X., Smirnova, I., Liu, M.Y., Van Huffel, C., Du, X., et al., 1998. Defective LPS signaling in C3H/HeJ and C57BL/10ScCr mice: mutations in Tlr4 gene. Science 2822085–2822088.

Qin, X.Y., Zhang, S.P., Cao, C., Loh, Y.P., Cheng, Y., 2016. Aberrations in peripheral inflammatory cytokine levels in Parkinson disease: a systematic review and meta-analysis. JAMA Neurol. 73, 1316–1324.

Rannikko, E.H., Weber, S.S., Kahle, P.J., 2015. Exogenous alpha-synuclein induces toll-like receptor 4 dependent inflammatory responses in astrocytes. BMC Neurosci. 16, 57.

Rassu, M., Del Giudice, M.G., Sanna, S., Taymans, J.M., Morari, M., Brugnoli, A., et al., 2017. Role of LRRK2 in the regulation of dopamine receptor trafficking. PLoS One 12, e0179082.

Recchia, A., Debetto, P., Negro, A., Guidolin, D., Skaper, S.D., Giusti, P., 2004. Alpha-synuclein and Parkinson's disease. FASEB J. 18, 617–626.

Reid, W.G., Hely, M.A., Morris, J.G., Loy, C., Halliday, G.M., 2011. Dementia in Parkinson's disease: a 20-year neuropsychological study (Sydney multicentre study). J. Neurol. Neurosurg. Psychiatry 82, 1033–1037.

Ribeiro, M., Rosenstock, T.R., Oliveira, A.M., Oliveira, C.R., Rego, A.C., 2014. Insulin and IGF-1 improve mitochondrial function in a PI-3K/Akt-dependent manner and reduce mitochondrial generation of reactive oxygen species in Huntington's disease knock-in striatal cells. Free Radic. Biol. Med. 74, 129–144.

Riera, H., Afonso, V., Collin, P., Lomri, A., 2015. A central role for JNK/AP-1 pathway in the pro-oxidant effect of pyrrolidine dithiocarbamate through superoxide dismutase 1 gene repression and reactive oxygen species generation in hematopoietic human cancer cell line U937. PLoS One 10, e0127571.

Rocha, E.M., Smith, G.A., Park, E., Cao, H., Brown, E., Hallett, P., et al., 2015. Progressive decline of glucocerebrosidase in aging and Parkinson's disease. Ann. Clin. Transl. Neurol. 2, 433–438.

Rott, R., Szargel, R., Haskin, J., Shani, V., Shainskaya, A., Manov, I., et al., 2008. Monoubiquitylation of alpha-synuclein by seven in absentia homolog (SIAH) promotes its aggregation in dopaminergic cells. J. Biol. Chem. 283, 3316–3328.

Rousseaux, M.W.C., de Haro, M., Lasagna-Reeves, C.A., De Maio, A., Park, J., Jafar-Nejad, P., et al., 2016. TRIM28 regulates the nuclear accumulation and toxicity of both alpha-synuclein and tau. eLife 5, e19809.

Ruiperez, V., Darios, F., Davletov, B., 2010. Alpha-synuclein, lipids and Parkinson's disease. Prog. Lipid. Res. 49, 420–428.

Sánchez-Alegría, K., Flores-León, M., Avila-Muñoz, E., Rodríguez-Corona, N., Arias, C., 2018. PI3K signaling in neurons: a central node for the control of multiple functions. Int. J. Mol. Sci. 19, pii: E3725.
Santiago, J.A., Potashkin, J.A., 2013. Shared dysregulated pathways lead to Parkinson's disease and diabetes. Trends Mol. Med. 19, 176–186.
Saponaro, C., Cianciulli, A., Calvello, R., Dragone, T., Iacobazzi, F., Panaro, M.A., 2012. The PI3K/Akt pathway is required for LPS activation of microglial cells. Immunopharmacol. Immunotoxicol. 34, 858–865.
Schernhammer, E., Hansen, J., Rugbjerg, K., Wermuth, L., Ritz, B., 2011. Diabetes and the risk of developing Parkinson's disease in Denmark. Diabetes Care 34, 1102–1008.
Schuh, A.F., Rieder, C.M., Rizzi, L., Chaves, M., Roriz-Cruz, M., 2011. Mechanisms of brain aging regulation by insulin: implications for neurodegeneration in late-onset Alzheimer's disease. ISRN Neurol. 2011, 306905.
Sengupta, U., Guerrero-Muñoz, M.J., Castillo-Carranza, D.L., Lasagna-Reeves, C.A., Gerson, J.E., Paulucci-Holthauzen, A.A., et al., 2015. Pathological interface between oligomeric α-synuclein and tau in synucleinopathies. Biol. Psychiatry 78, 672–683.
Serpell, L.C., 2000. Alzheimer's amyloid fibrils: structure and assembly. Biochim. Biophys. Acta 1502, 16–30.
Sharon, R., Bar-Joseph, I., Frosch, M.P., Walsh, D.M., Hamilton, J.A., Selkoe, D.J., 2003. The formation of highly soluble oligomers of alpha-synuclein is regulated by fatty acids and enhanced in Parkinson's disease. Neuron 37, 583–595.
Sheridan, G.K., Murphy, K.J., 2013. Neuron-glia crosstalk in health and disease: fractalkine and CX3CR1 take centre stage. Open Biol. 3, 130181.
Si, X., Pu, J., Zhang, B., 2017. Structure, distribution, and genetic profile of α-synuclein and their potential clinical application in Parkinson's disease. J. Mov. Disord. 10, 69–79.
Siddiqui, A., Chinta, S.J., Mallajosyula, J.K., Rajagopolan, S., Hanson, I., Rane, A., et al., 2012. Selective binding of nuclear alpha-synuclein to the PGC1alpha promoter under conditions of oxidative stress may contribute to losses in mitochondrial function: implications for Parkinson's disease. Free Radic. Biol. Med. 53, 993–1003.
Siddle, K., 2011. Signaling by insulin and IGF receptors: supporting acts and new players. J. Mol. Endocrinol. 47, R1–R10.
Simón-Sánchez, J., Schulte, C., Bras, J.M., Sharma, M., Gibbs, J.R., et al., 2009. Genome-wide association study reveals genetic risk underlying Parkinson's disease. Nat. Genet. 41, 1308–1312.
Smith, B.R., Santos, M.B., Marshall, M.S., Cantuti-Castelvetri, L., Lopez-Rosas, A., Li, G., et al., 2014. Neuronal inclusions of alpha-synuclein contribute to the pathogenesis of Krabbe disease. J. Pathol. 232, 509–521.
Sonntag, W.E., Delrae, M., Ingraham, J., Riddle, D.R., 2007. Regulation of cerebrovascular aging. In: Riddle, D.R. (Ed.), Brain Aging: Models, Methods, and Mechanisms. CRC Press/Taylor & Francis, Boca Raton, FL.
Spassieva, S., Bieberich, E., 2016. Lysosphingolipids and sphingolipidoses: psychosine in Krabbe's disease. J. Neurosci. Res. 94, 974–981.
Stojkovska, I., Krainc, D., Mazzulli, J.R., 2018. Molecular mechanisms of α-synuclein and GBA1 in Parkinson's disease. Cell Tissue Res. 373, 51–60.
Storz, P., 2011. Forkhead homeobox type O transcription factors in the responses to oxidative stress. Antioxid. Redox. Signal. 14, 593–605.
Streit, W.J., Xue, Q.S., 2016. Microglia in dementia with Lewy bodies. Brain Behav. Immun. 55, 191–201.
Stuendl, A., Kunadt, M., Kruse, N., Bartels, C., Moebius, W., Danzer, K.M., et al., 2016. Induction of α-synuclein aggregate formation by CSF exosomes from

patients with Parkinson's disease and dementia with Lewy bodies. Brain 139 (Pt 2), 481–494.

Sun, G.Y., Horrocks, L.A., Farooqui, A.A., 2007. The roles of NADPH oxidase and phospholipases A_2 in oxidative and inflammatory responses in neurodegenerative diseases. J. Neurochem. 103, 1–16.

Sun, X., Yao, H., Douglas, R.M., Gu, X.Q., Wang, J., Haddad, G.G., 2010. Insulin/PI3K signaling protects dentate neurons from oxygen-glucose deprivation in organotypic slice cultures. J. Neurochem. 112, 377–388.

Surendranathan, A., Rowe, J.B., O'Brien, J.T., 2015. Neuroinflammation in Lewy body dementia. Parkinsonism Relat. Disord. 21, 1398–1406.

Surguchov, A., 2008. Molecular and cellular biology of synucleins. Int. Rev. Cell Mol. Biol. 271, 227–282.

Surguchov, A., 2015. Intracellular dynamics of synucleins: "here, there and everywhere". Int. Rev. Cell Mol. Biol. 320, 103–169.

Swan, M., Saunders-Pullman, R., 2013. The association between ß-glucocerebrosidase mutations and parkinsonism. Curr. Neurol. Neurosci. Rep. 13, 368.

Tanaka, T., Kai, S., Matsuyama, T., Adachi, T., Fukuda, K., Hirota, K., 2013. General anesthetics inhibit LPS-induced IL-1beta expression in glial cells. PLoS One 8, e82930.

Tansey, M.G., McCoy, M.K., Frank-Cannon, T.C., 2007. Neuroinflammatory mechanisms in Parkinson's disease: potential environmental triggers, pathways, and targets for early therapeutic intervention. Exp. Neurol. 208, 1–25.

Taylor, R., 2012. Insulin resistance and type 2 diabetes. Diabetes 61, 778–779.

Tokuda, T., Qureshi, M.M., Ardah, M.T., Varghese, S., Shehab, S.A., et al., 2010. Detection of elevated levels of α-synuclein oligomers in CSF from patients with Parkinson disease2010 Neurology 75, 1766–1772.

Tolosa, E., Pont-Sunyer, C., 2011. Progress in defining the premotor phase of Parkinson's disease. J. Neurol. Sci. 310, 4–8.

Tong, Q., Zhang, L., Yuan, Y., Jiang, S., Zhang, R., Xu, Q., et al., 2015. Reduced plasma serotonin and 5-hydroxyindoleacetic acid levels in Parkinson's disease are associated with nonmotor symptoms. Parkinsonism Relat. Disord. 21, 882–887.

Truban, D., Hou, X., Caulfield, T.R., Fiesel, F.C., Springer, W., 2017. PINK1, parkin, and mitochondrial quality control: what can we learn about Parkinson's disease pathobiology? J. Parkinson Dis. 7, 13–29.

Tysnes, O.B., Storstein, A., 2017. Epidemiology of Parkinson's disease. J. Neural Transm. 124, 901–905.

Uversky, V.N., Eliezer, D., 2009. Biophysics of Parkinson's disease: structure and aggregation of α-synuclein. Curr. Protein Pept. Sci. 10, 483–499.

van Uem, J.M., Marinus, J., Canning, C., van Lummel, R., Dodel, R., Liepelt-Scarfone, I., et al., 2016. Health-related quality of life in patients with parkinson's disease—a systematic review based on the ICF model. Neurosci. Biobehav. Rev. 61, 26–34.

Vasquez, V., Mitra, J., Hegde, P.M., Pandey, A., Sengupta, S., Mitra, S., et al., 2017. Chromatin-bound oxidized alpha-synuclein causes strand breaks in neuronal genomes in *in vitro* models of Parkinson's disease. J. Alzheimer Dis. 60, S133–S150.

Vasudevaraju, P., Guerrero, E., Hegde, M.L., Collen, T.B., Britton, G.B., Rao, K.S., et al., 2012. New evidence on alpha-synuclein and Tau binding to conformation and sequence specific GC rich DNA: relevance to neurological disorders. J. Pharm. Bioallied Sci. 4, 112–117.

Veselý, B., Dufek, M., Thon, V., Brozman, M., Királová, S., Halászová, T., et al., 2018. Interleukin 6 and complement serum level study in Parkinson's disease. J. Neural Transm. 125, 875–881.

Videira, P.A.Q., Castro-Caldas, M., 2018. Linking glycation and glycosylation with inflammation and mitochondrial dysfunction in Parkinson's Disease. Front. Neurosci. 12, 381.
Videnovic, A., Golombek, D., 2013. Circadian and sleep disorders in Parkinson's disease. Exp. Neurol. 243, 45–56.
Virarkar, M., Alappat, L., Bradford, O.G., Awad, A.B., 2013. L-arginine and nitric oxide in CNS function and neurodegenerative diseases. Crit. Rev. Food Sci. 53, 1157–1167.
Volles, M.J., Lee, S.J., Rochet, J.C., Shtilerman, M.D., Ding, T.T., et al., 2001. Vesicle permeabilization by protofibrillar alpha-synuclein: implications for the pathogenesis and treatment of Parkinson's disease. Biochemistry 40, 7812–7819.
Wakabayashi, K., Takahashi, H., Takeda, S., Ohama, E., Ikuta, F., 1988. Parkinson's disease: the presence of Lewy bodies in Auerbach's and Meissner's plexuses. Acta Neuropathol. 76, 217–221.
Wakabayashi, K., Tanji, K., Mori, F., Takahashi, H., 2007. The Lewy body in Parkinson's disease: molecules implicated in the formation and degradation of alpha-synuclein aggregates. Neuropathology 27, 494–506.
Walker, Z., Possin, K.L., Boeve, B.F., Aarsland, D., 2015. Lewy body dementias. Lancet 386, 1683–1697.
Wang, Q., Cheng, X.-L., Zhang, D.-Y., Gao, X.-J., Zhou, L., 2013. Tectorigenin attenuates palmitate-induced endothelial insulin resistance via targeting ROS-associated inflammation and IRS-1 pathway. PLoS One 8, e66417.
Wang, Q., Liu, Y., Zhou, J., 2015. Neuroinflammation in Parkinson's disease and its potential as therapeutic target. Transl. Neurodegener. 4, 19.
Wills, J., Credle, J., Haggerty, T., Lee, J.H., Oaks, A.W., Sidhu, A., 2011. Tauopathic changes in the striatum of A53T α-synuclein mutant mouse model of Parkinson's disease. PLoS One 6, e17953.
Winterbourn, C.C., Hampton, M.B., 2008. Thiol chemistry and specificity in redox signaling. Free Radic. Biol. Med. 45, 549–561.
Wu, R., Liu, X.M., Sun, J.G., Chen, H., Ma, J., Dong, M., et al., 2017. DJ-1 maintains energy and glucose homeostasis by regulating the function of brown adipose tissue. Cell Discov. 3, 16054.
Xu, Q., Park, Y., Huang, X., Hollenbeck, A., Blair, A., Schatzkin, A., et al., 2011. Diabetes and risk of Parkinson's disease. Diabetes Care 34, 910–915.
Yamada, T., Kawamata, T., Walker, D.G., McGeer, P.L., 1992. Vimentin immunoreactivity in normal and pathological human brain tissue. Acta Neuropathol. 84, 157–162.
Yang, W., Yu, S., 2017. Synucleinopathies: common features and hippocampal manifestations. Cell. Mol. Life Sci. 74, 1485–1501.
Yang, Y., Gehrke, S., Haque, M.E., et al., 2005. Inactivation of Drosophila DJ-1 leads to impairments of oxidative stress response and phosphatidylinositol 3-kinase/Akt signaling. Proc. Natl. Acad. Sci. U.S.A. 102, 13670–13675.
Yang, L., Wang, H., Liu, L., Xie, A., 2018. The role of insulin/IGF-1/PI3K/Akt/GSK3β signaling in Parkinson's disease dementia. Front. Neurosci. 12, 73.
Yazıcı, D., Sezer, H., 2017. Insulin resistance, obesity and lipotoxicity, Advances in Experimental Medicine and Biology, 960. Springer, Cham, Switzerland, pp. 277–304.
Yu, B.P., 1994. Cellular defenses against damage from reactive oxygen species. Physiol. Rev. 74, 139–162.
Yu, S., Li, X., Liu, G., Han, J., Zhang, C., Li, Y., et al., 2007. Extensive nuclear localization of alpha-synuclein in normal rat brain neurons revealed by a novel monoclonal antibody. Neuroscience 145, 539–555.

Zeng, X.S., Geng, W.S., Jia, J.J., Chen, L., Zhang, P.P., 2018. Cellular and molecular basis of neurodegeneration in Parkinson disease. Front. Aging Neurosci. 10, 109.
Zhang, X., Gao, F., Wang, D., Li, C., Fu, Y., He, W., et al., 2018. Tau pathology in Parkinson's disease. Front. Neurol. 9, 809.
Zhao, M., Zhou, A., Xu, L., Zhang, X., 2014. The role of TLR4-mediated PTEN/PI3K/AKT/NF-κB signaling pathway in neuroinflammation in hippocampal neurons. Neuroscience 269, 93–101.

Further reading

Abdelkarim, H., Marshall, M.S., Scesa, G., Smith, R.A., Rue, E., et al., 2018. α-Synuclein interacts directly but reversibly with psychosine: implications for α-synucleinopathies. Sci. Rep. 8, 12462.
Adamowicz, D.H., Roy, S., Salmon, D.P., Galasko, D.R., Hansen, L.A., et al., 2017. Hippocampal α-synuclein in dementia with Lewy bodies contributes to memory impairment and is consistent with spread of pathology. J. Neurosci. 37, 1675–1684.
Alberio, T., McMahon, K., Cuccurullo, M., Gethings, L.A., Lawless, C., et al., 2014. Verification of a Parkinson's disease protein signature in T-lymphocytes by multiple reaction monitoring. J. Proteome Res. 13, 3554–3561.
Allen Reish, Standaert, D., 2005. Role of α-synuclein in inducing innate and adaptive immunity in Parkinson disease. Parkinson Dis. 5, 1–19.
Ariga, H., Takahashi-Niki, K., Kato, I., Maita, H., Niki, T., Iguchi-Ariga, S.M., 2013. Neuroprotective function of DJ-1 in Parkinson's disease. Oxid. Med. Cell. Longev. 2013, 683920.
Arshad, A.R., Sulaiman, S.A., Saperi, A.A., Jamal, R., Mohamed Ibrahim, N., et al., 2017. MicroRNAs and target genes as biomarkers for the diagnosis of early onset of Parkinson disease. Front. Mol. Neurosci. 10, 352.
Baba, Y., Kuroiwa, A., Uitti, R.J., Wszolek, Z.K., Yamada, T., 2005. Alterations of T-lymphocyte populations in Parkinson disease. Parkinsonism Relat. Disord. 11, 493–498.
Bandres-Ciga, S., Cookson, M.R., 2017. Alpha-synuclein triggers T-cell response. Is Parkinson's disease an autoimmune disorder? Mov. Disord. 32, 1327.
Barbour, R., Kling, K., Anderson, J., Banducci, K., Cole, T., et al., 2008. Red blood cells are the major source of α-synuclein in blood. Neurodegener. Dis. 5, 55–59.
Berg, D., 2008. Biomarkers for the early detection of Parkinson's and Alzheimer's disease. Neurodegener. Dis. 5, 133–136.
Brochard, V., Combadiere, B., Prigent, A., Laouar, Y., Perrin, A., et al., 2009. Infiltration of CD4 + lymphocytes into the brain contributes to neurodegeneration in a mouse model of Parkinson disease. J. Clin. Invest. 119, 182–192.
Burton, E.J., Barber, R., Mukaetova-Ladinska, E.B., Robson, J., Perry, R.H., et al., 2009. Medial temporal lobe atrophy on MRI differentiates Alzheimer's disease from dementia with Lewy bodies and vascular cognitive impairment: a prospective study with pathological verification of diagnosis. Brain 132, 195–203.
Bus, J.S., Gibson, J.E., 1984. Paraquat: model for oxidant-initiated toxicity. Environ. Health Perspect. 55, 37–46.
Canet-Aviles, R.M., Wilson, M.A., Miller, D.W., Ahmad, R., McLendon, C., Bandyopadhyay, S., et al., 2004. The Parkinson's disease protein DJ-1 is neuroprotective due to cysteine-sulfinic acid-driven mitochondrial localization. Proc. Natl. Acad. Sci. U.S.A 101, 9103–9108.

Cebrian, C., Zucca, F.A., Mauri, P., Steinbeck, J.A., Studer, L., et al., 2014. MHC-I expression renders catecholaminergic neurons susceptible to T-cell-mediated degeneration. Nat. Commun. 5, 3633.
Chan, N.C., Salazar, A.M., Pham, A.H., Sweredoski, M.J., Kolawa, N.J., Graham, R.L., et al., 2011. Broad activation of the ubiquitin-proteasome system by Parkin is critical for mitophagy. Hum. Mol. Genet. 20, 1726–1737.
Chauhan, A., Jeans, A.F., 2015. Is Parkinson's disease truly a prion-like disorder? An appraisal of current evidence. Neurol. Res. Int. 2015, 345285.
Chen, L., Periquet, M., Wang, X., Negro, A., McLean, P.J., Hyman, B.T., et al., 2009. Tyrosine and serine phosphorylation of alpha-synuclein have opposing effects on neurotoxicity and soluble oligomer formation. J. Clin. Invest. 119, 3257–3265.
Chung, C.Y., Koprich, J.B., Siddiqi, H., Isacson, O., 2009. Dynamic changes in presynaptic and axonal transport proteins combined with striatal neuroinflammation precede dopaminergic neuronal loss in a rat model of AAV alpha-synucleinopathy. J. Neurosci. 29, 3365–3373.
Clark, T.A., Lee, H.P., Rolston, R.K., et al., 2010. Oxidative stress and its implications for future treatments and management of Alzheimer disease. Int. J. Biomed. Sci. 6, 225–227.
Cookson, M.R., 2010. The role of leucine-rich repeat kinase 2 (LRRK2) in Parkinson's disease. Nat. Rev. Neurosci. 11, 791–797.
Cromarty, R.A., Elder, G.J., Graziadio, S., Baker, M., Bonanni, L., et al., 2016. Neurophysiological biomarkers for Lewy body dementias. Clin. Neurophysiol. 127, 349–359.
Dauer, W., Przedborski, S., 2003. Parkinson's disease: mechanisms and models. Neuron 39, 889–909.
Dawson, T.M., Dawson, V.L., 2003. Molecular pathways of neurodegeneration in Parkinson's disease. Science 302, 819–822.
DeMaagd, G., Philip, A., 2015. Parkinson's disease and its management: part 1: disease entity, risk factors, pathophysiology, clinical presentation, and diagnosis. P. T. 40, 504–532.
Desplats, P., Lee, H.-J., Bae, E.-J., Patrick, C., Rockenstein, E., et al., 2009. Inclusion formation and neuronal cell death through neuron-to-neuron transmission of α-synuclein. Proc. Natl. Acad. Sci. U.S.A. 106, 13010–13015.
Edison, P., Rowe, C.C., Rinne, J.O., Ng, S., Ahmed, I., et al., 2008. Amyloid load in Parkinson's disease dementia and Lewy body dementia measured with [11C]PIB positron emission tomography. J. Neurol. Neurosurg. Psychiatry 79, 1331–1338.
Emre, M., Aarsland, D., Brown, R., Burn, D.J., Duyckaerts, C., Mizuno, Y., et al., 2007. Clinical diagnostic criteria for dementia associated with Parkinson's disease. Mov. Disord. 22, 1689–1707.
Farooqui, A.A., 2012. Phytochemicals, Signal Transduction, and Neurological Disorders. Springer, New York.
Farooqui, A.A., 2014. Inflammation and Oxidative Stress in Neurological Disorders. Springer, New York.
Fasano, A., Visanji, N.P., Liu, L.W., Lang, A.E., Pfeiffer, R.F., 2015. Gastrointestinal dysfunction in Parkinson's disease. Lancet Neurol. 14, 625–639.
Floor, E., Wetzel, M.G., 1998. Increased protein oxidation in human substantia nigra pars compacta in comparison with basal ganglia and prefrontal cortex measured with an improved dinitrophenylhydrazine assay. J. Neurochem. 70, 268–275.
Gomperts, S.N., Rentz, D.M., Moran, E., Becker, J.A., Locascio, J.J., et al., 2008. Imaging amyloid deposition in Lewy body diseases. Neurology 71, 903–910.

Greenamyre, J.T., Betarbet, R., Sherer, T.B., 2003. The rotenone model of Parkinson's disease: genes, environment and mitochondria. Parkinsonism Relat. Disord. 9 (Suppl. 2), S59–S64.

Harms, A.S., Thome, A.D., Yan, Z., Schonhoff, A.M., Williams, G.P., et al., 2018. Peripheral monocyte entry is required for alpha-Synuclein induced inflammation and Neurodegeneration in a model of Parkinson disease. Exp. Neurol. 300, 179–187.

Hirsch, E.C., Vyas, S., Hunot, S., 2012. Neuroinflammation in Parkinson's disease. Parkinsonism Relat. Disord. 18, S210–S212.

Hsu, L.J., Sagara, Y., Arroyo, A., Rockenstein, E., Sisk, A., et al., 2000. Alpha-synuclein promotes mitochondrial deficit and oxidative stress. Am. J. Pathol. 157, 401–410.

Hu, Q., Wang, G., 2016. Mitochondrial dysfunction in Parkinson's disease. Transl. Neurodegener. 5, 14.

Jimenez-Ferrer, I., Jewett, M., Tontanahal, A., Romero-Ramos, M., Swanberg, M., 2017. Allelic difference in Mhc2ta confers altered microglial activation and susceptibility to alpha-synuclein-induced dopaminergic neurodegeneration. Neurobiol. Dis. 106, 279–290.

Jokinen, P., Scheinin, N., Aalto, S., Nagren, K., Savisto, N., et al., 2010. [(11)C]PIB-, [(18)F]FDG-PET and MRI imaging in patients with Parkinson's disease with and without dementia. Parkinsonism Relat. Disord. 16, 666–670.

Karpenko, M.N., Vasilishina, A.A., Gromova, E.A., Muruzheva, Z.M., Bernadotte, A., 2018. Interleukin-1β, interleukin-1 receptor antagonist, interleukin-6, interleukin-10, and tumor necrosis factor-α levels in CSF and serum in relation to the clinical diversity of Parkinson's disease. Cell. Immunol. 327, 77–82.

Katsumoto, A., Lu, H., Miranda, A.S., Ransohoff, R.M., 2014. Ontogeny and functions of central nervous system macrophages. J. Immunol. 193, 2615–2621.

Keeney, P.M., Xie, J., Capaldi, R.A., Bennett Jr., J.P., 2006. Parkinson's disease brain mitochondrial complex I has oxidatively damaged subunits and is functionally impaired and misassembled. J. Neurosci. 26, 5256–5564.

Kim, W.S., Kagedal, K., Halliday, G.M., 2014. Alpha-synuclein biology in Lewy body diseases. Alzheimer Res. Ther. 6, 73.

Klein, C., Westenberger, A., 2012. Genetics of Parkinson's disease. Cold Spring Harb. Perspect. Med. 2, a008888.

Kramer, M.L., Schulz-Schaeffer, W.J., 2007. Presynaptic alpha-synuclein aggregates, not Lewy bodies, cause neurodegeneration in dementia with Lewy bodies. J. Neurosci. 27, 1405–1410.

Krogsgaard, M., Davis, M.M., 2005. How T cells 'see' antigen. Nat. Immunol. 6, 239–245.

Lee, H.J., Suk, J.E., Patrick, C., Bae, E.J., Cho, J.H., et al., 2010. Direct transfer of α-synuclein from neuron to astroglia causes inflammatory responses in synucleinopathies. J. Biol. Chem. 285, 9262–9272.

Lee, H.J., Kang, S.J., Lee, K., Im, H., 2011. Human α-synuclein modulates vesicle trafficking through its interaction with prenylated rab acceptor protein 1. Biochem. Biophys. Res. Commun. 412, 526–531.

Lee, H., Brekelmans, G.J., Roks, G., 2015. The EEG as a diagnostic tool in distinguishing between dementia with Lewy bodies and Alzheimer's disease. Clin. Neurophysiol. 126, 1735–1739.

Li, S., Le, W., 2017. Biomarker discovery in Parkinson's disease: present challenges and future opportunities. Neurosci. Bull. 33, 481–482.

Linazasoro, G., 2007. Classical Parkinson disease versus Parkinson complex—reflections against staging and in favour of heterogeneity. Eur. J. Neurol. 14, 721–728.

Litvan, I., et al., 2011. MDS Task Force on mild cognitive impairment in Parkinson's disease: critical review of PD-MCI. Mov. Disord. 26, 1814–1824.

Liu, G.H., Qu, J., Suzuki, K., Nivet, E., Li, M., Montserrat, N., et al., 2012. Progressive degeneration of human neural stem cells caused by pathogenic LRRK2. Nature 491, 603–607.

Liu, J., Wang, X., Lu, Y., Duan, C., Gao, G., et al., 2017. Pink1 interacts with α-synuclein and abrogates α-synuclein-induced neurotoxicity by activating autophagy. Cell Death Dis. 8, e3056.

Logroscino, G., 2005. The role of early-life environmental risk factors in Parkinson disease: what is the evidence? Environ. Health Perspect. 113, 1234–1238.

Mak, E., Su, L., Williams, G.B., O'Brien, J.T., 2015. Neuroimaging correlates of cognitive impairment and dementia in Parkinson's disease. Parkinsonism Relat. Disord. 21, 862–870.

Martin, I., Dawson, V.L., Dawson, T.M., 2011. Recent advances in the genetics of Parkinson's disease. Annu. Rev. Genomics Hum. Genet. 12, 301–325.

McKeith, I., 2007. Dementia with Lewy bodies. In: Koller, W.C., Melamed, E. (Eds.), Parkinson's Disease and Related Disorders, Part II. Handbook of Clinical Neurology, vol. 84. An Imprint of Elsevier, San Diego, CA, pp. 531–548.

McKeith, I.G., Boeve, B.F., Dickson, D.W., Halliday, G., Taylor, J.P., Weintraub, D., et al., 2017. Diagnosis and management of dementia with Lewy bodies: fourth consensus report of the DLB consortium. Neurology 89, 88–100.

Meeus, B., Nuytemans, K., Crosiers, D., Engelborghs, S., Peeters, K., et al., 2010. Comprehensive genetic and mutation analysis of familial dementia with Lewy bodies linked to 2q35-q36. J. Alzheimer Dis. 20, 197–205.

Meulener, M.C., Xu, K., Thompson, L., Ischiropoulos, H., Bonini, N.M., 2006. Mutational analysis of DJ-1 in Drosophila implicates functional inactivation by oxidative damage and aging. Proc. Natl. Acad. Sci. U.S.A. 103, 12517–12522.

Mollenhauer, B., Cullen, V., Kahn, I., Krastins, B., Outeiro, T.F., Pepivani, I., et al., 2008. Direct quantification of CSF alpha-synuclein by ELISA and first cross-sectional study in patients with neurodegeneration. Exp. Neurol. 213, 315–325.

Mollenhauer, B., Schulz-Schaeffer, W., Schlossmacher, M., 2011. Synaptic alpha-synuclein pathology as the likely cause of Parkinson's disease dementia. Lancet Neurol. 10, 68–69.

Mosley, R.L., Hutter-Saunders, J.A., Stone, D.K., Gendelman, H.E., 2012. Inflammation and adaptive immunity in Parkinson's disease. Cold Spring Harb. Perspect. Med. 2, 009381.

Nalls, M.A., Duran, R., Lopez, G., Kurzawa-Akanbi, M., McKeith, I.G., et al., 2013. A multicenter study of glucocerebrosidase mutations in dementia with Lewy bodies. JAMA Neurol. 70, 727–735.

Nedelska, Z., Ferman, T.J., Boeve, B.F., Przybelski, S.A., Lesnick, T.G., et al., 2015. Pattern of brain atrophy rates in autopsy-confirmed dementia with Lewy bodies. Neurobiol. Aging 36, 452–461.

Nikolaus, S., Antke, C., Muller, H.W., 2009. In vivo imaging of synaptic function in the central nervous system: I. Movement disorders and dementia. Behav. Brain Res. 204, 1–31.

Ohrfelt, A., Grognet, P., Andreasen, N., Wallin, A., Vanmechelen, E., Blennow, K., et al., 2009. Cerebrospinal fluid alpha-synuclein in neurodegenerative disorders-a marker of synapse loss? Neurosci. Lett. 450, 332–335.

Pievani, M., de Haan, W., Wu, T., Seeley, W.W., Frisoni, G.B., 2011. Functional network disruption in the degenerative dementias. Lancet. Neurol. 10, 829–843.

Ramonet, D., Daher, J.P., Lin, B.M., Stafa, K., Kim, J., Banerjee, R., et al., 2011. Dopaminergic neuronal loss, reduced neurite complexity and autophagic abnormalities in transgenic mice expressing G2019S mutant LRRK2. PLoS One 6, e18568.

Reeve, A., Simcox, E., Turnbull, D., 2014. Ageing and Parkinson's disease: why is advancing age the biggest risk factor? Ageing Res. Rev. 14, 19–30.

Reynolds, A.D., Kadiu, I., Garg, S.K., Glanzer, J.G., Nordgren, T., et al., 2008. Nitrated alpha-synuclein and microglial neuroregulatory activities. J. Neuroimmune Pharmacol. 3, 59–74.

Reynolds, A.D., Stone, D.K., Mosley, R.L., Gendelman, H.E., 2009. Proteomic studies of nitrated alpha-synuclein microglia regulation by CD4 + CD25 + T cells. J. Proteome Res. 8, 3497–3511.

Roks, G., Korf, E.S., van der Flier, W.M., Scheltens, P., Stam, C.J., 2008. The use of EEG in the diagnosis of dementia with Lewy bodies. J. Neurol. Neurosurg. Psychiatry 79, 377–380.

Saunders, J.A., Estes, K.A., Kosloski, L.M., Allen, H.E., Dempsey, K.M., et al., 2012. CD4 + regulatory and effector/memory T cell subsets profile motor dysfunction in Parkinson's disease. J. Neuroimmune Pharmacol. 7, 927–938.

Savitt, J.M., Dawson, V.L., Dawson, T.M., 2006. Diagnosis and treatment of Parkinson disease: molecules to medicine. J. Clin. Invest. 116, 1744–1754.

Schlossmacher, M.G., Frosch, M.P., Gai, W.P., Medina, M., Sharma, N., Forno, L., et al., 2002. Parkin localizes to the Lewy bodies of Parkinson disease and dementia with Lewy bodies. Am. J. Pathol. 160, 1655–1667.

Schwab, A.J., Ebert, A.D., 2015. Neurite aggregation and calcium dysfunction in iPSC-derived sensory neurons with Parkinson's disease-related LRRK2 G2019S mutation. Stem Cell Rep. 5, 1039–1052.

Shi, M., Liu, C., Cook, T.J., Bullock, K.M., Zhao, Y., et al., 2014. Plasma exosomal α-synuclein is likely CNS-derived and increased in Parkinson's disease. Acta Neuropathol. 128, 639–650.

Sohma, H., Imai, S.-I., Takei, N., Honda, H., Matsumoto, K., et al., 2013. Evaluation of annexin A5 as a biomarker for Alzheimer's disease and dementia with Lewy bodies. Front. Aging Neurosci. 5, 15.

Srinivas, B.M.M., Mythri, R., Jagatha, B., Vali, S., 2008. Neuroprotective effect of curcumin against inhibition of mitochondrial complex I in vitro and in vivo. Implications for Parkinson's disease explained via in silico studies. J. Neurochem. 106, 9–28.

Stevens, C.H., Rowe, D., Morel-Kopp, M.C., Orr, C., Russell, T., et al., 2012. Reduced T helper and B lymphocytes in Parkinson's disease. J. Neuroimmunol. 252, 95–99.

Subramaniam, S.R., Chesselet, M.F., 2013. Mitochondrial dysfunction and oxidative stress in Parkinson's disease. Prog. Neurobiol. 106–107, 17–32.

Sui, Y.T., Bullock, K.M., Erickson, M.A., Zhang, J., Banks, W.A., 2014. α synuclein is transported into and out of the brain by the blood-brain barrier. Peptides 62, 197–202.

Sulzer, D., Alcalay, R.N., Garretti, F., Cote, L., Kanter, E., et al., 2017. T cells from patients with Parkinson's disease recognize alpha-synuclein peptides. Nature 546, 656–661.

Teixeira Dos Santos, M.C., Bell, R., da Costa, A.N., 2016. Recent developments in circulating biomarkers in Parkinson's disease: the potential use of miRNAs in a clinical setting. Bioanalysis 8, 2497–2518.

Tolö, J., Taschenberger, G., Leite, K., Stahlberg, M.A., Spehlbrink, G., et al., 2018. Pathophysiological consequences of neuronal α-synuclein overexpression: impacts on ion homeostasis, stress signaling, mitochondrial integrity, and electrical activity. Front. Mol. Neurosci. 11, 49.

Trancikova, A., Tsika, E., Moore, D.J., 2012. Mitochondrial dysfunction in genetic animal models of Parkinson's disease. Antioxid. Redox Signal. 16, 896–919.

Ueda, K., Fukushima, H., Masliah, E., Xia, Y., Iwai, A., et al., 1993. Molecular cloning of cDNA encoding an unrecognized component of amyloid in Alzheimer disease. Proc. Natl. Acad. Sci. U.S.A. 90, 11282–11286.

Van Der Flier, W.M., Pijnenburg, Y.A., Prins, N., Lemstra, A.W., Bouwman, F.H., et al., 2014. Optimizing patient care and research: the Amsterdam dementia cohort. J. Alzheimer Dis. 41, 313–327.

Vila, M., Vukosavic, S., Jackson-Lewis, V., Neystat, M., Jakowec, M., Przedborski, S., 2000. α-Synuclein up-regulation in substantia nigra dopaminergic neurons following administration of the parkinsonian toxin MPTP. J. Neurochem. 74, 721–729.

Wang, Y., Yang, Z., Le, W., 2017. Tiny but mighty: promising roles of MicroRNAs in the diagnosis and treatment of Parkinson's disease. Neurosci. Bull. 33, 543–551.

Williams-Gray, C.H., Foltynie, T., Brayne, C.E., Robbins, T.W., Barker, R.A., 2007. Evolution of cognitive dysfunction in an incident Parkinson's disease cohort. Brain 130, 1787–1798.

Yue, X., Li, H., Yan, H., Zhang, P., Chang, L., Li, T., 2016. Risk of Parkinson disease in diabetes mellitus: an updated meta-analysis of population-based cohort studies. Medicine. 95, e3549.

Zaltieri, M., Longhena, F., Pizzi, M., Missale, C., Spano, P., et al., 2015. Mitochondrial dysfunction and α-synuclein synaptic pathology in Parkinson's disease: who's on first? Parkinsons Dis. 2015, 108029.

Zhang, T., Shi, Z., Wang, Y., Wang, L., Zhang, B., Chen, G., et al., 2018. Akt3 deletion in mice impairs spatial cognition and hippocampal CA1 long long-term potentiation through downregulation of mTOR. Acta Physiol. e13167.

CHAPTER 8

Insulin resistance, dementia, and depression

Introduction

In developed countries, life expectancy is increasing with a constant rate (United Nations, 2015; WHO, 2016) due to advances in modern medicine, nutrition, hygiene, and safety standards (Aw et al., 2007; United Nations, 2015; WHO, 2016). It is predicted that in the United States, the number of seniors will increase from approximately 45 million currently to 70 million by the year 2030 (Ortman et al., 2014). Similarly, in the European Union the number of seniors over the age of 80 is expected to grow from 5% to 12% of the population (The 2015 Ageing Report, 2015). In 2016 Canada population had more persons over the age of 65 (16.9%) than under the age of 15 (16.6%) (Government of Canada, Statistics Canada, 2017). It is predicted that in the next 50 years, the elderly will comprise approximately 20% of the world population (Ellison et al., 2015). Therefore it is imperative that the scientific and medical communities must investigate approaches to minimize age-associated disease and maximize quality of life. Insulin-linked neurological disorders are debilitating diseases, which not only caregivers psychologically, but also have tremendous socioeconomic burden to the families. Changes in the immune system have long been recognized to occur with aging, and it is now appreciated that neuroinflammation likely contributes to age-associated and insulin-linked neurological diseases (Ransohoff, 2016). However, it is less well understood how specific changes in the immune system and blood–brain barrier (BBB) with aging and insulin resistance may affect brain functions by altering homeostasis between neurons and glial cells contributing to neurological diseases.

Increase in life expectancy is due to aging, a multifactorial and heterogeneous process, which is caused by a progressive decline in the physiological integrity of different organs of the human body leading to impaired body function and enhanced vulnerability to death. Increase in life expectancy is accompanied by increased susceptibility to age-related

Insulin Resistance as a Risk Factor in Visceral and Neurological Disorders
DOI: https://doi.org/10.1016/B978-0-12-819603-8.00008-0

diseases and, consequently, by the presence of multiple pathologies and comorbidities (Miwa et al., 2008; Farooqui, 2014), which are characterized by chronic processes, such as induction of oxidative stress and inflammation (Miwa et al., 2008; Farooqui, 2014). These processes in conjunction with immunosenescence result in a decline of multiple physiological systems, vulnerability, and the fear of functional dependence (Simen et al., 2011; Farooqui, 2014). This deterioration is a crucial risk factor for the many neurodegenerative diseases [Alzheimer's disease (AD), Parkinson's disease (PD), and Huntington's disease], neuropsychiatric diseases (dementia and depression), cardiovascular disorders (congestive heart failure, cardiomyopathy, and atrial fibrillation), cancer, dementia, depression, and visceral diseases (type 2 diabetes, metabolic syndrome, and sleep apnea) (López-Otín et al., 2013; Farooqui, 2013). Among abovementioned pathological conditions, dementia is the oldest syndrome since ancient times. However, it has been assumed a normal part of growing old. While the risk of becoming demented increases with age, not everybody becomes demented. Most of these pathological conditions including dementia are characterized by induction of oxidative stress, neuroinflammation, and insulin resistance in conjunction with immune senescence. As mentioned in Chapter 1, Insulin resistance and obesity, insulin resistance is characterized by a diminished ability of cells or tissues to respond to physiological levels of insulin. Genetic and environmental factors, including aging, obesity, lack of exercise, and stress, contribute to insulin resistance. Disorders of glucose and lipid metabolism cause defects in insulin signaling that are linked to various pathological conditions (Maciejczyk et al., 2019). Thus molecular and cellular mechanisms of insulin resistance are important in understanding the pathogenesis of various diseases mediated by insulin resistance. Insulin resistance is promoted not only by circulating high levels of free fatty acids, excessive energy intake, elevated plasma levels of diacylglycerol, and triacylglycerol in nonadipose tissue, including skeletal muscle, liver, heart, and β cells. In fact, lipid infusions and high-fat feeding in human subjects and rodents reduce insulin-stimulated glucose disposal (Han et al., 1997). These data suggest that defects in lipid metabolism may lead to the impairment of insulin signaling seem to be a major mechanism for insulin resistance (Dresner et al., 1999; Maciejczyk et al., 2019). Impaired insulin signaling not only affects insulin-stimulated glucose metabolism in skeletal muscle but also impairs other actions of insulin in diverse tissues including, liver, adipose tissue, heart, and the vasculature (Ouwens et al., 2005). In addition, high levels

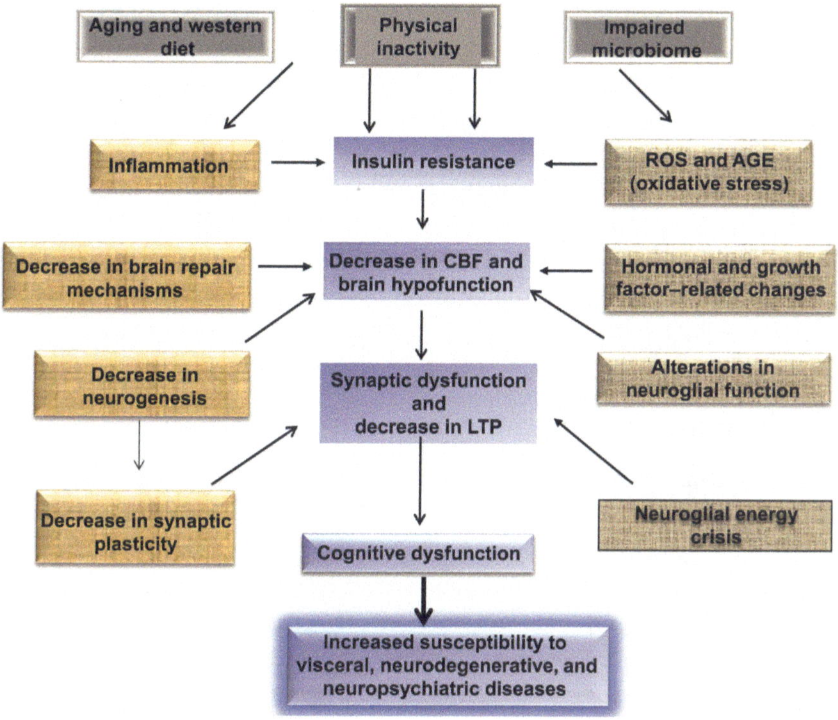

Figure 8.1 Factors contributing to insulin resistance and susceptibility to visceral, neurodegenerative, and neuropsychiatric diseases in humans.
AGEs, Advanced glycated end products; *LTP*, long-term potentiatation; *ROS*, reactive oxygen species.

of reactive oxygen species (ROS) and advanced glycation end products (AGEs) can also contribute to insulin resistance (Fig. 8.1) (Singh et al., 2001; Unoki and Yamagishi, 2008; Yamagishi et al., 2012). Furthermore, both ROS and AGEs may exert negative effects on tissues but interacting and upregulating nuclear factor-kappaB, a transcription factor, which is crucial mediator of neuroinflammation (Goldin et al., 2006). AGEs may also prevent the production of nitric oxide (NO) in the endothelium, thereby mitigating its vasodilatory effects. Moreover, a complex of AGE–RAGE may interfere with vascular structure, making it permeable to macromolecule invasion and resultant pathology (Goldin et al., 2006). It is also reported that inhibition of RAGE prevents the AGE-related changes in vascular permeability in diabetic rats (Wautier et al., 1996) supporting that view that a prominent feature of insulin resistance may be to promote and propagate an inflammatory cascade in the brain and vascular injury in blood vessels. These changes also occur in AD patients compared to age-matched controls and may ultimately contribute to AD pathology (Sasaki et al., 1998; Srikanth et al., 2011). Aging also depends upon the biological, the psychological, and the social aspects of human life. Institutionalization of the elderly results in changes in their lifestyle and is often accompanied by psychological and social deficiencies due to isolation from familiar surroundings (Paniz et al., 2007; Milà et al., 2012). The very fact of living in a public retirement home leads to a reduction in their autonomy and can result in a reduced quality of life (Estrada et al., 2011).

Normal aging and cognitive function

Cognitive function is defined as the ability of brain to process information about the thinking, memory, recall, mental flexibility, problem solving, and learning (Morrison and Baxter, 2012). Cognitive function is regulated not only by neurochemical and age-related intricate synaptic changes, but also by neuronal and glial interactions (Morrison and Baxter, 2012). Cognitive dysfunction refers to the loss of intellectual functions such as thinking, remembering, and reasoning that interfere with daily activities. Patients with cognitive dysfunction loss ability to learn, recall, concentrate, and problem solving. Neuropathologic changes associated with cognitive dysfunction include multifocal and/or diffuse disease and focal lesions: multiinfarct encephalopathy, white matter lesions or arteriosclerotic subcortical (leuko)encephalopathy, multilacunar state, mixed cortico-subcortical type, borderline/watershed lesions, rare granular cortical atrophy, postischemic encephalopathy, and hippocampal sclerosis (Farooqui, 2019). They result from systemic, cardiac, and local large or small vessel disease. Recent data indicate that cognitive dysfunction is commonly associated with widespread small ischemic/vascular lesions (microinfarcts, lacunes) throughout the brain with predominant involvement of subcortical and functionally important brain areas (Farooqui, 2019). Their pathogenesis is multifactorial, and their pathophysiology affects neuronal networks involved in cognition, memory, behavior, and executive functioning. Aging differently affects various aspects of episodic, explicit, semantic, short term, procedural, and perceptual memories (Nilsson, 2003). Explicit memory includes episodic memory involves the conscious recall of events and experiences. The semantic memory involves the conscious recall of facts and information (Tulving, 1987). Episodic memory is affected by aging much more than semantic memory. Cognitive dysfunction is one of the primary defect of aging process. Cognitive dysfunction is modulated by numerous factors including genes, age, lifestyle, and environmental factors. Several studies have indicated that hormonal changes also modulate cognitive function (Ebner et al., 2015; Kawas et al., 1997; Lupien et al., 2007). In brain, different regions respond to aging with different rate. Thus forebrain regions (superior frontal gyrus, entire cortex, and hippocampal CA1) are more susceptible to aging than other regions of the brain due to aging and hormonal changes (Berchtold et al., 2008; Zeier et al., 2011). Furthermore, in the dentate gyrus, neurogenesis declines dramatically with ageing and pathological situation, this may represent another possible cause of age-related impairment in hippocampal-dependent memory and cognitive decline. Above changes along with decrease in synaptic plasticity and long-term potentiation may promote synaptic dysfunction, which ultimately may be responsible for cognitive dysfunction (Fig. 8.1).

Finally, mounting evidence suggests that systemic immune activation also contributes to cognitive impairment in elderly. The brain monitors peripheral innate immune responses by several means that act in parallel. Two pathways: one involving afferent nerve and the other involves Toll-like receptors (TLRs) on microglial residing (Watkins et al., 1994; Romeo et al., 2001). The activation of microglial cells triggers the release of soluble mediators called proinflammatory cytokines [interleukin-1α and β (IL-1α and IL-1β), tumor necrosis factor-α (TNF-α), interleukin-6 (IL-6), and C-reactive protein (CRP)]. These mediators coordinate the local and systemic inflammatory response to microbial pathogens. These mediators act not only on peripheral tissue, but also on the brain to cause the aforementioned behavioral symptoms of inflammation. These release of proinflammatory cytokines not only results in the impairment of overall cognition (Schram et al., 2007), but also alters specific functions, such as reduced processing speed (Bettcher et al., 2014), executive function (Heringa et al., 2014), and memory (Teunissen et al., 2003). Now it is becoming increasingly evident that cognitive decline predisposes individuals for neurological and psychiatric disorders eventually affecting the quality of life. The intensity of cognitive decline is markedly increased not only in patients with diabetes, metabolic syndrome, atrial fibrillation, stroke and dementia, neurodegenerative diseases, but also in patients of neuropsychiatric diseases (Schuh et al., 2011; Farooqui, 2013; Johansson, 2015). Collective evidence suggests that cognitive decline during aging is a multifactorial process, which is controlled by several factors such as genes for oxidative stress, neuroinflammation, immune response, mitochondrial functions, growth factors, neuronal survival, and calcium homeostasis (Lu et al., 2004; Loerch et al., 2008).

Neurochemical aspects of dementia

Dementia is a chronic or progressive loss of cortical and subcortical functions resulting in cognitive decline, which is characterized by synaptic loss, impairment in memory and activities of daily living, resulting in altered mood, behavior, and personality. Dementia affects thinking, orientation, comprehension, calculation, learning capacity, language, and judgment (Farooqui, 2019). The prevalence of dementia increases with advancing age. Dementia incidence increases exponentially with age between the ages of 65 and 90 years and doubles approximately every 5 years (Jorm and Jolley, 1998). Whether this doubling of rates continues at older ages and whether the pattern is the same in very elderly men and women are not known (Fishman, 2017; Farooqui, 2019). The number of people aged 90 years and older was approximately 2 million in 2007 but will increase to 8.7 million by the middle of the 21st century (U.S. Census Bureau, 2008), making the oldest old the fastest growing segment of the US population. Precise estimates of dementia rates in the oldest old are therefore critical for accurate projection of the number of affected people and estimation of the social and economic impact of dementia in future years (Sosa-Ortiz et al., 2012; Fishman, 2017; Farooqui, 2019). All symptoms of dementia are not displayed by patients at one time. Only 75% oldest adults show not all but some symptoms of dementia (Lyketsos et al., 2002). A recent study estimated that 8.8% of adults aged 65 and older in the United States suffered dementia in 2012 (Langa et al., 2017). Predementia or mild cognitive impairment (MCI) is characterized by objective impairment in cognition that is not severe enough to require help with usual activities of daily living (Sosa-Ortiz et al., 2012; Rizzi et al., 2014). Moderate dementia involves loss of ability to communicate. In contrast, severe dementia is not only associated with all symptoms, but also need for full-time care. Simple tasks, such as sitting and holding one's head up become impossible along with loss of bladder control. Collective evidence suggests that dementia is a global health problem and burden of public health. Risk factors for dementia include advancing age, long-term consumption of western diet, physical and cognitive inactivity, and epigenetic and environmental factors (Farooqui, 2017). Other risk factors for dementia include (1) cardiovascular (e.g., hypertension, atrial fibrillation, diabetes, sleep apnea, insulin resistance, obesity) and cerebrovascular problems (e.g., stroke); (2) excessive alcohol consumption; (3) social isolation; (4) traumatic brain injury; (5) hearing loss; and (6) having one or two copies of the ApoE-ε4 genetic variant (Farooqui, 2017). A combination of all these factors is known to contribute to the pathogenesis and development of the dementia syndrome (Farooqui, 2017; Farooqui, 2019).

Classification of dementias

Several types dementia have been reported to occur in human population including AD type of dementia, vascular dementia, Lewy body dementia, frontotemporal dementia/degeneration (FTD), and infective dementia. The most common type of dementia is AD which accounts for 70% of cases; the next most common type is vascular dementia, which alone accounts for approximately 17% of cases (Alzheimer's Association (A.D.), 2014). The prevalence of Lewy body dementia, FTD, and infective dementia is 10%, 3%, and 5%, respectively (Fig. 8.2). AD type dementia is often accompanied by pathological conditions such as diabetes and stroke (Todd et al., 2013; Langa et al., 2017). Dementia on its own is associated

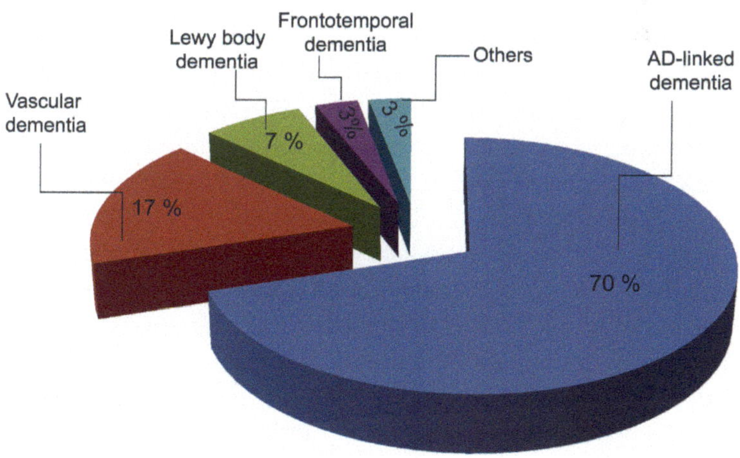

Figure 8.2 Proportions of various types of dementia found in human population. *AD*, Alzheimer's type of dementia; *FTD*, frontotemporal dementia; *ID*, infective dementia; *LBD*, Lewy body dementia; *MD*, mixed dementia; *VD*, vascular dementia.

with significantly increased risk of death, and other pathological conditions such as diabetes and stroke. These conditions increase the risk of morbidity (Todd et al., 2013). Secondary causes of dementia include AIDS and multiple sclerosis (Kabasakalian and Finney, 2009; Ironside and Bell, 2007).

Alzheimer's type of dementia

AD type of dementia is characterized clinically by progressive memory and orientation loss and other cognitive deficits, including impaired judgment and decision making, apraxia, and language disturbances. AD type of dementia is accompanied by the accumulation of Aβ peptide in the form of senile plaques (Farooqui, 2017) and deposition of hyperphosphorylated tau in the form of neurofibrillary tangles. The accumulation of senile plaques is thought to be one of the major contributors to dementia caused by AD (Villemagne et al., 2013). The number of NFTs is known to be paralleled with the severity of AD. Besides these lesions, sustained neuroinflammatory processes occur, involving notably micro- and astro-glial activation, which contribute to the progression of AD (Fig. 8.3). Furthermore, there also occurs an adaptive immune response during the progression of disease course involving vascular and parenchymal T-cell in AD patients' brain. The underlying mechanisms of this infiltration and its consequences with regard to Tau pathology are not known.

Insulin resistance is characterized by impaired insulin signal transduction, diminished glucose uptake, and dysregulated energy metabolism, is frequently preceded by glucose intolerance, and can lead to the development of type 2 diabetes. Numerous clinical, basic, and epidemiological studies have indicated that insulin resistance contributes to AD pathogenesis. For example, the Hisayama Study has shown that systemic insulin resistance is related to AD pathology (Matsuzaki et al., 2010), and ample evidences show that insulin resistance contributes to cognitive dysfunction (Farooqui, 2017). Actions of insulin in the brain include control of food intake (through insulin receptors located in the olfactory bulb and thalamus) and effects the cognitive functions, including memory. Insulin signaling modulates Aβ and Aβ metabolism and tau phosphorylation through phosphatidylinositol 3-kinase (PtdIns 3K)/Akt pathway (De Felice et al., 2009). Recent studies have indicated that glycogen synthase kinase-3β (GSK-3β) may be the potential link between type 2 diabetes and AD (Fig. 8.4). In type 2 diabetes, GSK-3β is the crucial enzyme of

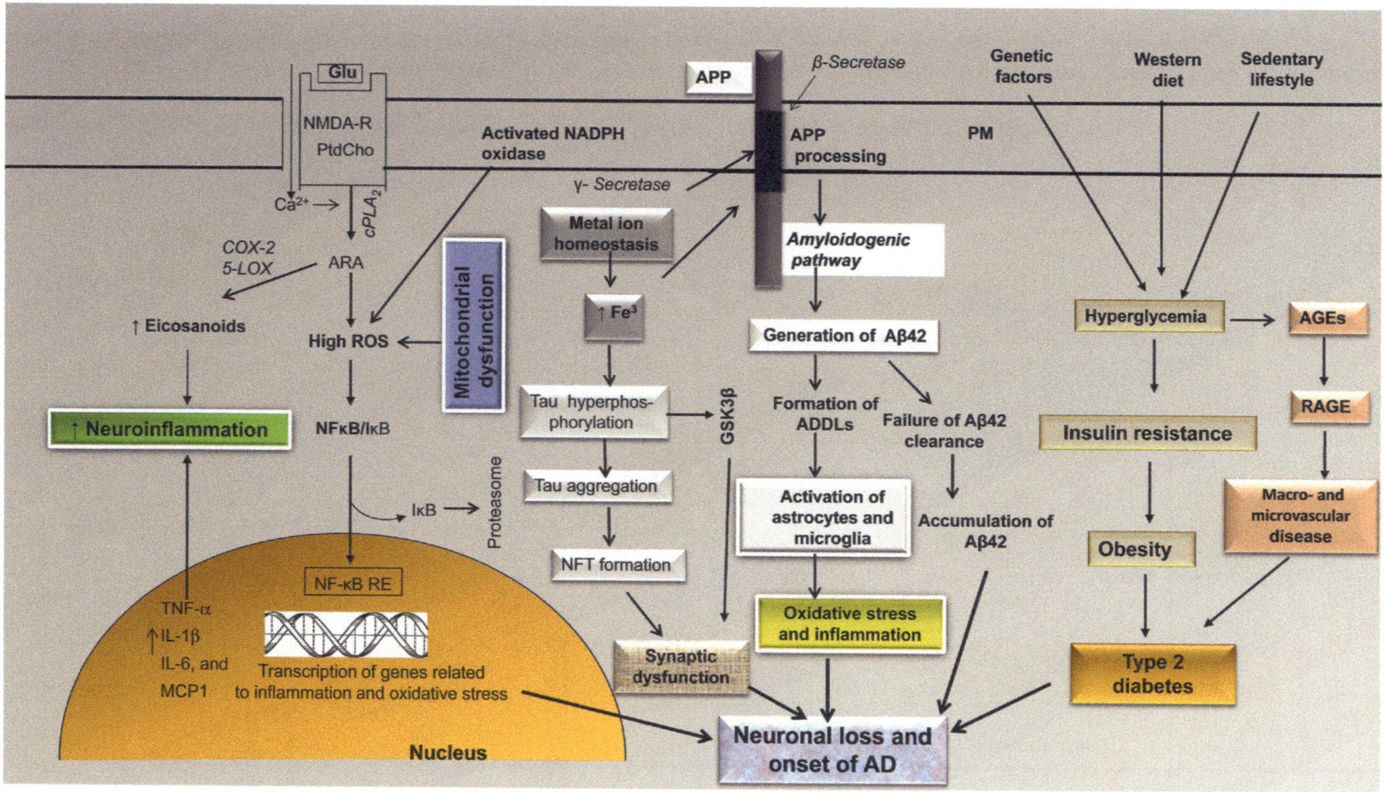

Figure 8.3 Hypothetical diagram showing molecular mechanisms contributing to insulin resistance and pathogenesis of AD.
APP, Amyloid precursor protein; *ARA*, arachidonic acid; *Bcl-2*, B-cell lymphoma 2; *COX-2*, cyclooxygenase-2; *$cPLA_2$*, cytosolic phospholipase A_2; *cyto-c*, cytochrome; *Glu*, glutamate; *I-κB*, inhibitory subunit of NF-κB; *IL-1β*, interleukin-1β; *IL-6*, interleukin-6; *5-LOX*, 5-lipoxygenase; *lyso-PtdCho*, lyso-phosphatidylcholine; *MCP-1*, monocyte chemoattractant protein-1; *NF-κB*, nuclear factor-κB; *NF-κB-RE*, nuclear factor-κB-response element; *NMDA-R*, NMDA receptor; *PtdCho*, phosphatidylcholine; *TNF-α*, tumor necrosis factor-α.

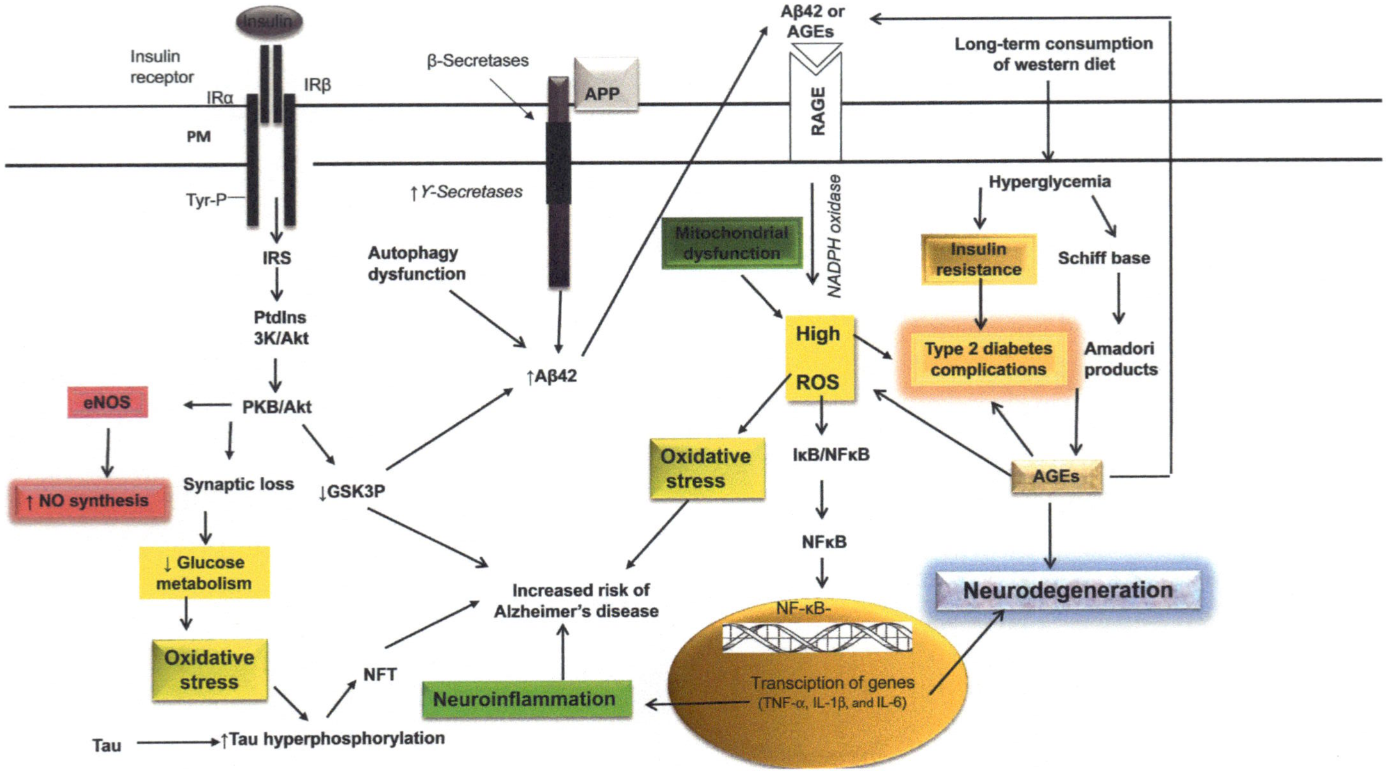

Figure 8.4 Hypothetical diagram showing the contribution of PtdIns 3K/Akt, AGE/RAGE signaling, and insulin resistance in the pathogenesis of AD type dementia in humans.
Akt, Protein kinase B; *GSK3*, glycogen synthetase 3 kinase; *IRS-1*, insulin receptor substrate-1; *PM*, plasma membrane; *PtdIns 3K*, phosphatidylinositol 3-kinase; *RAGE*, receptor for advanced glycated products.

glycogen synthesis, which plays a key role in regulating blood glucose. More importantly, GSK-3β is one of the key factors leading to insulin deficiency and insulin resistance, and these processes are important hallmark of the occurrence and development of type 2 diabetes. In AD, GSK-3β plays an important role in hyperphosphorylation of microtubule-associated protein tau (tau), which is one of the pathological features in AD. GSK-3β is one of the important kinases of tau phosphorylation and is involved in the insulin/PtdIns 3K/Akt signaling pathway. Dysfunction of the insulin/PtdIns 3K/Akt signaling pathway, which regulates glucose metabolism in the brain, can lead to tau hyperphosphorylation in the brain of AD patents. Additionally, insulin resistance in type 2 diabetes may cause β-amyloid (Aβ) deposition, which will be cleared by tau, but excessive phosphorylation of tau will further aggravate the neurotoxicity; then damage the brain and affect the cognitive function (Zhang et al., 2013; Kandimalla et al., 2017). These observations support the view that GSK-3β is a potential link between type 2 diabetes and AD (Zhang et al., 2013; Kandimalla et al., 2017).

Risk factors for AD include old age, expression of *ApoE* genes, disruption of BBB, and brain hypofunction (Bertram and Tanzi, 2008; Zlokovic, 2011). AD type of dementia is diagnosed by neuroimaging, analyzing cerebrospinal fluid (CSF) for the levels of Aβ, and phosphorylated tau protein. Recent evidence supports the contention that AD may be a slow-progressing brain metabolic disease. Individuals with type 2 diabetes and obesity have a higher risk of developing AD due to the development of hyperglycemia and insulin resistance, which is caused by impaired insulin signaling. As mentioned above, insulin signaling regulates Aβ and tau, and Aβ has negative effects on insulin signaling; therefore, dysfunctional insulin signaling can enhance Aβ and tau pathology, and increased Aβ production can further exacerbate insulin resistance. On the basis of these studies, it is suggested that intranasal insulin is only effective treatment for early AD. However, individuals with the ApoE-ε4 allele do not respond well.

Neurochemical aspects of vascular dementia

The onset of vascular dementia occurs when the blood supply to the brain is reduced by various cerebrovascular pathologies, such as hypoperfusions or hemorrhages, angiopathies (affecting both large and small arteries) resulting from hypertension, smoking, and others (cerebral amyloid angiopathy) causing disruption of the BBB and neurovascular units, usually in hemispheric white matter (Iadecola, 2013) leading to a progressive decline in memory and cognitive function. Pathological features of vascular dementia are diffuse myelin pallor, astrocytic gliosis, and the loss of oligodendrocytes leading to rarefaction, vacuolization, and the loss of myelin and axons without definite necrosis, ultimately culminating in white matter lesions and lacunes (Román et al., 2002). The hardening of cerebral arteries in vascular dementia leads to additional reduction in blood flow, a process, which is a major contributor of cognitive decline. The lesion pattern of "pure" vascular dementia is related to arteriosclerosis and microangiopathies. The lesion pattern of vascular dementia differs from that in mixed-type dementia (AD with vascular encephalopathy), more often shows large infarcts, which suggests different pathogenesis of both types of lesions. Due to the high variability of cerebrovascular pathology and its causative factors, no validated neuropathologic criteria for vascular dementia are available, and a large variability across laboratories still exists in the procedures for morphologic examination and histology techniques (Farooqui, 2017). Furthermore, sustained cerebral hypoperfusion in vascular dementia may cause of white matter attenuation, a key feature common to both AD and vascular dementia along with involvement of cerebral small vessel disease. Chronic hypoperfusion contributes to white matter rarefaction, glial activation, and axon damage resulting in diffused ischemic-neuronal loss (Libon et al., 2006). It is proposed that insulin resistance and cerebral hypoperfusion are the common pathophysiological mechanism which contributes to cognitive decline and degenerative processes leading to vascular dementia and cerebral small vessel disease. White matter changes are also closely associated with increased risk for stroke, dementia, and disability (Duncombe et al., 2017). Collective evidence suggests that vascular dementia is a heterogeneous construct with cerebrovascular pathology that can range from large vessel infarction, to small vessel ischemic disease and microvascular injury. Although AD and cerebrovascular pathology may occur in isolation and independently, recent autopsy and neuroimaging studies show that the lines between AD and vascular dementia are often blurred (Schneider et al., 2004; Cullen et al., 2006). For many patients, markers of vascular injury coexist with traditional AD hallmarks to increase the risk for dementia (Schneider et al., 2004). In some cases, AD pathology may be promoted by a specific form of vascular injury (e.g., lacunar or microvascular infarction, cerebral microbleeds, BBB dysfunction) which affects Aβ transport between brain and periphery (Iliff et al., 2012; Carare et al., 2008; Weller et al., 2009), and thereby contributes to parenchymal and neurovascular Aβ deposition in or around the neurovascular unit. Conversely, AD pathology may cause vascular injury, as Aβ deposits in the vessel wall and inducing inflammation that damages the endothelium (Bell and Zlokovic, 2009). Recent AD research suggests that cardiovascular and metabolic risk factors (often referred to as "cardiometabolic risk factors") can affect various forms of dementia pathology and their progression over time via multiple interacting pathways. Converging evidence suggests that pathological features of vascular dementia are diffuse myelin pallor, astrocytic gliosis, and the loss of oligodendrocytes leading to rarefaction, vacuolization, and the loss of myelin and axons without definite necrosis, ultimately culminating in white matter lesions and lacunes (Román et al., 2002). These findings are supported by structural neuroimaging studies (Risacher and Saykin, 2013; Valkanova and Ebmeier, 2014). The onset of dementia syndrome may occur in several stages including MCI, mild dementia, moderate dementia, and severe dementia (Sosa-Ortiz et al., 2012; Rizzi et al., 2014; Hornykiewicz, 2008).

Neurochemical aspects of Lewy body dementia

Neuropathologically, Lewy body dementia is characterized by the accumulation of Lewy bodies and the presence of Lewy neurites in the brainstem, limbic system, and cortical areas (Hashimoto and Masliah, 1999; Fujishiro et al., 2008; Ince, 2011). Lewy body dementia is characterized by the abnormal accumulation of α-synuclein not only in striatonigral system but also in the limbic areas, the insula, frontal cortex, and subcortical nuclei (Hurtig et al., 2000; Marui et al., 2002; Jankovic, 2008). Neurochemically, Lewy body dementia is characterized by mitochondrial dysfunction, induction of oxidative stress, onset of excitotoxicity, energy failure, neuroinflammation, misfolding and aggregation of α-synuclein, impairment of protein clearance pathways, cell-autonomous mechanisms, and deficits in proteasomal function or autophagy-lysosomal degradation of defective proteins (e.g., α-synuclein) (Fig. 8.5) (Michel et al., 2016; Si et al., 2017; Maiti et al., 2017; Franco-Iborra et al., 2016; Truban et al., 2017; Moors et al., 2016).

High incidence of glucose intolerance ($>50\%$) has been observed in patients with Lewy body dementia and PD (Sandyk, 1993). In addition, it has been shown that patients with Lewy body dementia and PD exhibit

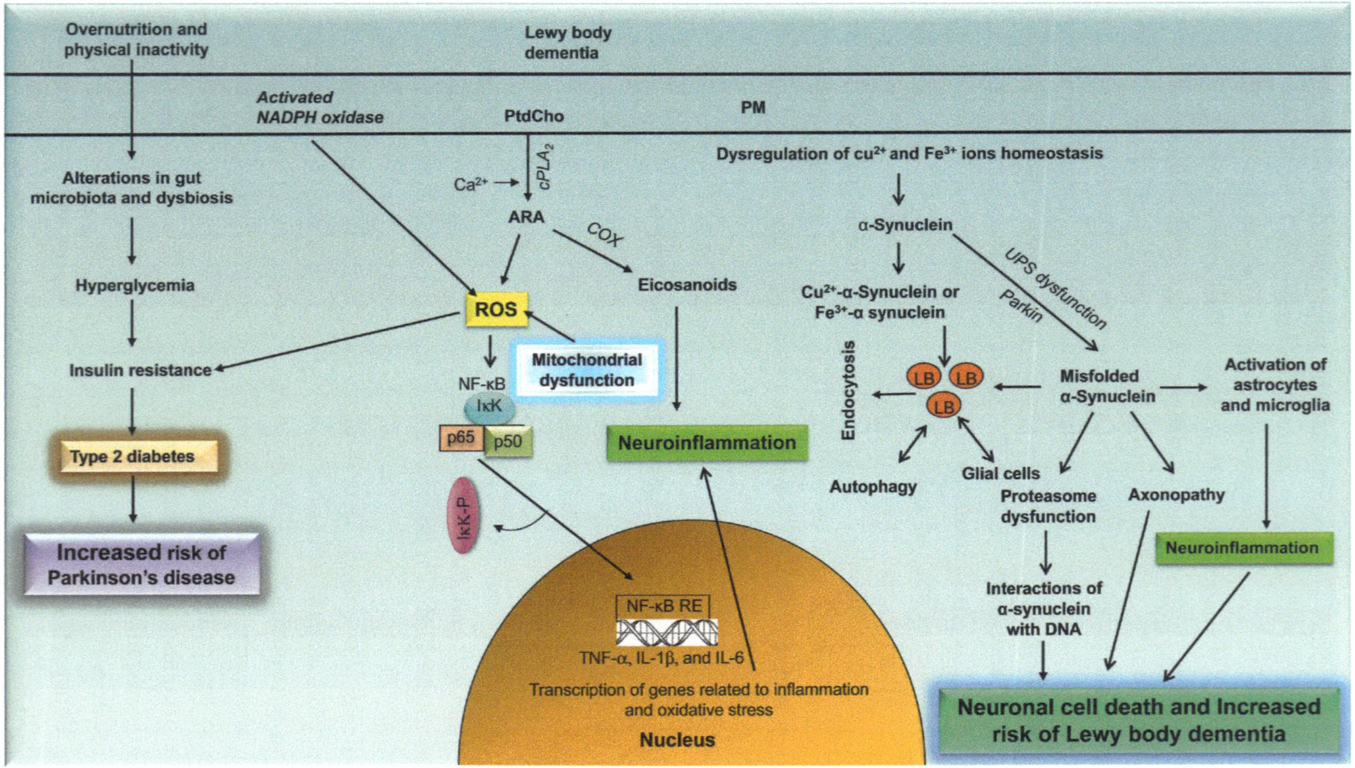

Figure 8.5 Hypothetical diagram showing contribution of insulin resistance in the pathogenesis of Pakinson's-linked dementia. *ARA*, Arachidonic acid; *ARE*, antioxidant response element; *cPLA*$_2$, cytosolic phospholipase A$_2$; *γ-GCL*, γ-glutamate cystein ligase; *HO-1*, heme oxygenase; *Keap1*, kelch-like ECH-associated protein 1; *LB*, Lewy body; *Maf*, small leucine zipper proteins; *NQO-1*, NADPH quinine oxidoreductase; *Nrf2*, nuclear factor E2-related factor 2; *PtdCho*, phosphatidylcholine.

increased autoimmune reactivity to insulin (Wilhelm et al., 2007), and that PD patients with diabetes have increased risk of PD (Papapetropoulos et al., 2004). Comorbidity in PD and type 2 diabetes patients correlates with a significant increase in the cost of care of affected individuals (Pressley et al., 2003). Hyperglycemia has been suggested to decrease the effectiveness of levodopa (L-DOPA) therapy and increase motor dyskinesias experienced by PD patients (Sandyk, 1993). Further, L-DOPA therapy may exacerbate hyperinsulinemia and hyperglycemia in PD patients (Boyd et al., 1971; Sirtori et al., 1972), possibly by diminishing peripheral glucose disposal in skeletal muscle (Smith et al., 2004). An early marker for the development of peripheral glucose intolerance may be insulin resistance in neural tissues. Hypothalamic parasympathetic nerves affect insulin release by beta cells, while sympathetic nerves act directly on the liver to affect hepatic glucose metabolism (Uyama et al., 2004). Thus it is important to understand the mechanistic link between nigrostriatal DA depletion and CNS insulin signaling. The molecular mechanisms contributing to the pathogenesis of Lewy body dementia is not known. However, it is suggested that impairment of insulin signaling increases the risk of PD (Morris et al., 2008; Bosco et al., 2012; Ashraghi et al., 2016; Pang et al., 2016). A case control study indicates that PD patients have a higher risk of having abnormal glucose metabolism and insulin resistance than nondemented PD patients (Bosco et al., 2012). Another study found that patients with PD and comorbid type 2 diabetes exhibited increased impairment in attentional function and executive function deficits than PD patients without type 2 diabetes (Ashraghi et al., 2016). These studies indicate that the impairment of insulin signaling may play a role in the pathogenesis Lewy body dementia and may initiate or accelerate the development of PD. Insulin treatment has positive effects on nervous system development and growth, and can alleviate and repair damage caused by the inflammation response. It is also reported that insulin treatment normalizes dopamine production and ameliorate motor impairments in the 6-OHDA-induced rat PD model (Hölscher, 2014; Pang et al., 2016). However, studies on insulin therapy in PD and Lewy body dementia patients have not yet been assessed.

Recent studies have indicated that dysregulation of the brain–gut–microbiota axis may also contribute to Lewy body dementia and gastrointestinal dysfunction, which is present in over 80% of PD subjects (Pfeiffer, 2011). These patients also show weight loss, dental deterioration, salivary excess, dysphagia, gastroparesis, and anorectal dysfunction

(Pfeiffer, 2011; Liddle, 2018). Moreover, the dysregulation of brain–gut–microbiota axis and alterations in microbiota composition (dysbiosis) may also be associated with the pathogenesis of PD itself, supporting the view that the pathological process in PD spreads from the gut to the brain (Braak et al., 2006; Lebouvier et al., 2009). This suggestion is supported by following findings. Clinically, gastrointestinal symptoms of PD often appear in patients before other neurological signs in the brain and aggregation of α-synuclein occurs in enteric nerves of PD patients (Liddle, 2018). Importantly, patients undergoing vagotomy have a reduced risk of developing PD. Furthermore, experimentally, aggregation of α-synuclein occurs in enteric nerves before it appears in the brain and injection of abnormal α-synuclein into the wall of the intestine spreads to the vagus nerve. Ingestion of toxins and alterations in gut microbiota can promote the aggregation of α-synuclein aggregation in PD. Thus microbial dysbiosis induces atypical immune signaling, imbalance in host homeostasis, and even CNS disease progression. It is proposed that crosscommunication between commensal microorganisms and different components of brain along with immune signaling may be involved in this complex crosstalk between microbiota and host tissues (Ma et al., 2019). It is not known how PD starts. Recent studies have also indicated that onset of dysbiosis contributes to hyperglycemic conditions, which in combination with oxidative stress promote insulin resistance. Furthermore, TLRs are activated by conserved motifs primarily found in microorganisms and dysregulation of their signaling may aid the pathogenesis of PD (Caputi and Giron, 2018). Following engagement with ligands of microbial origin, TLR4 on dendritic cells undergo maturation, which in turn activates the adaptive immune response. On the basis of several studies, it is suggested that an overstimulation of the innate immune system due to gut dysbiosis may provoke local and systemic inflammation and enteric neuroglial activation promoting the development of α-synuclein pathology.

Patients with Lewy body dementia not only show the accumulation of Lewy bodies, which are enriched in α-synuclein, but also signs of cerebral angiopathy, and deposition of Aβ and hyperphosphorylation of tau. Postural instability is a major component of functional mobility. Balance problems and resulting falls are major factors determining quality of life, morbidity, and mortality in individuals with Lewy body dementia (Stolze et al., 2004). Neurodegeneration associated with Lewy body dementia involves multiple brain areas including both dopaminergic and cholinergic neurons and for these reasons, it is often misdiagnosed as AD or other forms of dementia. Moreover, oxidative stress is significantly involved in the pathology of Lewy body dementia (DLB) (Mao, 2013). In particular, α-synuclein accumulation causes mitochondrial degeneration, which leads to the induction of oxidative stress followed by neurodegeneration. Falls contribute to the increased risk of bone fracture, impairment of mobility, low body mass index, and low bone mineral density in Lewy body patients (Sato et al., 2001). These problems start with the loss of dopaminergic neuronal loss in substantia nigra pars compacta. The molecular mechanisms of dopaminergic neuronal loss are not fully understood. However, it is suggested that the loss of dopaminergic neurons in the substantia nigra pars compacta may be due to monoamine oxidase (MAO)-mediated abnormal dopamine metabolism and hydrogen peroxide generation leading to oxidative stress. However, only 5%–10% of PD patients are known to have monogenic forms of PD. The majority of patients are sporadic, which may be induced by complex interactions among genetic factors, environmental exposures to toxins (paraquat, rotenone, herbicide, and insecticide), and aging of genetic variants with environmental risk factors (Lesage and Brice, 2009). There is a large variability in the onset and course of the disease. Current treatment approaches for vascular dementia are aimed at preventing future vascular insults by controlling the major risk factors such as hypertension, hypercholesterolemia, and type 2 diabetes (O'Brien and Thomas, 2015).

Neurochemical aspects of frontotemporal dementia

The FTDs are a group of heterogeneous neurodegenerative disorders characterized by progressive deterioration of behavior or language and associated pathology in the frontal or temporal lobes. Six clinical subtypes of FTD have been described in the literature. They are (1) behavioral variant of FTD, (2) semantic variant primary progressive aphasia, (3) nonfluent agrammatic variant primary progressive aphasia, (4) corticobasal syndrome, (5) progressive supranuclear palsy, and (6) FTD associated with motor neuron disease (Finger, 2016; Olney et al., 2017). Some FTD-related disorders include FTD with motor neuron disease, progressive supranuclear palsy syndrome, and corticobasal syndrome.

Very little is known about insulin resistance in FTD. However, it is well known that a range of eating behaviors and metabolic changes occur in FTD and amyotrophic lateral sclerosis (ALS) patients. In ALS, metabolic changes have been linked to disease progression and prognosis.

Furthermore, eating behavior changes that affect metabolism have been incorporated into the diagnostic criteria for FTD supporting the view that there is some clinical and pathological overlap between FTD and ALS (Hamdy et al., 2018). It is very likely that eating behavior changes may be related to that insulin resistance.

Insulin resistance, stress, and depression

Depression is a multisystem and multifactorial mental disorder characterized by poor (low) mood, changes in weight (decrease or increase), sleep disturbances (insomnia or hypersomnia), psychomotor retardation or agitation, fatigue, feelings of worthlessness or guilt, diminished cognitive functioning, and recurrent thoughts of death. Other symptoms of depression include psychomotor change, loss of energy, concentration difficulties/indecisiveness, irritability, and low self-esteem (Davidson et al., 2002). Not all individuals show all of the symptoms of depression (Anisman and Matheson, 2005). Depression afflicts approximately 20%–25% of women and 10%–17% of men within their lifetime (Levinson, 2006). Current estimates of the recurrence of depression suggest that 50%–60% of individuals who experience one depressive episode go on to experience a second one, with 70%–80% of these eventually experiencing a third episode, and 90% of individuals with three past episodes going on to experience a fourth (American Psychiatric Association, 2000; Burcusa et al., 2003). It is becoming increasingly evident that there is a relationship between depression and stress (Hammen, 2006). Two types of stresses (acute stress and chronic stress) are known to occur in humans and animals. Acute stress is accompanied by several biological, cognitive, and behavioral changes (Shonkoff et al., 2009). In contrast, chronic stress affects the operation of the organism's adaptive biological systems, causing illness (Shonkoff et al., 2009). Proper response to stress facilitates survival at the individual level and species propagation at the population level. Despite this necessity, prolonged stress responses become maladaptive. Chronic stress, for example, leads to a host of adverse health consequences, including cardiovascular disease, obesity, depression, pain, inflammation, and exacerbation of neurodegeneration (Topic et al., 2013; Lloyd et al., 2012; Ohayon and Schatzberg, 2003; Slavich and Irwin, 2014; McEwen, 2005). In humans, chronic psychosocial stress is a well-known risk factor for the onset and development of depression (Gold, 2015; Pittenger and Duman, 2008). Increase in neuronal plasticity is essential for adaptive intracellular changes during the normal stress response, which promotes dendritic growth, new synapse formation, and facilitates neuronal protein synthesis in the face of an acute challenge. In addition, a successful stress response requires continuity of the response to ensure normal brain function and promote survival (Gold, 2015; Radley et al., 2015). On the one hand, moderate stress enhances neural function in most cases, while severe or chronic stress is detrimental and can disrupt the ability of the brain to maintain its normal stress response, eventually inducing depression (Radley et al., 2015; McEwen, 2005). Furthermore, it has been shown that significant but brief stressful events (acute stress) result in the differentiation of stem cells into new nerve cells that improve the mental performance of rats (Kirby et al., 2013). On the other hand, chronic stress increases the levels of the stress hormone glucocorticoid and suppresses the production of new neurons in the hippocampus. This response results in decreased dendritic spine density and synapse number and impaired memory (Duman and Duman, 2015; Qiao et al., 2014).

Although depression has been studied for decades, its cellular and molecular mechanisms still remain largely unknown (Marsden, 2013). However, involvement of elevations in insulin resistance, induction of low-grade inflammation, and alterations in leptin homeostasis have been repeatedly reported (Kan et al., 2013). More severe cases of depression have been associated with activation of the hypothalamic–pituitary–adrenocortical (HPA) axis and sympathetic nervous system which is known to lead to an increased release of catecholamines which in turn inhibit insulin-induced uptake of glucose in adipocytes (Haring et al., 1986). At the molecular level, involvement of synaptic proteins such as Kalirin-7, spinophilin, Homer1, cofilin, Rac-1, cadherin, p-Akt, p-GSK-3β, p-Erk1/2, PKC, NCAM, SNAP-25, SNAP-29, synaptophysin, synapsin 1, GluR1, GluR2, NR1, NR2A, NR2B, PSD95, αCaMKII, melanocortin 4 receptors, and CRH receptor 1 in the regulation of synaptic plasticity and pathogenesis of depression has also been reported. The chronic stress alters the expression of synaptic proteins in the brain. These proteins play a key role in chronic stress–induced both depression-like behaviors and in alterations in spines. In addition, chronic stress is accompanied not only by changes in several signal transduction pathways including cAMP–PKA–CREB, cAMP–ERK1/2–CREB, cAMP–PKA, Ras–ERK, PtdIns 3K–Akt, TNFα–NF-κB, GSK-3β, mTOR, and CREB, but also by chronic stress–induced spine loss or increase in certain brain areas (Marsden, 2013; Wang et al., 2015; Gray et al., 2014). Postmortem studies have shown that the levels of NR2A, NR2B, mGLuR5, PSD-95, and mTOR as well as the levels of S6K, eIF4B, and p-eIF4B, the core downstream signaling targets of mTOR, are decreased in the PFC of depressed patients (Jernigan et al., 2011).

Detailed investigations have indicated that insulin plays important roles in modulation of feeding behavior and energy maintenance not only in the hypothalamus, but also in memory-related processes in the hippocampus (Talbot et al., 2012; Bruning et al., 2000; De Felice and Benedict, 2015; Grillo et al., 2015). Insulin receptors are expressed in regions classically involved with mood regulation, such as the nucleus accumbens, the ventral tegmental area, the amygdala, and the raphe nuclei (Figlewicz et al., 2003; Woods et al., 2016). The knockdown of insulin receptors in the hypothalamus of rats triggered depressive and anxiety-like behaviors in mice (Grillo et al., 2011). Anxiety and depressive-like behaviors have been further reported in mice with neuronal-specific knockout of insulin receptors (NIRKO). NIRKO mice also exhibited mitochondrial dysfunction, oxidative stress and increased MAO expression, and dopamine turnover in the mesolimbic system. Interestingly, altered behavior has been detected in 17-month-old NIRKO mice, but not in younger animals. It is important to note that by this age, these animals display increased white adipose tissue and plasma leptin concentration (Bruning et al., 2000), raising the possibility of the behavior response being a secondary effect to the absence of insulin signaling in neurons. Other mechanisms underlying depression include endocrine disturbances (Holsboer, 2000) or immunological alterations (Rubin et al., 1987). Oxidative stress has also been associated with depression because of its regulation through the imbalance between antioxidant defense (uric acid, vitamin E, glutathione, and coenzyme Q10) and free-radical production (superoxides, peroxynitrite, and NO) (Palta et al., 2014) has been associated with depression (Kodydkova et al., 2009; Khanzode et al., 2003). Among abovementioned antioxidant, uric acid protects axons from damage, attenuates myelin vacuolization and demyelination, and inhibits nitrotyrosine formation (Touil et al., 2001). Exposure to stress causes oxidative/nitrosative damage to the brain tissue in general and prefrontal cortex and hippocampal area in particular. These changes are mediated by catecholamines, glucocorticoids, and glutamate (Glu), and are shown to be balanced by antiinflammatory pathways in the brain. In particular, acute restraint stress is followed by the upregulation of cyclooxygenase-2 and subsequent generation of proinflammatory prostaglandin, PGE_2 release in the cortex.

Among abovementioned possibilities, involvement of insulin signaling in the neuronal dysfunction and cognitive decline in depression is a major hypothesis (Rasgon and McEwen, 2016). Thus, in children, the severity of depressive symptoms is linked with the development of insulin resistance (Shomaker et al., 2011) and a recent report has demonstrated that worst insulin resistance correlated with more pronounced depressive symptoms and dysfunction of the anterior cingulate cortex-hippocampal motivational network in a cohort of obese depressed youths (Singh et al., 2018). Several mechanisms contribute to depression-mediated disruptions of glucose metabolism and central adiposity. These processes via the activation of the HPA axis lead to type 2 diabetes, a condition, which is a risk factor not only for the development of cardiovascular diseases, but also to neurological and psychiatric disorders, including AD (Svenningsson et al., 2012; Lyra et al., 2019). The precise diagnostic parameters for the diagnosis of depression are >2 weeks: sad/anxious mood, hopelessness, helplessness, decreased energy, appetite/weight changes, headaches, sleep changes, feelings of guilt, loss of interest, decreased concentration, psychomotor retardation, and suicide attempts (National Institute of Mental Health Depression Basics, 2016). The molecular mechanisms linking depression with metabolic dysregulation have not been fully elucidated. However, on the basis of several studies, it has been suggested that increase in insulin resistance, induction of low-grade inflammation, and elevation in leptin levels contribute to depression mediated metabolic dysregulation (Kan et al., 2013; Liu et al., 2012). In addition, increase in expression of proinflammatory cytokines such as TNF-α and IL-1 activates several stress kinases in the brain, including IκB kinase β, c-Jun N-terminal kinase, and protein kinase RNA-activated may also contribute to the pathogenesis of low-grade inflammation and depression. The underlying mechanism between inflammation and depression may not only involve complex interactions among inflammation, psychological factors, but also pathophysiological and behavioral mechanisms. First, inflammation may lead to depression. Several studies have indicated that proinflammatory cytokines (TNF-α and IL-1) may contribute to the development of depression through the activation of indoleamine-2,3-dioxygenase, a process, which leads to decrease in production of serotonin and increase in production of kynurenic and quinolinic acids (Miller et al., 2009; Haroon et al., 2012; Capuron and Miller, 2011). Thus decrease in serotonin levels is an important factor in the pathogenesis of depression. Also, increase in production of kynurenic and quinolinic acids may leads to increase in the release of Glu and thereby to decrease in production of trophic factors, including brain-derived neurotrophic factor (BDNF), a trophic factor associated with depression (Haroon et al., 2012; Hashimoto, 2009). Furthermore, there is an increase in levels of CRP, a marker of systemic inflammation. CRP does not cross the BBB. However, elevated CRP levels probably indicate elevated levels of proinflammatory cytokines, which can cross the BBB. Secondly, depression may also lead to neuroinflammation. Psychological stress activates the HPA and sympathetic nervous system, which releases stress hormones (Kyrou and Tsigos, 2009). These hormones together with released cytokine initiate the acute-phase response triggering neuroinflammation. Furthermore, depression may lead to inflammation mediated by weight gain (Miller et al., 2003). Expansion of adipose tissue increases synthesis of leptin, which in turn increases levels of the proinflammatory cytokine IL-6, stimulating the production of acute-phase proteins, including CRP (Gabay and Kushner, 1999).

In terms of involvement insulin resistance in depression, the activation of stress kinase pathways may lead to the phosphorylation of insulin receptor substrate-1 at serine residues, impacting insulin signaling response (Fig. 8.6). Central insulin signaling may also contribute to hippocampal neurogenesis, synaptic plasticity, HPA axis response, and the reward system (Lyra et al., 2019). Furthermore, the HPA axis is sensitive not only to physical factors (increase in intake of alcohol and smoking), but also to psychosocial and socioeconomic factors (divorce, unemployment, work-related stress, poor education, and poverty). These factors provide the basis for depression through the activation of the HPA axis. Although the HPA axis functions as a protective mechanism to maintain allostasis, intense chronic activation is believed to lead to permanent derangements of the HPA axis and increased susceptibility to disease. Studies on primates have shown that exposure to moderate psychological stress is followed by a depressive reaction and the development of adverse metabolic indicators, including abdominal fat accumulation and insulin resistance (Fig. 8.7). Similar perturbations in the HPA axis associated with low socioeconomic status and leading to visceral obesity have been reported in humans (Rosmond and Björntorp, 2000). It is interesting to know that there is a relationship between depression and type 2 diabetes. This relationship can be explained by lifestyle factors associated with depression, including physical inactivity and poor dietary habits that increase the risk of developing insulin resistance (Weber-Hamann et al., 2002). Common to both

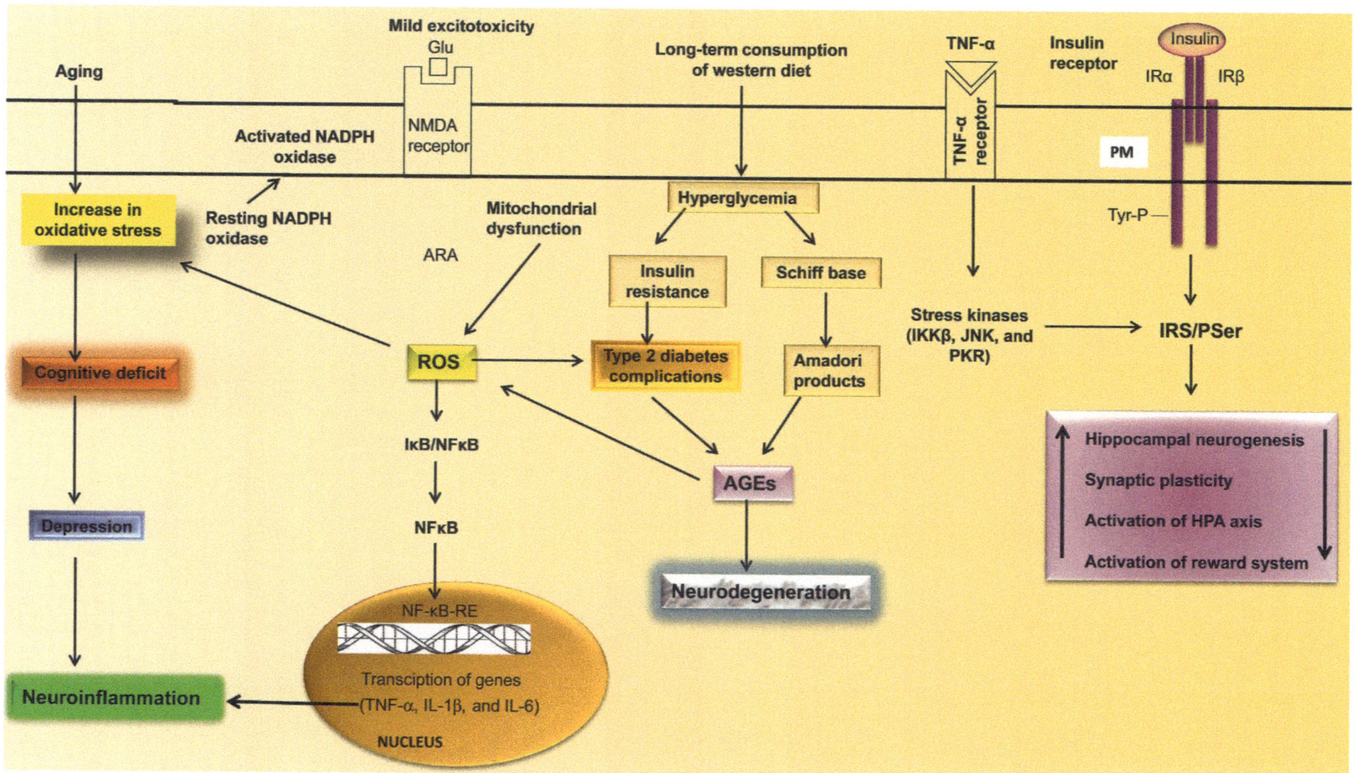

Figure 8.6 Hypothetical diagram showing the contribution of insulin receptor, insulin resistance, and stress kinases in the pathogenesis of depression.

IKKs, IkappaB kinases; *JNK*, c-jun kinase; *PKR*, protein kinase RNA-activated.

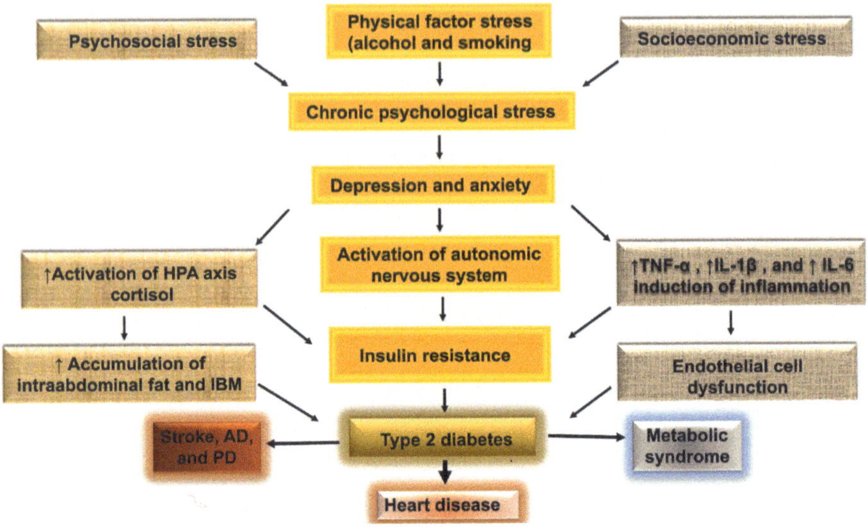

Figure 8.7 Hypothetical diagram showing the contribution of stress insulin resistance, and diabetes in the pathogenesis of depression, metabolic syndrome, neurodegenetative [Alzheimer's disease (AD) and Parkinson's disease (PD)]; and heart disease; tumor necrosis factor-α (TNF-α); interleukin-1β (IL-1β); interleukin-6 (IL-6); hypothalamic-pituitary-adrenal (HPA) axis.

of these pathways is obesity, which is a significant risk factor for increasing insulin resistance and type 2 diabetes. One hypothesis suggests that hyperactivity of the HPA axis associated with depression promotes intraabdominal fat accumulation and there is some support for this in the literature (Weber-Hamann et al., 2002). The clinical features of depression also include reduction in plasma levels of BDNF (Sen et al., 2008), elevated blood levels of IL-1β, IL-6, and TNF-α (Dowlati et al., 2010), low levels of magnesium, over-activity of HPA axis (Varghese and Brown, 2001), alterations in cerebral structures such as an increased ventricle/brain ratio and localized atrophy of the prefrontal cortex, cingulated gyrus, ventral striatum, amygdale, cerebellum, and hippocampus (aan het Rot et al., 2009). Furthermore, depression is also accompanied by changes in neurotransmitters (Glu, γ-aminobutyric acid, and monoaminergic systems), neuropeptides (vasopressin), biogenic amine (dopamine, norepinephrine, and serotonin), and alterations in gene—environmental interactions along with modification in neuronal synaptic density in the brain (Fig. 8.7) (Krystal et al., 2002; Cryan and Slattery, 2010). Collectively these studies suggest that the relationship among depression, insulin resistance, and type 2 diabetes can be explained not only by lifestyle factors, but also by physical inactivity genes, and epigenetic factors. Common to above factors is obesity, a pathological condition which is an important risk factor for insulin resistance and type 2 diabetes (Lyra Silva et al., 2019).

Effects of diet and exercise on dementia and depression

Exercise is defined as activities that improve one or more aspects of physical conditioning, including cardiovascular endurance, muscular strength, flexibility, balance, and fine motor control (Farooqui, 2014). It is well established that regular physical exercise is an important part of a healthy lifestyle in all age groups. Exercise slows not only on brain aging (decrease in brain atrophy), but also delays the onset of depression. The underlying mechanisms associated with slowing of depression by exercise are not fully understood. However, exercise produces changes the brain structural integrity by enhancing neurogenesis and angiogenesis and secretions of growth factors promoting synaptic plasticity (Gomez-Pinilla et al., 2008; Farooqui, 2014). These processes improve cognitive function by increasing the gray matter volume (Hillman et al., 2008) and inducing neurogenesis in the dentate gyrus. Neurogenesis is coupled with angiogenesis, which in turn is related to cerebral blood volume (van Praag et al., 1999). It is hypothesized that measurement of cerebral blood volume may provide an in vivo correlation between neurogenesis and increased cerebral blood flow due to exercise. At the molecular level, many molecules contribute to exercise-induced neurogenesis, angiogenesis, and cerebral blood flow. These molecules include vascular endothelial growth factor, BDNF, insulin-like growth factor-1, catechol-O-methyltransferase, endorphins, and NO (Neeper et al., 1996; Stroth et al., 2010; Camargo et al., 2013). In addition, exercise also modulates gene involved in insulin-like signaling, energy metabolism, neurogenesis, and synaptic plasticity along with learning and memory (Reagan, 2007; van Praag et al., 2005). Also, long-term exercise treatment reduces oxidative stress in the hippocampus of aging rats (Marosi et al., 2012). A number of studies have also described that a single bout of exercise leads to immediate changes in the methylation pattern of certain genes in DNA and affects the proteins that these genes express (Hamer et al., 2013). Exercise-related methylation change appears to be stronger among older people, with age accounting for 30% of the methylation variation (Brown, 2015). In a study of 90-year-old male physicians in the United States, it is reported that maintaining active exercise is an important contributor to good quality ageing for physicians who reached 90 years and beyond, in good health (Yates et al., 2008). Collectively, these studies strongly support the view that exercise improves brain health by contributing to disease prevention and helping recovery from illness. It influences our cognition and psychological health, and reduces our risk of developing illnesses such as diabetes, heart disease, and stroke (Barnes and Yaffe, 2011).

Conclusion

Dementia and depression syndrome are disorders of cognitive impairment that interferes with everyday life such as memory, thinking, orientation, comprehension, calculation, learning capacity, language, and judgment. The prevalence of dementia and depression is increasing exponentially with increasing age and chances of developing dementia and depression increase with age after age 70–75 years. Age-related cognitive changes in early stages of dementia and depression are accompanied by alterations in neuronal structure with neuronal death, loss of synapses, and dysfunction of neuronal networks. Several types of dementia and depression are known to occur in human population. For example, AD type of dementia, vascular dementia, mixed dementia (vascular dementia plus AD), Lewy body dementia, FTD, and depression. Molecular mechanisms of dementia and depression are based on several molecular changes in the brain such as increase in oxidative stress, dysregulation in calcium homeostasis, activation of microglia and astrocytes, induction of neuroinflammation along with early synaptic disconnection and late apoptotic cell death. Healthy lifestyle in dementia and depression has been reported to decrease the rate of cognitive decline and help delay the onset of cognitive symptoms in the setting of age-associated diseases.

References

aan het Rot, M., Mathew, S.J., Charney, D.S., 2009. Neurobiological mechanisms in major depressive disorder. CMAJ 180, 305–313.

Alzheimer's Association (A.D.), 2014. Alzheimer's disease facts and figures. Alzheimer Demen 2014 (10), e47–e92.

American Psychiatric Association, 2000. Diagnostic and Statistical Manual of Mental Disorders, Text Revision – Fourth ed. American Psychiatric Association, Washington, DC.

Anisman, H., Matheson, K., 2005. Stress, depression, and anhedonia: caveats concerning animal models. Neurosci. Biobehav. Rev. 29, 525–546.

Ashraghi, M.R., Pagano, G., Polychronis, S., Niccolini, F., Politis, M., 2016. Parkinson's disease, diabetes and cognitive impairment. Recent Pat. Endocr. Metab. Immune Drug Discov. 10, 11–21.

Aw, D., Silva, A.B., Palmer, D.B., 2007. Immunosenescence: emerging challenges for an ageing population. Immunology 120, 435–446.

Bell, R.D., Zlokovic, B.V., 2009. Neurovascular mechanisms and blood-brain barrier disorder in Alzheimer's disease. Acta Neuropathol. 118, 103–113.

Barnes, D.E., Yaffe, K., 2011. The projected impact of risk factor reduction on Alzheimer's disease prevalence. Lancet Neurol. 10, 819–828.

Berchtold, N.C., Cribbs, D.H., Coleman, P.D., Rogers, J., Head, E., et al., 2008. Gene expression changes in the course of normal brain aging are sexually dimorphic. Proc. Natl. Acad. Sci. U.S.A. 105, 15605–15610.

Bertram, L., Tanzi, R.E., 2008. Thirty years of Alzheimer's disease genetics: the implications of systematic meta-analyses. Nat. Rev. Neurosci. 9, 768–778.

Bettcher, B.M., Watson, C.L., Walsh, C.M., Lobach, I.V., Neuhaus, J., Miller, J.W., et al., 2014. Interleukin-6, age and corpus callosum integrity. PLoS One 9, e106521.

Bosco, D., Plastino, M., Cristiano, D., Colica, C., Ermio, C., De, B.M., 2012. Dementia is associated with insulin resistance in patients with Parkinson's disease. J. Neurol. Sci. 2012 (315), 39–43.

Boyd 3rd, A.E., Lebovitz, H.E., Feldman, J.M., 1971. Endocrine function and glucose metabolism in patients with Parkinson's disease and their alternation by L-Dopa. J. Clin. Endocrinol. Metab. 33, 829–837.

Braak, H., de Vos, R.A., Bohl, J., Del Tredici, K., 2006. Gastric alpha-synuclein immunoreactive inclusions in Meissner's and Auerbach's plexuses in cases staged for Parkinson's disease-related brain pathology. Neurosci. Lett. 396, 67–72.

Brown, W.M., 2015. Exercise-associated DNA methylation change in skeletal muscle and the importance of imprinted genes: a bioinformatics meta-analysis. Br. J. Sports Med. 49, 1567–1578.

Bruning, J.C., Gautam, D., Burks, D.J., Gillette, J., Schubert, M., Orban, P.C., et al., 2000. Role of brain insulin receptor in control of body weight and reproduction. Science 289, 2122–2125.

Burcusa, S.L., Iacono, W.G., McGue, M., 2003. Adolescent twins discordant for major depressive disorder: shared familial liability to externalizing and other internalizing disorders. J. Child Psychol. Psychiatry 44, 997–1005.

Caputi, V., Giron, M.C., 2018. Microbiome-gut-brain axis and toll-like receptors in Parkinson's disease. Int. J. Mol. Sci. 19 (6), pii: E1689.

Capuron, L., Miller, A.H., 2011. Immune system to brain signaling: neuropsychopharmacological implications. Pharmacol. Ther. 130, 226–238.

Camargo, L.H., Alves, F.H., Biojones, C., Correa, F.M., Resstel, L.B., et al., 2013. Involvement of N-methyl-D-aspartate glutamate receptor and nitric oxide in cardiovascular responses to dynamic exercise in rats. Eur. J. Pharmacol. 713, 16–24.

Carare, R.O., Bernardes-Silva, M., Newman, T.A., Page, A.M., Nicoll, J.A., Perry, V.H., et al., 2008. Solutes, but not cells, drain from the brain parenchyma along basement membranes of capillaries and arteries: significance for cerebral amyloid angiopathy and neuroimmunology. Neuropathol. Appl. Neurobiol. 34, 131–144.

Cryan, J.F., Slattery, D.A., 2010. GABA B receptors and depression: current status. Adv. Pharmacol. 58, 427–451.

Cullen, K.M., Kócsi, Z., Stone, J., 2006. Microvascular pathology in the aging human brain: evidence that senile plaques are sites of microhaemorrhages. Neurobiol. Aging 27, 1786–1796.

Davidson, R.J., Lewis, D.A., Alloy, L.B., Amaral, D.G., Bush, G., Cohen, J.D., et al., 2002. Neural and behavioral substrates of mood and mood regulation. Biol. Psychiatry 52, 478–502.
De Felice, F.G., Benedict, C., 2015. A key role of insulin receptors in memory. Diabetes 64, 3653–3655.
De Felice, F.G., Vieira, M.N., Bomfim, T.R., Bomfim, T.R., Decker, H., Velasco, P.T., et al., 2009. Protection of synapses against Alzheimer's-linked toxins: insulin signaling prevents the pathogenic binding of Abeta oligomers. Proc. Natl. Acad. Sci. U.S.A. 106, 1971–1976.
Dowlati, Y., Herrmann, N., Swardfager, W., Liu, H., Sham, L., Reim, E.K., et al., 2010. A meta-analysis of cytokines in major depression. Biol. Psychiatry 67, 446–457.
Dresner, A., Laurent, D., Marcucci, M., Griffin, M.E., Dufour, S., Cline, G.W., et al., 1999. Effects of free fatty acids on glucose transport and IRS-1-associated phosphatidylinositol 3-kinase activity. J. Clin. Invest. 103, 253–259.
Duman, C.H., Duman, R.S., 2015. Spine synapse remodeling in the pathophysiology and treatment of depression. Neurosci. Lett. 601, 20–29.
Duncombe, J., Kitamura, A., Hase, Y., Ihara, M., Kalaria, R.N., Horsburgh, K., 2017. Chronic cerebral hypoperfusion: a key mechanism leading to vascular cognitive impairment and dementia. Closing the translational gap between rodent models and human vascular cognitive impairment and dementia. Clin. Sci. (Lond) 131, 2451–2468.
Ebner, N.C., Kamin, H., Diaz, V., Cohen, R.A., MacDonald, K., 2015. Hormones as "difference makers" in cognitive and socioemotional aging processes. Front. Psychol. 5, 1–16.
Ellison, D., White, D., Farrar, F.C., 2015. Aging population. Nurs. Clin. North Am. 50, 185–213.
Estrada, A., Cardona, D., Segura, A.M., Chavarriaga, L.M., Ordóñez, J., Osorio, J.J., 2011. Quality of life in institutionalized elderly people of medellín. Biomedica 31, 492–502.
Farooqui, A.A., 2013. Metabolic Syndrome: An Important Risk Factor for Stroke, Alzheimer Disease, and Depression. Springer International Publishing, Switzerland.
Farooqui, A.A., 2014. Inflammation and Oxidative Stress in Neurological Disorders. Springer International Publishing, Switzerland.
Farooqui, A.A., 2017. Neurochemical Aspects of Alzheimer Disease: Risk Factors, Pathogenesis, Biomarkers, and Treatment Strategies. Elsevier, New York.
Farooqui, A.A., 2019. Molecular Mechanisms of Dementias: Biomarkers, Neurochemistry, and Therapy. Academic Press, an imprint of Elsevier, San Diego, CA.
Figlewicz, D.P., Evans, S.B., Murphy, J., Hoen, M., Baskin, D.G., 2003. Expression of receptors for insulin and leptin in the ventral tegmental area/substantia nigra (VTA/SN) of the rat. Brain Res. 964, 107–115.
Finger, E.C., 2016. Frontotemporal Dementias. Continuum (Minneap Minn) 22 (2 Dementia), 464–489.
Fishman, E., 2017. Risk of developing dementia at older ages in the United States. Demography 54, 1897–1919.
Franco-Iborra, S., Vila, M., Perier, C., 2016. The Parkinson disease mitochondrial hypothesis: where are we at? Neuroscientist 22, 266–277.
Fujishiro, H., Ferman, T.J., Boeve, B.F., Smith, G.E., Graff-Radford, N.R., et al., 2008. Validation of the neuropathologic criteria of the third consortium for dementia with Lewy bodies for prospectively diagnosed cases. J. Neuropathol. Exp. Neurol. 67, 649–656.
Gabay, C., Kushner, I., 1999. Acute-phase proteins and other systemic responses to inflammation. N. Engl. J. Med. 340, 448–454.
Gold, P.W., 2015. The organization of the stress system and its dysregulation in depressive illness. Mol. Psychiatry 20, 32–47.

Goldin, A., Beckman, J.A., Schmidt, A.M., Creager, M.A., 2006. Advanced glycation end products: sparking the development of diabetic vascular injury. Circulation 114, 597–605.

Gomez-Pinilla, F., Vaynman, S., Ying, Z., 2008. Brain-derived neurotrophic factor functions as a metabotrophin to mediate the effects of exercise on cognition. Eur. J. Neurosci. 28, 2278–2287.

Government of Canada, Statistics Canada, 2017. Census in brief: recent trends for the population aged 15 to 64 in Canada, census year 2016. Available from: <http://www12.statcan.gc.ca/census-recensement/2016/as-sa/98-200-x/2016003/98-200-x2016003-eng.cfm> (accessed 22.06.17.).

Gray, J.D., Rubin, T.G., Hunter, R.G., McEwen, B.S., 2014. Hippocampal gene expression changes underlying stress sensitization and recovery. Mol. Psychiatry 19, 1171–1178.

Grillo, C.A., Piroli, G.G., Kaigler, K.F., Wilson, S.P., Wilson, M.A., Reagan, L.P., 2011. Downregulation of hypothalamic insulin receptor expression elicits depressive-like behaviors in rats. Behav. Brain. Res. 222, 230–235.

Grillo, C.A., Piroli, G.G., Lawrence, R.C., Wrighten, S.A., Green, A.J., Wilson, S.P., et al., 2015. Hippocampal insulin resistance impairs spatial learning and synaptic plasticity. Diabetes 64, 3927–3936.

Hamdy, O., Ashrafzadeh, S., Mottalib, A., 2018. Weight management in patients with type 2 diabetes: a multidisciplinary real-world approach. Curr. Diab. Rep. 18, 66.

Hamer, M., Lavoie, K.L., Bacon, S.L., 2013. Taking up physical activity in later life and healthy ageing: the English longitudinal study of ageing. Br. J. Sports Med. 48, 239–243.

Hammen, C., 2006. Stress generation in depression: Reflections on origins, research, and future directions. J. Clin. Psychol. 62, 1065–1082.

Han, X., Ploug, T., Galbo, H., 1997. Effect of diet on insulin- and contraction-mediated glucose transport and uptake in rat muscle. Am. J. Physiol. 269, R544–R551.

Haring, H., Kirsch, D., Obermaier, B., Ermel, B., Machicao, F., 1986. Decreased tyrosine kinase activity of insulin receptor isolated from rat adipocytes rendered insulin-resistant by catecholamine treatment in vitro. Biochem. J. 234, 59–66.

Haroon, E., Raison, C.L., Miller, A.H., 2012. Psychoneuroimmunology meets neuropsychopharmacology: translational implications of the impact of inflammation on behavior. Neuropsychopharmacology 37, 137–162.

Hashimoto, K., 2009. Emerging role of glutamate in the pathophysiology of major depressive disorder. Brain Res. Rev. 61, 105–123.

Hashimoto, M., Masliah, E., 1999. Alpha-synuclein in Lewy body disease and Alzheimer's disease. Brain Pathol. 9, 707–720.

Heringa, S.M., van den Berg, E., Reijmer, Y.D., Nijpels, G., Stehouwer, C.D., Schalkwijk, C.G., et al., 2014. Markers of low-grade inflammation and endothelial dysfunction are related to reduced information processing speed and executive functioning in an older population - the Hoorn Study. Psychoneuroendocrinology 40, 108–118.

Hillman, C.H., Erickson, K.I., Kramer, A.F., 2008. Be smart, exercise your heart: exercise effects on brain and cognition. Nat. Rev. Neurosci. 9, 58–65.

Holsboer, F., 2000. The corticosteroid receptor hypothesis of depression. Neuropsychopharmacology 23, 477–501.

Hölscher, C., 2014. Insulin, incretins and other growth factors as potential novel treatments for Alzheimer's and Parkinson's diseases. Biochem. Soc. Trans. 42, 593–599.

Hornykiewicz, O., 2008. Basic research on dopamine in Parkinson's disease and the discovery of the nigrostriatal dopamine pathway: the view of an eyewitness. Neurodegener. Dis. 5, 114–117.

Hurtig, H., Trojanowski, J., Galvin, J., Ewbank, D., Schmidt, M., et al., 2000. Alpha-synuclein cortical Lewy bodies correlate with dementia in Parkinson's disease. Neurology 54, 1916–1921.

Iadecola, C., 2013. The pathobiology of vascular dementia. Neuron 80, 844–866.

Iliff, J.J., Wang, M., Liao, Y., Plogg, B.A., Peng, W., Gundersen, G.A., et al., 2012. A paravascular pathway facilitates CSF flow through the brain parenchyma and the clearance of interstitial solutes, including amyloid β. Sci. Transl. Med. 4, 147ra111.

Ince, P.G., 2011. Dementia with Lewy bodies and Parkinson's disease with dementia. In: Dickson, D.W., Weller, R.O. (Eds.), Neurodegeneration, the Molecular Pathology of Dementia and Movement Disorders. Wiley-Blackwell, Oxford, pp. 224–237.

Ironside, J.W., Bell, J.E., 2007. Infective dementias. Panminerva Med. 49, 209–225.

Jankovic, J., 2008. Parkinson's disease: clinical features and diagnosis. J. Neurol. Neurosurg. Psychiatry 79, 368–376.

Jernigan, C.S., Goswami, D.B., Austin, M.C., et al., 2011. The mTOR signaling pathway in the prefrontal cortex is compromised in major depressive disorder. Prog. Neuro-Psychopharmacol. Biol. Psychiatry 35, 1774–1779.

Jorm, A.F., Jolley, D., 1998. The incidence of dementia: a meta-analysis. Neurology 51, 728–733.

Kabasakalian, A., Finney, G.R., 2009. Reversible dementias. Int. Rev. Neurobiol. 84, 283–302.

Kan, C., Silva, N., Golden, S.H., Rajala, U., Timonen, M., Stahl, D., et al., 2013. A systematic review and meta-analysis of the association between depression and insulin resistance. Diabetes Care 36, 480.

Kandimalla, R., Thirumala, V., Reddy, P.H., 2017. Is Alzheimer's disease a type 3 diabetes? A critical appraisal. Biochim. Biophys. Acta Mol Basis Dis. 1863, 1078–1089.

Kawas, C., Resnick, S., Morrison, A., Brookmeyer, R., Corrada, M., Zonderman, A., et al., 1997. A prospective study of estrogen replacement therapy and the risk of developing Alzheimer's disease: the Baltimore longitudinal study of aging. Neurology 48, 1517–1521.

Khanzode, S.D., Dakhale, G.N., Khanzode, S.S., Saoji, A., Palasodkar, R., 2003. Oxidative damage and major depression: the potential antioxidant action of selective serotonin re-uptake inhibitors. Redox. Rep. 8, 365–370.

Kirby, E.D., Muroy, S.E., Sun, W.G., et al., 2013. Acute stress enhances adult rat hippocampal neurogenesis and activation of newborn neurons via secreted astrocytic FGF2. eLife 2, 00362.

Kodydkova, J., Vavrova, L., Zeman, M., Jirak, R., Macasek, J., Stankova, B., et al., 2009. Antioxidative enzymes and increased oxidative stress in depressive women. Clin. Biochem. 42, 1368–1374.

Krystal, J.H., Sanacora, G., Blumberg, H., Anand, A., Charney, D.S., et al., 2002. Glutamate and GABA systems as targets for novel antidepressant and mood-stabilizing treatments. Mol. Psychiatry 7, S71–S80.

Kyrou, I., Tsigos, C., 2009. Stress hormones: physiological stress and regulation of metabolism. Curr. Opin. Pharmacol. 9, 787–793.

Langa, K.M., Larson, E.B., Crimmins, E.M., Faul, J.D., Levine, D.A., et al., 2017. A comparison of the prevalence of dementia in the United States in 2000 and 2012. JAMA Intern. Med. 177, 51–58.

Lebouvier, T., Chaumette, T., Paillusson, S., Duyckaerts, C., Bruley des Varannes, S., Neunlist, M., et al., 2009. The second brain and Parkinson's disease. Eur. J. Neurosci. 30, 735–741.

Lesage, S., Brice, A., 2009. Parkinson's disease: from monogenic forms to genetic susceptibility factors. Hum. Mol. Genet. 18, R48–R59.

Levinson, S.S., 2006. Inflammatory and long-term risk markers. Clin. Lab. Med. 26, 553–570.
Libon, D.J., Price, C.C., Heilman, K.M., Grossman, M., 2006. Alzheimer's "other dementia". Cogn. Behav. Neurol. 19, 112–116.
Liddle, R.A., 2018. Parkinson's disease from the gut. Brain Res. 1693 (Pt B), 201–206.
Liu, Y., Ho, R.C., Mak, A., 2012. Interleukin (IL)-6, tumour necrosis factor alpha (TNF-alpha) and soluble interleukin-2 receptors (sIL-2R) are elevated in patients with major depressive disorder: a meta-analysis and meta-regression. J. Affect Disord. 139, 230.
Lloyd, C.E., Roy, T., Nouwen, A., Chauhan, A.M., 2012. Epidemiology of depression in diabetes: international and cross-cultural issues. J. Affect Disord. 142 (Suppl.), S22–S29.
Loerch, P.M., Lu, T., Dakin, K.A., Vann, J.M., Isaacs, A., et al., 2008. Evolution of the aging brain transcriptome and synaptic regulation. PLoS One 3, e3329.
López-Otín, C., Blasco, M.A., Partridge, L., Serrano, M., Kroemer, G., 2013. The hallmarks of aging. Cell 153, 1194–1217.
Lu, T., Pan, Y., Kao, S.Y., Li, C., Kohane, I., Chan, J., et al., 2004. Gene regulation and DNA damage in the ageing human brain. Nature 429, 883–891.
Lupien, S.J., Maheu, F., Tu, M., Fiocco, A., Schramek, T.E., 2007. The effects of stress and stress hormones on human cognition: implications for the field of brain and cognition. Brain Cogn. 65, 209–237.
Lyketsos, C.G., Lopez, O., Jones, B., Fitzpatrick, A.L., Breitner, J., Dekosky, S., 2002. Prevalence of neuropsychiatric symptoms in dementia and mild cognitive impairment: results from the cardiovascular health study. JAMA 288, 1475–1483.
Lyra, E., Silva, N.M., Lam, M.P., Soares, C.N., Munoz, D.P., Milev, R., et al., 2019. Insulin resistance as a shared pathogenic mechanism between depression and type 2 diabetes. Front. Psychiatry 10, 57.
Ma, Q., Xing, C., Long, W., Wang, H.Y., Liu, Q., Wang, R.F., 2019. Impact of microbiota on central nervous system and neurological diseases: the gut-brain axis. J. Neuroinflamm. 16, 53.
Maciejczyk, M., Żebrowska, E., Chabowski, A., 2019. Insulin resistance and oxidative stress in the brain: what's new? Int. J. Mol. Sci. 20. pii: E874.
Maiti, P., Manna, J., Dunbar, G.L., 2017. Current understanding of the molecular mechanisms in Parkinson's disease: targets for potential treatments. Transl. Neurodegener. 6, 28.
Mao, P., 2013. Oxidative Stress and Its Clinical Applications in Dementia. J. Neurodegener. Dis. 2013, 319898.
Marosi, K., Bori, Z., Hart, N., Sárga, L., Koltai, E., Radák, Z., et al., 2012. Long-term exercise treatment reduces oxidative stress in the hippocampus of aging rats. Neuroscience 226, 21–28.
Marsden, W.N., 2013. Synaptic plasticity in depression: molecular, cellular and functional correlates. Prog. Neuro-Psychopharmacol. Biol. Psychiatry 43, 168–184.
Marui, W., Iseki, E., Nakai, T., Miura, S., Kato, M., Ueda, K., et al., 2002. Progression and staging of Lewy pathology in brains from patients with dementia with Lewy bodies. J. Neurol. Sci. 195, 153–159.
Matsuzaki, T., Sasaki, K., Tanizaki, Y., Hata, J., Fujimi, K., Matsui, Y., et al., 2010. Insulin resistance is associated with the pathology of Alzheimer disease: the Hisayama Study. Neurology 75, 764–770.
McEwen, B.S., 2005. Glucocorticoids, depression, and mood disorders: structural remodeling in the brain. Metabolism 54, 20–23.
Michel, P.P., Hirsch, E.C., Hunot, S., 2016. Understanding dopaminergic cell death pathways in Parkinson disease. Neuron 90, 675–691.

Milà, V.R., Abellana, S.R., Padró, M.L., Farran, C.A., 2012. Assessment of food consumption, energy and protein intake in the meals offered in four Spanish nursing homes. Nutr. Hosp. 27, 914–921.
Miller, G.E., Freedland, K.E., Carney, R.M., Stetler, C.A., Banks, W.A., 2003. Pathways linking depression, adiposity, and inflammatory markers in healthy young adults. Brain Behav. Immun. 17, 276–285.
Miller, A.H., Maletic, V., Raison, C.L., 2009. Inflammation and its discontents: the role of cytokines in the pathophysiology of major depression. Biol. Psychiatry 65, 732–741.
Miwa, H., Fukaya, M., Watabe, A.M., Watanabe, M., Manabe, T., 2008. Functional contributions of synaptically localized NR2B subunits of the NMDA receptor to synaptic transmission and long-term potentiation in the adult mouse CNS. J. Physiol. 586, 2539–2550.
Moors, T., Paciotti, S., Chiasserini, D., Calabresi, P., Parnetti, L., et al., 2016. Lysosomal dysfunction and alpha-synuclein aggregation in Parkinson's disease: diagnostic links. Mov. Disord. 31, 791–801.
Morris, J.K., Zhang, H., Gupte, A.A., Bomhoff, G.L., Stanford, J.A., et al., 2008. Measures of striatal insulin resistance in a 6-hydroxydopamine model of Parkinson's disease. Brain Res. 1240, 185–195.
Morrison, J.H., Baxter, M.G., 2012. The aging cortical synapse: hallmarks and implications for cognitive decline. Nat. Rev. Neurosci. 13, 240–250.
National Institute of Mental Health Depression Basics, 2016. <https://www.nimh.nih.gov/health/publications/depression/index.shtml>.
Neeper, S.A., Gomez-Pinilla, F., Choi, J., Cotman, C.W., 1996. Physical activity increases mRNA for brain-derived neurotrophic factor and nerve growth factor in rat brain. Brain Res. 726, 49–56.
Nilsson, L.G., 2003. Memory function in normal aging. Acta Neurol. Scand. Suppl. 179, 7–13.
O'Brien, J.T., Thomas, A., 2015. Vascular dementia. Lancet 386, 1698–1706.
Ohayon, M.M., Schatzberg, A.F., 2003. Using chronic pain to predict depressive morbidity in the general population. Arch. Gen. Psychiatry 60, 39–47.
Olney, N.T., Spina, S., Miller, B.L., 2017. Frontotemporal Dementia. Neurol. Clin. 35, 339–374.
Ortman, J.M., Velkoff, V.A., Hogan, H., 2014. An Aging Nation: The Older Population in the United States. US Census Bureau, Washington, DC, pp. 25–1140.
Ouwens, D.M., Boer, C., Fodor, M., de Galan, P., Heine, R.J., Maassen, J.A., et al., 2005. Cardiac dysfunction induced by high-fat diet is associated with altered myocardial insulin signalling in rats. Diabetologia 48, 1229–1237.
Palta, P., Samuel, L.J., Miller 3rd, E.R., Szanton, S.L., 2014. Depression and oxidative stress: results from a meta-analysis of observational studies. Psychosom. Med. 76, 12–19.
Pang, Y., Lin, S., Wright, C., Shen, J., Carter, K., Bhatt, A., et al., 2016. Intranasal insulin protects against substantia nigra dopaminergic neuronal loss and alleviates motor deficits induced by 6-OHDA in rats. Neuroscience 318, 157–165.
Paniz, C., Bairros, A., Valentini, J., Charão, M., Bulcão, R., Moro, A., et al., 2007. The influence of the serum vitamin C levels on oxidative stress biomarkers in elderly women. Clin. Biochem. 40, 1367–1372.
Papapetropoulos, S., Ellul, J., Argyriou, A.A., Talelli, P., Chroni, E., Papapetropoulos, T., 2004. The effect of vascular disease on late onset Parkinson's disease. Eur. J. Neurol. 11, 231–235.
Pfeiffer, R.F., 2011. Gastrointestinal dysfunction in Parkinson's disease. Parkinsonism Relat. Disord. 17, 10–15.

Pittenger, C., Duman, R.S., 2008. Stress, depression, and neuroplasticity: a convergence of mechanisms. Neuropsychopharmacology 33, 88–109.
Pressley, J.C., Louis, E.D., Tang, M.X., Cote, L., Cohen, P.D., Glied, S., et al., 2003. The impact of comorbid disease and injuries on resource use and expenditures in parkinsonism. Neurology 60, 87–93.
Qiao, H., An, S.C., Ren, W., Ma, X.M., 2014. Progressive alterations of hippocampal CA3-CA1 synapses in an animal model of depression. Behav. Brain. Res. 275, 191–200.
Radley, J., Morilak, D., Viau, V., Campeau, S., 2015. Chronic stress and brain plasticity: mechanisms underlying adaptive and maladaptive changes and implications for stress-related CNS disorders. Neurosci. Biobehav. Rev. 58, 79–91.
Ransohoff, R.M., 2016. How neuroinflammation contributes to neurodegeneration. Science 353, 777–783.
Rasgon, N.L., McEwen, B.S., 2016. Insulin resistance—a missing link no more. Mol. Psychiatry 21, 1648–1652.
Reagan, L.P., 2007. Insulin signaling effects on memory and mood. Curr. Opin. Pharmacol. 7, 633–637.
Risacher, S.L., Saykin, A.J., 2013. Neuroimaging biomarkers of neurodegenerative diseases and dementia. Semin. Neurol. 33, 386–416.
Rizzi, L., Rosset, I., Roriz-Cruz, M., 2014. Global epidemiology of dementia: Alzheimer's and vascular types. Biomed. Res. Int. 2014, 908915.
Román, G.C., Erkinjuntti, T., Wallin, A., Pantoni, L., Chui, H.C., 2002. Subcortical ischaemic vascular dementia. Lancet Neurol. 1, 426–436.
Romeo, H.E., Tio, D.L., Rahman, S.U., Chiappelli, F., Taylor, A.N., 2001. The glossopharyngeal nerve as a novel pathway in immune-to-brain communication: relevance to neuroimmune surveillance of the oral cavity. J. Neuroimmunol. 115, 91–100.
Rosmond, R., Björntorp, P., 2000. Occupational status, cortisol secretory pattern, and visceral obesity in middle-aged men. Obes. Res. 8, 445–450.
Rubin, R.T., Poland, R.E., Lesser, I.M., Winston, R.A., Blodgett, A.L., 1987. Neuroendocrine aspects of primary endogenous depression. I. Cortisol secretory dynamics in patients and matched controls. Arch. Gen. Psychiatry 44, 328–336.
Sandyk, R., 1993. The relationship between diabetes mellitus and Parkinson's disease. Int. J. Neurosci. 69, 125–130.
Sasaki, N., Fukatsu, R., Tsuzuki, K., Hayashi, Y., Yoshida, T., Fujii, N., et al., 1998. Advanced glycation end products in Alzheimer's disease and other neurodegenerative diseases. Am. J. Pathol. 153, 1149–1155.
Sato, Y., Kaji, M., Tsuru, T., et al., 2001. Risk factors for hip fracture among elderly patients with Parkinson's disease. J. Neurol. Sci. 182, 89–93.
Schneider, J.A., Wilson, R.S., Bienias, J.L., Evans, D.A., Bennett, D.A., 2004. Cerebral infarctions and the likelihood of dementia from Alzheimer disease pathology, Neurology, 62. pp. 1148–1155.
Schram, M.T., Euser, S.M., De Craen, A.J.M., Witteman, J.C., Frölich, M., Hofman, A., et al., 2007. Systemic markers of inflammation and cognitive decline in old age. J. Am. Geriatr. Soc. 55, 708–716.
Schuh, A.F., Rieder, C.M., Rizzi, L., Chaves, M., Roriz-Cruz, M., 2011. Mechanisms of brain aging regulation by insulin: implications for neurodegeneration in late-onset Alzheimer's disease. ISRN Neurol. 2011, 306905.
Sen, S., Duman, R., Sanacora, G., 2008. Serum brain-derived neurotrophic factor, depression, and antidepressant medications: meta-analyses and implications. Biol. Psychiatry 64, 527–532.

Shomaker, L.B., Tanofsky-Kraff, M., Stern, E.A., Miller, R., Zocca, J.M., Field, S.E., et al., 2011. Longitudinal study of depressive symptoms and progression of insulin resistance in youth at risk for adult obesity. Diabetes. Care 34, 2458–2463.

Shonkoff, J.P., Boyce, W.T., McEwen, B.S., 2009. Neuroscience, molecular biology, and the childhood roots of health disparities: building a new framework for health promotion and disease prevention. JAMA 301, 2252–2259.

Si, X., Pu, J., Zhang, B., 2017. Structure, distribution, and genetic profile of α-synuclein and their potential clinical application in Parkinson's disease. J. Mov. Disord. 10, 69–79.

Simen, A.A., Bordner, K.A., Martin, M.P., Moy, L.A., Barry, L.C., 2011. Cognitive dysfunction with aging and the role of inflammation. Therap. Adv. Chron. Dis 2, 175–195.

Singh, R., Barden, A., Mori, T., Beilin, L., 2001. Advanced glycation end-products: a review. Diabetologia 44, 129–146.

Singh, M.K., Leslie, S.M., Packer, M.M., Zaiko, Y.V., Phillips, O.R., Weisman, E.F., et al., 2018. Brain and behavioral correlates of insulin resistance in youth with depression and obesity. Horm. Behav 17, 30501–30509.

Sirtori, C.R., Bolme, P., Azarnoff, D.L., 1972. Metabolic responses to acute and chronic L-dopa administration in patients with parkinsonism. N. Engl. J. Med. 287, 729–733.

Slavich, G.M., Irwin, M.R., 2014. From stress to inflammation and major depressive disorder: a social signal transduction theory of depression. Psychol. Bull. 140, 774–815.

Smith, J.L., Ju, J.S., Saha, B.M., Racette, B.A., Fisher, J.S., 2004. Levodopa with carbidopa diminishes glycogen concentration, glycogen synthase activity, and insulin-stimulated glucose transport in rat skeletal muscle. J. Appl. Physiol. 97, 2339–2346.

Sosa-Ortiz, A.L., Acosta-Castillo, I., Prince, M.J., 2012. Epidemiology of dementias and Alzheimer's disease. Arch. Med. Res. 43, 600–608.

Srikanth, V., Maczurek, A., Phan, T., Steele, M., Westcott, B., Juskiw, D., et al., 2011. Advanced glycation endproducts and their receptor RAGE in Alzheimer's disease. Neurobiol. Aging 32, 763–777.

Stolze, H., Klebe, S., Zechlin, C., Baecker, C., Friege, L., et al., 2004. Falls in frequent neurological diseases--prevalence, risk factors and aetiology. J. Neurol. 251, 79–84.

Stroth, S., Reinhardt, R.K., Thöne, J., Hille, K., Schneider, M., et al., 2010. Impact of aerobic exercise training on cognitive functions and affect associated to the COMT polymorphism in young adults. Neurobiol. Learn Memory 94, 364–372. 2010.

Svenningsson, I., Bjorkelund, C., Marklund, B., Gedda, B., 2012. Anxiety and depression in obese and normal-weight individuals with diabetes type 2: a gender perspective. Scand. J. Caring. Sci. 26, 349–354.

Talbot, K., Wang, H.Y., Kazi, H., Han, L.Y., Bakshi, K.P., Stucky, A., et al., 2012. Demonstrated brain insulin resistance in Alzheimer's disease patients is associated with IGF-1 resistance, IRS-1 dysregulation, and cognitive decline. J. Clin. Invest. 122, 1316–1338.

Teunissen, C.E., van Boxtel, M.P.J., Bosma, H., Bosmans, E., Delanghe, J., De Bruijn, C., et al., 2003. Inflammation markers in relation to cognition in a healthy aging population. J. Neuroimmunol. 2003 (134), 142–150.

The 2015 Ageing Report, 2015. The 2015 Ageing Report: Economic and Budgetary Projections for the 28 EU Member States (2013–2060). Publications Office, Luxembourg, <http://bookshop.europa.eu/uri?target = EUB:NOTICE: KCAR15003:EN:HTML>.

Todd, S., Barr, S., Roberts, M., Passmore, A.P., 2013. Survival in dementia and predictors of mortality: a review. Int. J. Geriatr. Psychiatry 28, 1109–1124.

Topic, R., Milicic, D., Stimac, Z., Loncar, M., Velagic, V., Marcinko, D., et al., 2013. Somatic comorbidity, metabolic syndrome, cardiovascular risk, and CRP in patients with recurrent depressive disorders. Croat. Med. J. 54, 453–459.

Touil, T., Deloire-Grassin, M.S., Vital, C., Petry, K.G., Brochet, B., 2001. In vivo damage of CNS myelin and axons induced by peroxynitrite. Neuroreport 12, 3637–3644.

Truban, D., Hou, X., Caulfield, T.R., Fiesel, F.C., Springer, W., 2017. PINK1, parkin, and mitochondrial quality control: what can we learn about Parkinson's disease pathobiology? J. Parkinsons Dis. 7, 13–29.

Tulving, E., 1987. Multiple memory systems and consciousness. Hum. Neurobiol. 6, 67–80.

United Nations, 2015. Department of Economic and Social Affairs, Population Division. World Population Ageing 2015 (ST/ESA/SER.A/390). United Nations, New York.

Unoki, H., Yamagishi, S., 2008. Advanced glycation end products and insulin resistance. Curr. Pharm. Des. 14, 987–989.

U.S. Census Bureau. 2008. National Population Projections. 2008. Available at http://www.census.gov/population/www/projections/2008projections.html (accessed September 22, 2009).

Uyama, N., Geerts, A., Reynaert, H., 2004. Neural connections between the hypothalamus and the liver. Anat. Rec. A Discov. Mol. Cell. Evol. Biol. 280, 808–820.

Valkanova, V., Ebmeier, K.P., 2014. Neuroimaging in dementia. Maturitas 79, 202–208.

van Praag, H., Kempermann, G., Gage, F.H., 1999. Running increases cell proliferation and neurogenesis in the adult mouse dentate gyrus. Nat. Neurosci. 2, 266–270.

van Praag, H., Shubert, T., Zhao, C., Gage, F.H., 2005. Exercise enhances learning and hippocampal neurogenesis in aged mice. J. Neurosci. 25, 8680–8685.

Varghese, F.P., Brown, E.S., 2001. The hypothalamic-pituitary-adrenal axis in major depressive disorder: a brief primer for primary care physicians. Prim. Care Campanion J. Clin. Psychiatry 3, 151–155.

Villemagne, V.L., Burnham, S., Bourgeat, P., Brown, B., Ellis, K.A., Salvado, O., et al., 2013. Amyloid β deposition, neurodegeneration, and cognitive decline in sporadic Alzheimer's disease: a prospective cohort study. Lancet Neurol. 12, 357–367.

Wang, Z.Z., Yang, W.X., Zhang, Y., et al., 2015. Phosphodiesterase-4D knock-down in the prefrontal cortex alleviates chronic unpredictable stress-induced depressive-like behaviors and memory deficits in mice. Sci. Rep. 5, 11332.

Watkins, L.R., et al., 1994. Neurocircuitry of illness-induced hyperalgesia. Brain Res. 639, 283–299.

Wautier, J.L., Zoukourian, C., Chappey, O., Wautier, M.P., Guillausseau, P.J., Cao, R., et al., 1996. Receptor-mediated endothelial cell dysfunction in diabetic vasculopathy. Soluble receptor for advanced glycation end products blocks hyperpermeability in diabetic rats. J. Clin. Invest. 97, 238–243.

Weber-Hamann, B., Hentschel, F., Kniest, A., Deuschle, M., Colla, M., Lederbogen, F., et al., 2002. Hypercortisolemic depression is associated with increased intra-abdominal fat. Psychosom. Med. 64, 274–277.

Weller, R.O., Boche, D., Nicoll, J.A., 2009. Microvasculature changes and cerebral amyloid angiopathy in Alzheimer's disease and their potential impact on therapy. Acta Neuropathol. 118, 87–102.

WHO, 2016. Dementia. <http://www.who.int/mediacentre/factsheets/fs362/en/>.

Wilhelm, K.R., Yanamandra, K., Gruden, M.A., Zamotin, V., Malisauskas, M., Casaite, V., et al., 2007. Immune reactivity towards insulin, its amyloid and protein S100B in blood sera of Parkinson's disease patients. Eur. J. Neurol. 14, 327–334.

Woods, C.A., Guttman, Z.R., Huang, D., Kolaric, R.A., Rabinowitsch, A.I., Jones, K.T., et al., 2016. Insulin receptor activation in the nucleus accumbens reflects nutritive value of a recently ingested meal. Physiol. Behav. 159, 52–63.

Yamagishi, S., Maeda, S., Matsui, T., Ueda, S., Fukami, K., Okuda, S., 2012. Role of advanced glycation end products (AGEs) and oxidative stress in vascular complications in diabetes. Biochim. Biophys. Acta 1820, 663–671.

Yates, L.B., Djousse, L., Kurth, T., Buring, J.E., Gaziano, M., 2008. Exceptional longevity in men: modificable factors associated with survival and function to age 90 years. Arch. Intern. Med. 168, 284–290.

Zhang, Z., Hayward, M.D., Yu, Y.L., 2013. Life course pathways to racial disparities in cognitive impairment among older Americans*. J. Health Soc. Behav. 57, 184–199.

Zeier, Z., Madorsky, I., Xu, Y., Ogle, W.O., Notterpek, L., Foster, T.C., 2011. Gene expression in the hippocampus: regionally specific effects of aging and caloric restriction. Mech. Ageing Dev. 132, 8–19.

Zlokovic, B.V., 2011. Neurovascular pathways to neurodegeneration in Alzheimer's disease and other disorders. Nat. Rev. Neurosci. 12, 723–738.

Further reading

Arnau-Soler, A., Macdonald-Dunlop, E., Adams, M.J., Clarke, T.K., MacIntyre, D.J., Milburn, K., et al., 2019. Generation scotland; major depressive disorder working group of the psychiatric genomics consortium. Transl. Psychiatry 9, 14.

Athilingam, P., Moynihan, J., Chen, L., D'Aoust, R., Groer, M., Kip, K., 2013. Elevated levels of interleukin 6 and C-reactive protein associated with cognitive impairment in heart failure. Congest. Heart Fail. 19, 92–98.

Cerejeira, J., Lagarto, L., Mukaetova-Ladinska, E.B., 2012. Behavioral and psychological symptoms of dementia. Front. Neurol. 3, 73.

Charlton, R.A., Lamar, M., Zhang, A., Ren, X., Ajilore, O., Pandey, G.N., et al., 2018. Associations between pro-inflammatory cytokines, learning and memory in late-life depression and healthy aging. Int. J. Geriatr. Psychiatry 33, 104–112.

Marsland, A.L., Gianaros, P.J., Abramowitch, S.M., Manuck, S.B., Hariri, A.R., 2008. Interleukin-6 covaries inversely with hippocampal grey matter volume in middle-aged adults. Biol. Psychiatry 64, 484–490.

Marsland, A.L., Gianaros, P.J., Kuan, D.C.H., Sheu, L.K., Krajina, K., Manuck, S.B., 2015. Brain morphology links systemic inflammation to cognitive function in midlife adults. Brain Behav. Immun. 48, 195–204.

Plassman, B.L., Langa, K.M., McCammon, R.J., Fisher, G.G., Potter, G.G., Burke, J.R., et al., 2011. Incidence of dementia and cognitive impairment, not dementia in the United States. Ann. Neurol. 70, 418–426.

Tegeler, C., O'Sullivan, J.L., Bucholtz, N., Goldeck, D., Pawelec, G., Steinhagen-Thiessen, E., et al., 2016. The inflammatory markers CRP, IL-6 and IL-10 are associated with cognitive function—data from the Berlin aging study II. Neurobiol. Aging 38, 112–117.

CHAPTER 9

Use of phytochemicals for the treatment of insulin resistance–linked visceral and neurological disorders

Introduction

Insulin resistance–linked visceral and neurological disorders are multifactorial chronic diseases, which include type 1 and type 2 diabetes, metabolic syndrome (MetS), sleep apnea, cardiovascular diseases (CVDs), and various types of cancers. Neurological disorders which involve insulin resistance include stroke, Alzheimer's disease (AD), Parkinson's disease (PD), and Huntington's disease (HD). Among neurological disorders, stroke is acute metabolic trauma, whereas AD, PD, and HD are chronic neurodegenerative diseases (Farooqui, 2016a,b). The overproduction of oxidants [reactive oxygen species (ROS) and reactive nitrogen species] in the human body is closely associated with the pathogenesis of many insulin resistance–linked chronic diseases, such as type 2 diabetes, MetS, CVD, neurological disorders, and various types of cancers. On one hand, type 2 diabetes is linked with MetS and CVD; and on the other hand, it is a risk factor for stroke, AD, PD, and other types of dementias (Farooqui, 2013). All abovementioned diseases are global health problems. They cause death and disability to millions of people throughout the world. It has been demonstrated that consumption of fruits, vegetables, and grains exerts a protective effect against the development of insulin resistance–mediated visceral and neurological disorders (Farooqui, 2008; Mursu et al., 2014; Kruk, 2014; Kyro et al., 2013). This protective effect of fruits, vegetables, and grains is mainly due to the presence phytochemicals, which are a heterogeneous nonnutritive group of chemical compounds with numerous biological effects in animals and men. In plants, phytochemicals are used for the survival of plants from various types of (1) environmental stresses, including ultraviolet (UV) light, heat and cold stresses, osmotic stress and high salinity, extreme pH, water deficit and dehydration, and nutrient

Insulin Resistance as a Risk Factor in Visceral and Neurological Disorders
DOI: https://doi.org/10.1016/B978-0-12-819603-8.00009-2

deprivation; (2) phytochemicals protect plants from viral, bacterial, yeast, and fungal infections; (3) phytochemicals defend plants from invading insects, herbivorous animals, and competitor plant species; (4) phytochemicals provide plants with a protection from environmental pollutants; (5) phytochemicals attract pollinators and other symbiotes; (6) phytochemicals attract the natural predators of herbivorous insects and animals (Kennedy and Wightman, 2011; Hooper et al., 2010; Tahara, 2007; Murakami, 2013); and finally (7) phytochemicals promote the growth of nonpathogenic endophytic microorganisms (Fig. 9.1). These microorganisms promote the survival of the host plants by protecting them from being eaten by herbivorous insects and animals as well as by defending them from many environmental stresses and infections by pathogenic microorganisms (Verma et al., 2009; Aly et al., 2013).

Phytochemicals are divided into 11 major classes: (1) phenolic compounds, including flavonoids, phenolic acids, hydroxycinnamic acids, lignans, tyrosol esters, stilbenoids, and alkylresorcinols; (2) terpenes, including carotenoids, monoterpenes, saponins, some modified lipid species, and triterpenoids; (3) betalains, including betacyanins and betaxanthins; (4) polysulfides; (5) organosulfur compounds; (6) indole compounds; (7) some protease inhibitors; (8) oxalic and anacardic organic acids; (9) modified purines; (10) quinones; and (11) polyamines (Farooqui, 2008; Kennedy and Wightman, 2011; Si and Liu, 2014). Examples of phytochemicals include flavonoids, curcumin, catechins, resveratrol, ginkgo biloba, and sulfur compounds found in garlic (Fig. 9.2). Today, approximately 6 in 10 Americans regularly consume some type of phytochemical, and approximately 1 in 6 Americans reports using herbal

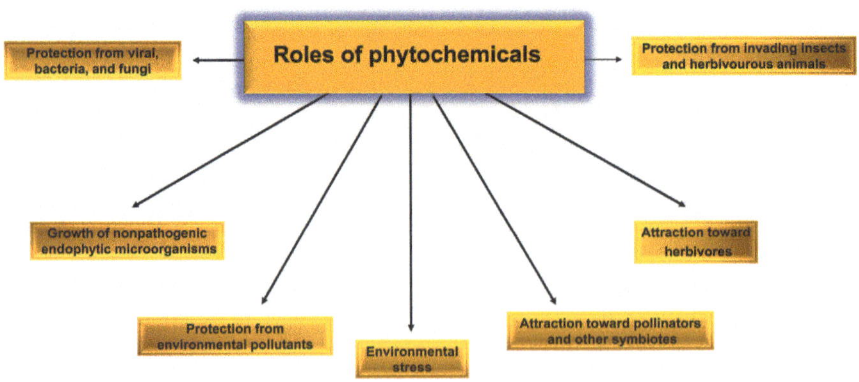

Figure 9.1 Roles of phytochemicals in plants.

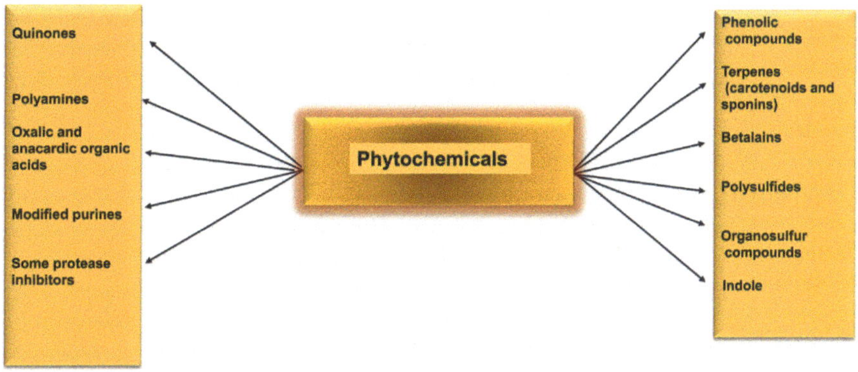

Figure 9.2 Classes of phytochemicals found in plants.

remedies on a regular basis (Gershwin et al., 2010). Unlike drugs (pharmaceuticals), which undergo extensive clinical trials in animals and humans prior to federal drug administration (FDA) approval, dietary supplements are not tested for their efficacy and safety. Phytochemicals contain complicated mixtures of organic chemicals, the levels of which may vary substantially depending upon many factors related to the growth of plants, harvesting, production, and storage conditions. These factors may cause alterations in chemical compositions of dietary supplement resulting in batch variation (Bent and Ko, 2004; Gershwin et al., 2010). Phytochemicals are the back bone of traditional medicinal system, which is still used in India, China, Egypt, and other African countries (Farooqui, 2008). In recent years, studies on phytochemicals have increased all over the world and new terms such as "dietary supplements," "functional food," and "nutraceutical" have been introduced. These terms illustrate the high expectations associated with current research on phytochemicals (Farooqui, 2008; Bellik et al., 2012). At present, "dietary supplements" compose a $32 billion industry, with over 50% of the American public regularly using dietary supplements (Kantor et al., 2016; Garcia-Cazarin et al., 2014). The precise molecular mechanisms of beneficial effects and specific actions of phytochemicals still remain the subject of intense research. The bioavailability of most phytochemicals to visceral organs is relatively higher than the brain not only because of the presence of blood–brain barrier (BBB), but also due to rapid metabolism and elimination of phytochemicals in the urine (Farooqui, 2008). The BBB controls the entrance of plasma components, red blood cells, and leukocytes into the brain, while exporting the neurotoxic molecules from the brain to the blood. The brain also contains two additional sites, which make up a barrier between the blood and cerebrospinal fluid (CSF), for example, the arachnoid epithelium constituting the central layer of the meninges and choroid plexus epithelium. These distinctive BBB membranes encompass a group of physical, transport, and metabolic barriers that separate the neural environment from the blood (Abbott et al., 2010). Health benefits of phytochemicals on visceral and brain tissues are due to their antiinflammatory, antioxidant, anticarcinogenic, antiproliferative, hypocholesterolemic, and cellular repair properties (Farooqui, 2008; Yoo et al., 2018). Regular consumption of phytochemicals from childhood to old-age may reduce risks of insulin resistance–linked visceral, neurodegenerative, and neuropsychiatric diseases. Antioxidant and antiinflammatory properties of phytochemicals mitigate the damaging effect of oxidative stress, neuroinflammation, and apoptosis. The chemical structures of phytochemicals have been used for synthesizing their analogues. These analogues have improved pharmacological activities through optimizing their bioavailability and pharmacokinetic profiles. The effects of phytochemicals on insulin resistance–linked visceral and brain diseases can be conducive, additive, synergistic, and antagonistic (Liu, 2003; Farooqui, 2008). Among phytochemicals, flavonoids represent a major group of secondary metabolites which are extensively distributed in nature especially in green plants. Majority of natural flavonoids are pigments and are usually allied with some vital pharmacological functions. Flavonoids differ from each other on the basis of differences in the aglycon ring structure and state of oxidation/reduction. Moreover, based on the extent of hydroxylation of aglycon, positions of the hydroxyl groups, saturation of pyran ring, and differences in the derivatization of the hydroxyl groups are major differentiating features among the various classes of flavonoids (Manach et al., 2004; Farooqui, 2008). Common flavonoids include quercetin, kaempferol (flavonols), and myricetin. These flavonoids are predominantly present in the onions, leeks, and broccoli. Other common types of flavonoids include isoflavones (daidzein, genistein). They are naturally distributed in soy and soy products. Flavanones include naringenin and hesperetin. They are present in the citrus fruits and tomatoes. Flavanols are present in epigallocatechin gallate (EGCG), catechin, epicatechin (EC), and epigallocatechin (EGC). They are mainly sequestered in the green tea, red wine, and chocolate, whereas anthocyanidins including malvidin, pelargonidin, and cyanidinare are widely distributed in the berry fruits and red wine (Manach et al., 2005). Flavonoids act by scavenging ROS, singlet molecular oxygen, and peroxyl radicals generated during lipid peroxidation (Farooqui, 2008; Upadhyay and Dixit, 2015). ROS are generated during the aerobic respiration and are counteracted by the bodily diverse system of antioxidants. When the ROS are generated in excess, they lead to oxidative stress and thus disturb the functions of different proteins, lipids and essential body elements (Farooqui, 2008). Besides their role in several disease processes, ROS are implicated in the inflammatory damage to neurons and development of AD (Farooqui, 2008). In addition, the use of polyphenols and flavonoids may not only result in improvements of memory acquisition and consolidation, but also in storage and retrieval of memory (Vauzour, 2012). These phytochemicals are highly effective in producing antiaging effects and reversing age-related declines in memory via their ability to interact with the cellular and molecular architecture of the brain responsible for memory related processes (Vauzour, 2012).

Phytochemicals (including flavonoids) induce their effects not only through their ability to modulate synaptic plasticity, but also by inducing neurogenesis in the hippocampus. The ability of many phytochemicals to activate the extracellular signal-regulated kinase (ERK1/2) and the protein kinase B (PKB/Akt) signaling pathways leads to the activation of the cAMP response element binding protein (CREB), a transcription factor responsible for increasing the expression of a number of growth factors (neurotrophins), which play important in defining memory, a process by which knowledge is encoded, stored, and later retrieved (Farooqui, 2008; Farooqui and Farooqui, 2018). So far, about 12,000 phytochemicals have been identified, and still a large percentage remains unknown (Barbosa et al., 2013). As mentioned above, the protective role of phytochemicals may be associated not only their antioxidant activity but also their antiaging effects (Si and Liu, 2014; Leonov et al., 2015). The purpose of this chapter is to provide information on beneficial effects of phytochemicals on insulin resistance and insulin resistance–linked chronic diseases with the hope that this discussion will jumpstart more studies on molecular mechanisms and signal transduction processes associated with the beneficial effects of phytochemicals in chronic diseases in humans.

Effects of various types of diets on insulin resistance

Major components of diet (carbohydrate, lipids, and proteins) not only provide energy but also modulate genes associated with protection against acute and chronic diseases associated with aging (Farooqui, 2013). If diet intake does not meet cellular nutritional needs, cellular metabolism and function slows down or even stop resulting into death (Farooqui, 2015).

Western diet is composed of refined grains (simple sugars), saturated and omega-6 fatty acids, proteins of animal origins, high salt, and low in fiber. Long-term consumption of western diet increases obesity, a pathological condition, mainly caused by a positive energy balance for several months to years. Obesity is accompanied by hyperglycemia, increase in insulin resistance, dyslipidemia, and hypertension (Fig. 9.3). Consumption of western diet not only impairs feedback control of the hypothalamic–pituitary–adrenal axis and elevates basal levels of plasma corticosterone, but also increases glucocorticoid receptor immunoreactivity (McNeilly et al., 2015). In rodents, long-term consumption of western diet blunts leptin and insulin anorexigenic signaling in the hypothalamus and produces apoptotic cell death in hypothalamic neurons with reduction in synaptic inputs to the arcuate nucleus and lateral hypothalamus. Collective evidence suggests that long consumption of a western diet not only results in hyperglycemia, weight gain obesity, and insulin resistance, but also reduction in brain-derived neurotrophic factor (BDNF) in the hippocampus and prefrontal cortex (Kanoski et al., 2007) leading to impairment in optimal

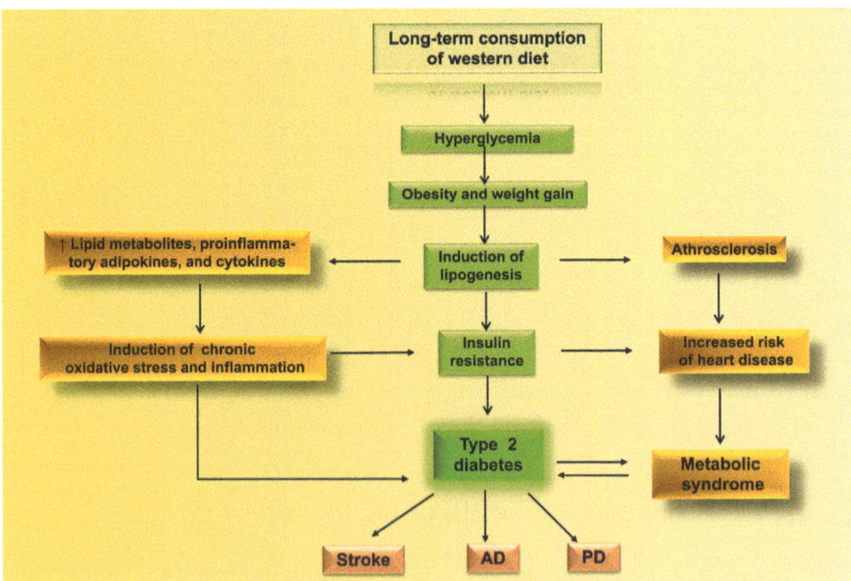

Figure 9.3 Effects of long-term consumption of western diet on human health. *AD*, Alzheimer's disease; *PD*, Parkinson's disease.

brain development and function by decreasing synaptic plasticity and changes in dendritic morphology (Granholm et al., 2008). This may lead to adverse cognitive and emotional outcomes later in life (Stranahan et al., 2008; Farooqui, 2013; Sellbom and Gunstad, 2012; Farooqui, 2013). In addition, consumption of western diet also results in alterations in blood vessel structure (Freeman and Granholm, 2012) and increase in neuroinflammation in the hippocampus (Pistell et al., 2010). The molecular mechanisms contributing to above processes are not fully established. However, it is suggested that insulin resistance, oxidative stress, genetic and environmental factors may contribute to the pathogenesis of chronic visceral and neurological disorders. In addition, impairments in the leptin system is another link that may connect insulin resistance, type 2 diabetes, MetS, and obesity, and CVD with stroke, AD, and PD (Morley and Banks, 2010).

Mediterranean diet is composed of vegetables, legumes, fruits, whole grains, fish, olive oil, fresh garlic, low levels of dairy products (cheese and yogurt), nuts, and red wine (Fig. 9.4) (Wang et al., 2006). Fresh fruits and vegetables in Mediterranean diet contain vitamins, carotenoids, flavonoids, fiber, potassium, magnesium, and other minerals. Olive oil contains

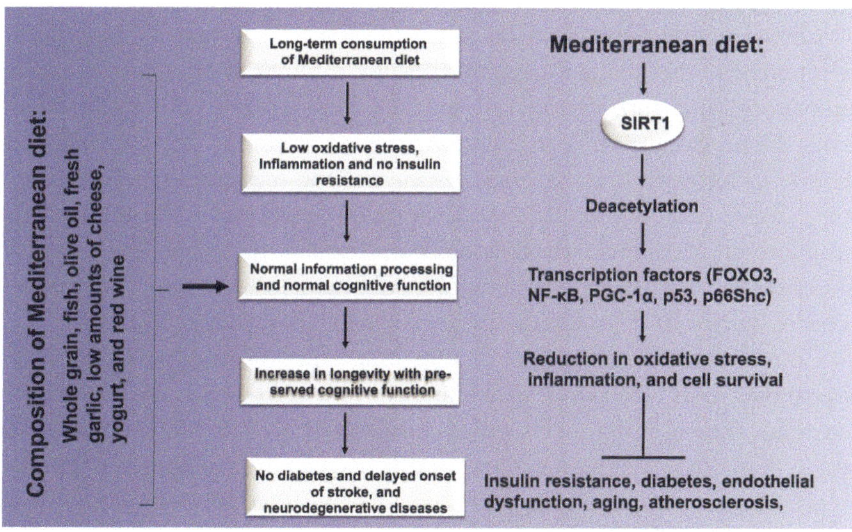

Figure 9.4 Effects of long-term consumption of Mediterranean diet.
FoxO3, Transcription factor forkhead box O-3; *NF-κB,* nuclear factor kappa B; *PGC-1α,* peroxisome proliferator-activated receptor-gamma coactivator-1α; *p53,* tumor suppressor protein p53; *p66Shc,* adaptor protein.

phytochemicals including tyrosol, hydroxytyrosols, oleocanthal, and oleuropein. Garlic is enriched in allicin, alliin, diallyl sulfide (DAS), diallyl disulfide (DADS), diallyl trisulfide (DATS), *S*-allylcysteine, dithiins, ajoene. Fish is enriched in omega-3 fatty acids such as eicosapentaenoic acid (EPA; 20:5n3) and docosahexaenoic acid (DHA; 22:6n3). The enzymic oxidation of these fatty acids produces resolvins, neuroprotectins, and maresins (Farooqui, 2011). These mediators produce antioxidant, antiinflammatory, and antiapoptotic effects. Red wine in Mediterranean diet contains resveratrol, a nonflavonoids polyphenolic compound, which not only produces cardioprotective, anticancer, antiinflammatory, and antioxidant effects, but also increases life span of yeast, worms, flies, and rodents (Baur et al., 2006) by stimulating sirtuin (SIRT) proteins, NAD^+-dependent deacetylase involved in the regulation of metabolism, apoptosis, mitochondrial biogenesis, inflammation, fatty acid metabolism, and glucose homeostasis (Lagouge et al., 2006a,b; Wang et al., 2009). In addition, cardioprotective and neuroprotective effects of resveratrol may be not only due to its free radical scavenging and metal chelation properties, but also due to antiamyloid action. Converging evidence suggests that resveratrol as a component of Mediterranean diet may reduce the detrimental effects of oxidative stress and neuroinflammation not only by normalizing mitochondrial dysfunction, decreasing insulin resistance, and improving endothelial function, but also by increasing cognitive function, on signal transduction processes. In addition, Mediterranean diet reduces cognitive decline through its antioxidant, antiinflammatory, and antiapoptotic effects (Fig. 9.4) (Baur et al., 2006; Wang et al., 2006). Long-term consumption of Mediterranean diet also results in increase in longevity through maintenance of the telomeres length (Riviere et al., 2007; Calabrese et al., 2007). Collective evidence suggests that consumption of Mediterranean diet decreases insulin resistance, diabetes risk, and cardiovascular health complications indicating that the type and composition of dietary fat is important for good health (Kastorini et al., 2011).

Components of vegetarian diet include fresh vegetables, fruits, chick peas, pulses, nuts, curcumin, garlic, and soy products (soy milk, soy yogurt, and tofu). These components produce healthy effects in visceral and brain tissues (Fig. 9.5). Vegetarian diets usually exceed human protein requirements (Willcox et al., 2014; Millward, 1999). Vegetables and fruits have adequate levels of magnesium, potassium, and nitrate. Thus vegetarian diet meets all current recommendations for most of micro and macronutrients. A vegetarian diet contains low levels of saturated fat and cholesterol, but high amounts of dietary fiber along with many health-promoting phytochemicals.

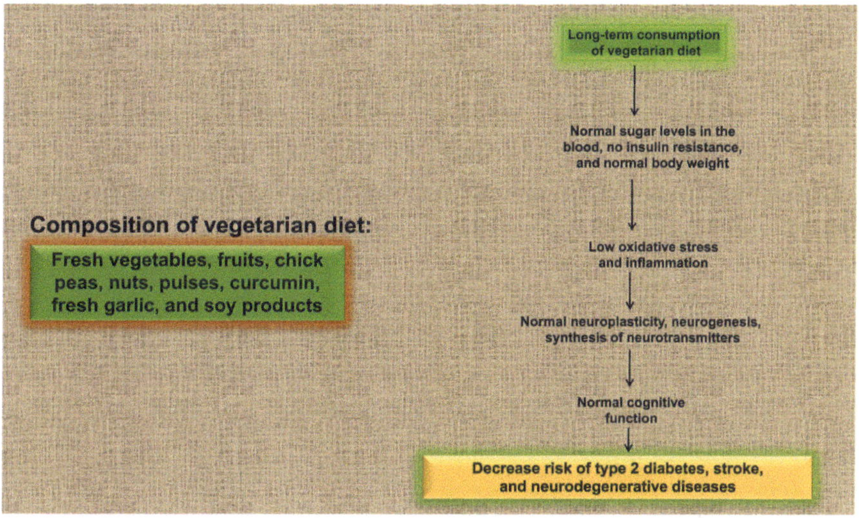

Figure 9.5 Effects of vegetarian diet consumption on human health.

Polyphenols (flavonoids, phenolic acids, proanthocyanidins, and tannins) in vegetarian diet inhibit hyperglycemia by inhibiting carbohydrate digestion, reducing carbohydrate absorption in the intestines, stimulating the release of insulin from pancreatic β-cells, and modulating hepatic glucose output (Rahimi et al., 2005). Vegetarian diet also contains significant amount of soluble and insoluble fibers. Fiber in vegetarian diet not only impacts intestinal tract time and absorption of macronutrients, but also alters the action of digestive enzymes and secretion of gastrointestinal and pancreatic hormones (Anderson, 1986). Insoluble fiber can decrease intestinal tract time, potentially reducing time for the carbohydrates to be absorbed in the jejunum (Montonen et al., 2003). Soluble fiber delays gastric emptying slowing the absorption and digestion of carbohydrates potentially delaying the insulin response (Montonen et al., 2003). In addition, fermentation of nondigestive fiber by the microbiota of the colon produces short-chain fatty acids (SCFAs). The production of SCFAs also impacts on carbohydrate metabolism (Thorburn et al., 1993). Vegetarian diet contains high levels of magnesium, an element which is deficient in at least 70%–80% Americans. Magnesium plays a critical role in visceral and brain tissue (Volpe, 2013). The deficiency of magnesium is a risk factor for a number of chronic diseases including migraine headaches, AD, stroke, hypertension, CVD, and type 2 diabetes (Volpe, 2013). On the basis of the abovementioned information, it is proposed that vegetarians typically have a lower body mass index (BMI)

(Spencer et al., 2003), and an improved lipid profile (Quiles et al., 2014). Adherence to a vegetarian diet reduces the risk of type 2 diabetes (Kahleova and Pelikanova, 2015; Olfert and Wattick, 2018), insulin resistance (Kim and Bae, 2015), hypertension (Yokoyama et al., 2014), coronary heart disease (Fraser, 2004), and dementia (Giem et al., 1993).

Components of Okinawan diet include Satsamu sweet potato, green and yellow vegetables, soybean-based foods, fruit, and low amounts of meat (fish and pork). In addition, Okinawan diet also contain a wide array of other plant foods including seaweed (especially konbu), soy, green tea, and kohencha tea (Willcox et al., 2009; Willcox et al., 2014). Among these components, soybeans are a rich source of isoflavones (phytoestrogens). These plant-derived, estrogen-like substances may partly suppress or inhibit normal estrogen secretion or normal estrogen activity in estrogen-responsive tissues such as breast (possibly by competing with endogenous estrogens for receptor sites in breast tissue), while themselves being less estrogenic to breast tissue, thereby reducing breast cancer risk (Wu et al., 2003). Okinawan diet improves metabolism and reduces insulin resistance and inflammation. Consumption of Okinawan diet promotes cardioprotective and neuroprotective effects (Fig. 9.6). The life expectancy of Okinawan peoples is significantly higher than the western countries.

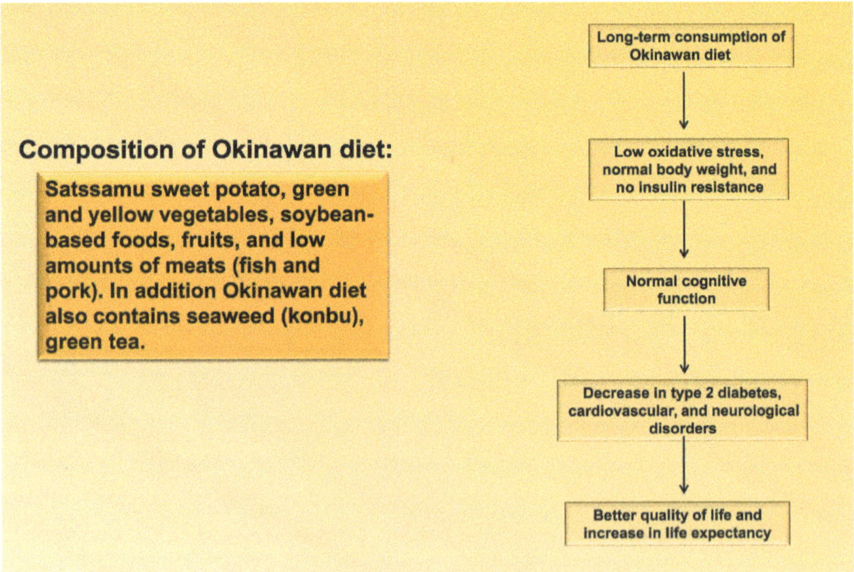

Figure 9.6 Effects of Okinawan diet on human health.

Much of the longevity advantage in Okinawan peoples is probably related to a traditional lifestyle (Suzuki et al., 2001) in which people consume low calories yet nutritionally dense, particularly with regard to vitamins, minerals, and phytonutrients (Willcox et al., 2007)

The molecular mechanisms by which Okinawan diet exerts cardioprotective and neuroprotective effects involves the beneficial effects of fresh fruits, vegetables, and n-3 fatty acids. Among these components, fresh fruits and vegetable provide vitamins and fiber, and n-3 fatty acids are precursors for resolvins, neuroprotectins, and maresins. These mediators produce antioxidant, anti-inflammatory, and antiapoptotic effects (Farooqui, 2011). Furthermore, seaweeds, an important component of Okinawan diet, have been reported to retards chronic diseases due to the presence of different compounds and proteins (phlorotannins, lipids, and minerals are macroalgae's lectins, phycobiliproteins, peptides, and amino acids), polyphenols, and polysaccharides, which are not present in terrestrial food sources (Jiménez-Escrig et al., 2011). Purported benefits of Okinawan diet also include antiviral, anticancer, and anticoagulant effects as well as the ability to modulate gut health and risk factors for obesity and diabetes. Although most of studies on seaweed effects have been performed in cell and animal models, there is evidence of the beneficial effect of seaweed and seaweed components on markers of human health and disease status (Jiménez-Escrig et al., 2011). No information is available on the effect of Okinawan diet on neurological disorders. However, based on components of Okinawan diet, it can be suggested that like Mediterranean diet, Okinawan diet may produce neuroprotective effects in humans. As mentioned above that the overproduction of oxidants (ROS, AGEs, and reactive nitrogen species) results in redox imbalance in the human body. These processes are closely associated with the pathogenesis of many insulin resistance–linked chronic visceral diseases, such as type 2 diabetes, Mets, CVD, and neurological disorders. It is well known that type 2 diabetes is a complex metabolic disorder characterized by impaired insulin secretion, β-cell dysfunction, and, aberrant insulin signaling, and insulin resistance (Stumvoll et al., 2005). Hyperglycemia, insulin resistance, hyperinsulinemia, hyperlipidemia (in particular elevated free fatty acids), and hyperhomocysteinemia are important pathophysiological components of type 2 diabetes that trigger systemic inflammation and impair nitric oxide (NO) bioavailability, with consequent impaired endothelial function (Kampoli et al., 2011; Tang et al., 2014). In type 2 diabetes, impairment in NO production, increase in oxidative stress and changes in function of endothelial progenitor cells contributing to the accelerated atherosclerotic process. Collective evidence suggests that type 2 diabetes on one side is associated with MetS and CVD and on the other side, it increases the risk factor for stroke, AD, PD, and other types of dementias (Fig. 9.7) (Farooqui, 2013; Goudis et al., 2015). All these disorders are linked primarily to high blood glucose levels. Converging evidence suggests that insulin resistance and type 2 diabetes are closely associated with the pathogenesis of CVD due to a complex combination of various traditional and nontraditional risk factors, which not only play important role in endothelial cell dysfunction, but are also associated with the evolution of atherosclerosis over its long natural history in clinical events (Fonseca et al., 2004). The clustering of vascular risk observed in association with insulin resistance has led to the view that cardiovascular risk appears early, before the development of type 2 diabetes whereas the solid interactions between hyperglycemia and microvascular disease suggests that this risk is not appear until hyperglycemic conditions appears. These notions highlight the progressive nature of both type 2 diabetes and related cardiovascular risk which contribute to specific challenges in diverse stages of the life of a subject with type 2 diabetes (Rydén et al., 2013), but do diabetic patients have specific risk factors which can explain the observed increase in CVD, or have all cardiovascular risk factors, traditional and nontraditional, the same strength?

Collective evidence suggests that at the molecular levels, insulin and glucagon contribute to maintenance of homeostasis in glucose and lipids metabolites through signaling cascades. These hormones stimulate translocation of the glucose transporter isoform 4 (GLUT4) from an intracellular location to the cell surface and facilitate the rapid insulin-dependent storage of glucose in muscle and fat cells (Simpson et al., 2001). Details of insulin signaling have been described in earlier chapters. Through insulin signaling, insulin controls energy metabolism and regulates cells growth and survival, as well as uptake, synthesis and hydrolysis of glycogen, proteins, and lipids (Saltiel and Kahn, 2001). Three mechanisms contribute to the transport of glucose in the cells. These include mammalian target of rapamycin complex 1 and ERK pathways, but these pathways are not enough to satisfy glucose need in the cell (Lazar et al., 1995; Polak et al., 2008). An alternative pathway, which involves adaptor protein with pleckstrin homology and Src homology 2 (SH2) domain (APS) and a phosphoinositide 3-kinase (PtdIns 3K)-dependent insulin signaling pathways facilitates glucose transport in adipose tissue and brain. Src (abbreviated form

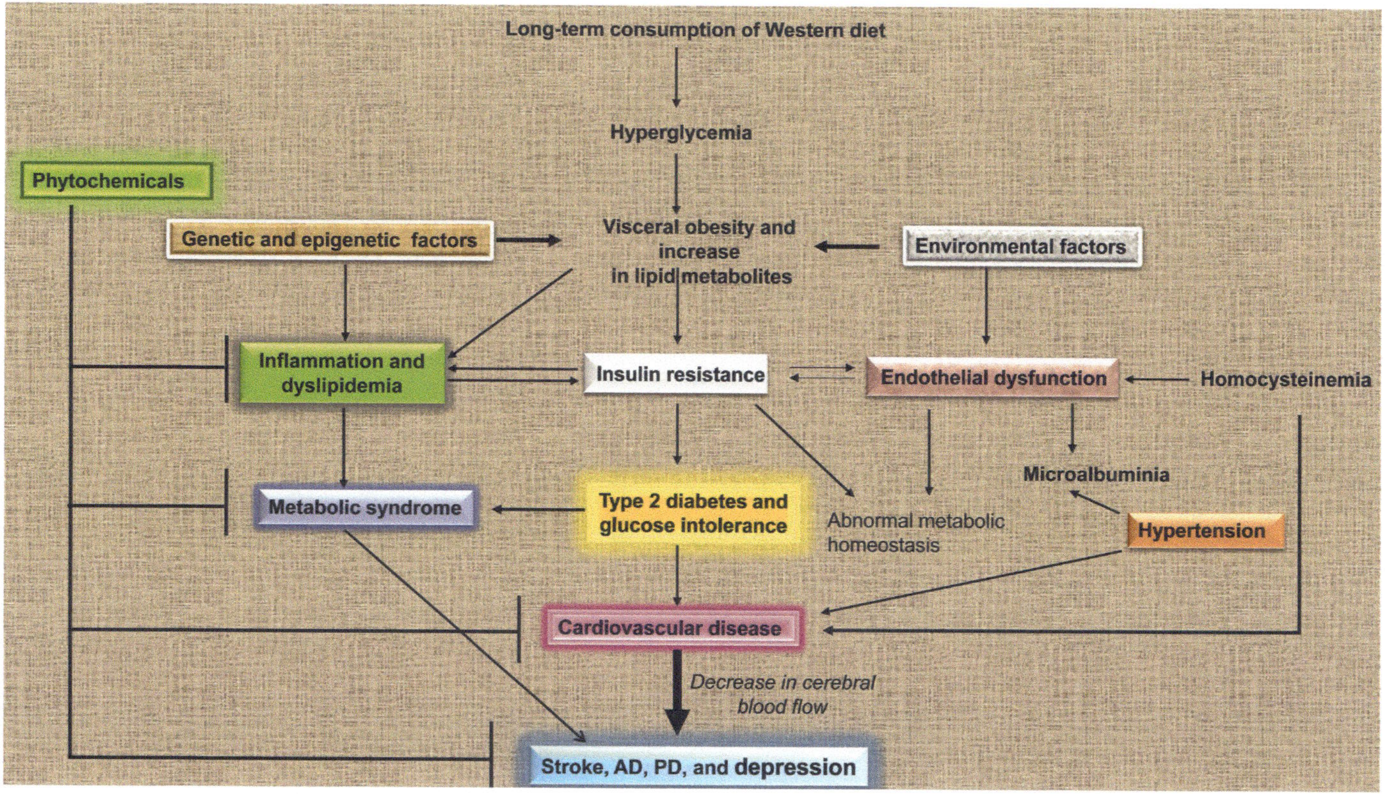

Figure 9.7 Effects of long-term consumption of western diet on insulin resistance and its effects on cardiovascular disease and neurological disorders.

of sarcoma) is a proto-oncogene encoding one of tyrosine kinases (Larsen, 1989). Of note, the PtdIns 3K dependent insulin signaling pathway for glucose uptake in muscle is fairly well known, although the necessity for the APS-dependent insulin signaling pathway in this particular tissue is still elusive. All these pathways together, clarify the delivery efficiency of GLUT4 to the cell surface by assembling signaling stage at the plasma membrane that are comprised of protein kinases, lipids, adaptor proteins, small GTPases (enzymes that hydrolyze guanosine triphosphate) and lipid kinases. These signaling pathways involve the compartmentalizing mechanism in regulation of GLUT4 cycling. It is interesting to note that GLUT4 cycling can be overstimulated in various types of cancers, like plasma cell malignancy, multiple myeloma, etc.

Effects of phytochemicals on insulin resistance–linked visceral and neurological disorders

Phytochemicals have been used to treat type 2 diabetes, heart disease, and neurological disorders (stroke, AD, and PD) from ancient times (Hu et al., 2002; Parveen et al., 2018; Jiang et al., 2007; Goudis et al., 2015; Firdous, 2014; Laddha and Kulkarni, 2019; Sayem et al., 2018). Phytochemicals may be more effective in treating these diseases and have less side effects than the modern-days medications (Farooqui, 2012). One positive aspect of phytochemical supplementation is that large amounts of phytochemicals are consumed in the daily diet from ancient time. The list of phytochemical supplements includes curcumin, n-3 fatty acids, resveratrol, cinnamon, olive oil, green tea, ginsenoside, and garlic (Fig. 9.8). Although the possible mechanisms of actions of these phytochemicals are not absolutely understood, numerous studies are being conducted to reveal the different signaling pathways by which abovementioned phytochemicals and phytochemical-enriched diet act to retard insulin resistance–linked chronic diseases (McEvoy et al., 2012). Although many phytochemicals undergo detoxification and are poorly absorbed with rapid excretion, they still exert antiinflammatory, antioxidant, and anticarcinogenic effects at low doses. The effects of phytochemicals are mediated through their abilities to counteract, reduce and repair damage resulting from oxidative stress and neuroinflammation—processes that are modulated by the transcription factors, such nuclear factor-kappaB (NF-κB) and nuclear factor erythroid 2–related factor 2 (Nrf2). Phytochemicals not only stimulate the synthesis of adaptive enzymes and proteins that protect against cellular stress (Farooqui, 2008; Virgili and Marino, 2008), but also modulate age-related

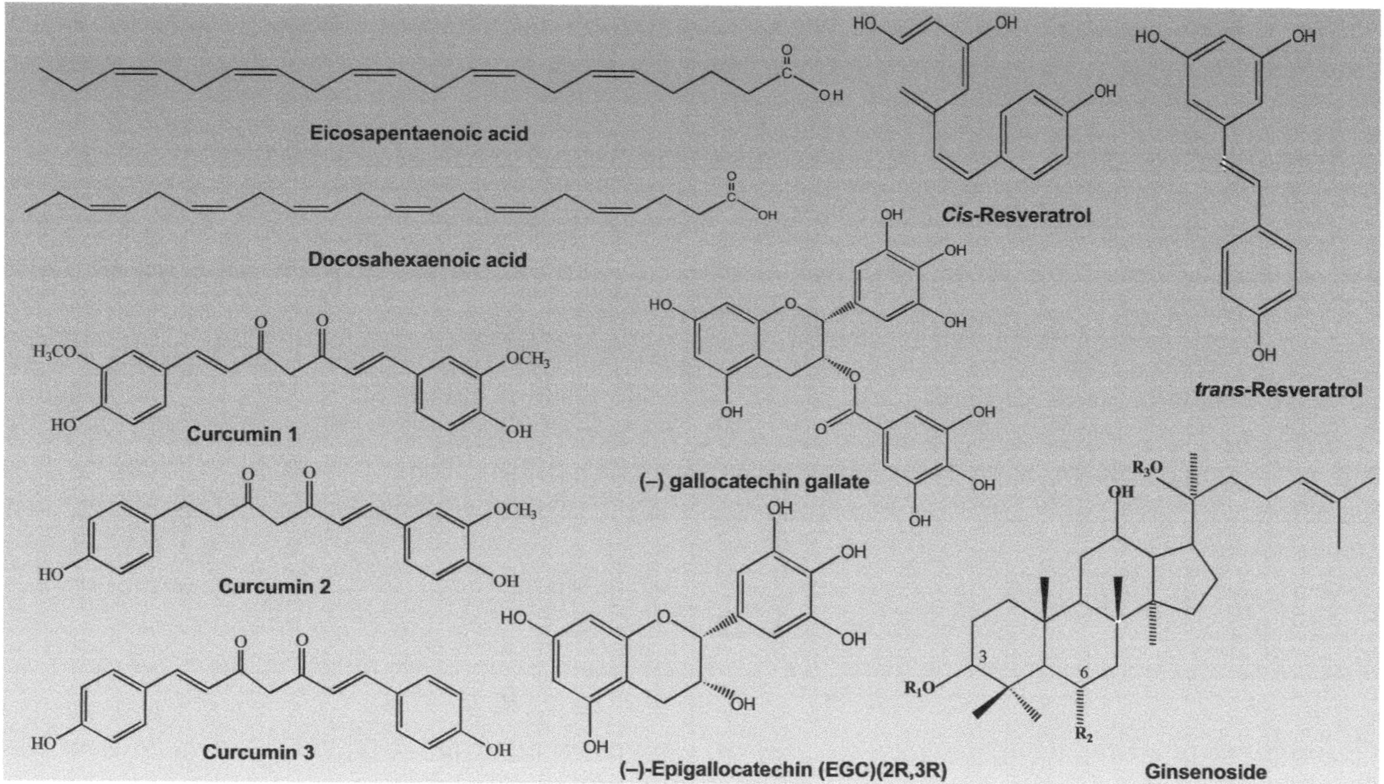

Figure 9.8 Chemical structures of phytochemicals found in plants.

decline in memory by upregulating signaling pathways that control synaptic plasticity. They activate the ERK1/2 and the PKB/Akt signaling pathway and CREB, enzymes and transcription factor that upregulate the expression of several neurotrophins that facilitate memory formation (Spencer et al., 2009). Low levels of phytochemicals can cross the BBB and induce the synthesis of heat shock proteins (HSPs). These proteins promote the production of antiinflammatory cytokines. In this scenario, responses to HSPs production are considered an attempt by the immune system to correct the inflammatory condition. These highly integrated and regulated processes are controlled by redox sensitive genes called vitagenes. These genes code for HSPs, thioredoxin, and sirtuin protein systems and modulate a complex network of intracellular signaling pathways relevant to life span and preservation of cellular homeostasis under stressful conditions (Farooqui, 2008).

Effects of curcumin on insulin resistance and insulin resistance–linked diseases

Curcumin is a yellow-orange pigment obtained from the plant Curcuma longa, which belongs to the family Zingiberaceae. It provides beneficial effects on insulin resistance and glucose intolerance in in vitro conditions and in preclinical studies (Weisberg et al., 2008; Na et al., 2011). These effects result in not only in decreasing of low-grade inflammation, but also by downregulation of NF-κB leading to cytoprotection of pancreatic β-cells by increasing the concentrations of antioxidant enzymes (Weisberg et al., 2008). In mice models of diabetes, curcumin increases the expression of 5′ adenosine monophosphate–activated protein kinase (AMPK), a key regulator of glucose and lipid homeostasis (Kim et al., 2009; Kang et al., 2013).

A single injection of curcumin (1 and 2 mg/kg, i.v.) 30 minutes after focal cerebral ischemic/reperfusion injury in rats not only diminishes infarct volume, improves neurological deficit, and decreases mortality, but also reduces the water content of the brain, and the extravasates Evans blue dye in ipsilateral hemisphere in a dose-dependent manner (Jiang et al., 2007). In cultured astrocytes, curcumin significantly inhibits inducible nitric oxide synthase (iNOS) expression and NO(x) (nitrites/nitrates contents) formation mediated by lipopolysaccharide (LPS)/tumor necrosis factor-α (TNF-α). Based on in vivo and in vitro observations, it is proposed that curcumin ameliorates cerebral ischemia/reperfusion injury not only by preventing peroxynitrite-mediated BBB, but also by decreasing lipid peroxidation–mediated damage and retarding apoptotic cell death as well as glial cell activation (Jiang et al., 2007). In addition, curcumin stimulates transcription factor and Nrf2 and induces the expression of enzymes involved not only in detoxification of free radicals, but also in the maintenance of redox potential (Fig. 9.9) (Yang et al., 2009). Nrf2 pathway is the main antioxidant defense pathway present in visceral and neural cells. It involves antioxidant response element (ARE) (Raghunath et al., 2018). This element contributes to the dynamic expression of phase II genes. The detoxification of toxic chemicals three phases namely, phase I, II, and III (Xu et al., 2005). Phase I enzymes primarily consist of cytochrome P450 superfamily of enzymes and metabolizes the xenobiotics (Lewis, 2003). Phase II conjugating enzymes include glutathione (GSH) *S*-transferases which catalyzes the conjugation of reactive electrophile species with GSH thereby it attenuates the toxic potential of xenobiotics (Strange et al., 2000). Phase III transporters eliminate the GSH conjugates and offer protection against deleterious chemicals (Mizuno et al., 2003). Finally, curcumin up- and downregulates different kinds of miRNA (Gupta et al., 2013) and takes part in epigenetic changes by inhibiting DNA methyltransferases and regulating histone modifications via effects on histone acetyltransferases and histone deacetylases (Boyanapalli and Kong, 2015). However, in our opinion, this is not a disadvantage but, quite to the contrary, an advantage when one compound can affect diverse biological processes, such as the redox state, inflammation, proliferation, migration, apoptosis, wound healing and as a consequence positively affect memory, postpone aging and age-related diseases such as atherosclerosis (Kunnumakkara et al., 2017).

Epidemiological studies have indicated that in India, where curcumin is widely used in daily diet, there is a significant reduction in the prevalence of AD (in patients between 70 and 79 years of age is 4.4-fold less than that of the United States) (Chandra et al., 2001). In addition, curcumin has been reported to decrease oxidative damage and β-amyloid (Aβ) deposition in a transgenic mouse model of AD, and to reverse Aβ-induced cognitive deficits and neuropathological changes in rats (Yang et al., 2005). It is suggested that curcumin acts through its pleiotropic effects.

AD and PD are accompanied by reduced expression of insulin receptor and in the brain and impairment of insulin signaling. These changes are associated with decrease in metabolism of the glucose, reduction in

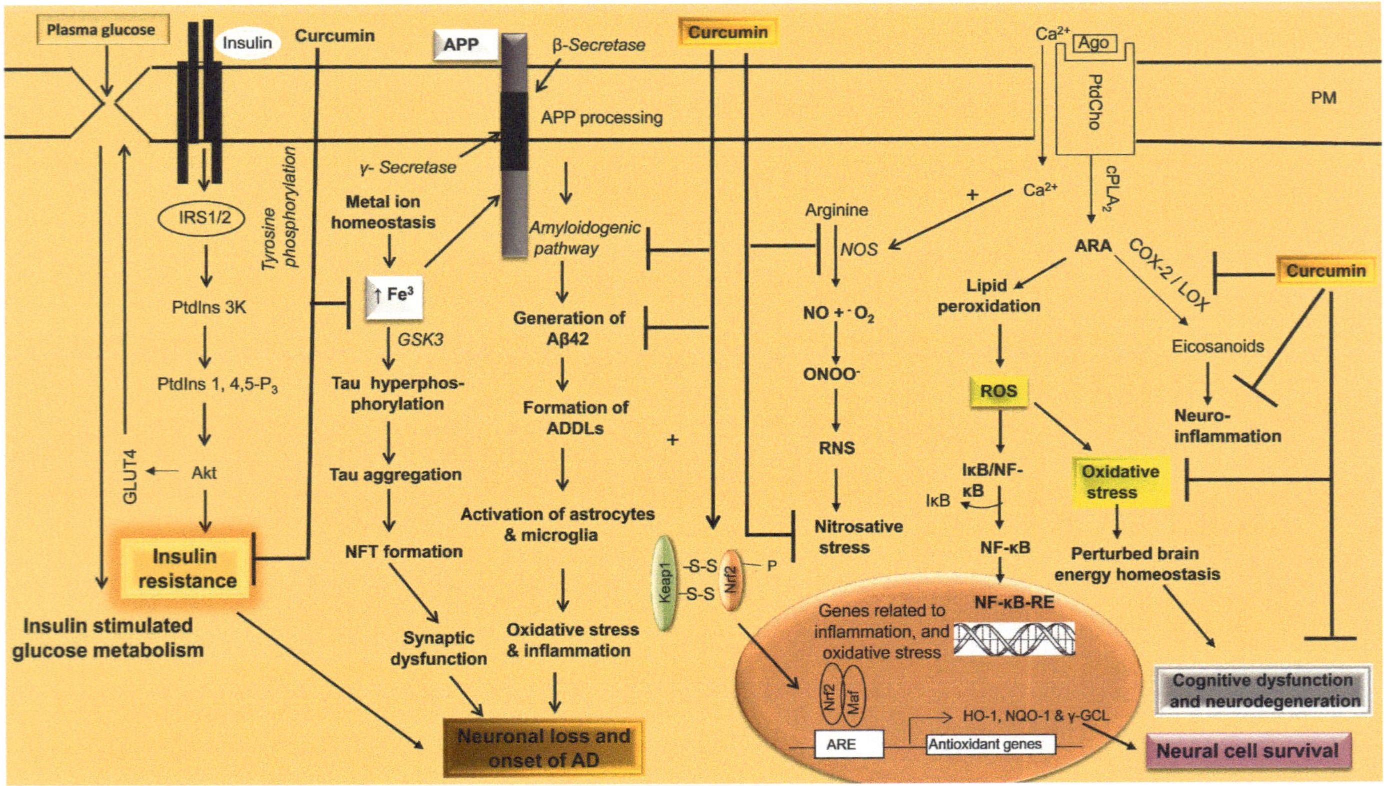

Figure 9.9 Hypothetical diagram showing effects of curcumin on neurochemical changes in Alzheimer's disease. In AD, induction of insulin resistance decreases entry of glucose into the cell. Curcumin inhibits insulin resistance and stimulates GLUT4 through the stimulation of Akt activity. *Aβ*, β-Amyloid; *ADDL*, Aβ-derived diffusible ligand; *Ago*, agonist; *APP*, amyloid precursor protein; *ARA*, arachidonic acid; *ARE*, antioxidant response element; *COX-2*, cyclooxygenase-2; *cPLA₂*, cytosolic phospholipase A_2; *γ-GCL*, glutamate cystein ligase; *HO-1*, heme oxygenase; *I-κB*, inhibitory subunit of NF-κB; *IL-1β*, interleukin-1β; *IL-6*, interleukin-6; *Keap1*, kelch-like ECH-associated protein 1; *LOX*, lipoxygenase; *Maf*, small leucine zipper proteins; *NF-κB-RE*, nuclear factor-κB-response element; *NQO-1*, NADPH quinine oxidoreductase; *Nrf2*, nuclear factor E2-related factor 2; *PtdCho*, phosphatidylcholine; *ROS*, reactive oxygen species; *TNF-α*, tumor necrosis factor-α.

development of neurons, and decrease in learning and memory (Moroo et al., 1994; Enrique et al., 2014). In scopolamine-induced model of AD, neurochemical changes involve decrease in glucose metabolism, hyperinsulinemia, and impairment of insulin signaling (Das et al., 2019). This animal model of AD is not only characterized not only by dysfunction of insulin signaling proteins (IR-, IGF-1, IRS1, and IRS-2) and marked decrease in Akt activity, but also by reduction in GLUT3 and GLUT4 levels. Treatment of rats with curcumin results significantly increases glucose levels in plasma and in brain. However, insulin levels are decreased in plasma and are increased in scopolamine-injected rat brain. Moreover, GLUT3 and GLUT4 levels are significantly increased in curcumin-treated AD rats. Collectively, this study suggests that curcumin ameliorates the altered insulin signaling, reduces hyperinsulinemia, and improves glucose levels in scopolamine-injected models of AD (Das et al., 2019). More studies are needed to better understand the mechanism of action of curcumin in the treatment of AD.

Curcumin produces antiamyloidogenic effects by inhibiting Aβ aggregation and downregulating BACE1 expression. It also possesses antitau function by inhibiting tau hyperphosphorylation and plays a key role in tau tangle clearance. In a randomized, placebo-controlled double-blinded study, Baum et al. reported that curcumin exhibits neuroprotective effects in AD patients via several mechanisms, including the inhibition of Aβ aggregation, the inflammatory pathways, and free radical–induced neurodegeneration (Baum et al., 2008). The targets of curcumin action also include transcription factors (NF-κB, AP-1, PPARγ, and Nrf2), enzymes [COX-2, 5-lipoxygenase (5-LOX), iNOS, HO-1, and Jun kinase (JNK)], and proinflammatory cytokines (TNF-α, IL-1β, IL-6) (Shishodia et al., 2005) (Fig. 9.9). Modulation of above transcription factors, enzymes, and proinflammatory cytokines by curcumin leads to neuroprotection through antiinflammatory, antioxidant, and antiprotein aggregative and neurogenic effects of curcumin in AD models (Yang et al., 2005; Cole and Frautschy, 2007; Ma et al., 2009; Farooqui, 2016a,b).

Curcumin retards oxidative stress and significantly reduces the cytotoxicity mediated by extracellular or intracellular α-synuclein (α-Syn) aggregates, suggesting that curcumin may be a useful therapeutic agent for treating PD. Since extracellularly added curcumin provides protection even against intracellularly induced α-Syn toxicity, it is suggested that there is a significant extracellular or cell surface component to α-Syn-induced neurotoxicity in PD models (Liu et al., 2011). This observation is consistent with recently published reports of interneuronal transmission of extracellular α-Syn pathology in neuronal cells (Desplats et al., 2009). α-Syn is mainly located in synaptic terminals. The conversion of α-Syn into oligomers and fibrils is the hallmark of PD and dementia with Lewy bodies. α-Syn is disordered in solution but can adopt an α-helical conformation upon binding to lipid membranes (Alza et al., 2019). This lipid–protein interaction plays an important role in its proposed biological function, that is, synaptic plasticity, but can also entail the aggregation of the protein. Both the chemical properties of the lipids and the lipid–protein ratio have been reported to modulate the aggregation propensity of α-syn (Alza et al., 2019).

Curcumin also protects against A53T α-Syn-mediated cell death in a dose-dependent manner. In mitochondrial permeability transition pore (MPTP)-induced model of PD, systemic administration of curcumin and tetrahydrocurcumin significantly reverses the MPTP-induced depletion of dopamine and 3,4-dihydroxy phenyl acetic acid supporting the view that curcumin and tetrahydrocurcumin exert neuroprotective effects against MPTP-induced neurotoxicity (Rajeswari and Sabesan, 2008). Collective evidence suggests that curcumin can be used for treating stroke, AD, and PD in animal models. Curcumin has ability to cross BBB and neural cell membranes and induces its antioxidant, antiinflammatory, antiamyloidogenic, antiapoptotic, and metal chelating effects leading to retardation of signal transduction pathways associated with oxidative stress and neuroinflammation (Fig. 9.9). On the basis of this information, it is suggested that curcumin fulfills the characteristics for an ideal neuroprotective agent for stroke, AD, and PD with its low toxicity, affordability, and easy accessibility. However, poor bioavailability of curcumin is the major hurdle for its more widespread use in animals and humans. The bioavailability of curcumin can be increased by encapsulation of curcumin into liposomes, cyclodextrin, curcumin conjugate with PLGA, complexation with phospholipids, and synthesis of curcumin analogues.

Effect of green tea on insulin resistance–linked diseases

Green tea is one of the most popular drinks consumed worldwide. It is produced mainly in China and Japan from the leaves of the *Camellia sinensis* plant. Health benefits of green tea have been widely studied (Reygaert, 2018; Farooqui, 2008). Green tea contains four main catechins. These include EC, (−)-epicatechin-3-gallate, EGC, and EGCG. Of these catechins, EGCG and EGC are found in the highest amounts in green tea (Fig. 9.8). In addition, green tea also contains gallic acid, chlorogenic acid, and caffeic acid, and flavonols such as kaempferol, myricetin, and quercetin (USDA Database, 2003; Wang and Ho, 2009). Human studies indicate that green tea not only contributes to a reduction in the risk of CVD and some forms of cancer, but also induces antihypertensive effects by suppressing angiotensin I converting enzyme, body weight control by suppressing the appetite, antibacterial and antivirasic effects, solar UV protection, bone mineral density increase, antifibrotic effects, and neuroprotective effects. The beneficial effects of green tea are due to decrease in blood pressure (Henry and Stephens-Larson, 1984) and blood sugar (Matsumoto et al., 1993). Lipid metabolism studies in animals have indicated that green tea catechins reduce triacylglycerol and total cholesterol concentrations (Chan et al., 1999), inhibit hepatic and body fat accumulation (Ishigaki et al., 1991), and stimulate thermogenesis (Dulloo et al., 2000). In addition, green tea boosts metabolism and improves immune function. Several studies have indicated that drinking green tea increases insulin sensitivity and glucose levels in diabetic (*db/db*) mice (Ortsater et al., 2012; Tsuneki et al., 2008). Similarly, in humans, epidemiological studies indicate that long-term consumption of green tea associated with a reduction of the incidence of diabetes (Oba et al., 2010; Toolsee et al., 2013). A number of randomized controlled trials have reported that daily consumption of green tea extract may enhance oral glucose tolerance in healthy people as well as reduce fasting plasma glucose and glycosylated hemoglobin A1c (HbA1c) levels in people at risk of diabetes (Basu et al., 2011; Hsu et al., 2011). These observations support the view that consuming green tea and green tea extract may play an important role not only in reduction of insulin resistance, but also in improvement of glycemic control in people with type 2 diabetes (Fig. 9.10).

Mechanisms by which EGCG influences insulin resistance and blood pressure are not fully understood. However, it is proposed that green tea extract may act by stimulating endothelial production of NO via the phosphotidylinositol-3-kinase/Akt pathway (Lorenz et al., 2004; Kim et al., 2007). In addition, green tea extract is known to produce inhibitory effect on I-kappa kinase activity (Chen et al., 2002; Kim et al., 2005; Pan et al., 2000). Green tea extract is known to impact the endothelial cell function through the production of vasoconstrictors and vasodilators which subsequently impact on smooth muscle cell function and vascular tone. Indeed, EGCG induces dose-dependent vasodilation in

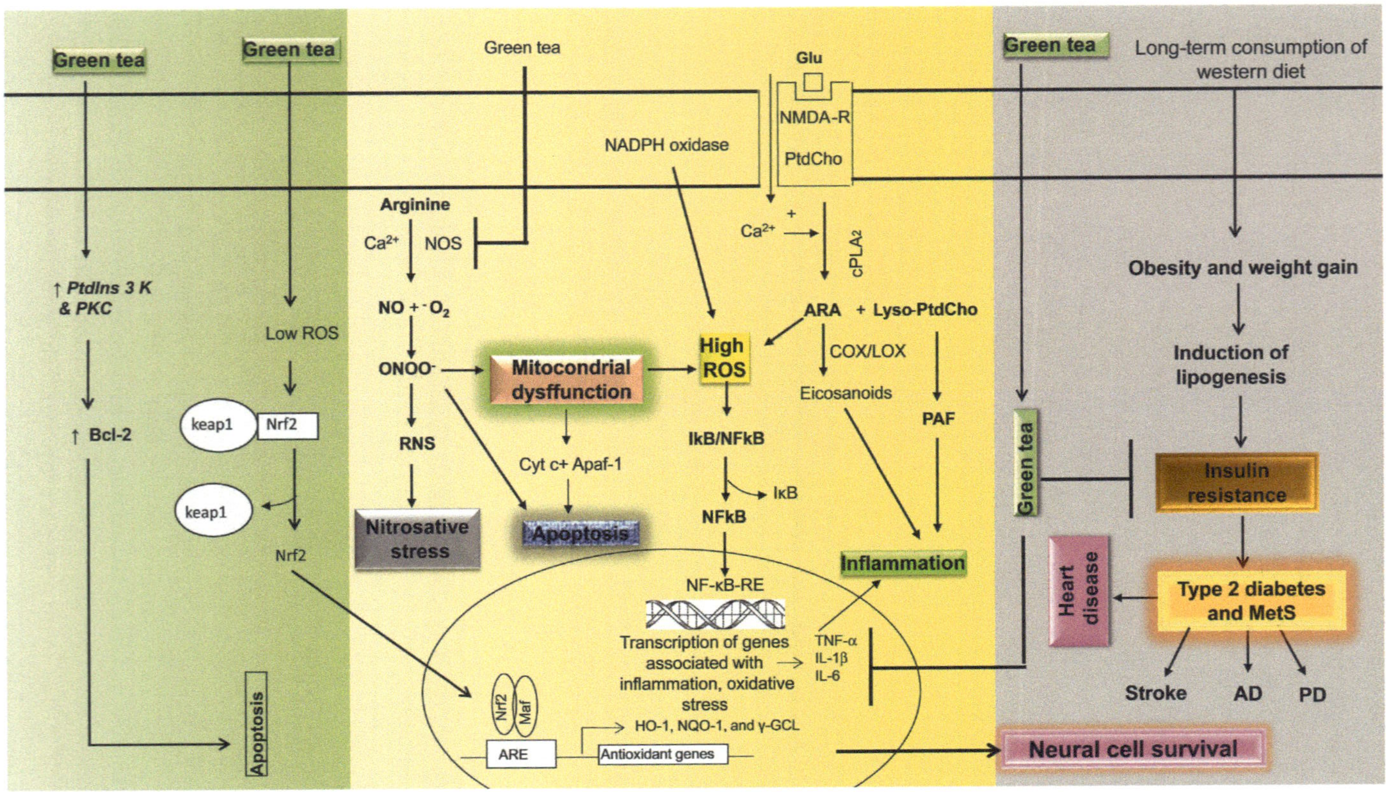

Figure 9.10 Hypothetical diagram showing effects of green tea catechins on neurochemical changes in neurological disorders. *Bcl-2*, B-cell lymphoma 2; *Glu*, glutamate; *lyso-PtdCho*, lyso-phosphatidylcholine; *MCP-1*, monocyte chemoattractant protein 1; *NMDA-R*, NMDA receptor; *NO*, nitric oxide; *ONOO−*, peroxynitrite; *PAF*, platelet activating factor. For other abbreviations, refer caption of Fig. 9.9.

precontracted rat aortic rings (Lorenz et al., 2004) and in mesenteric vascular beds isolated from spontaneously hypertensive rats (Potenza et al., 2007). These effects can be observed at concentrations as low as 1 μM, albeit weakly. In cultured endothelial cells, EGCG has been demonstrated to activate endothelial NOS and stimulate production of NO, which is a potent vasodilator (Lorenz et al., 2004; Persson et al., 2006). Collective evidence suggests that supplementation diet with EGCG may affect insulin resistance and other associated metabolic risk factors such as obesity and may reduce tea antihypertensive effect. This may produce beneficial effects on cardiovascular function associated with habitual green tea consumption.

Effect of resveratrol on insulin resistance and insulin resistance–linked diseases

Resveratrol (3,5,4′-trihydroxystilbene), a natural polyphenolic compound that occurs naturally in nuts, berries, and the skin of grapes (Fig. 9.8). Resveratrol occurs in two isoforms: *trans*-resveratrol and *cis*-resveratrol. The *trans* isomer possesses a greater biological activity due to the presence of the 4′-hydroxystyryl group. Resveratrol possesses antioxidant, antiinflammatory, antiapoptotic, and antiobesity activities. The health-promoting effects of resveratrol are not only due to the stimulation of the AMPK and the sirtuins, but also calorie restriction, which putatively affects metabolic pathways that modulate molecular damage (Poulsen et al., 2013; Most et al., 2017). In addition, resveratrol not only produces positive effects on longevity (age) and lipid levels, but also decreases obesity and mediates beneficial effects against certain cancers and viral infections.

Antiobesity activity of resveratrol is due to the activation of brown adipose tissue and the induction of white adipose tissue (WAT) browning (Wang et al., 2015; Liao et al., 2018). However, the mechanism by which dietary resveratrol affects adipose tissue remains unclear. However, it is proposed that resveratrol may affect not only composition of gut microbiota, but also increases levels of gut microbiota–derived metabolites (beta-amino isobutyric acid, gamma amino butyric acid, PPARY, exercise, irisin, and mico-rRNAs) (Jeremic et al., 2017). This may affect food intake, and other factors such as glucagon-like peptide-1 (GLP-1) and peptide YY (PYY) secreted by L cells in the intestinal epithelium. SCFAs, GLP-1, and PYY are known to influence energy intake and energy metabolism to affect the onset of obesity by activating various pathways through calcium signaling pathways, including neuronal excitability, islet cell proliferation and apoptosis, insulin and glucagon secretion, adipose tissue metabolism, and gastrointestinal tract activity (Song et al., 2019). Increase in these metabolites has also been reported to influence and regulate the browning of WAT as well as thermogenesis (Bird et al., 2017). Thus in rodents, resveratrol modifies the relative Bacteroidetes:Firmicutes ratio and reverses the gut microbial dysbiosis induced by a high-fat diet. By upregulating the expression of genes involved in maintaining tight junctions between intestinal cells, resveratrol contributes to gut barrier integrity. The composition of the gut microbiome and rapid metabolism of resveratrol determines the production of resveratrol metabolites, which are found at greater concentrations in humans after ingestion than their parent molecule and can have similar biological effects.

Resveratrol provides protection against CVDs, currently the leading cause of death, and a relevant health problem worldwide (Markus and Morris, 2008). At the concentration obtainable physiologically by the consumption of red wine (resveratrol) increases the expression of endothelial NO synthase in human vascular endothelial cells, which is responsible for synthesizing the potent vasodilator, NO (Nicholson et al., 2008); it also decreases the expression of the potent vasoconstrictor, endothelin (Nicholson et al., 2008). Other mechanisms underlying the cardioprotective effects of resveratrol include the inhibition of platelet aggregation, similar to aspirin, and its antioxidant effects on cholesterol metabolism (Markus and Morris, 2008). Resveratrol also produces its effect on endothelial cells injury. It maintains a balance between vasodilators, such as NO, and vasoconstrictors, such as endothelin-1 (Kinlay et al., 2001), which, together, provide thrombus resistance and prevent atherogenesis (Davingnon and Ganz, 2004).

In animal model of stroke, intravenous administration of resveratrol attenuates deleterious effects of ischemia/reperfusion injury (Shigematsu et al., 2003). Resveratrol has also been shown to attenuate the proinflammatory effects invoked by platelet activating factor. Resveratrol also confers vasculoprotection by regulating the expression of proinflammatory and proatherogenic genes in endothelial cells of cerebrovascular and cardiovascular systems. Resveratrol decreases neural and endothelial cells VCAM and ICAM-1 expression (Carluccio et al., 2003). Because of the potent antiinflammatory action of resveratrol, regulates the expression of inflammatory mediators, such as adhesion molecules, cytokines (e.g., TNF-α,

IL-1β, IL-6), and iNOS through transcriptional mechanisms that include C/EBP, fos/jun, AP-1, and NF-κB. Resveratrol also increases levels of MMP-2 and VEGF, which may be contribute to neuroprotective effects by inducing angiogenesis (Dong et al., 2008).

The production of free radicals contributes to insulin resistance, which leads to cognitive impairment observed in AD. Several studies have indicated that supplementation of resveratrol not only ameliorates cognitive impairment and elevates GSH levels, but also scavenges intracellular free radicals and improves the cell viability (Savaskan et al., 2003; Kumar et al., 2007). Also, supplementation of resveratrol in diet reduces the high malondialdehyde and nitrite levels observed in rat model of AD (Kumar et al., 2007; Al-Bishri et al., 2017). Furthermore, in vitro studies show that resveratrol improves the functional recovery and reduces the DNA fragmentation and apoptosis through the modulation of glutamate uptake activity and S100B secretion (Kutuk et al., 2004; de Almeida et al., 2007). Additionally, resveratrol inhibits the expression of proteins that lead to oxidative stress, such as cyclooxygenase (COX), LOX (Fig. 9.11), or NOS, and glycogen synthase kinase 3β (GSK-3β) (Bastianetto et al., 2000).

Oxidative stress and mitochondrial dysfunction play crucial role in the dopaminergic neurons death, which is the pathological feature of PD.

Resveratrol also produces neuroprotective effects in animal models of PD. Numerous in vitro and in vivo studies have shown that resveratrol produces neuroprotective effects (Zeng et al., 2017). Thus resveratrol significantly alleviated 1-methyl-4-phenylpyridinium (MPP +)-induce cytotoxicity and restores MPP + -induced mitochondrial dysfunction in SN4741 cells. Resveratrol supplementation rescues MPP + induced a decline in the level of p-Akt, p-GSK-3β, and the ratio of Bcl-2/Bax. Supplementation with resveratrol also elevates the expression of Bax and caspase-3, caspase-9. However, inhibition GSK-3β activity clearly abolished the protective effects of resveratrol. Taken together, these results suggest that resveratrol attenuates MPP + -induced mitochondrial dysfunction and cell apoptosis, and these protections may be achieved through Akt/GSK-3β pathway (Zeng et al., 2017).

The Nrf2 is another target protein of resveratrol. The stimulation of Nrf2-keap complex by resveratrol at lower concentrations (1 μM) promote the release of free Nrf2, which migrates to the nucleus and interacts with ARE triggering the expression of phase II and antioxidant defense enzymes, like heme oxygenase-1 (HO-1) (Fig. 9.11) (Ungvari et al., 2010). AMPK is

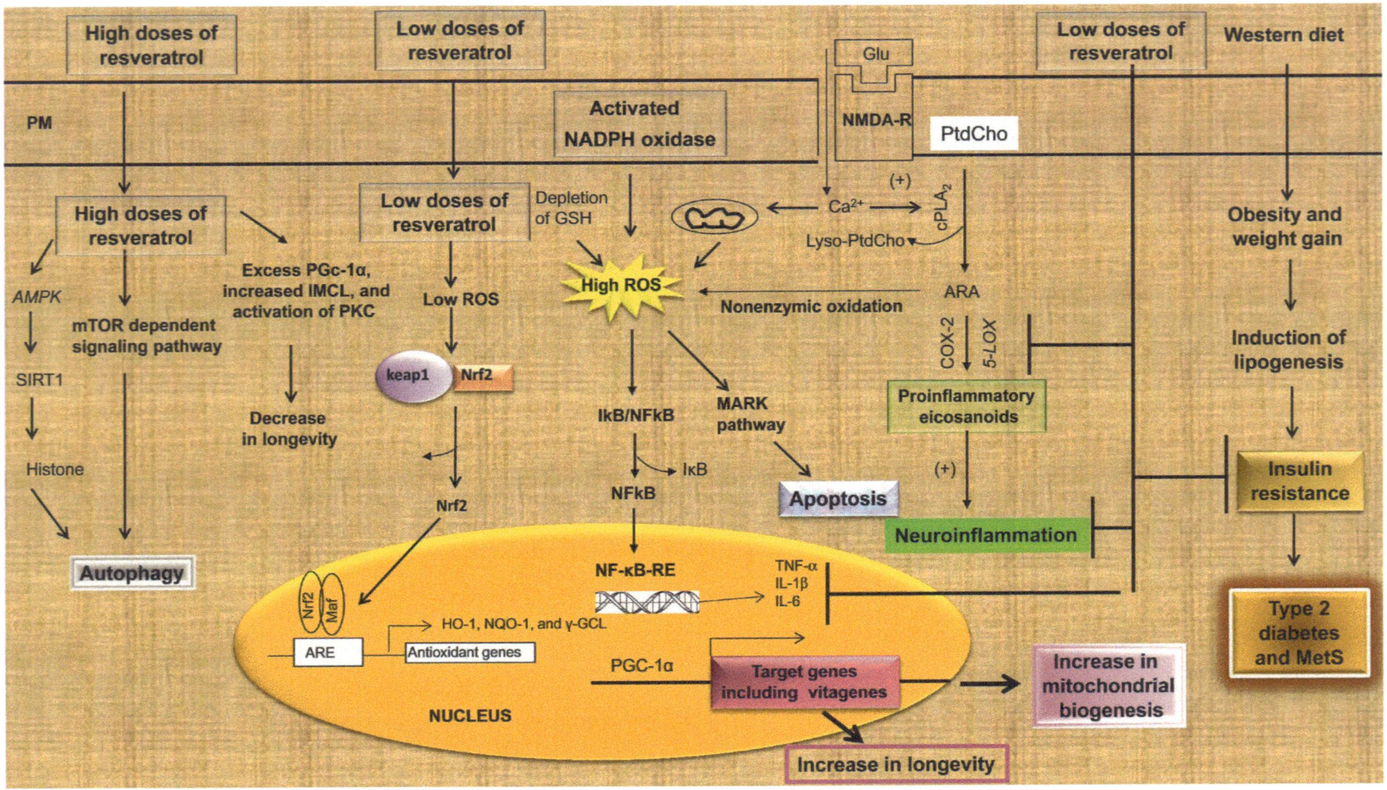

Figure 9.11 Hypothetical diagram showing effects of resveratrol on neurochemical changes on human health. *AMPK*, adenosine monophosphate–activated protein kinase; *IMCL*, intramyocellular lipid; *PKC*, protein kinase C. For other abbreviations, refer caption of Fig. 9.9.

another target of resveratrol. The activation of AMPK by resveratrol results in reduction of the intracellular ATP level (Dasgupta and Milbrandt, 2007). Interest in studies on resveratrol has skyrocketed in recent years, not only from its association with the health benefits of red wine (the "French paradox"), but also its in vitro effects on anticancer activity (Jang et al., 1997). Resveratrol at low doses produces beneficial effects not only on glucose and lipid homeostasis, but also reduces body fat accumulation (Haghighatdoost and Hariri, 2018). Resveratrol modulates apoptotic cell death by upregulating the expression of apoptosis-related proteins. In addition, resveratrol simultaneously downregulates the expression of apoptosis retarding proteins (Bcl-2, Bcl-XL, and Mcl-1) (Shankar et al., 2007). It is proposed that these proteins exert their effect mainly at the level of mitochondria. Furthermore, resveratrol facilitates the translocation of p53 and Bax to mitochondria where these proteins may interact with other Bcl-2 family members to cause permeabilization of outer mitochondrial membrane and release of mitochondrial proteins leading to caspase activation and apoptosis. Resveratrol also regulates G1 and G1/S phases of cell cycle by modulating the expression of cyclin-dependent kinase inhibitors p21/WAF1/CIP1 and p27/KIP1. It reduces inflammation not only by inhibiting COX-2 activity and blocking synthesis of prostaglandin, but also by downregulating NF-κB activity.

As mentioned above, resveratrol has been suggested to modulate cellular processes by activating key metabolic sensors/effectors, including AMPK, SIRT1, and peroxisome proliferator-activated receptor-gamma coactivator-1α (PGC-1α) (Fig. 9.11) (Vingtdeux et al., 2010; Um et al., 2010). Resveratrol not only produces its effect on mitochondrial functions, but also acts as an activator of SIRT1, increases NAD^+/NADH ratio, and enhances the clearance of mutant proteins associated with neurodegenerative diseases via mTOR-dependent or independent manner to promote neuronal survival (Wang et al., 2016). Resveratrol has also been reported as the antiaging agent to enhance life span through proautophagic mechanisms (Morselli et al., 2010; Tavernarakis et al., 2008; Hawley et al., 2010). In addition, AMPK is a Ser/Thr protein kinase predominantly expressed in neural tissue and is activated by different upstream kinases such as Ca^{2+}/CaM-dependent protein kinase kinase β. Resveratrol is a potent activator of AMPK in cultured cells or mice (Vingtdeux et al., 2010; Hawley et al., 2010). Moreover, resveratrol and its analogues, resveratrol A314 and resveratrol A405, show protective effect against AD by activating AMPK. Mechanically, resveratrol activates

AMPK by increasing intracellular Ca^{2+} levels and promoting AMPK phosphorylation at Thr172 site, thus resulting in mTOR inhibition and increasing autophagic and lysosomal clearance of Aβ. However, recent studies have indicated that resveratrol directly interacts with SIRT1 in vitro (Borra et al., 2005; Pacholec et al., 2010). SIRT1, AMPK, and mTOR signal pathways are involved in the pathogenesis of AD. Collective evidence suggests that resveratrol has therapeutic potential on AD not only by activating autophagy via controlling SIRT1-mediated transcriptional regulation, but also by modulating mTOR-dependent signal pathway (Lagouge et al., 2006a,b). More importantly, resveratrol via oral administration can pass through BBB, which will motivate the exploration of resveratrol metabolites/analogues with improved potency and brain penetration properties as antineurodegenerative molecules. Resveratrol is known to counter high-fat diet—mediated metabolic deterioration with a mechanism which relayed albeit in part, on its beneficial effects on mitochondrial health and function (Haohao et al., 2015). On the basis of these findings, it has been hypothesized that the beneficial effects of resveratrol are due either to its antioxidant properties, positive effects on glucose metabolism, insulin signaling, and lipid metabolism. Resveratrol-mediated activation of SIRT1 may also be involved in responding to molecular damage and restoring metabolic imbalances (Borra et al., 2005; Baur and Sinclair, 2006).

In the brain, resveratrol also enhances SIRT1-dependent cellular processes such as axonal protection (Araki et al., 2004) and fat mobilization (Picard et al., 2004). Resveratrol increases life span, cell survival, and neuroprotection not only by downregulating p53 activity; but as mentioned above, by deacetylating PGC-1α, increasing mitochondrial size and number, and restoring energy metabolism in the brain tissue (Borra et al., 2005; Howitz et al., 2003; Lagouge et al., 2006a,b; Baur and Sinclair, 2006). Resveratrol also effects the metabolomic profile, the metabolites of long-chain polyunsaturated fatty acids (PUFAs) (n-3 and n-6) in adipose tissue. This may explain the increased conversion of α-linolenic acid and linoleic acid into long-chain PUFAs. Additionally, glycolysis, gluconeogenesis, and pyruvate metabolism are also affected by the increase in important metabolites of the glycolytic pathway, such as glucose-6-phosphate, dihydroxyacetone phosphate, 3-phosphoglycerate, and phosphoenolpyruvate (Korsholm et al., 2017). Collective evidence suggests that resveratrol is an antioxidant that has the ability to reverse dyslipidemia and obesity. Resveratrol has ability to attenuate hyperglycemia and hyperinsulinemia promotes, protects, maintains endothelial cell function (Bertelli and Das, 2009). It also contributes to modulation of vascular function, SIRT1, Nrf2, and the AMPK. Resveratrol produces positive effects on insulin sensitivity and insulin resistance–linked chronic diseases in human clinical trials (Korsholm et al., 2017; Méndez-del Villar et al., 2014). A randomized, double-blind, placebo-controlled clinical trial has indicated that resveratrol produces positive effects on metabolic parameters, such as BMI, fat mass, waist circumference as well as insulin signaling and insulin secretion (Méndez-del Villar et al., 2014; Chen et al., 2015).

Effect of n-3 fatty acids in insulin resistance and insulin resistance–linked diseases

Many factors predispose one to the development of insulin resistance. These factors include environmental factors, such as the lack of physical activity (sedentary lifestyle), improper diet, as well as genetic and epigenetic factors. As mentioned in Chapter 1, Insulin resistance and obesity, and Chapter 2, Insulin resistance, diabetes, and metabolic syndrome, that the consumption of western diet leads to the induction of insulin resistance (Farooqui, 2015), whereas a diet rich in PUFAs may exert a positive influence on human health, including cardiovascular system, brain function, insulin resistance, and prevention of inflammation (Farooqui, 2015). It has been shown that the consumption of a western diet, which is poor in n-3 PUFAs, EPA, and DHA (Fig. 9.8) and rich in n-6 PUFAs (arachidonic acid), leads to an increase in the n-6:n-3 ratio, to a range from 10:1 to 20:1 (Farooqui, 2009; Simopoulos, 2016). Western diet contributes to low-grade inflammation, oxidative stress, and hyperactivation of stress-sensitive Ser/Thr kinases, such as JNK and IkB kinase (IKKβ). These parameters in turn inhibit the insulin receptor/insulin receptor substrate 1 (IRS1) axis. The etiopathogenesis of insulin resistance involves a multitude of metabolic pathways including production of ROS and lipid accumulation due to mitochondrial dysfunction. In addition, mitochondria are functionally and structurally linked to ER, which undergoes stress in conditions of chronic overnutrition, activating the unfolded protein response, which in turn activates the principal inflammatory pathways that impair insulin action (Lepretti et al., 2018). Among the nutrients, saturated fatty acids play key roles in insulin resistance onset. However, not all dietary fats exert the same effects on cellular energy metabolism. Dietary n-3

PUFAs have been suggested to not only counteract insulin resistance development by modulating mitochondrial bioenergetics and ER stress, but also retardation of low-grade inflammation. Converging evidence suggests that supplementation of fish oil (EPA and DHA) in diet may reduce insulin resistance–related diseases such as type 2 diabetes, MetS, heart disease, AD, and PD (Farooqui, 2009). These diseases represent major complications of hyperglycemia, obesity and related metabolic complications, which are clustered in the MetS (namely, dyslipidemia, impaired insulin sensitivity, and hypertension) (Farooqui, 2013).

The fundamental problem in the prevention and treatment of insulin resistance–linked diseases is a maintenance and restoration of insulin sensitivity in target tissues, which is disturbed already in predisease state. Optimal strategies in the prevention of insulin resistance–linked diseases are always based on a healthy lifestyle, including increased physical activity and proper nutrition. These manipulations are sufficient to lower the incidence of type 2 diabetes in patients with impaired glucose tolerance by 60% (Flachs et al., 2014; Tuomilehto et al., 2001; Knowler et al., 2002). Mechanisms of actions of n-3 fatty acids may include hypolipidemic, antioxidant, antiinflammatory, antiapoptotic, and immunomodulatory effects. In addition, n-3 fatty acids promote the stimulation of trophic factors production leading to improvement of various metabolic characteristics (Cole et al., 2009). In rat model of AD, dietary intake of DHA significantly reduces the levels of Aβ40, cholesterol, and saturated fatty acids (Hashimoto et al., 2006, 2008). Chronic preadministration of DHA also prevents Aβ-induced impairment of an avoidance ability-related memory function in a rat model of AD (Hashimoto et al., 2006), and protects mice from synaptic loss and dendritic pathology in another model of AD (Calon et al., 2004). DHA and its metabolite, neuroprotectin D_1 not only decrease Aβ secretion from aging brain cells, but also prevent apoptosis (Lukiw et al., 2005). DHA also inhibits c-Jun N-terminal kinase and the phosphorylation of adaptor protein IRS1 and tau in cultured hippocampal neurons (Ma et al., 2009).

n-3 fatty acids also produce neuroprotective effects in animal models of PD. Thus supplementation of n-3 fatty acids in the diet of animals protects rodents from MPTP-induced neurotoxicity (Bousquet et al., 2008). The dietary supplementation of n-3 fatty acids not only prevents the MPTP-mediated decrease in tyrosine hydroxylase-labeled nigral cells, but also downregulates Nurr1 mRNA, and dopamine transporter mRNA levels in the substantia nigra (Bousquet et al., 2008). Similarly, in

6-hydroxydopamine-induced model of PD, n-3 fatty acids act by increasing dopamine turnover in the surviving neurons without modifying neuronal population (Delattre et al., 2009). Although the molecular mechanisms associated with neuroprotective effects of n-3 fatty acids are not fully understood. However, it is suggested that n-3 fatty acids inhibit the synthesis and release of proinflammatory cytokines such as TNF-α, IL-1β, and IL-2 (Farooqui, 2009). In addition, n-3 fatty acids have antidepressant effect in PD patients and improve their quality of life.

Effects of cinnamon on insulin resistance and insulin resistance–linked diseases

Cinnamon (*Cinnamomum verum* and *Cinnamomum zeylanicum*) and cassia (*Cinnamomum aromaticum*) has been used as a spice and as traditional herbal medicine for centuries in South East Asia. In addition to containing polyphenolic compounds (proanthocyanidins and procyanidin), EC, camphene, eugenol, phenol, salicylic acid and tannins, cinnamon also contains manganese, dietary fiber, iron, and calcium. Cinnamon contains four major compounds (cinnamaldehyde, cinnamyl acetate, cinnamyl alcohol, and eugenol) (Fig. 9.12). These components produce antioxidant effects by scavenging ROS and free radicals, suppressing lipid peroxidation, and reducing malondialdehyde (Vanschoonbeek et al., 2006; Qin et al., 2010). Cinnamon also produces antiinflammatory effects by reducing inflammatory mediators such as prostaglandin-E2, interleukin (IL)-6, and NO production (Fig. 9.13) (Tung et al., 2008). In chow diet-fed rats, cinnamon extracts enhance insulin-stimulated insulin receptor β and IRS1 tyrosine phosphorylation levels and IRS1/PtdIns 3K in skeletal muscle. Extensive in vitro studies have also indicated that cinnamon improves insulin resistance by inhibiting and reversing impairments in insulin signaling in skeletal muscle. In adipose tissues, cinnamon acts by increasing the regulation of genes related to insulin signaling and blocking lipogenesis in adipose tissue (Sheng et al., 2008; Qin et al., 2009, 2010). Extensive in vitro studies have indicated that cinnamon improves insulin resistance by preventing and reversing impairments in insulin signaling in skeletal muscle (Ziegenfuss et al., 2006; Roussel et al., 2009). In adipose tissue, cinnamon increases the expression of peroxisome proliferator-activated receptors including PPARγ (Sheng et al., 2008; Qin et al., 2009, 2010). In addition, in mouse 3T3-L1 adipocytes, cinnamon extracts also downregulate the expression of genes encoding insulin signaling pathway proteins

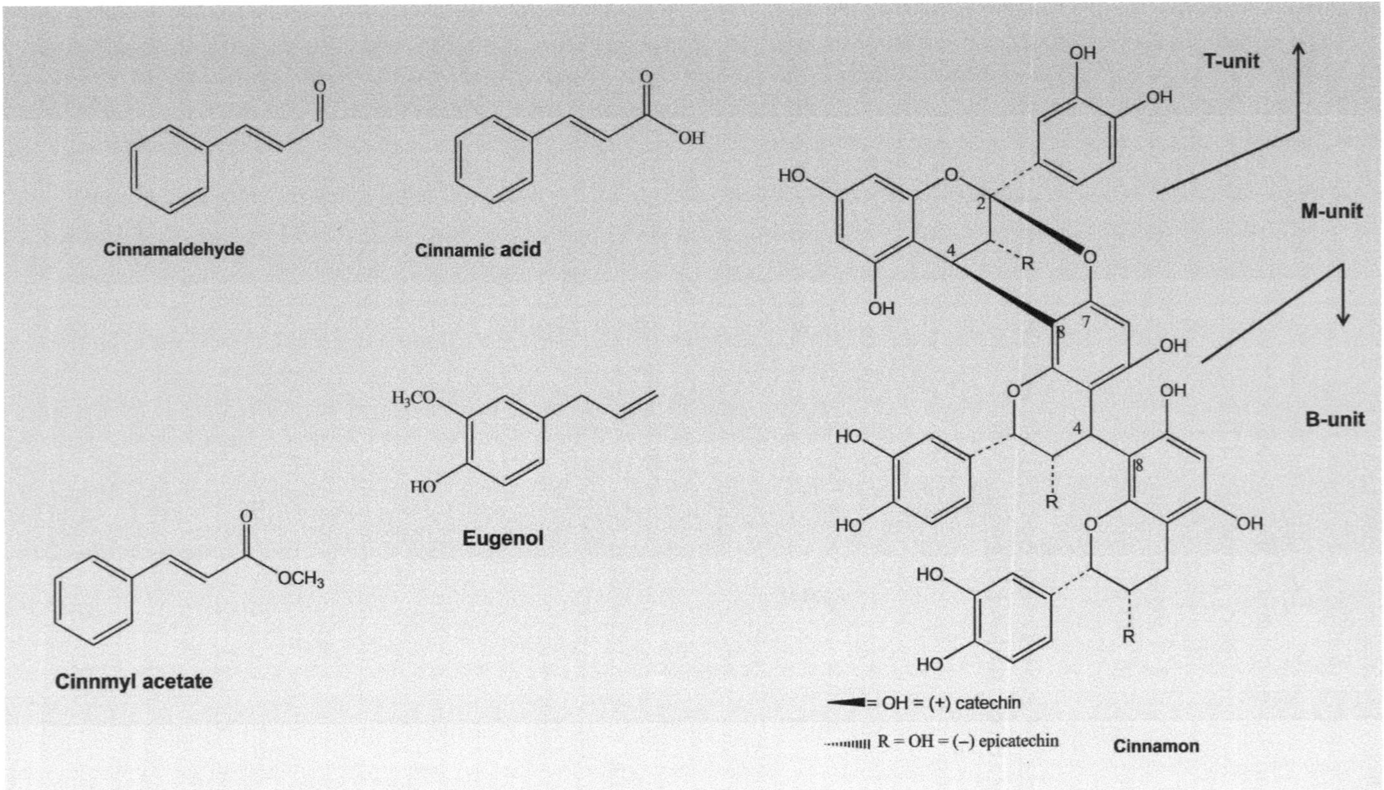

Figure 9.12 Chemical structures of cinnamon and its derivatives.

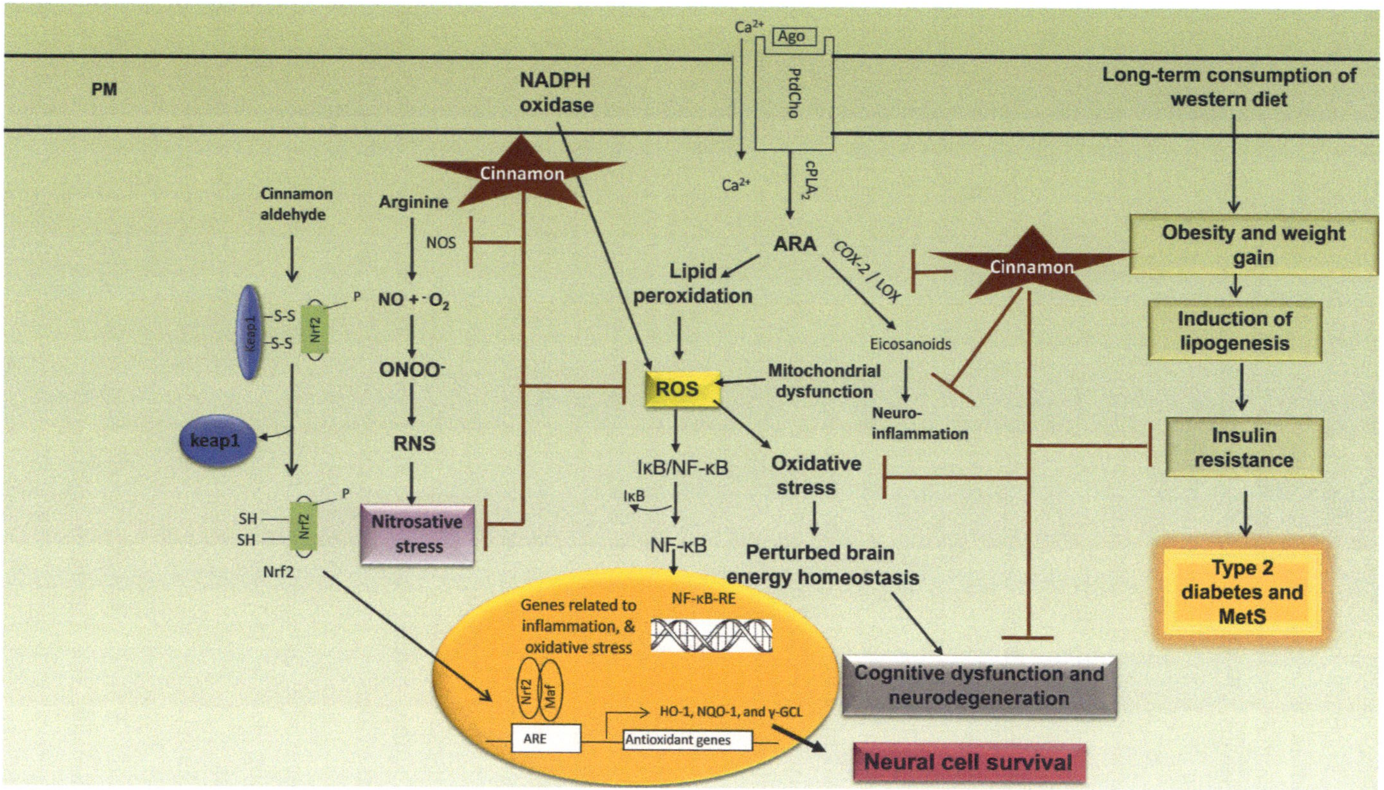

Figure 9.13 Hypothetical diagram showing the effect of cinnamon on insulin resistance.

including GSK3β, IGF1R, IGF2R, and PtdIns 3K, which may contribute to the potential health benefits of this phytochemical (Cao et al., 2010). Cinnamon also improves cholesterol levels, systolic blood pressure, and postprandial glucose levels in rodents. In type 2 diabetes patients, cinnamon lowers HbA1c by 0.83% compared with usual care alone lowering HbA1c by 0.37% in a randomized, controlled trial (Crawford, 2009; Khan et al., 2003; Ziegenfuss et al., 2006; Qin et al., 2010). Cinnamaldehyde in cinnamin preparation exhibits glucolipid lowering effects in diabetic animals by increasing glucose uptake and improving insulin sensitivity in adipose and skeletal muscle tissues, improving glycogen synthesis in liver, restoring pancreatic islets dysfunction, slowing gastric emptying rates, and improving diabetic renal and brain disorders. Cinnamaldehyde exerts these effects through multiple signaling pathways, including PPARs, AMPK, and Nrf2 pathways. Collective evidence indicates that cinnamon polyphenol produces antiinflammatory, antimicrobial, antioxidant, antitumor, cardiovascular, cholesterol lowering, and immunomodulatory effects (Zhu et al., 2017). As mentioned in Chapter 1, Insulin resistance and obesity, and Chapter 2, Insulin resistance, diabetes, and metabolic syndrome, type 2 diabetes is accompanied by a chronic state of hyperglycemia, systemic redox imbalance, inflammation, advanced glycation end products, dyslipidemia, and endothelial dysfunction. Type 2 diabetes patients are two to three times more likely to have CVD than nondiabetics (International Diabetes Atlas, 2017) and exhibit an increased rate of atherosclerotic plaque development (Matheus et al., 2013). Revascularization procedures for occluded arteries include balloon angioplasty with or without stent placement. The arterial injury response after revascularization often results in vessel reocclusion or restenosis, which in turn requires further intervention. Type 2 diabetes also accelerates restenosis as evidenced by lower patency rates after revascularization in diabetic patients (Lexis et al., 2009; DeRubertis et al., 2008; Singh et al., 2011). Recent studies have indicated that local redox imbalance is a major contributing factor in vascular smooth muscle cell (VSMC) pathology (Buglak et al., 2018). Specifically, ROS downstream of nicotinamide adenine dinucleotide phosphate (NADPH) oxidase 1, superoxide, and subsequently hydrogen peroxide are overproduced after vascular injury and promote VSMC growth and migration (Streeter et al., 2013). Additionally, the family of platelet-derived growth factors have been implicated as one of the major growth signals upstream of NADPH oxidase (Jawien et al., 1992). Small molecules that target the redox imbalance have shown promise as potential therapeutics for preventing the VSMC phenotypic changes (Pan et al., 2017). This process may involve the activation of Nrf2 pathway.

Effects of garlic on insulin resistance and insulin resistance–linked diseases

Garlic (*Allium sativum*) belongs to the family of Lilliaceae. Garlic preparations induce cardioprotective, antioxidant, antineoplastic, and antimicrobial effects in humans (Rahman and Lowe, 2006; Tattelman, 2005). Garlic preparations are widely used for the prevention of cardiovascular and metabolic diseases, such as atherosclerosis, arrhythmia, hyperlipidemia, thrombosis, hypertension, and type 2 diabetes (Banerjee and Moulik, 2000). Furthermore, garlic preparations also produce significant antiarrhythmic effect in both ventricular and supraventricular arrhythmias (Schwingshackl et al., 2016). Garlic preparations contain high concentrations of sulfur-containing compounds such as allicin (a thiosulfinate), DAS, DADS, DATS, and dipropyl sulfide, ajoene, as well as adenosine, vitamins E and C, minerals, polyphenols, inhibitors of adenosine deaminase, and cyclic AMP phosphodiesterase (Fig. 9.14) (Santhosha et al., 2013). Unlike antibiotics, garlic does not weaken the immune system but promotes it. It stimulates humoral and cell responses of the immune system. Allicin is the most active compound found in garlic. Allicin (thio-2-propene-1-sulfinic acid *S*-allyl ester) is formed when alliin (+*S*-allyl-L-cysteine sulfoxide), a sulfur-containing amino acid, comes into contact with the enzyme alliinase when raw garlic is chopped, crushed, or chewed. Allicin, unstable in aqueous solution, rapidly decomposes mainly to DAS, DADS, DATS, and ajoene (Amagase, 2006). Most medicinal effects of garlic preparations (antimicrobial, hypolipidemic, vasodilatory, antihypertensive, antioxidant, antineoplastic, and antithrombotic) are related to allicin and other breakdown products (Amagase, 2006; Santhosha et al., 2013).

In insulin-linked chronic diseases such as type 2 diabetes, MetS, and heart disease, oral administration of raw garlic for a period of 8 weeks shows significant reduction of blood glucose, improvement of insulin sensitivity and endothelial cell function in rats (Atkin et al., 2016). Late complications in type 2 diabetic patients are commonly associated with accelerated development of atherosclerosis. In type 2 diabetes, nonenzymic glycosylation of apo-B is caused by hyperglycemia. This process is an

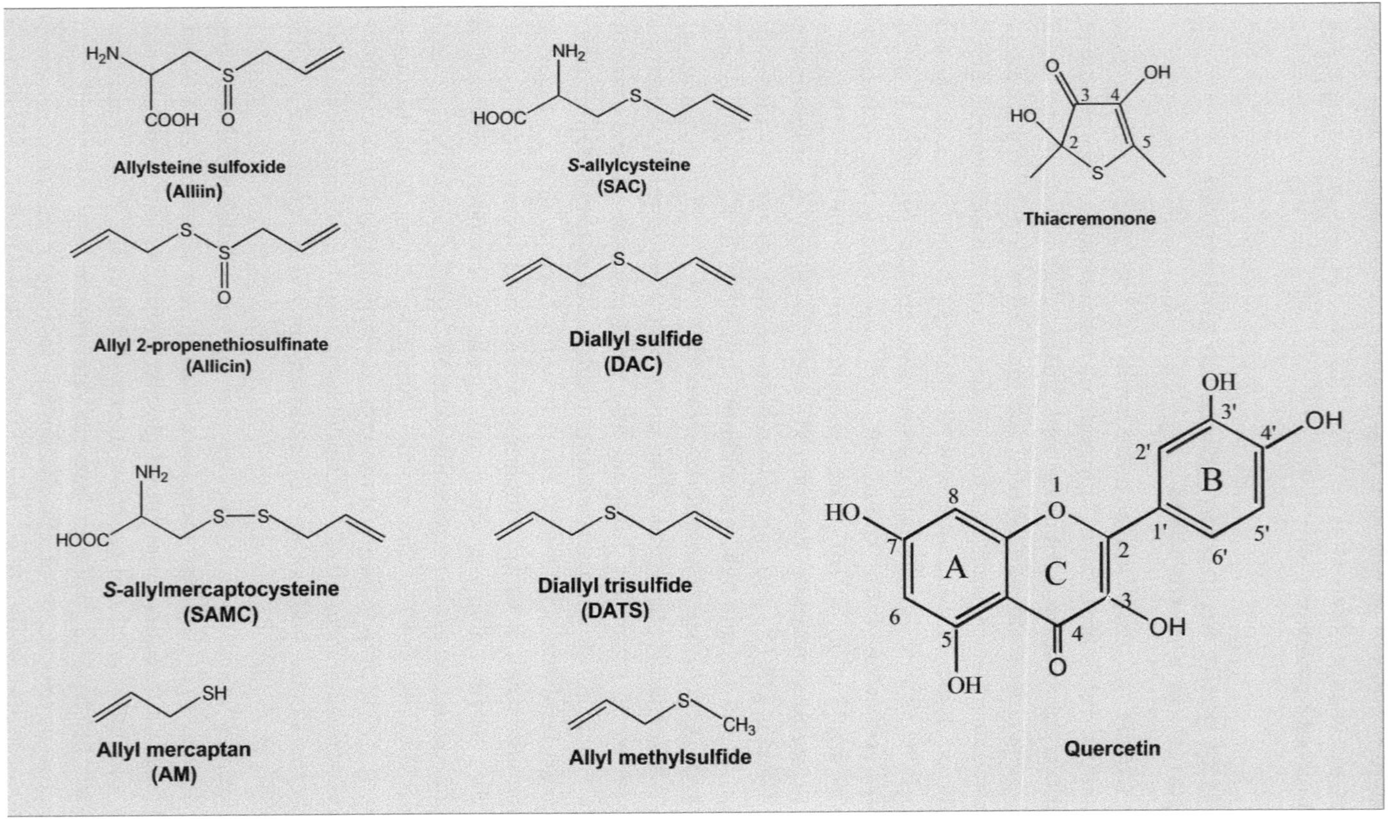

Figure 9.14 Chemical structures of sulfur compounds found in garlic.

efficient biochemical way of atherogenic modification of low-density lipoprotein. Studies on the effect of time-released garlic powder tablet Allicor (double-blinded placebo-controlled study) on the parameters of metabolic control and plasma lipids in 60 type 2 diabetes patients have indicated that Allicor treatment results in better metabolic control due to the lowering of fasting blood glucose, serum fructosamine, and serum triglyceride levels supporting the view that Allicor treatment may lead to the reduction of cardiovascular risk in type 2 diabetic patients (Sobenin et al., 2008). Other metabolic complications of type 2 diabetes and MetS, such as increased serum triacylglycerol, insulin, and uric acid levels, are also normalized after garlic administration. It is suggested that lowering serum uric acid and triacylglycerol after garlic administration may be responsible for improving insulin resistance in fructose fed rats (Nakagawa et al., 2006). Fructose fed rats show increase in serum levels of NO and decrease in H_2S levels compare to normal subjects and control rats (Padiya et al., 2011). In high fructose fed model of MetS, marked increase in TBARS levels is accompanied by reduction in GSH levels in liver in comparison to the control group. However, administration of raw garlic homogenate normalizes not only increase in TBARS, but also decreases in GSH levels in diabetic liver, supporting the view that oral administration of raw garlic homogenate increases insulin sensitivity and reduces metabolic complications along with oxidative stress in fructose fed rat model of MetS (Padiya et al., 2011). Long-term, double-blind studies are needed on large human population to establish the role of garlic in controlling MetS and its complications.

Effect of quercetin on insulin resistance and insulin resistance–related diseases

Quercetin is a polyphenolic bioflavonoid (Fig. 9.14), which is abundantly found in kales, onions, berries, apples, red grapes, broccoli, citrus fruits, parsley, berries, cherries, as well as green tea and red wine (Anand David et al., 2016). Quercetin exhibits antioxidant and antiinflammatory activities, as well as antitumor activities (Russo et al., 2012). Quercetin and its metabolites are known to enter rat brain after oral administration of quercetin (De Boer et al., 2005). Quercetin produces beneficial effects in diabetes (Kobori et al., 2009) and AD through regulation of antioxidative stress enzymes via the action of Nrf2 and the antioxidant effect of paraoxonase 2 expression (Costa et al., 2016). Quercetin has also been shown to improve on memory and cognition function. Improvement may be associated with ISR regulation (Nakagawa et al., 2017). In RAW 264.7 cells, quercetin is known to inhibit LPS-mediated NO, PGE_2, iNOS, COX-2, TNF-α, IL-1β, IL-6 and GM-CSF mRNA and related protein expressions. It also facilitates heme oxygenase (HO-1) induction in a dose- and time-dependent manner. Quercetin not only suppresses I-κB-phosphorylation, NF-κB translocation, AP-1, NF-κB DNA binding, and reporter gene transcription, but also attenuates p38(MAPK) and JNK1/2 but not ERK1/2 activations. Moreover, quercetin retards PtdIns 3K, PDK1, and Akt activation in a time- and dose-dependent manner. Quercetin also disrupts LPS-mediated p85 association to TLR4/MyD88 complex and then limits the activation of IRAK1, TRAF6, and TAK1 with a subsequent reduction in p38 and JNK activations, and suppression in IKKα/β-mediated I-κB phosphorylation (Endale et al., 2013). Many studies have indicated that consumption of flavonoids not only reduce the risk of CVDs, metabolic disorders (type 2 diabetes and MetS), but also certain types of cancer (Salvamani et al., 2014). These effects are due to the reduction of oxidative stress, inhibition of low-density lipoproteins oxidation and platelet aggregation, and vasodilatory effects in blood vessels (Salvamani et al., 2014; Anand David et al., 2016). At the molecular level, flavonoids act on the SIRT1−AMPK−PGC-1α axis (Davis et al., 2009). Indeed, this polyphenol activates the AMPK and SIRT1. These enzymes are pivotal regulators of mitochondrial oxidative metabolism (Hawley et al., 2010). Furthermore, quercetin can stimulate mitochondria oxidative metabolism by directly decreasing ATP:AMP ratio, which in turn results in the activation of AMPK and its downstream catabolic pathways (Dorta et al., 2005). Finally, despite quercetin being able to modulate pivotal pathways involved in mitochondria biogenesis and oxidative metabolism in rodents, its effect on mitochondria function and fatty acid catabolism in human skeletal muscle remains to be fully elucidated (Davis et al., 2009). Quercetin also increases the antioxidant capacity of the body by regulating levels of reduced GSH. This is because generation of free radicals and their detoxification involves superoxide dismutase, the enzyme, which transforms free radicals into H_2O_2 into the nontoxic H_2O (Fig. 9.15). This reaction requires GSH as a hydrogen donor. Collectively, animal and cellular studies have indicated that quercetin induces the synthesis of GSH (Perron and Brumaghim, 2009). Oxidative stress and mitochondrial dysfunction are the key players in triggering neurodegeneration in neurodegenerative diseases. Supplementation of food containing quercetin may reduce the risk of neurodegenerative disorders (AD and PD), cancer,

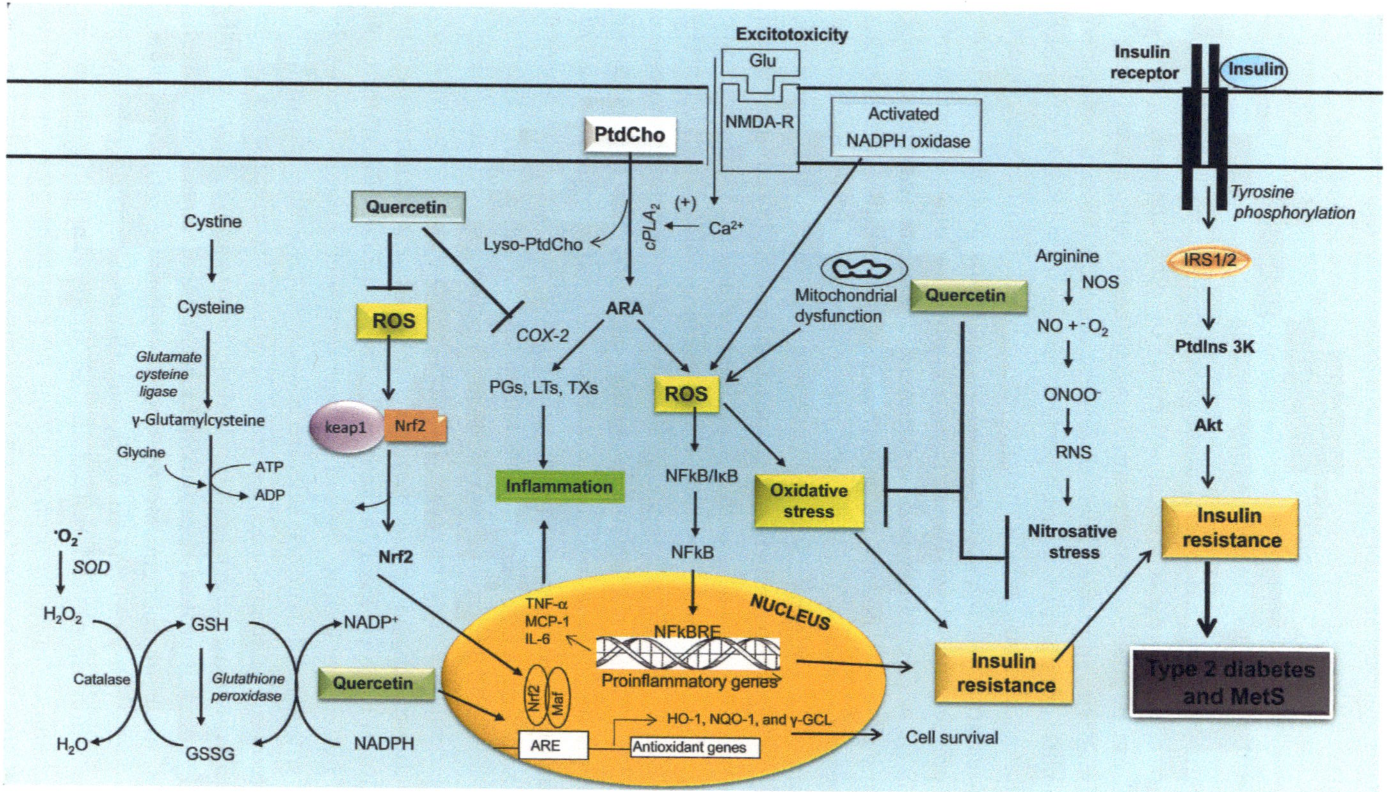

Figure 9.15 Hypothetical diagram showing the effect of quercetin on insulin resistance.
Akt, serine/threonine protein kinase; *GSH*, reduced glutathione; *GSK3*, glycogen synthase kinase 3; *GSSG*, oxidized glutathione; *IRS*, insulin receptor substrate; *PtdIns 3K*, phosphatidylinositol 3-kinase. For other abbreviations, refer caption of Fig. 9.9.

CVDs, allergic disorders, thrombosis, atherosclerosis, hypertension, and arrhythmia.

Conclusion

For centuries, humans have been using phytochemicals to improve their health. In addition, the chemical structure of phytochemicals (parent structure) can be used to develop structural analogues, which can be used as safer drugs with improved pharmacological activities through optimized bioavailability and pharmacokinetic profiles, and lower side effects. Unlike drugs (pharmaceuticals), which undergo extensive clinical trials in animals and humans prior to FDA approval, dietary phytochemicals are not tested for their efficacy and safety. Phytochemicals contain complicated mixtures of organic chemicals, the levels of which may vary substantially depending upon many factors related to the growth of plants, harvesting, production, and storage conditions. These factors may cause alterations in chemical compositions of dietary supplement resulting in batch variation. Significant US population believe that phytochemicals are economical and the use of phytochemicals is better than drugs, which are more expensive than preparations of phytochemicals. The precise molecular mechanisms through which specific phytochemicals exert their beneficial biological effects still remain the subject of intense investigations. The hypoglycemic effects of phytochemicals are not only due to lowering of the uptake of carbohydrates in the intestine, affecting the glucose metabolism by altering enzyme activities, improving β-cell function and insulin action, initiating insulin release and antioxidant as well as antiinflammatory effects. Insulin-linked visceral and neurological disorders respond to phytochemicals through the stimulation of receptors located on cellular membranes. The stimulation of these receptors induces cell survival through the stimulation of Ca^{2+} influx and PtdIns 3K and mitogen-activated protein kinases. In addition, visceral and neuronal cells are also sensitive to other phytochemicals (curcumin, catechins, and resveratrol), which are absorbed through the gut and distributed through the blood stream to various organs including brain, which is protected by BBB. As mentioned above, the bioavailability of most phytochemicals to visceral organs is relatively higher than the brain not only due to BBB, but also due to rapid metabolism and elimination of phytochemicals in the urine. To enter the brain, a phytochemical must be either highly lipid-soluble or subjected to uptake transport processes through ATP-binding cassette transporter

(ABC transporter). Thus there occurs a complex interplay between the physicochemical properties and active ABC transporters. When phytochemicals reach in the tissue, they play a vital role in restoring mitochondrial function, inhibiting oxidative stress, and inflammation associated with chronic insulin resistance–linked visceral and neurological diseases. It is becoming increasingly evident that beneficial effects of phytochemicals are due to not only the induction of many endogenous antioxidant defense mechanisms, such as GSH and Nrf2, but also due to abilities of phytochemicals to block the activation of transcription factor, NF-κB. Phytochemicals may also induce the synthesis and secretion of growth factors (protein chaperones and BDNF) and stimulate the induction of genetic and epigenetic mechanisms.

References

Abbott, N.J., Patabendige, A.A., Dolman, D.E., Yusof, S.R., Begley, D.J., 2010. Structure and function of the blood–brain barrier. Neurobiol. Dis. 37, 13–25.

Al-Bishri, W.M., Hamza, A.H., Farran, S.K., 2017. Resveratrol treatment attenuates amyloid beta, Tau protein and markers of oxidative stress, and inflammation in Alzheimer's disease rat model. Int. J. Pharm. Res. Allied Sci. 6, 71–78.

Aly, A.H., Debbab, A., Proksch, P., 2013. Fungal endophytes—secret producers of bioactive plant metabolites. Pharmazie 68, 499–505.

Alza, N.P., Iglesias González, P.A., Conde, M.A., Uranga, R.M., Salvador, G.A., 2019. Lipids at the crossroad of α-synuclein function and dysfunction: biological and pathological implications. Front. Cell Neurosci 13, 175.

Amagase, H., 2006. Clarifying the real bioactive constituents of garlic. J. Nutr 136, 716S–725S.

Anand David, A.V., Arulmoli, R., Parasuraman, S., 2016. Overviews of biological importance of quercetin: a bioactive flavonoid. Pharmacogn. Rev. 10, 84–89.

Anderson, J.W., 1986. Fiber and health: an overview. Am. J. Gastroenterol 81, 892–897.

Araki, T., Sasaki, Y., Milbrandt, J., 2004. Increased nuclear NAD biosynthesis and SIRT1 activation prevent axonal degeneration. Science 305, 1010–1013.

Atkin, M., Laight, D., Cummings, M.H., 2016. The effects of garlic extract upon endothelial function, vascular inflammation, oxidative stress and insulin resistance in adults with type 2 diabetes at high cardiovascular risk. A pilot double blind randomized placebo controlled trial. J. Diabetes Complications 30, 723–727.

Banerjee, K.S., Moulik, S.K., 2000. Effect of garlic on cardiovascular disorders: a review. Nutr. J. 1, 1–14.

Barbosa, A., Silveira, G.D., de Menezes, I., Neto, J., Bitencurt, J., Estavam, C.D., et al., 2013. Antidiabetic effect of the *Chrysobalanus icaco* L. aqueous extract in rats. J. Med. Food. 16, 538–543.

Bastianetto, S., Zheng, W.H., Quirion, R., 2000. Neuroprotective abilities of resveratrol and other red wine constituents against nitric oxide-related toxicity in cultured hippocampal neurons. Br. J. Pharmacol 131, 711–720.

Basu, A., Du, M., Sanchez, K., Leyva, M.J., Betts, N.M., Blevins, S., et al., 2011. Green tea minimally affects biomarkers of inflammation in obese subjects with metabolic syndrome. Nutrition 27, 206–213.

Baum, L., Lam, C.W.K., Cheung, S.K., Kwok, T., Lui, V., Tsoh, J., et al., 2008. Six-month randomized, placebo-controlled, double-blind, pilot clinical trial of curcumin in patients with Alzheimer disease. J. Clin. Psychopharmacol. 28, 110–113.

Baur, J.A., Sinclair, D.A., 2006. Therapeutic potential of resveratrol: the in vivo evidence. Nat. Rev. Drug Discov 5, 493–506.

Baur, J.A., Pearson, K.J., Price, N.L., 2006. Resveratrol improves health and survival of mice on a high-calorie diet. Nature 444, 337–342.

Bellik, Y., Boukraâ, L., Alzahrani, H.A., Bakhotmah, B.A., Abdellah, F., Hammoudi, S. M., et al., 2012. Molecular mechanism underlying anti-inflammatory and anti-allergic activities of phytochemicals: an update. Molecules 18, 322–353.

Bent, S., Ko, R., 2004. Commonly used herbal medicines in the United States: a review. Am. J. Med. 116, 478–485.

Bertelli, A.A., Das, D.K., 2009. Grapes, wines, resveratrol, and heart health. J. Cardiovasc. Pharmacol. 54, 468–476.

Bird, J.K., Raederstorff, D., Weber, P., Steinert, R.E., 2017. Cardiovascular and antiobesity effects of resveratrol mediated through the gut microbiota. Adv. Nutr. 8, 839–849.

Borra, M.T., Smith, B.C., Denu, J.M., 2005. Mechanism of human SIRT1 activation by resveratrol. J. Biol. Chem 280, 17187–17195.

Bousquet, M., Saint-Pierre, M., Julien, C., Salem Jr, N., Cicchetti, F., Calon, F., 2008. Beneficial effects of dietary omega-3 polyunsaturated fatty acid on toxin-induced neuronal degeneration in an animal model of Parkinson's disease. FASEB J. 22, 1213–1225.

Boyanapalli, S.S.S., Kong, A.N.T., 2015. "Curcumin, the king of spices": epigenetic regulatory mechanisms in the prevention of cancer, neurological, and inflammatory diseases. Curr. Pharmacol. Rep 1, 129–139.

Buglak, N.E., Jiang, W., Bahnson, E.S.M., 2018. Cinnamic aldehyde inhibits vascular smooth muscle cell proliferation and neointimal hyperplasia in Zucker Diabetic Fatty rats. Redox Biol. 19, 166–178.

Calabrese, E.J., Bachmann, K.A., Bailer, A.J., Bolger, P.M., Borak, J., et al., 2007. Biological stress response terminology: integrating the concepts of adaptive response and preconditioning stress within a hormetic dose-response framework. Toxicol. Appl. Pharmacol. 222, 122–128.

Calon, F., Lim, G.P., Yang, F.S., Morihara, T., Teter, B., Ubeda, O., et al., 2004. Docosahexaenoic acid protects from dendritic pathology in an Alzheimer's disease mouse model. Neuron 43, 633–645.

Cao, H., Graves, D.J., Anderson, R.A., 2010. Cinnamon extract regulates glucose transporter and insulin-signaling gene expression in mouse adipocytes. Phytomedicine 17, 1027–1032.

Carluccio, M.A., Siculella, L., Ancora, M.A., Massaro, M., Scoditti, E., Storelli, C., et al., 2003. Olive oil and red wine antioxidant polyphenols inhibit endothelial activation: antiatherogenic properties of Mediterranean diet phytochemicals. Arterioscler. Thromb. Vasc. Biol. 23, 622–629.

Chan, P.T., Fong, W.P., Cheung, Y.L., Huang, Y., Ho, W.K.K., Chen, Z.Y., 1999. Jasmine green tea epicatechins are hypolipidemic in hamsters (*Mesocricetus auratus*) fed a high fat diet. J. Nutr. 129, 1094–1101.

Chandra, V., Pandav, R., Dodge, H.H., Johnston, J.M., Belle, S.H., DeKosky, S.T., et al., 2001. Incidence of Alzheimer's disease in a rural community in India: the Indo-US study. Neurology 57, 985–989.

Chen, P.C., Wheeler, D.S., Malhotra, V., Odoms, K., Denenberg, A.G., Wong, H.R., 2002. A green tea-derived polyphenol, epigallocatechin-3-gallate, inhibits I kappa B kinase activation and IL-8 gene expression in respiratory epithelium. Inflammation 26, 233–241.

Chen, S., Zhao, X., Ran, L., Wan, J., Wang, X., Qin, Y., et al., 2015. Resveratrol improves insulin resistance, glucose and lipid metabolism in patients with non-alcoholic fatty liver disease: a randomized controlled trial. Dig. Liver Dis. 47, 226–232.

Cole, G.M., Frautschy, S.A., 2007. The role of insulin and neurotrophic factor signaling in brain aging and Alzheimer's disease. Exp. Gerontol. 42, 10–21.

Cole, G.M., Ma, Q.L., Frautschy, S.A., 2009. Omega-3 fatty acids and dementia. Prostaglandins Leukot. Essent. Fatty Acids 81, 213–221.

Costa, L.G., Garrick, J.M., Roque, P.J., Pellacani, C., 2016. Mechanisms of neuroprotection by quercetin: counteracting oxidative stress and more. Oxid. Med. Cell. Longev. 2016:2986796.

Crawford, P., 2009. Effectiveness of cinnamon for lowering hemoglobin A1C in patients with type 2 diabetes: a randomized, controlled trial. J. Am. Board Family Med 22, 507–512.

Das, T.K., Chakrabarti, S.K., Zulkipli, I.N., Abdul Hamid, M.R.W., 2019. Curcumin ameliorates the impaired insulin signaling involved in the pathogenesis of Alzheimer's disease in rats. J. Alzheimer Dis. Rep 3, 59–70.

Dasgupta, B., Milbrandt, J., 2007. Resveratrol stimulates AMP kinase activity in neurons. Proc. Natl. Acad. Sci. U.S.A. 104, 7217–7222.

Davingnon, J., Ganz, P., 2004. Role of endothelial dysfunction in atherosclerosis. Circulation. 109, III27–III32.

Davis, J.M., Murphy, E.A., Carmichael, M.D., Davis, B., 2009. Quercetin increases brain and muscle mitochondrial biogenesis and exercise tolerance. Am. J. Phys. Regul. Integr. Comp. Phys 296, R1071–R1077.

de Almeida, L.M., Pineiro, C.C., Leite, M.C., Brolese, G., Tramontina, F., Feoli, A.M., et al., 2007. Resveratrol increases glutamate uptake, glutathione content, and S100B secretion in cortical astrocyte cultures. Cell. Mol. Neurobiol. 27, 661–668.

De Boer, V.C., Dihal, A.A., van der Woude, H., Arts, I.C., Wolffram, S., Alink, G.M., et al., 2005. Tissue distribution of quercetin in rats and pigs. J. Nutr 135, 1718–1725.

Delattre, A.M., Kiss, A., Szawka, R.E., Anselmo-Franci, J.A., Bagatini, P.B., Xavier, L.L., et al., 2009. Evaluation of chronic omega-3 fatty acids supplementation on behavioral and neurochemical alterations in 6-hydroxydopamine-lesion model of Parkinson's disease. Neurosci. Res. 66, 256–264.

DeRubertis, B.G., Pierce, M., Ryer, E.J., Trocciola, S., Kent, K.C., Faries, P.L., 2008. Reduced primary patency rate in diabetic patients after percutaneous intervention results from more frequent presentation with limb-threatening ischemia. J. Vasc. Surg. 47, 101–108.

Desplats, P., Lee, H.J., Bae, E.J., Patrick, C., Rockenstein, E., Crews, L., et al., 2009. Inclusion formation and neuronal cell death through neuron-to-neuron transmission of alpha-synuclein. Proc. Natl. Acad. Sci. U.S.A 106, 13010–13015.

Dong, W., Li, N., Gao, D., Zhen, H., Zhang, X., Li, F., 2008. Resveratrol attenuates ischemic brain damage in the delayed phase after stroke and induces messenger RNA and protein express for angiogenic factors. J. Vasc. Surg 48, 709–714.

Dorta, D.J., Pigoso, A.A., Mingatto, F.E., Rodrigues, T., Prado, I.M., Helena, A.F., et al., 2005. The interaction of flavonoids with mitochondria: effects on energetic processes. Chem. Biol. Interact. 152, 67–78.

Dulloo, A.G., Seydoux, J., Girardier, L., Chantre, P., Vandermander, J., 2000. Green tea and thermogenesis: interactions between catechin-polyphenols, caffeine and sympathetic activity. Int. J. Obes. Relat. Metab. Disord 24, 252–258.

Endale, M., Park, S.C., Kim, S., Kim, S.H., Yang, Y., Cho, J.Y., et al., 2013. Quercetin disrupts tyrosine-phosphorylated phosphatidylinositol 3-kinase and myeloid differentiation factor-88 association, and inhibits MAPK/AP-1 and IKK/NF-κB-induced

inflammatory mediators production in RAW 264.7 cells. Immunobiology 218, 1452–1467.
Enrique, B., Esther, V., Verónica, H.C., Juan, M.R.A., 2014. Insulin in the brain: its pathophysiological implications for states related with central insulin resistance, type 2 diabetes and Alzheimer's disease. Front. Endocrinol 5, 1–21.
Farooqui, A.A., 2008. Phytochemicals, Signal Transduction, and Neurological Disorders. Springer Science + Business Media, New York.
Farooqui, A.A., 2009. Beneficial Effects of Fish Oil on Human Brain. Springer, New York.
Farooqui, A.A., 2011. Lipid Mediators and Their Metabolism in the Brain. Springer Science + Business Media LLC, New York.
Farooqui, A.A., 2012. Phytochemicals, Signal Transduction, and Neurological Disorders. Springer Science + Business Media LLC, New York.
Farooqui, A.A., 2013. Metabolic Syndrome: An Important Risk Factor for Stroke, Alzheimer Disease, and Depression. Springer, New York.
Farooqui, A.A., 2015. High Calorie Diet and the Human Brain-Metabolic Consequences of Long-Term Consumption. Springer International Publishing, Switzerland.
Farooqui, A.A., 2016a. Therapeutic Potentials of Curcumin for Alzheimer's Disease. Springer International Publishing, Switzerland.
Farooqui, A.A., 2016b. Neurochemical aspects of neurological disorders. In: Farooqui, T., Farooqui, A.A. (Eds.), Trace Amines and Neurological Disorders. Academic Press, An Imprint of Elsevier, San Diego, CA, pp. 237–256.
Farooqui, A.A., Farooqui, T., 2018. Effects of western, Mediterranean, vegetarian, and Okinawan diet patterns on human brain. In: Farooqui, T., Farooqui, A.A. (Eds.), Role of the Mediterranean Diet in the Brain and Neurodegenerative Diseases. Academic Press, An Imprint of Elsevier, San Diego, CA.
Firdous, A.N., 2014. Phytochemicals for treatment of diabetes. EXCLI J 13, 451–453.
Flachs, P., Rossmeisl, M., Kopecky, J., 2014. The effect of n-3 fatty acids on glucose homeostasis and insulin sensitivity. Physiol. Res 63 (Suppl. 1), S93–S118.
Fonseca, V., Desouza, C., Asnani, S., Jialal, I., 2004. Nontraditional risk factors for cardiovascular disease in diabetes. Endocr. Rev 25, 153–175.
Fraser, G., 2004. A comparison of first event coronary heart disease rates in two contrasting California populations. J. Nutr. Health Aging 9, 53–58.
Freeman, L.R., Granholm, A.C., 2012. Vascular changes in rat hippocampus following a high saturated fat and cholesterol diet. J. Cereb. Blood Flow Metab. 32, 643–653.
Garcia-Cazarin, M.L., Wambogo, E.A., Regan, K.S., Davis, C.D., 2014. Dietary supplement research portfolio at the NIH, 2009-2011. J. Nutr 144, 414–418.
Gershwin, M.E., Borchers, A.T., Keen, C.L., Hendler, S., Hagie, F., Greenwood, M.R., 2010. Public safety and dietary supplementation. Ann. N.Y. Acad. Sci 1190, 104–117.
Giem, P., Beeson, W.L., Fraser, G.E., 1993. The incidence of dementia and intake of animal products: preliminary findings from the Adventist Health Study. Neuroepidemiology 12, 28–36.
Goudis, C.A., Korantzopoulos, P., Ntalas, I.V., Kallergis, E.M., Liu, T., Ketikoglou, D. G., 2015. Diabetes mellitus and atrial fibrillation: pathophysiological mechanisms and potential upstream therapies. Int. J. Cardiol 184, 617–622.
Granholm, A.C., Bimonte-Nelson, H.A., Moore, A.B., Nelson, M.E., Freeman, L.R., 2008. Effects of a saturated fat and high cholesterol diet on memory and hippocampal morphology in the middle-aged rat. J. Alzheimer Dis 14, 133–145.
Gupta, S.C., Kismali, G., Aggarwal, B.B., 2013. Curcumin, a component of turmeric: from farm to pharmacy. BioFactors 39, 2–13.

Haghighatdoost, F., Hariri, M., 2018. Effect of resveratrol on lipid profile: an updated systematic review and meta-analysis on randomized clinical trials. Pharmacol. Res. 129, 141–150.
Haohao, Z., Guijun, Q., Juan, Z., Wen, K., Lulu, C., 2015. Resveratrol improves high-fat diet induced insulin resistance by rebalancing subsarcolemmal mitochondrial oxidation and antioxidation. J. Physiol. Biochem. 71, 121–131.
Hashimoto, M., Hossain, S., Shimada, T., Shido, O., 2006. Docosahexaenoic acid-induced protective effect against impaired learning in amyloid β-infused rats is associated with increased synaptosomal membrane fluidity. Clin. Exp. Pharmacol. Physiol. 33, 934–939.
Hashimoto, M., Shahdat, H.M., Yamashita, S., Katakura, M., Tanabe, Y., Fujiwara, H., et al., 2008. Docosahexaenoic acid disrupts in vitro amyloid beta fibrillation and concomitantly inhibits amyloid levels in cerebral cortex of Alzheimer's disease model rats. J. Neurochem 107, 1634–1646.
Hawley, S.A., Ross, F.A., Chevtzoff, C., Green, K.A., Evans, A., et al., 2010. Use of cells expressing γ subunit variants to identify diverse mechanisms of AMPK activation. Cell Metab. 11, 554–565.
Henry, J.P., Stephens-Larson, P., 1984. Reduction of chronic psychosocial hypertension in mice by decaffeinated tea. Hypertension 6, 437–444.
Hooper, P.L., Hooper, P.L., Tytell, M., Vígh, L., 2010. Xenohormesis: health benefits from an eon of plant stress response evolution. Cell Stress Chaperones 15, 761–770.
Howitz, K.T., Bitterman, K.J., Cohen, H.Y., Lamming, D.W., Lavu, S., Wood, J.G., et al., 2003. Small molecule activators of sirtuins extend *Saccharomyces cerevisiae* lifespan. Nature 425, 191–196.
Hsu, C.H., Liao, Y.L., Lin, S.C., Tsai, T.H., Huang, C.J., Chou, P., 2011. Does supplementation with green tea extract improve insulin resistance in obese type 2 diabetics? A randomized, double-blind, and placebo-controlled clinical trial. Altern. Med. Rev 16, 157–163.
Hu, F.B., Bronner, L., Willett, W.C., Stampfer, M.J., Rexrode, K.M., Albert, C.M., et al., 2002. Fish and omega-3 fatty acid intake and risk of coronary heart disease in women. JAMA 287, 1815–1821.
International Diabetes Federation Diabetes Atlas, eighth edition, 2017.
Ishigaki, A., Tonooka, F., Matsumoto, N., Hara, Y., 1991. Suppression of the Accumulation of Body and Liver Fat by Tea Catechin. Organizing Committee of International Symposium on Tea Science, Shizuoka, Japan, pp. 309–313.
Jang, M., Cai, L., Udeani, G.O., et al., 1997. Cancer chemopreventive activity of resveratrol, a natural product derived from grapes. Science 275, 218–220.
Jawien, A., Bowen-Pope, D.F., Lindner, V., Schwartz, S.M., Clowes, A.W., 1992. Platelet-derived growth factor promotes smooth muscle migration and intimal thickening in a rat model of balloon angioplasty. J. Clin. Investig. 89, 507–511.
Jeremic, N., Chaturvedi, P., Tyagi, S.C., 2017. Browning of white fat: novel insight into factors, mechanisms, and therapeutics. J. Cell. Physiol 232, 61–68.
Jiang, J., Wang, W., Sun, Y., Hu, M., Li, F., Zhu, D.Y., 2007. Neuroprotective effect of curcumin on focal cerebral ischemic rats by preventing blood-brain barrier damage. Eur. J. Pharmacol 561, 54–62.
Jiménez-Escrig, A., Gómez-Ordóñez, E., Rupérez, P., 2011. Seaweed as a source of novel nutraceuticals: sulfated polysaccharides and peptides. Adv. Food. Nutr. Res. 64, 325–337.
Kahleova, H., Pelikanova, T., 2015. Vegetarian diets in the prevention and treatment of type 2 diabetes. J. Am. Coll. Nutr. 34, 448–458.
Kampoli, A.M., Tousoulis, D., Briasoulis, A., Latsios, G., Papageorgiou, N., et al., 2011. Potential pathogenic inflammatory mechanisms of endothelial dysfunction induced by type 2 diabetes mellitus. Curr. Pharm. Des 17, 4147–4158.

Kang, O.H., Kim, S.B., Seo, Y.S., Joung, D.K., Mun, S.H., Choi, J.G., et al., 2013. Curcumin decreases oleic acid-induced lipid accumulation via AMPK phosphorylation in hepatocarcinoma cells. Eur. Rev. Med. Pharmacol. Sci 17, 2578–2586.

Kanoski, S.E., Meisel, R.L., Mullins, A.J., Davidson, T.L., 2007. The effects of energy-rich diets on discrimination reversal learning and on BDNF in the hippocampus and prefrontal cortex of the rat. Behav. Brain Res 182, 57–66.

Kantor, E.D., Rehm, C.D., Du, M., White, E., Giovannucci, E.L., 2016. Trends in dietary supplement use among US adults from 1999-2012. JAMA 316, 1464–1474.

Kastorini, C.M., Milionis, H.J., Esposito, K., Giugliano, D., Goudevenos, J.A., Panagiotakos, D.B., 2011. The effect of Mediterranean diet on metabolic syndrome and its components: a meta-analysis of 50 studies and 534,906 individuals. J. Am. Coll. Cardiol 57, 1299–1313.

Kennedy, D.O., Wightman, E.L., 2011. Herbal extracts and phytochemicals: plant secondary metabolites and the enhancement of human brain function. Adv. Nutr 2, 32–50.

Khan, A., Safdar, M., Ali Khan, M.M., Khattak, K.N., Anderson, R.A., 2003. Cinnamon improves glucose and lipids of people with type 2 diabetes. Diabetes Care 26, 3215–3218.

Kim, M.-H., Bae, Y.-J., 2015. Comparative study of serum leptin and insulin resistance levels between Korean postmenopausal vegetarian and non-vegetarian women. Clin. Nutr. Res. 4, 175–181.

Kim, F., Tysseling, K.A., Rice, J., Gallis, B., Haji, L., Giachelli, C.M., et al., 2005. Activation of IKK beta by glucose is necessary and sufficient to impair insulin signaling and nitric oxide production in endothelial cells. J. Mol. Cell. Cardiol 39, 327–334.

Kim, J.A., Formoso, G., Li, Y.H., Potenza, M.A., Marasciulo, F.L., Montagnani, M., et al., 2007. Epigallocatechin gallate, a green tea polyphenol, mediates NO-dependent vasodilation using signaling pathways in vascular endothelium requiring reactive oxygen species and Fyn. J. Biol. Chem 282, 13736–13745.

Kim, T., Davis, J., Zhang, A.J., He, X., Mathews, S.T., 2009. Curcumin activates AMPK and suppresses gluconeogenic gene expression in hepatoma cells. Biochem. Biophys. Res. Commun 388, 377–382.

Kinlay, S., Creager, M.A., Fukumoto, M., Hikita, H., Fang, J.C., Selwyn, A.P., et al., 2001. Endothelium-derived nitric oxide regulates arterial elasticity in human arteries in vivo. Hypertension 38, 1049–1053.

Knowler, W.C., Barrett-Connor, E., Fowler, S.E., Hamman, R.F., Lachin, J.M., Walker, E.A., et al., 2002. Reduction in the incidence of type 2 diabetes with lifestyle intervention or metformin. N. Engl. J. Med. 346, 393–403.

Kobori, M., Masumoto, S., Akimoto, Y., Takahashi, Y., 2009. Dietary quercetin alleviates diabetic symptoms and reduces streptozotocin-induced disturbance of hepatic gene expression in mice. Mol. Nutr. Food Res. 53, 859–868.

Korsholm, A., Kjær, T., Ornstrup, M., Pedersen, S., 2017. Comprehensive metabolomic analysis in blood, urine, fat, and muscle in men with metabolic syndrome: a randomized, placebo-controlled clinical trial on the effects of resveratrol after four months' treatment. Int. J. Mol. Sci. 18, 554.

Kruk, J., 2014. Association between vegetable, fruit and carbohydrate intake and breast cancer risk in relation to physical activity. Asian Pac. J. Cancer Prev. 15, 4429–4436.

Kumar, A., Naidu, P.S., Seghal, N., Padi, S.S.V., 2007. Neuroprotective effects of resveratrol against intracerebroventricular colchicine-induced cognitive impairment and oxidative stress in rats. Pharmacology 79, 17–26.

Kunnumakkara, A.B., Bordoloi, D., Padmavathi, G., Monisha, J., Roy, N.K., Prasad, S., et al., 2017. Curcumin, the golden nutraceutical: multitargeting for multiple chronic diseases. Br. J. Pharmacol. 174, 1325–1348.

Kutuk, O., Adli, M., Poli, G., Basaga, H., 2004. Resveratrol protects against 4-HNE induced oxidative stress and apoptosis in Swiss 3T3 fibroblasts. Biofactors 20, 1–10.

Kyro, C., Skeie, G., Loft, S., Landberg, R., Christensen, J., Lund, E., et al., 2013. Intake of whole grains from different cereal and food sources and incidence of colorectal cancer in the Scandinavian HELGA cohort. Cancer Causes Control 24, 1363–1374.

Laddha, A.P., Kulkarni, Y.A., 2019. Tannins and vascular complications of diabetes: an update. Phytomedicine 56, 229–245.

Lagouge, M., Argmann, C., Gehart-Hines, Z., 2006a. Resveratrol improves mitochondrial function and protects against metabolic disease by activating SIRT1 and PRG-1. Cell 127, 1–14.

Lagouge, M., Argmann, C., Gerhart-Hines, Z., Meziane, H., Lerin, C., Daussin, F., et al., 2006b. Resveratrol improves mitochondrial function and protects against metabolic disease by activating SIRT1 and PGC-1alpha. Cell 127, 1109–1122.

Larsen, C.J., 1989. The Nobel Prize in physiology and medicine 1989. J. Michael Bish. Harold E. Varmus. Pathol. Biol. (Paris) 37, 1077–1078.

Lazar, D.F., Wiese, R.J., Brady, M.J., Mastick, C.C., Waters, S.B., Yamauchi, K., et al., 1995. Mitogen-activated protein kinase kinase inhibition does not block the stimulation of glucose utilization by insulin. J. Biol. Chem. 270, 20801–20807.

Leonov, A., Arlia-Ciommo, A., Piano, A., Svistkova, V., Lutchman, V., Medkour, Y., et al., 2015. Longevity extension by phytochemicals. Molecules 20, 6544–6572.

Lepretti, M., Martucciello, S., Burgos Aceves, M.A., Putti, R., Lionetti, L., 2018. Omega-3 fatty acids and insulin resistance: focus on the regulation of mitochondria and endoplasmic reticulum stress. Nutrients 10, pii: E350.

Lewis, D.F., 2003. Human cytochromes P450 associated with the phase 1 metabolism of drugs and other xenobiotics: a compilation of substrates and inhibitors of the CYP1, CYP2 and CYP3 families. Curr. Med. Chem. 10, 1955–1972.

Lexis, C.P., Rahel, B.M., Meeder, J.G., Zijlstra, F., van der Horst, I.C., 2009. The role of glucose lowering agents on restenosis after percutaneous coronary intervention in patients with diabetes mellitus. Cardiovasc. Diabetol. 8, 41.

Liao, W., Yin, X., Li, Q., Zhang, H., Liu, Z., Zheng, X., et al., 2018. Resveratrol-induced white adipose tissue browning in obese mice by remodeling fecal microbiota. Molecules 23, E3356.

Liu, R.H., 2003. Health benefits of fruit and vegetables are from additive and synergistic combinations of phytochemicals. Am. J. Clin. Nutr 78, 517S–520S.

Liu, Z., Yu, Y., Li, X., Ross, C.A., Smith, W.W., 2011. Curcumin protects against A53T alpha-synuclein-induced toxicity in a PC12 inducible cell model for Parkinsonism. Pharmacol. Res. 63, 439–444.

Lorenz, M., Wessler, S., Follmann, E., Michaelis, W., Dusterhoft, T., Baumann, G., et al., 2004. A constituent of green tea, epigallocatechin-3-gallate, activates endothelial nitric oxide synthase by a phosphatidylinositol-3-OH-kinase-, cAMP-dependent protein kinase-, and Akt-dependent pathway and leads to endothelial-dependent vasorelaxation. J. Biol. Chem 279, 6190–6195.

Lukiw, W.J., Cui, J.G., Marcheselli, V.L., Bodker, M., Botkjaer, A., Gotlinger, K., et al., 2005. A role for docosahexaenoic acid-derived neuroprotectin D1 in neural cell survival and Alzheimer disease. J. Clin. Invest. 115, 2774–2783.

Ma, Q.L., Yang, F., Rosario, E.R., Ubeda, O.J., Beech, W., Gant, D.J., et al., 2009. Beta-amyloid oligomers induce phosphorylation of tau and inactivation of insulin receptor substrate via c-Jun N-terminal kinase signaling: suppression by omega-3 fatty acids and curcumin. J. Neurosci. 29, 9078–9089.

Manach, C., Scalbert, A., Morand, C., Rémésy, C., Jiménez, L., 2004. Polyphenols: food sources and bioavailability. Am. J. Clin. Nutr. 79, 727–747.

Manach, C., Williamson, G., Morand, C., Scalbert, A., Rémésy, C., 2005. Bioavailability and bioefficacy of polyphenols in humans. I. Review of 97 bioavailability studies. Am. J. Clin. Nutr 81, 230S–242S.
Markus, M.A., Morris, B.J., 2008. Resveratrol in prevention and treatment of common clinical conditions of aging. Clin. Interv. Aging 3, 331–339.
Matheus, A.S., Tannus, L.R., Cobas, R.A., Palma, C.C., Negrato, C.A., Gomes, M.B., 2013. Impact of diabetes on cardiovascular disease: an update. Int. J. Hypertens. 2013:653789.
Matsumoto, N., Ishigaki, F., Ishigaki, A., Iwashin, H., Hara, Y., 1993. Reduction of blood glucose levels by tea catechin. Biosci. Biotech. Biochem. 57, 525–527.
McEvoy, C.T., Temple, N., Woodside, J.V., 2012. Vegetarian diets, low-meat diets and health: a review. Public Health Nutr 15, 2287–2294.
McNeilly, A.D., Stewart, C.A., Sutherland, C., Balfour, D.J., 2015. High fat feeding is associated with stimulation of the hypothalamic-pituitary-adrenal axis and reduced anxiety in the rat. Psychoneuroendocrinology 52, 272–280.
Méndez-del Villar, M., González-Ortiz, M., Martínez-Abundis, E., Pérez-Rubio, K.G., Lizárraga-Valdez, R., 2014. Effect of Resveratrol Administration on Metabolic Syndrome, Insulin Sensitivity, and Insulin Secretion. Metab. Syndr. Relat. Disord. 12, 497–501.
Millward, D.J., 1999. The nutritional value of plant-based diets in relation to human amino acid and protein requirements. Proc. Nutr. Soc 58, 249–260.
Mizuno, N., Niwa, T., Yotsumoto, Y., Sugiyama, Y., 2003. Impact of drug transporter studies on drug discovery and development. Pharmacol. Rev. 55, 425–461.
Montonen, J., Knekt, P., Järvinen, R., Aromaa, A., Reunanen, A., 2003. Whole-grain and fiber intake and the incidence of type 2 diabetes. Am. J. Clin. Nutr. 77, 622–629.
Morley, J.E., Banks, W.A., 2010. Lipids and cognition. J. Alzheimer Dis 20, 737–747.
Morselli, E., Maiuri, M.C., Markaki, M., Megalou, E., Pasparaki, A., Palikaras, K., et al., 2010. The life span-prolonging effect of sirtuin-1 is mediated by autophagy. Autophagy 6, 186–188.
Murakami, A., 2013. Modulation of protein quality control systems by food phytochemicals. J. Clin. Biochem. Nutr. 52, 215–227.
Moroo, I., Yamada, T., Makino, H., Tooyama, I., McGeer, P.L., McGeer, E.G., et al., 1994. Loss of insulin receptor immunoreactivity from the substantia nigra pars compacta neurons in Parkinson's disease. Acta Neuropathol. 87, 343–348.
Most, J., Tosti, V., Redman, L.M., Fontana, L., 2017. Calorie restriction in humans: an update. Ageing Res. Rev. 39, 36–45.
Mursu, J., Virtanen, J.K., Tuomainen, T.P., Nurmi, T., Voutilainen, S., 2014. Intake of fruit, berries, and vegetables and risk of type 2 diabetes in Finnish men: the Kuopio Ischaemic Heart Disease Risk Factor Study. Am. J. Clin. Nutr 99, 328–333.
Na, L.X., Zhang, Y.L., Li, Y., Liu, L.Y., Li, R., Kong, T., et al., 2011. Curcumin improves insulin resistance in skeletal muscle of rats. Nutr. Metab. Cardiovasc. Dis 21, 526–533.
Nakagawa, T., Hu, H., Zharikov, S., Tuttle, K.R., Short, R.A., Glushakova, O., et al., 2006. A causal role for uric acid in fructose-induced metabolic syndrome. Am. J. Physiol. 290, 625–631.
Nakagawa, K., Ueda, M., Itoh, M., Islam, S., Tana Nakagawa, T., 2017. Dietary quercetin ameliorates memory impairment in a murine model of Alzheimer's disease with obesity and diabetes, suppressing ATF4 expression. J. Neurol. Neurosci. 8, 234.
Nicholson, S.K., Tucker, G.A., Brameld, J.M., 2008. Effects of dietary polyphenols on gene expression in human vascular endothelial cells. Proc. Nutr. Soc. 67, 42–47.

Oba, S., Nagata, C., Nakamura, K., Fujii, K., Kawachi, T., Takatsuka, N., et al., 2010. Consumption of coffee, green tea, oolong tea, black tea, chocolate snacks and the caffeine content in relation to risk of diabetes in Japanese men and women. Br. J. Nutr 103, 453–459.

Ortsater, H., Grankvist, N., Wolfram, S., Kuehn, N., Sjoholm, A., 2012. Diet supplementation with green tea extract epigallocatechin gallate prevents progression to glucose intolerance in db/db mice. Nutr. Metab. (Lond.) 9, 11.

Olfert, M.D., Wattick, R.A., 2018. Vegetarian diets and the risk of diabetes. Curr. Diab. Rep 18, 101.

Pacholec, M., Bleasdale, J.E., Chrunyk, B., Cunningham, D., Flynn, D., Garofalo, R.S., et al., 2010. SRT1720, SRT2183, SRT1460, and resveratrol are not direct activators of SIRT1. J. Biol. Chem. 285, 8340–8351.

Padiya, R., Khatua, T.N., Bague, R., Kuncha, M., Banerjee, S.K., 2011. Garlic improves insulin sensitivity and associated metabolic syndromes in fructose fed rats. Nutr. Metab. (Lond.) 8, 53.

Pan, M.H., Lin-Shiau, S.Y., Ho, C.T., Lin, J.H., Lin, J.K., 2000. Suppression of lipopolysaccharide-induced nuclear factor-kappa B activity by theaflavin-3,3′-digallate from black tea and other polyphenols through down-regulation of I kappa B kinase activity in macrophages. Biochem. Pharmacol. 59, 357–367.

Pan, S., Lin, H., Luo, H., Gao, F., Meng, L., Zhou, C., 2017. Folic acid inhibits dedifferentiation of PDGF-BB-induced vascular smooth muscle cells by suppressing mTOR/P70S6K signaling. Am. J. Transl. Res. 9, 1307–1316.

Parveen, A., Kim, J.H., Oh, B.G., Subedi, L., Khan, Z., Kim, S.Y., 2018. Phytochemicals: target-based therapeutic strategies for diabetic retinopathy. Molecules 23, pii: E1519.

Perron, N.R., Brumaghim, J.L., 2009. A review of the antioxidant mechanisms of polyphenol compounds related to iron binding. Cell Biochem. Biophys 53, 75–100.

Persson, J., Nyberg, L., Lind, J., Larsson, A., Nilsson, L.G., Ingvar, M., et al., 2006. Structure-function correlates of cognitive decline in aging. Cereb Cortex 16, 907–915.

Picard, F., Kurtev, M., Chung, N., Topark-Ngarm, A., Senawong, T., Machado De Oliveira, R., et al., 2004. Sirt1 promotes fat mobilization in white adipocytes by repressing PPAR-gamma. Nature 429, 771–776.

Pistell, P.J., Morrison, C.D., Gupta, S., Knight, A.G., Keller, J.N., et al., 2010. Cognitive impairment following high fat diet consumption is associated with brain inflammation. J. Neuroimmunol 219, 25–32.

Polak, P., Cybulski, N., Feige, J.N., Auwerx, J., Rüegg, M.A., Hall, M.N., 2008. Adipose-specific knockout of raptor results in lean mice with enhanced mitochondrial respiration. Cell Metab. 8, 399–410.

Potenza, M.A., Marasciulo, F.L., Tarquinio, M., Tiravanti, E., Colantuono, G., Federici, A., et al., 2007. EGCG, a green tea polyphenol, improves endothelial function and insulin sensitivity, reduces blood pressure, and protects against myocardial I/R injury in SHR. Am. J. Physiol. Endocrinol. Metab 292, E1378–E1387.

Poulsen, M.M., Jorgensen, J.O., Jessen, N., Richelsen, B., Pedersen, S.B., 2013. Resveratrol in metabolic health: an overview of the current evidence and perspectives. Ann. N.Y. Acad. Sci. 1290, 74–82.

Qin, B., Polansky, M.M., Anderson, R.A., 2009. Cinnamon extract regulates plasma levels of adipose-derived factors and expression of multiple genes related to carbohydrate metabolism and lipogenesis in adipose tissue of fructose-fed rats. Horm. Metab. Res 42, 187–193.

Qin, B., Panickar, K.S., Anderson, R.A., 2010. Cinnamon: potential role in the prevention of insulin resistance, metabolic syndrome, and type 2 diabetes. J. Diabetes Sci. Technol 4, 685–693.

Quiles, L., Portolés, O., Sorlí, J.V., Corella, D., 2014. Short term effects on lipid profile and glycaemia of a low-fat vegetarian diet. Nutr. Hosp. 32, 156–164.

Raghunath, A., Sundarraj, K., Nagarajan, R., Arfuso, F., Bian, J., Kumar, A.P., 2018. Antioxidant response elements: discovery, classes, regulation and potential applications. Redox Biol. 17, 297–314.

Rahimi, R., Nikfar, S., Larijiani, B., Abdollahi, M., 2005. A review on the role of antioxidants in the management of diabetes and its complications. Biomed. Pharmacother. 59, 365–373.

Rahman, K., Lowe, G.M., 2006. Garlic and cardiovascular disease: a critical review. J. Nutr. 136, 736S–740S.

Rajeswari, A., Sabesan, M., 2008. Inhibition of monoamine oxidase-B by the polyphenolic compound, curcumin and its metabolite tetrahydrocurcumin, in a model of Parkinson's disease induced by MPTP neurodegeneration in mice. Inflammopharmacology 16, 96–99.

Reygaert, W.C., 2018. Green tea catechins: their use in treating and preventing infectious diseases. Biomed. Res. Int 2018:9105261.

Riviere, C., Richard, T., Quentin, L., Krisa, S., Merillon, J.M., Monti, J.P., 2007. Inhibitory activity of stilbenes on Alzheimer's beta-amyloid fibrils in vitro. Bioorg. Med. Chem. 15, 1160–1167.

Roussel, A.M., Hininger, I., Benaraba, R., Ziegenfuss, T.N., Anderson, R.A., 2009. Antioxidant effects of a cinnamon extract in people with impaired fasting glucose that are overweight or obese. J. Am. Coll. Nutr 28, 16–21.

Russo, M., Spagnuolo, C., Tedesco, I., Bilotto, S., Russo, G.L., 2012. The flavonoid quercetin in disease prevention and therapy: facts and fancies. Biochem. Pharmacol. 83, 6–15.

Rydén, L., Grant, P.J., Anker, S.D., Berne, C., Cosentino, F., Danchin, N., 2013. The task force on diabetes, pré-diabetes and cardiovascular diseases of the European Society of Cardiology (ESC) and developed in collaboration with the European Association for the Study of Diabetes (EASD). ESC Guidelines on diabetes, pre-diabetes and cardiovascular diseases, developed in collaboration with the EASD. Eur. Heart J 34, 3035–3087.

Saltiel, A.R., Kahn, C.R., 2001. Insulin signaling and the regulation of glucose and lipid metabolism. Nature. 414, 799–806.

Salvamani, S., Gunasekaran, B., Shaharuddin, N.A., Ahmad, S.A., Shukor, M.Y., 2014. Antiartherosclerotic effects of plant flavonoids. Biomed. Res. Int. 2014:480258.

Santhosha, S.G., Jamuna, P., Prabhavathi, S.N., 2013. Bioactive components of garlic and their physiological role in health maintenance: a review. Food Biosci 3, 59–74.

Savaskan, E., Olivieri, G., Meier, F., Seifritz, E., Wirz-Justice, A., Müller-Spahn, F., 2003. Red wine ingredient resveratrol protects from β-amyloid neurotoxicity. Gerontology 49, 380–383.

Sayem, A.S.M., Arya, A., Karimian, H., Krishnasamy, N., Ashok Hasamnis, A., Hossain, C.F., 2018. Action of phytochemicals on insulin signaling pathways accelerating glucose transporter (GLUT4) protein translocation. Molecules 23, pii: E258.

Schwingshackl, L., Missbach, B., Hoffmann, G., 2016. An umbrella review of garlic intake and risk of cardiovascular disease. Phytomedicine 23, 1127–1133.

Sellbom, K.S., Gunstad, J., 2012. Cognitive function and decline in obesity. J. Alzheimer Dis 30 (Suppl. 2), S89–S95.

Shankar, S., Singh, G., Srivastava, R.K., 2007. Chemoprevention by resveratrol: molecular mechanisms and therapeutic potential. Front. Biosci. 12, 4839–4854.

Sheng, X., Zhang, Y., Gong, Z., Huang, C., Zang, Y.Q., 2008. Improved insulin resistance and lipid metabolism by cinnamon extract through activation of peroxisome proliferator-activated receptors. PPAR Res. 581348.

Shigematsu, S., Ishida, S., Hara, M., Takahashi, N., Yoshimatsu, H., Sakata, T., et al., 2003. Resveratrol, a red wine constituent polyphenol, prevents superoxide-dependent inflammatory responses induced by ischemia/reperfusion, platelet-activating factor, or oxidants. Free Radic. Biol. Med. 34, 810–817.

Shishodia, S., Sethi, G., Aggarwal, B.B., 2005. Curcumin: getting back to the roots. Ann. N.Y. Acad. Sci 1056, 206–217.

Si, H., Liu, D., 2014. Dietary antiaging phytochemicals and mechanisms associated with prolonged survival. J. Nutr. Biochem. 25, 581–591.

Simopoulos, A.P., 2016. An increase in the omega-6/omega-3 fatty acid ratio increases the risk for obesity. Nutrients 8, 128.

Simpson, F., Jonathan, P.W., David, E.J., 2001. GLUT4—at the cross roads between membrane trafficking and signal transduction. Traffic 2, 2–11.

Singh, M., Arora, R., Kodumuri, V., Khosla, S., Jawad, E., 2011. Coronary revascularization in diabetic patients: current state of evidence. Exp. Clin. Cardiol. 16, 16–22.

Sobenin, I.A., Nedosugova, L.V., Filatova, L.V., Balabolkin, M.I., Gorchakova, T.V., Orekhov, A.N., 2008. Metabolic effects of time-released garlic powder tablets in type 2 diabetes mellitus: the results of double-blinded placebo-controlled study. Acta Diabetol. 45, 1–6.

Song, Z., Wang, Y., Zhang, F., Yao, F., Sun, C., 2019. Calcium signaling pathways: key pathways in the regulation of obesity. Int. J. Mol. Sci 20, pii: E2768.

Spencer, E.A., Appleby, P.N., Davey, G.K., Key, T.J., 2003. Diet and body mass index in 38000 EPIC-Oxford meat-eaters, fish-eaters, vegetarians and vegans. Int. J. Obes 27, 728–734.

Spencer, J.P., Vauzour, D., Rendeiro, C., 2009. Flavonoids and cognition: the molecular mechanisms underlying their behavioural effects. Arch Biochem. Biophys. 492, 1–9.

Stranahan, A., Norman, E., Lee, K., Cutler, R., Telljohann, R., et al., 2008. Diet-induced insulin resistance impairs hippocampal synaptic plasticity and cognition in middle-aged rats. Hippocampus 18, 1085–1088.

Strange, R.C., Jones, P.W., Fryer, A.A., 2000. Glutathione S-transferase: genetics and role in toxicology. Toxicol. Lett. 112–113, 357–363.

Streeter, J., Thiel, W., Brieger, K., Miller Jr., F.J., 2013. Opportunity nox: the future of NADPH oxidases as therapeutic targets in cardiovascular disease. Cardiovasc. Ther. 31, 125–137.

Stumvoll, M., Goldstein, B.J., van Haeften, T.W., 2005. Type 2 diabetes: principles of pathogenesis and therapy. Lancet 365, 1333–1346.

Suzuki, M., Willcox, B.J., Willcox, D.C., 2001. Implications from and for food cultures for cardiovascular disease: longevity. Asia Pac. J. Clin. Nutr 10, 165–171.

Tahara, S., 2007. A journey of twenty-five years through the ecological biochemistry of flavonoids. Biosci. Biotechnol. Biochem. 71, 1387–1404.

Tang, X., Luo, Y.-X., Chen, H.-Z., Liu, D.P., 2014. Mitochondria, endothelial cell function, and vascular diseases. Front. Physiol. 5, 175.

Tattelman, E., 2005. Health effects of garlic. Am. Fam. Physician. 72, 103–106.

Tavernarakis, N., Pasparaki, A., Tasdemir, E., Maiuri, M.C., Kroemer, G., et al., 2008. The effects of p53 on whole organism longevity are mediated by autophagy. Autophagy 4, 870–873.

Thorburn, A., Muir, J., Proietto, J., 1993. Carbohydrate fermentation decreases hepatic glucose output in healthy subjects. Metabolism 42, 780–785.

Toolsee, N.A., Aruoma, O.I., Gunness, T.K., Kowlessur, S., Dambala, V., Murad, F., et al., 2013. Effectiveness of green tea in a randomized human cohort: relevance to diabetes and its complications. Biomed. Res. Int 2013:412379.

Tsuneki, H., Murata, S., Anzawa, Y., Soeda, Y., Tokai, E., Wada, T., et al., 2008. Age-related insulin resistance in hypothalamus and peripheral tissues of orexin knockout mice. Diabetologia 51, 657–667.

Tung, Y.T., Chua, M.T., Wang, S.Y., Chang, S.T., 2008. Anti-inflammation activities of essential oil and its constituents from indigenous cinnamon (*Cinnamomum osmophloeum*) twigs. Bioresour. Technol. 99, 3908–3913.

Tuomilehto, J., Lindström, J., Eriksson, J.G., Valle, T.T., Hämäläinen, H., Ilanne-Parikka, P., et al., 2001. Prevention of type 2 diabetes mellitus by changes in lifestyle among subjects with impaired glucose tolerance. N. Engl. J. Med. 344, 1343–1350.

Um, J.H., Park, S.J., Kang, H., Yang, S., Foretz, M., McBurney, M.W., et al., 2010. AMP-activated protein kinase-deficient mice are resistant to the metabolic effects of resveratrol. Diabetes 59, 554–563.

Ungvari, Z., Bagi, Z., Feher, A., Recchia, F.A., Sonntag, W.E., Pearson, K., et al., 2010. Resveratrol confers endothelial protection via activation of the antioxidant transcription factor Nrf2. Am. J. Physiol. Heart Circ. Physiol. 299, H18–H24.

Upadhyay, S., Dixit, M., 2015. Role of polyphenols and other phytochemicals on molecular signaling. Oxid. Med. Cell. Longev. 1–15. 2015.

USDA, 2003. USDA Database for the Flavonoid Contents of Selected Foods. US Department of Agriculture, Beltsville, MD.

Vanschoonbeek, K., Thomassen, B.J., Senden, J.M., Wodzig, W.K., van Loon, L.J., 2006. Cinnamon supplementation does not improve glycemic control in postmenopausal type 2 diabetes patients. J. Nutr 136, 977–980.

Vauzour, D., 2012. Dietary polyphenols as modulators of brain functions: biological actions and molecular mechanisms underpinning their beneficial effects. Oxid. Med. Cell. Longev 2012:914273.

Verma, V.C., Kharwar, R.N., Strobel, G.A., 2009. Chemical and functional diversity of natural products from plant associated endophytic fungi. Nat. Prod. Commun. 4, 1511–1532.

Vingtdeux, V., Giliberto, L., Zhao, H., Chandakkar, P., Wu, Q., Simon, J.E., et al., 2010. AMP-activated protein kinase signaling activation by resveratrol modulates amyloid-beta peptide metabolism. J. Biol. Chem. 285, 9100–9113.

Virgili, F., Marino, M., 2008. Regulation of cellular signals from nutritional molecules: a specific role for phytochemicals, beyond antioxidant activity. Free Radic. Biol. Med. 45, 1205–1216.

Volpe, S.L., 2013. Magnesium in disease prevention and overall health. Adv. Nutr. 4, 378S–383SS.

Wang, Y., Ho, C.T., 2009. Polyphenolic chemistry of tea and coffee: a century of progress. J. Agric. Food Chem. 57, 8109–8114.

Wang, J., Ho, L., Zhao, Z., Seror, I., Humala, N., Dickstein, D.L., et al., 2006. Moderate consumption of cabernet sauvignon attenuates Abeta neuropathology in a mouse model of Alzheimer's disease. FASEB J. 20, 2313–2320.

Wang, G.L., Fu, Y.C., Xu, W.C., Feng, Y.Q., Fang, S.R., Zhou, X.H., 2009. Resveratrol inhibits the expression of SREBP1 in cell model of steatosis via Sirt1-FOXO1 signaling pathway. Biochem. Biophys. Res. Commun 380, 644–664.

Wang, S., Liang, X., Yang, Q., Fu, X., Rogers, C.J., Zhu, M., et al., 2015. Resveratrol induces brown-like adipocyte formation in white fat through activation of AMP-activated protein kinase (AMPK) alpha1. Int. J. Obes 39, 967–976.

Wang, S.F., Wu, M.Y., Cai, C.Z., Li, M., Lu, J.H., 2016. Autophagy modulators from traditional Chinese medicine: mechanisms and therapeutic potentials for cancer and neurodegenerative diseases. J. Ethnopharmacol. 194, 861–876.

Weisberg, S.P., Leibel, R., Tortoriello, D.V., 2008. Dietary curcumin significantly improves obesity-associated inflammation and diabetes in mouse models of diabesity. Endocrinology. 149, 3549–3558.

Willcox, B.J., Willcox, D.C., Todoriki, H., Fujiyoshi, A., Yano, K., He, Q., et al., 2007. Caloric restriction, the traditional Okinawan diet and healthy aging: the diet of the world's longest-lived people and its potential impact on morbidity and lifespan. Ann. N.Y. Acad. Sci. 1114, 434–455.

Willcox, D.C., Willcox, B.J., Todoriki, H., Suzuki, M., 2009. The Okinawan diet: health implications of a low-calorie, nutrient-dense, antioxidant-rich dietary pattern low in glycemic load. J. Am. Coll. Nutr. 28 (Suppl.), 500S–516S.

Willcox, D.C., Scapagnini, G., Willcox, B.J., 2014. Healthy aging diets other than the Mediterranean: a focus on the Okinawan diet. Mech. Ageing Dev 136, 148–162.

Wu, A.H., Yu, M.C., Tseng, C.C., Hankin, J., Pike, M.C., 2003. Green tea and risk of breast cancer in Asian Americans. Int. J. Cancer 106, 574–579.

Xu, C., Li, C.Y., Kong, A.N., 2005. Induction of phase I, II and III drug metabolism/transport by xenobiotics. Arch. Pharm. Res. 28, 249–268.

Yang, F., Lim, G.P., Begum, A.N., Ubeda, O.J., Simmons, M.R., Ambegaokar, S.S., et al., 2005. Curcumin inhibits formation of amyloid beta oligomers and fibrils, binds plaques, and reduces amyloid in vivo. J. Biol. Chem 280, 5892–5901.

Yang, C., Zhang, X., Fan, H., Liu, Y., 2009. Curcumin upregulates transcription factor Nrf2, HO-1 expression and protects rat brains against focal ischemia. Brain Res. 1282, 133–141.

Yokoyama, Y., Nishimura, K., Barnard, N.D., Takegami, M., Watanabe, M., et al., 2014. Vegetarian diets and blood pressure: a meta-analysis. JAMA Int. Med. 174, 577–587.

Yoo, S., Kim, K., Nam, H., Lee, D., 2018. Discovering health benefits of phytochemicals with integrated analysis of the molecular network, chemical properties and ethnopharmacological evidence. Nutrients. 10, pii: E1042.

Zeng, W., Zhang, W., Lu, F., Gao, L., Gao, G., 2017. Resveratrol attenuates MPP^+-induced mitochondrial dysfunction and cell apoptosis via AKT/GSK-3beta pathway in SN4741 cells. Neurosci. Lett. 637, 50–56.

Zhu, R., Liu, H., Liu, C., Wang, L., Ma, R., Chen, B., et al., 2017. Cinnamaldehyde in diabetes: a review of pharmacology, pharmacokinetics and safety. Pharmacol. Res. 122, 78–89.

Ziegenfuss, T.N., Hofheins, J.E., Mendel, R.W., Landis, J., Anderson, R.A., 2006. Effects of a water-soluble cinnamon extract on body composition and features of the metabolic syndrome in pre-diabetic men and women. J. Int. Soc. Sports Nutr 3, 45–53.

Further reading

Armour, S.M., Baur, J.A., Hsieh, S.N., Land-Bracha, A., Thomas, S.M., Sinclair, D.A., 2009. Inhibition of mammalian S6 kinase by resveratrol suppresses autophagy. Aging (Albany NY) 1, 515.

Cardoso, S.M., Pereira, O.R., Seca, A.M., Pinto, D.C., Silva, A.M., 2015. Seaweeds as preventive agents for cardiovascular diseases: from nutrients to functional foods. Mar. Drugs. 13, 6838–6865.

Chuengsamarn, S., Rattanamongkolgul, S., Luechapudiporn, R., Phisalaphong, C., Jirawatnotai, S., 2012. Curcumin extract for prevention of type 2 diabetes. Diabetes Care 35, 2121–2127.

Francis, H., Stevenson, R., 2013. The longer-term impacts of Western diet on human cognition and the brain. Appetite. 63, 119–128.

Gil, A., Olza, J., Gil-Campos, M., Gomez-Llorente, C., Aguilera, C.M., 2011. Is adipose tissue metabolically different at different sites? Int. J. Pediatr. Obes 6, 13–20.

Irvine, M.A., Scholey, A., King, R., Gillings, R., Vauzour, D., Demichele, S.J., et al., 2018. The cognitive ageing, nutrition and neurogenesis (CANN) trial: design and progress. Alzheimer Dement. (NY). 4, 591–601.

Na, L.-X., Li, Y., Pan, H.-Z., Zhou, X.-L., Sun, D.-J., Meng, M., et al., 2013. Curcuminoids exert glucose-lowering effect in type 2 diabetes by decreasing serum free fatty acids: a double-blind, placebo-controlled trial. Mol. Nutr. Food Res 57, 1569–1577.

Nanjo, F., Hara, Y., Kikuchi, Y., 1994. Effects of tea polyphenols on blood rheology in rats fed a high-fat diet. In: Ho, C.T. (Ed.), Food Phytochemicals for Cancer Prevention: Teas, Spices and Herbs. American Chemical Society, Washington, DC, pp. 76–82.

Panahi, Y., Hosseini, M.S., Khalili, N., Naimi, E., Majeed, M., Sahebkar, A., 2015. Antioxidant and anti-inflammatory effects of curcuminoid-piperine combination in subjects with metabolic syndrome: a randomized controlled trial and an updated meta-analysis. Clin. Nutr. 34, 1101–1108.

Silva Figueiredo, P., Inada, A.C., Ribeiro Fernandes, M., Granja Arakaki, D., Freitas, K. C., et al., 2018. An overview of novel dietary supplements and food ingredients in patients with metabolic syndrome and non-alcoholic fatty liver disease. Molecules. 23, pii: E877.

Wei, C., Bajpai, R., Sharma, H., Heitmeier, M., Jain, A.D., Matulis, S.M., et al., 2017. Development of GLUT4-selective antagonists for multiple myeloma therapy. Eur. J. Med. Chem. 139, 573–586.

Wu, S., Ding, Y., Wu, F., Li, R., Hou, J., Mao, P., 2015. Omega-3 fatty acids intake and risks of dementia and Alzheimer's disease: a meta-analysis. Neurosci. Biobehav. Rev 48, 1–9.

Xiao, X., Shi, D., Liu, L., Wang, J., Xie, X., Kang, T., et al., 2011. Quercetin suppresses cyclooxygenase-2 expression and angiogenesis through inactivation of P300 signaling. PLoS One 6, e22934.

Xu, D., Hu, M.J., Wang, Y.Q., Cui, Y.L., 2019. Antioxidant activities of quercetin and its complexes for medicinal application. Molecules. 24, pii: E1123.

Yu, H., Zhang, L., Li, L., Zheng, C., Guo, L., Li, W., et al., 2010. Recent developments and future prospects of antimicrobial metabolites produced by endophytes. Microbiol. Res. 165, 437–449.

CHAPTER 10

Summery and perspective for future research on insulin resistance and insulin resistance–linked visceral and neurological disorders

Introduction

Chronic overconsumption of energy from diet in the absence of adequate physical activity leads to weight gain (obesity) and excess intra-abdominal fat. This factor strongly predisposes one to insulin resistance (Farooqui, 2013). Adipose tissue also releases adipocytokines (adipokines) such as leptin and adiponectin into the circulation. At the molecular level, leptin is known to modulate food consumption and energy expenditure via neuroendocrine signaling in the hypothalamus (Zhang et al., 1994). In contrast, adiponectin improves and promotes insulin sensitivity, and mice with adiponectin-deficiency are severely insulin resistant (Tateya et al., 2013). Thus adipokines are modulators of insulin sensitivity in liver, adipose tissue, and skeletal muscle.

As a component of the development of obesity and related metabolic dysfunction, immune cells such as macrophages accumulate in the adipose tissue, secreting proinflammatory cytokines that impact glucose and lipid metabolism (Tateya et al., 2013). The entry of excessive numbers of macrophages into adipose tissue results in increase production and release of proinflammatory cytokines [tumor necrosis factor-α (TNF-α) and interleukin-6 (IL-6)], which can reduce lipoprotein lipase enzyme activity, thus increasing blood lipid levels. In parallel, hormone-sensitive lipase activity can be enhanced in adipose tissue by TNF-α, which further increases the release of nonesterified fatty acids into the blood, while concomitantly reducing insulin-stimulated glucose uptake via impaired insulin signaling. These processes may lead to hyperglycemia and dyslipidemia, which are important parameters indicative of onset of insulin resistance,

Insulin Resistance as a Risk Factor in Visceral and Neurological Disorders
DOI: https://doi.org/10.1016/B978-0-12-819603-8.00010-9

obesity, and type 2 diabetes (Tateya et al., 2013). Insulin resistance and type 2 diabetes are also characterized by defects in uptake and oxidation of glucose, a decrease in glycogen synthesis, and, to a lesser extent, the ability to suppress lipid oxidation (Maciejczyk et al., 2019). At the clinical level, insulin resistance is induced by excessive energy intake, circulating high levels of free fatty acids (FFAs), diacylglycerol, triacylglycerol, and ceramides in nonadipose tissue, including skeletal muscle, liver, heart, β-cells, and central nervous system (CNS). Converging evidence suggests that the accumulation of abovementioned lipid mediators and defect in lipid metabolism may be major mechanisms for insulin resistance (Maciejczyk et al., 2019). Impaired insulin signaling not only affects insulin-stimulated glucose metabolism in skeletal muscle but also impairs actions of insulin in diverse tissues including, liver, adipose tissue, heart, and the vasculature (Ouwens et al., 2005). In these tissues, the production of proinflammatory cytokines (IL-1β, IL-6, and TNF-α) is also increased. This process activates several serine-threonine kinases (e.g., JNK, IKK) phosphorylating the serine residues of the insulin receptor substrate 1 (IRS-1), which in turn blocks the insulin signaling proteins (e.g., PI3-K–Akt, GSK3β, AMPK) (Maciejczyk et al., 2019). Additionally, the toxic effects of hyperglycemia not only involve the activation of the polyol/protein kinase C pathway, but also auto-oxidation of glucose, as well as accumulation of advanced glycation end product (AGE; Farooqui, 2013), whereas the excessive availability of FFAs inhibits glycolysis and impairs the functioning of the mitochondrial respiratory chain. All of these factors result in the activation of nuclear factor kappa B (NF-κB) pathway; and therefore, both lipotoxicity and glucotoxicity are an important source of free radicals and inflammation (Farooqui, 2013).

At the subcellular level, insulin resistance is accompanied by the decrease in the number of mitochondria, abnormal mitochondrial morphology, lower levels of mitochondrial oxidative enzymes, depletion of mitochondrial DNA, decrease in activity of respiratory chain complexes (Pérez-Carreras et al., 2003), and lower ATP synthesis both in vivo (Ritov et al., 2005; Petersen et al., 2004) and ex vivo in human muscle biopsies (Kim et al., 2000). Several mechanisms link obesity with insulin resistance–linked visceral and neurological disorders (Farooqui, 2013; Farooqui, 2017). One candidate mechanism is induction of oxidative stress. This process has been implicated in vascular complications of insulin resistance–linked visceral and neurological disorders. Levels of oxidative stress are markedly increased in obese subjects and in patients with insulin resistance–linked visceral and neurological disorders. Treating obese mice with antioxidant agents attenuates the development of type 2 diabetes in rodents. In addition, high levels of reactive oxygen species (ROS) and AGEs, which originate from simple carbohydrates (glucose and fructose) metabolism, may also promote the development of insulin resistance (Unoki and Yamagishi, 2008; Yamagishi et al., 2012). Furthermore, ROS and AGEs through their interactions with transcription factor NF-κβ can increase the expression of proinflammatory cytokines. These cytokines induce low-grade inflammation (Goldin et al., 2006).

Different tissues respond to insulin resistance in different ways (Martin and McGee, 2014). The liver is a vital organ, which not only controls metabolic homeostasis but also contributes to glucose uptake-storage-generation and processing. About one-third of consumed glucose is utilized by the liver and it is a key target for insulin action. Insulin controls lipogenesis (fatty acid and triglycerides biosynthesis) and restrains hepatic gluconeogenesis (glucose production) in the liver suggesting that insulin sensitivity is closely associated with rates of hepatic gluconeogenesis and lipid accumulation (Bechmann et al., 2012). Insulin resistance can also be induced by different environmental factors, including dietary habits. Thus vitamin D deficiency (i.e., hypovitaminosis D) is associated with increased insulin resistance, impaired insulin secretion, and poorly controlled glucose homeostasis (Leung, 2016).

In the liver, suppression of glucose output contributes to insulin resistant state, due to impaired suppression of gluconeogenesis and glycogenolysis (Hone and Pacini, 2008). In contrast, in the white adipose tissue, the impaired suppression of lipolysis contributes to the hyperlipidemia and insulin resistant states (Boden, 2011). Insulin resistance in skeletal muscle, fat, and liver, as well as cardiovascular tissues are central to the pathogenesis of the metabolic syndrome (MetS) and cardiovascular disease (CVD) (DeFronzo and Tripathy, 2009; Lark et al., 2012; Martin and McGee, 2014). It is well known that glucose and lipid metabolism are largely dependent on mitochondria to generate energy in cells. When oxidation of nutrients is inefficient, the ratio of ATP production/oxygen consumption is low. This process leads to the increased production of ROS particularly superoxide anions. ROS formation has maladaptive consequences, which not only increases the rate of mutagenesis, but also stimulates the expression of proinflammatory adipokines and cytokines. In the brain, the production of ROS from arachidonic acid (ARA) is accompanied by the generation of 4-hydroxynonenal, a nine carbon α,β-unsaturated aldehyde, which promotes mitochondrial dysfunction. This process contributes to metabolic and cardiovascular abnormalities and subsequent increases in CVD. Furthermore, interventions that improve mitochondrial dysfunction also improve insulin resistance. Collective evidence suggests that mitochondrial dysfunction is the central cause of insulin resistance and associated complications. At present specific biomarker for insulin resistance is not known. However, several potential biomarkers including adiponectin, RBP4, chemerin, A-FABP, FGF21, fetuin-A, myostatin, IL-6, and irisin have been used as potential biomarkers for insulin sensitivity (insulin resistance) (Fig. 10.1) (Suganami et al., 2012; Srikanth et al., 2011; Farooqui, 2013; Park et al., 2015).

In heart disease, insulin resistance not only promotes atherothrombotic conditions early in life and reduces fibrinolytic balance, but also impairs endothelial function, which may contribute to future episodes of cardiovascular events in adulthood (International Diabetes Federation, 2011). Atherosclerosis develops gradually, over the course of many years along with several silent dysfunctional changes in the endothelium (Martín-Timón et al., 2014). The increase in mortality during the development of atherosclerosis is related to the presence, duration, and intensity of risk factors in the lifespan. As mentioned above, inappropriate accumulations of lipids in muscle and liver due to abnormal fatty acid metabolism are the major cause of insulin resistance. The resulting dyslipidemia strongly

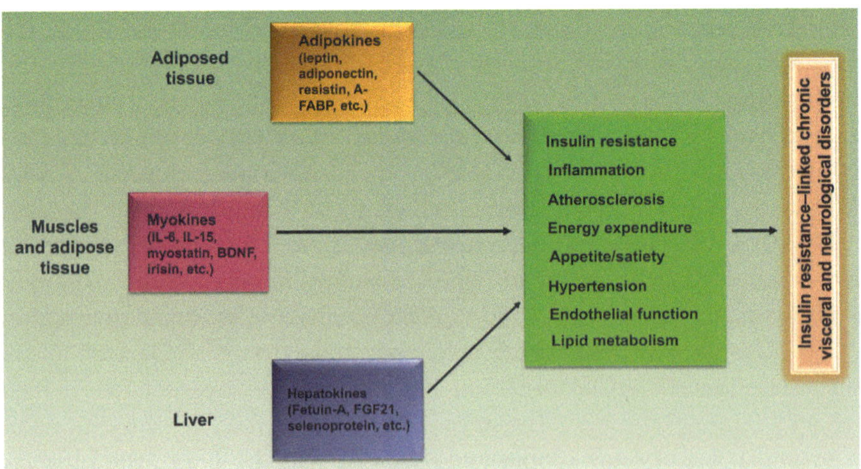

Figure 10.1 Potential biomarkers (adipokines, myokines, hepatokines, and cytokines) of insulin resistance and their involvement in the pathogenesis of insulin resistance.

correlates with increased CVD risk (Muntoni and Muntoni, 2011). Visceral fat is also infiltrated with inflammatory cells and secretes proinflammatory cytokines such as IL-6 and TNF-α (Espinola-Klein et al., 2011), which are implicated in the development of inflammation and insulin resistance. Thus cardiovascular health is closely associated with the pathogenesis of atherosclerosis (Lusis, 2000). Furthermore, in heart and brain tissues, oxidative stress and neuroinflammation are thought to be linked with impaired cellular function which may have direct effects on neuronal structure and integrity as well as neurodegeneration (Floyd and Hensley, 2002). Oxidative stress and inflammation therefore contribute to the pathogenesis of CVDs and a key intermediate mechanistic link between these two conditions and its risk factors may be represented by the NO pathway given the pleiotropic roles of NO in the regulation of vascular, metabolic, immune, and cognitive functions (Hirst and Robson, 2011).

Chronic insulin resistance: a common link between visceral and neurological disorders

Chronic insulin resistance is a common link among type 2 diabetes, MetS, sleep apnea, and CVDs. In addition, insulin resistance and MetS are also a major risk factor for neurological disorders such as stroke, Alzheimer's disease (AD), Parkinson's disease (PD), and other types of dementias (Fig. 10.2) (Farooqui, 2013). Several pathways and processes link peripheral metabolism with brain activities and functions. These pathways and processes involve transfer of proteins and metabolites from peripheral tissues to the brain and from brain to peripheral tissues across blood–brain barrier (BBB) along with induction of inflammation (Schwartz and Porte, 2005; Pavlov and Tracey, 2012). In addition, brain also regulates feeding behavior and energy intake and expenditure leading to metabolic homeostasis (Pavlov and Tracey, 2012; Kaelberer et al., 2018). Vagus nerve (the 10th cranial nerve), which contains fibers carrying ascending sensory signals to the brain and descending motor signals to the visceral organs, plays an important role in the abovementioned processes (Pavlov et al., 2018). The communication between gut microbiota and the brain is bidirectional. It involves the autonomic nervous system. The vagus nerve, because of its role in interoceptive awareness, is able to sense the microbiota-derived metabolites through its afferents, to transfer this gut information to the brain where it is integrated in the central autonomic

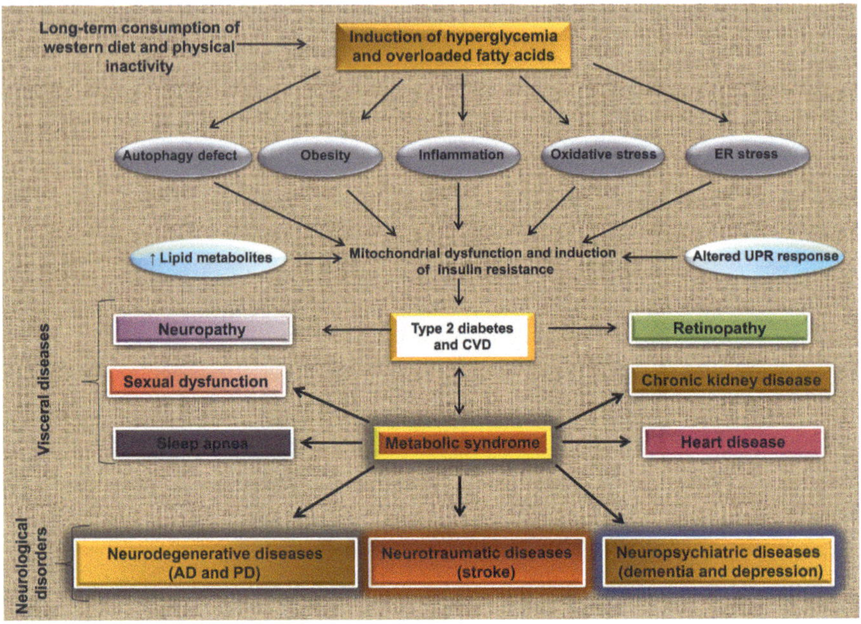

Figure 10.2 Hypothetical diagram showing involvement of western diet consumption, physical inactivity, and hyperglycemia in the pathogenesis of insulin resistance.

network, and then to generate an adapted or inappropriate response (Bonaz et al., 2018). A cholinergic antiinflammatory pathway has been described through vagus nerve fibers. This pathway is able to dampen peripheral inflammation and to decrease intestinal permeability leading to thus modulation of microbiota composition. Stress produces deleterious effects on the gastrointestinal tract by altering microbiota composition leading to gastrointestinal disorders such as irritable bowel syndrome (IBS) and inflammatory bowel disease (IBD). Both these disorders are characterized by a dysbiosis. A low vagal tone has been described in IBD and IBS patients thus favoring peripheral inflammation. Targeting the vagus nerve, for example, through vagus nerve stimulation may produce not only antiinflammatory effects, but also restore homeostasis in the microbiota–gut–brain axis (Bonaz et al., 2018).

It is well established that the immune system defends against infection and injury by inducing inflammation, which neutralizes invading pathogens and promotes tissue repair. If unresolved, however, inflammation can also be deleterious as in examples the onset of inflammatory and autoimmune disorders. Thus brain integrates biological functions and provides a nearly instantaneous homeostatic control mechanism by releasing of neurotransmitters and other regulatory molecules. Collective evidence suggests that brain communicates with the immune system not only by providing physiological homeostasis, but also by neutralizing threats against body tissues (Dantzer, 2018; Pavlov et al., 2018). Studies in rodents and humans have indicated that brain and peripheral tissues synthesize metabolites and molecular components, which induce, maintain, and facilitate the communication between the two systems. In addition, both immune cells and neurons express pattern recognition receptors, including Toll-like receptors, and cytokine receptors. These receptors act as molecular substrate for simultaneous modulation of immune and neuronal function through pathogen-associated molecular patterns, cytokines, and other immune molecules (de Lartigue et al., 2011; Steinberg et al., 2016; Park et al., 2014; Xu et al., 2015). The expression of neurotransmitter receptors, including acetylcholine receptors and adrenergic receptors has been identified on macrophages, dendritic cells, T cells, and other immune cells, facilitating neural regulation of immune responses (Kawashima et al., 2012, 2015). Immune cells also synthesize and release substances classically designated as neurotransmitters and neuromodulators, including acetylcholine, dopamine, and other catecholamines with a role in local immune regulation and in more complex neuroimmunomodulatory circuitry (Kawashima et al., 2012; Rosas-Ballina et al., 2008; Marino and Cosentino, 2013). In addition, there is a growing body of evidence that the entry of activated CD4 and CD8 T cells into the brain and dysfunctional "cross talk" between the brain and the peripheral immune system make a significant contribution to the genesis and/or exacerbation of pathology in neurodegenerative disease (Dá Mesquita et al., 2016). The purpose of these common metabolites is to integrate immune and neural communication and function through neural circuits triggered by a stimulus (e.g., infection or injury) and culminating in a response regulating immune function (e.g., inhibiting TNF-α or stimulating dendritic cells). In addition, vagus nerve also plays an important role in communication with immune system. Thus vagus nerve regulates inflammatory responses through inflammatory reflex (Pavlov et al., 2003). Furthermore, studies on electrical stimulation of vagus nerve have indicated that vagus nerve controls the release of proinflammatory cytokines and aberrant inflammation in many conditions (Chavan et al., 2017; Pavlov and Tracey, 2017). Studies on molecular mechanisms of the inflammatory reflex have indicated the involvement of alpha7 nicotinic acetylcholine receptor

(α7nAChR) signaling in its efferent arm. Cholinergic compounds, including α7nAChR agonists and centrally acting acetylcholinesterase inhibitors (AChE), have been reported to alleviate inflammation and metabolic derangements in insulin resistance, obesity, and MetS (Pavlov and Tracey, 2012). One of these drugs, the centrally acting AChE inhibitor galantamine is in clinical use for counteracting cognitive impairment in AD (Hampel et al., 2018). A recent clinical trial has revealed that galantamine produces antiinflammatory and beneficial metabolic effects in MetS patients (Consolim-Colombo et al., 2017) with MetS. Recent clinical studies have also shown the utility of bioelectronic stimulation of vagus nerve in rheumatoid arthritis and IBD—conditions characterized by immune and metabolic dysregulation (Bonaz et al., 2016; Koopman et al., 2016).

Development and onset of insulin resistance, type 2 diabetes, and MetS markedly effects brain function at the cellular and subcellular levels. As mentioned in earlier chapters that brain is a highly metabolic organ. It utilizes glucose-derived energy and oxygen for its optimal function. Among various subcellular structures, mitochondria are the major bioenergetic subcellular structure, which are highly susceptible to insulin resistance, type 2 diabetes, and MetS-mediated metabolic damage. Long-term consumption of high calorie diet (western diet), physical inactivity, and over-production of oxidants (ROS and reactive nitrogen species) contribute to the pathogenesis of many abovementioned insulin resistance–linked chronic visceral (type 2 diabetes, MetS, sleep apnea, and CVD) and neurological disorders (stroke, AD, and PD). Among chronic diseases, common risk factors include visceral obesity, dyslipidemia, low levels of high-density lipoprotein cholesterol, hypertriglyceridemia and increase in small dense low-density lipoprotein particle levels. These parameters promote and facilitate insulin resistance. Changes in adipose tissue mass and metabolism may link insulin resistance to visceral obesity, type 2 diabetes, MetS, heart disease, and neurological disorders (Lebovitz, 2006). Obesity contributes to type 2 diabetes by decreasing insulin sensitivity in adipose tissue, liver, and skeletal muscle, and subsequently impairing beta-cell function. So, the question arises: if insulin resistance is key in developing type 2 diabetes, what are the contributing factors in its development? Recently a study has described eight possible pathophysiological derangements that contribute to the development of type 2 diabetes (Defronzo, 2009), supporting the multifactorial nature of type 2 diabetes. In visceral tissues, induction of insulin resistance causes imbalance in glucose metabolism promoting chronic hyperglycemia, which in turn triggers oxidative stress and induces inflammatory responses leading to cell damage (Malin, 2019). Insulin resistance also alters systemic lipid metabolism which then leads to the development of dyslipidemia due to synthesis of high levels of plasma triglycerides, low levels of high-density lipoprotein, and the appearance of small dense low-density lipoproteins. As mentioned above, in heart disease, onset of insulin resistance in the myocardium generates damage by at least three different mechanisms: (1) signal transduction alteration, (2) impaired regulation of substrate metabolism, and (3) altered delivery of substrates to the myocardium. These characteristics, along with endothelial dysfunction, which can lead to not only aberrant insulin signaling, but also to formation of atherosclerotic plaques (Ormazabal et al., 2018). It should be noted that in type 2 diabetes, MetS, and heart disease, insulin resistance sets in long before any symptoms and sign of abovementioned diseases appear. It is important to recognize and treat individuals with insulin resistance as early as possible, because hyperinsulinemia remains undiagnosed for a long period, thereby increasing the risk of the development of other components of visceral diseases.

In neurological disorders (stroke, AD, and PD), onset of insulin resistance promotes the aggregation of Aβ and α-synuclein, induction of oxidative stress, along with onset of low-grade inflammation. These processes contribute to the induction of neurodegeneration. The underlying mechanism of insulin resistance in the brain involves the desensitization of insulin signaling (Fig. 10.3). In AD and PD patients, insulin signaling is desensitized in the brain, even without the induction of type 2 diabetes (Farooqui, 2017; Hölscher, 2019). As mentioned in Chapter 1, Insulin resistance and obesity, insulin is an important growth hormone, which through PtdIns 3K/Akt cascade is linked with multiple downstream pathways, including mTORC1, glycogen synthase kinase 3β (GSK-3β), and the FOXO family of transcription factors (Boucher et al., 2014). Many of these pathways have been shown to play pivotal roles in normal brain function such as regulation of neural cell growth, energy utilization, mitochondrial function and replacement, autophagy, oxidative stress management, synaptic plasticity, and cognitive function (Vandal et al., 2014; Farooqui, 2017). Induction of insulin desensitization due to the loss of above may enhance the risk of developing AD and PD in later life. In addition to brain, accumulation of Aβ, hyperphosphorylation of tau, and α-synuclein may also have roles in pancreatic β-cell dysfunction by reducing insulin sensitivity and glucose uptake by peripheral tissues such as liver,

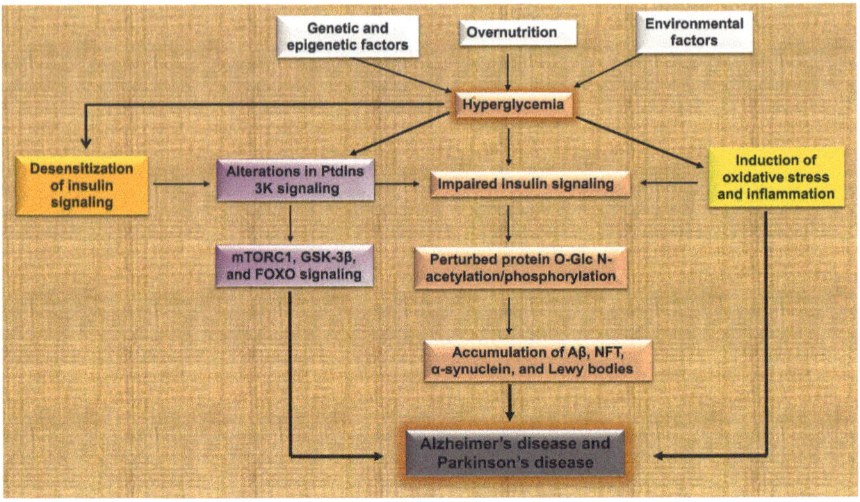

Figure 10.3 Effect of genetic factors, overnutrition, and environmental factors on insulin signaling.

skeletal muscle, and adipose tissue (Bharadwaj et al., 2017) supporting the view that Aβ and hyperphosphorylated tau may have a role in promoting type 2 diabetes. The accumulation of the islet amyloid polypeptide and amylin within islet β-cells is a major pathological feature of the pancreas in patients with chronic type 2 diabetes. Cosecreted with insulin, amylin accumulates over time. This process may contribute to Aβ-cell toxicity, ultimately leading to reduced insulin secretion and onset of overt (insulin dependent) type 2 diabetes (Bharadwaj et al., 2017). Recent evidence also suggests that accumulation of amylin in the brain of AD patients may exacerbate the neurodegenerative process (Bharadwaj et al., 2017). It should be noted that in the brain diminished responsiveness to insulin has different consequences than in peripheral tissues. Recent published data show that peripheral insulin and glucose tolerance were comparable between aged wild type amyloid precursor protein/presenilin (APP/PS) mice (an animal model of AD), while levels of serine phosphorylated IRS-1 are increased in the brain of APP/PS1 mice (Denver et al., 2018). This provides some support to the view that insulin resistance in brain may indeed exist as a distinct phenomenon, separate from insulin signaling in the visceral organs and one that can distinguish AD-related neurodegeneration from normal aging. In brain, insulin reverses the high-fat diet–induced increase in Aβ and improves memory in an animal model of AD (Vandal et al., 2014). Furthermore, the accumulation of soluble Aβ oligomers, known as amyloid beta-derived diffusible ligands also contributes to insulin resistance in AD by modifying synapse conformation. This altered shape conformation is responsible for reduced affinity of synaptic insulin receptor for its ligand (Mittal and Katare, 2016).

Long-term consumption of high calorie diet, which is low in eicosapentaenoic acid (EPA), and docosahexaenoic acid (DHA) and rich in ARA, leads not only to an increase in the n−6:n−3 ratio, but also to induction of low-grade inflammation, oxidative stress, and hyperactivation of stress-sensitive Ser/Thr kinases, such as Jun kinase (JNK) and IkB kinase (IKKβ). These parameters in turn inhibit the insulin receptor/IRS-1 axis (Farooqui, 2013). The etiopathogenesis of insulin resistance involves a multitude of metabolic pathways including production of ROS and lipid accumulation due to mitochondrial dysfunction. In addition, mitochondria are functionally and structurally linked to ER (Fig. 10.4). These organelles play important roles in cellular homeostasis and control of protein quality. Unfolded protein response (UPR) is a cellular response to ER stress, which

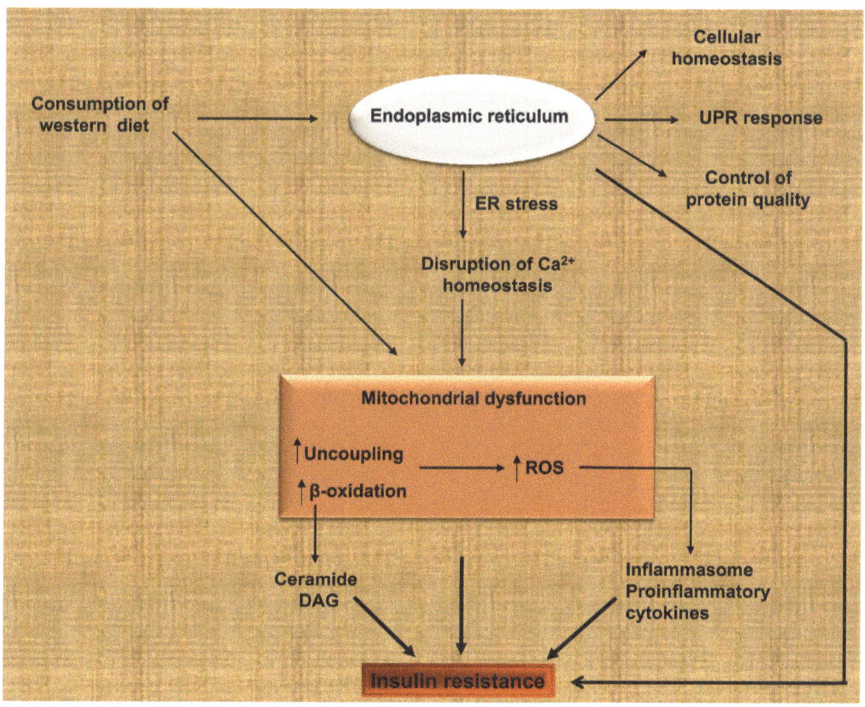

Figure 10.4 Effect of western diet on endoplasmic reticulum (ER) stress-mediated insulin resistance.

promotes cell survival. Severe or prolonged ER stress activates apoptosis signaling to trigger cell death. In mammals, the UPR is initiated by three major ER stress sensors, including inositol-requiring transmembrane kinase 1, double-stranded RNA-activated protein kinase-like ER kinase and activating transcription factor 6 (Xiang et al., 2017). UPR dysfunction plays an important role in the pathogenesis of neurodegenerative diseases such as AD, PD, and HD. These diseases are characterized by the accumulation and aggregation of misfolded proteins. During overnutrition, ER stress activates the UPR, which in turn activates the principal inflammatory pathways that impair insulin signaling leading to insulin resistance (Fig. 10.4) (Lepretti et al., 2018).

The composition of diet regulates the population of microbiota in the intestine. Many factors influence gut microbiota (Fig. 10.5). Thus, western diet, which is enriched in refined sugar, high fat, and low in fiber, increases the abundance of bile-tolerant microorganisms (*Alistipes*, *Bilophila*, and *Bacteroides*) and decreases the levels of Firmicutes. These conditions promote obesity along with changes in metabolic and immune biomarkers (Cani et al., 2009). Consumption of western diet promotes the development of inflammation not only in the hypothalamus, but also

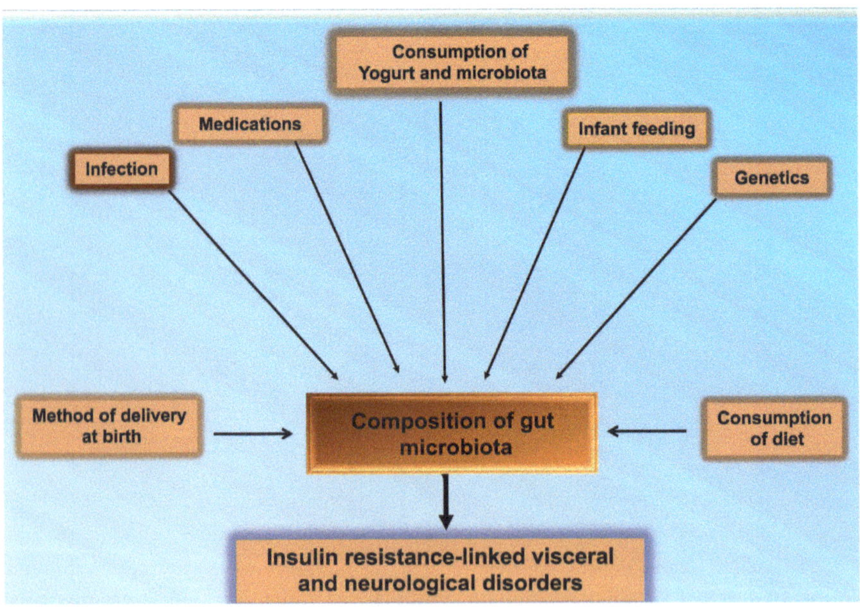

Figure 10.5 Factors affecting composition of microbiota in intestine.

in the peripheral tissues including the liver, adipose tissue, skeletal muscle, and intestine (Fig. 10.6) (Sanmiguel et al., 2015; Guillemot-Legris et al., 2016). In addition, consumption of western diet not only increases insulin resistance and obesity, but also decreases brain-derived neurotrophic factor (BDNF) in hippocampus. In contrast, low calorie plant-based diet increases that population of *Roseburia*, *Eubacterium rectale*, and *Ruminococcus bromii* (Martínez et al., 2015). Long-term consumption of Mediterranean diet retards development of obesity and insulin resistance (Farooqui, 2013). Dietary fiber induces prebiotic effects that may restore imbalances in the gut microbiota; however, no clinical trials have been reported in patients with metabolic diseases. In a study, six obese subjects with type 2 diabetes and/or hypertension were fed a strict vegetarian diet for 1 month. Determination of blood biomarkers such as glucose and lipid metabolites, fecal microbiota using 454-pyrosequencing of 16S ribosomal RNA genes, fecal lipocalin-2, and short-chain fatty acids indicates that consumption of vegetarian diet not only reduces body weight, concentrations of triglycerides, total cholesterol, low-density lipoprotein cholesterol, and hemoglobin A1c, but also improves fasting glucose and postprandial glucose levels. Plant-based diet reduces the *Firmicutes*-to-*Bacteroidetes* ratio in the gut microbiota, but did not alter enterotypes. This study underscores the benefits of dietary fiber for improving the risk factors of metabolic diseases and shows that increase in fiber intake reduces gut inflammation by changing the gut microbiota (Kim et al., 2013). Vegetarian diet increases that population of *Roseburia*, *E. rectale*, and *R. bromii*. These microorganisms stimulate nuclear receptors enzymes leading not only downregulation of oxidative metabolism, but also downregulation of the synthesis of proinflammatory molecules, and restoration of healthy symbiotic gut microbiota (Riccio and Rossano, 2015). In addition, the consumption of plant-based diet also increases hippocampal and prefrontal BDNF levels, proliferative cells, and neuron numbers in dentate gyrus, thus positively affecting spatial memory in adulthood (Kaptan et al., 2015). Moreover, the consumption of plant-based diet decreases serum glucose levels and level of lipid peroxidation in serum and hippocampus, suggesting the importance of nutrition in cognitive function in adulthood and old age (Kaptan et al., 2015). Recent studies have indicated the existence of a cross talk between the host and the commensal microbiota within the gut. During cross talk, nutrients play an important role either by directly interacting with the host via the epithelium or the intestinal immune system or by indirectly modulating the composition of the commensal

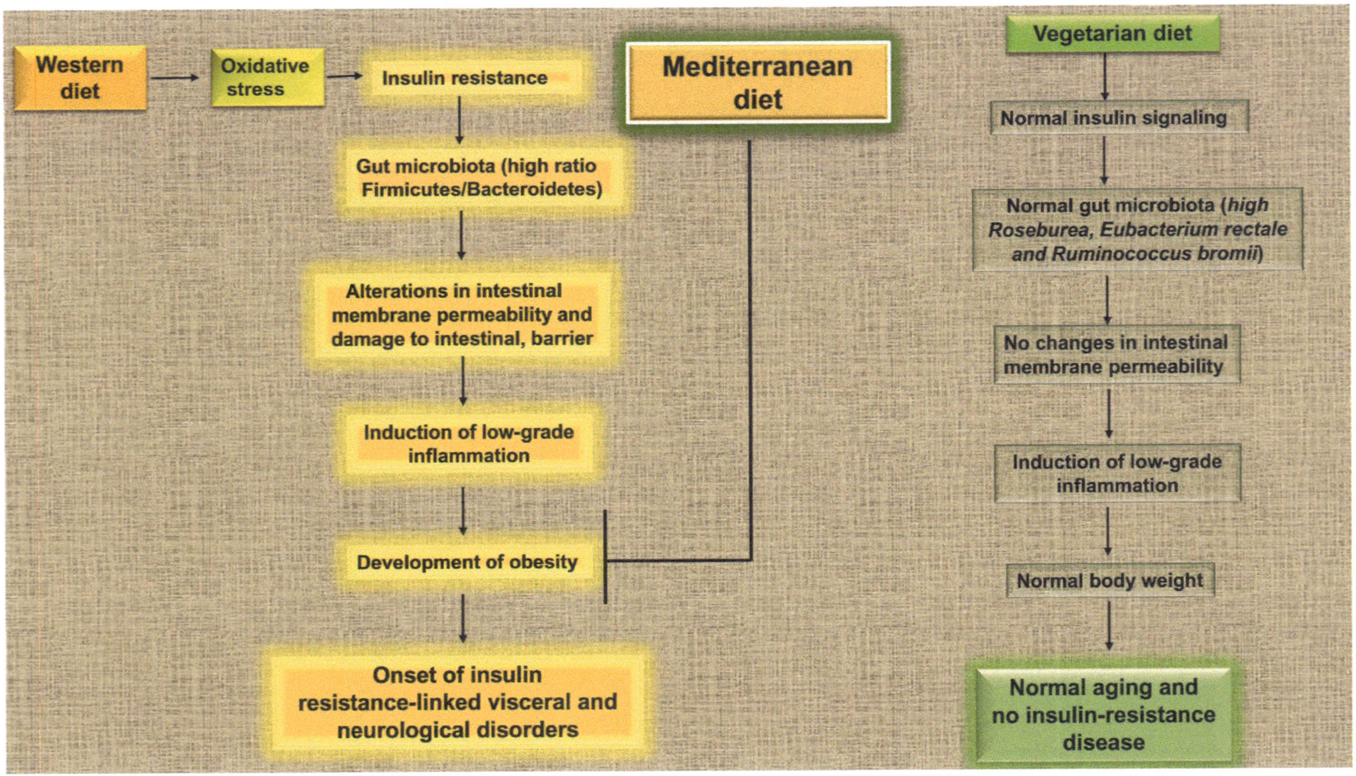

Figure 10.6 Effect of genetic factors, western diet and vegetarian diet on insulin resistance–linked visceral and neurological disorders.

microbiota which in turn will interact with the immune system, and vice versa.

Dysbiosis is a state in which the homeostasis of the gut microbiome is disrupted, often leading to health problems. One of the causes of dysbiosis is diet, and it is becoming increasingly evident that composition of diet may change the gut microbiota and contribute to insulin resistance, obesity, type 2 diabetes, and MetS (Matsuzawa-Nagata et al., 2008; Wu et al., 2011; Muegge et al., 2011). Over 80% of patients with type 2 diabetes and MetS in the Western world are overweight. Insulin resistance, obesity, type 2 diabetes, and MetS are characterized by an altered gut microbiota, inflammation, and gut barrier disruption (Gómez-Ambrosi et al., 2011; Everard et al., 2013; Zhang and Zhang, 2013). The gut microbiome interacts with the brain and regulates many brain functions through the maintenance of "gut–brain axis" homeostasis (Hamilton and Raybould, 2016), modulation of bidirectional neurohumoral communication, production of neuroactive molecules, and regulation of the circulating levels of some cytokines. The other mechanism of interaction between microbiome and brain is through the involvement of BBBs. Microbiota and their release factors may enter into the systemic circulation from the gut. Once in the blood, the microbiota-derived factors can alter peripheral immune cells. This process may facilitate interactions with the BBB and ultimately with other elements of the neurovascular unit. Based on these studies, it is suggested that gut microbiota-derived factors can cross the BBB not only by altering BBB integrity, changing BBB transport rates, but also by inducing the release of neuroimmune substances from the BBB cells (Logsdon et al., 2018). Metabolic products produced by the microbiota, such as short-chain fatty acids, can cross the BBB to affect brain function (Logsdon et al., 2018).

Direction for future research

As mentioned above, insulin resistance contributes to the pathophysiology of insulin-linked visceral diseases (type 2 diabetes obesity, MetS, sleep apnea, and heart disease) and neurological disorders (stroke, AD, PD, and various types of dementia). Therefore quantifying insulin sensitivity/resistance in abovementioned diseases in humans and animal models of these diseases is of great importance not only in the epidemiological, clinical, and basic science investigations, but also in clinical practice (Muniyappa et al., 2008; Khan et al., 2019). To this end, many methods available for estimation of insulin resistance which range from complex techniques down to simple indices. For all methods of assessing insulin resistance, it is essential that their validity and reliability is established before using them as investigations. The reference techniques of hyperinsulinaemic euglycemic clamp and its alternative the frequently sampled intravenous glucose tolerance test are the most reliable methods available for estimating insulin resistance (Borai et al., 2007). However, many simple methods, from which indices can be derived, have been assessed and validated, for example, homeostasis model assessment (HOMA), quantitative insulin sensitivity check index (QUICKI) (Borai et al., 2011). In addition, it has been observed that measurement of individual biochemical protein markers such as insulin-like growth factor binding protein-1 can provide useful information about the status of insulin resistance (Borai et al., 2009; Maddux et al., 2006). Among these tests, the quantitative determination of insulin sensitivity by QUICKI has been validated extensively against the reference standard glucose clamp method. QUICKI is a simple, robust, accurate, reproducible method that appropriately predicts changes in insulin sensitivity after therapeutic interventions as well as the onset of insulin-linked visceral and neurological disorders (Muniyappa et al., 2008; Gutch et al., 2015; Khan et al., 2019). In contrast, HOMA-IR has been used to assess longitudinal changes in insulin resistance in persons with type 2 diabetes of various ethnic groups in order to examine the natural history of diabetes and to assess the effects of treatment (Wallace et al., 2004). It can also be utilized in nondiabetic populations as it allows (1) comparisons of insulin sensitivity among persons with abnormal glucose tolerance and (2) the longitudinal assessment of persons who later develop abnormal glucose tolerance (Wallace et al., 2004). HOMA-IR has been validated against the hyperinsulinaemic euglycaemic clamp in multiple studies across several populations ($r = 0.5-0.8$) (Pacini and Mari, 2003). Other indices that utilize fasting glucose and insulin data (e.g., QUICKI have questionable superiority to HOMA) (Duncan et al., 2001; Katsuki et al., 2002).

In addition to above tests, one should also have a clear and full understanding of molecular aspects of glucose and lipid metabolism and homeostasis in insulin-linked chronic visceral and neurological disorders in vivo. Metabolomics is a method for the quantification of small-molecule metabolites in body fluids and body tissues. It aims to identify and quantify metabolites in the sample by either using mass spectrometry or nuclear magnetic resonance spectroscopy (Roberts et al., 2014; Savolainen et al., 2017;

Del Coco et al., 2019). Studies on fasting plasma samples from 399 nondiabetic healthy subjects have indicated the presence of α-hydroxybutyrate, an early marker for both insulin resistance and impaired glucose regulation (Gall et al., 2010). Among the 485 candidate biomarkers identified, plasma α-hydroxybutyrate levels were inversely related to insulin sensitivity and this association was independent of age, sex, and BMI. It is suggested that underlying biochemical mechanisms may involve increased lipid oxidation and oxidative stress. In another study, other metabolites such as linoleoyl-glycerophosphocholine, glycine, and creatine were also determined and their levels were highly correlated with insulin sensitivity (Cobb et al., 2013). Levels of branched chain amino acids (BCAAs) were significantly increase in obese compared to lean subjects and a BCAA-based index correlated with HOMA (Newgard et al., 2009). The elevation of levels of BCAA in subjects with impaired fasting glucose and type 2 diabetes have been confirmed in subsequent studies (Menni et al., 2013). Collective evidence suggests that metabolomic studies are very promising since they can measure hundreds of metabolites in a very small sample. However, the pricing, technology, and access precludes the use of this technology in clinical practice. Further studies using this approach are necessary in larger more heterogeneous cohorts to replicate and validate surrogate insulin resistance markers derived through metabolomics.

Conclusion

It is well established that chronic overconsumption of foods rich in carbohydrates and various saturated lipids affects insulin secretion and onset of obesity and type 2 diabetes. Both these conditions contribute to an increased prevalence of dementia and cognitive decline along with insulin resistance, which is proposed to play a critical role in linking metabolic disorders with CNS impairment. As mentioned in earlier chapters, aging, hypertension, type 2 diabetes, hypoxia/obstructive sleep apnea, insulin resistance, and obesity synergistically promote diverse pathological mechanisms including cerebral hypoperfusion, glucose hypometabolism, impaired insulin secretion, and signaling leading to insulin resistance. These risk factors trigger aberrant redox regulation, neuroinflammation, and oxidative-nitrosative stress that in turn decrease nitric oxide and enhance endothelin, Aβ deposition, cerebral amyloid angiopathy, and BBB disruption. Proinflammatory cytokines, endothelin-1, and oxidative-nitrosative stress trigger several pathological feedforward and feedback loops. These upstream factors persist in the brain for decades, upregulating Aβ and tau, before the cognitive decline. The hyperinsulinemic euglycemic glucose clamp is the gold standard method for the determining insulin sensitivity, but is impractical as a routine test because it is expensive, laborious, and time-consuming. A number of surrogate indices have therefore been employed to simplify and improve the determination of insulin resistance (Singh and Saxena, 2010). Assessment of insulin resistance in type 2 diabetes, MetS, AD, and PD may aid in studying the pathogenesis, etiology, and consequences of above disorders. The methods, which reduce the risk of complications of insulin resistance (regular physical activity and adherence to a low fat and carbohydrate diet), show beneficial effects in reducing risk the risk of type 2 diabetes, MetS, AD, and PD.

References

Bechmann, L.P., Hannivoort, R.A., Gerken, G., Hotamisligil, G.S., Trauner, M., Canbay, A., 2012. The interaction of hepatic lipid and glucose metabolism in liver diseases. J. Hepatol. 56, 952–964.

Bharadwaj, P., Wijesekara, N., Liyanapathirana, M., Newsholme, P., Ittner, L., Fraser, P., et al., 2017. The link between type 2 diabetes and neurodegeneration: roles for amyloid-β, amylin, and tau proteins. J. Alzheimer Dis. 59, 421–432.

Boden, G., 2011. Obesity, insulin resistance and free fatty acids. Curr. Opin. Endocrinol. Diabetes. Obes. 18, 139–143.

Bonaz, B., Sinniger, V., Hoffmann, D., Clarencon, D., Mathieu, N., Dantzer, C., et al., 2016. Chronic vagus nerve stimulation in Crohn's disease: a 6-month follow-up pilot study. Neurogastroenterol. Motil. 28, 948–953.

Bonaz, B., Bazin, T., Pellissier, S., 2018. The vagus nerve at the interface of the microbiota-gut-brain axis. Front. Neurosci. 12, 49.

Borai, A., Livingstone, C., Ferns, G., 2007. The biochemical assessment of insulin resistance. Ann. Clin. Biochem. 44, 324–342.

Borai, A., Livingstone, C., Zarif, H., Ferns, G., 2009. Serum insulin-like growth factor binding protein-1: an improvement over other simple indices of insulin sensitivity in the assessment of subjects with normal glucose tolerance. Ann. Clin. Biochem. 46 (Pt 2), 109–113.

Borai, A., Livingstone, C., Kaddam, I., Ferns, G., 2011. Selection of the appropriate method for the assessment of insulin resistance. BMC Med. Res. Methodol. 11, 158.

Boucher, J., Kleinridders, A., Kahn, C.R., 2014. Insulin receptor signaling in normal and insulin-resistant states. Cold Spring Harb. Perspect. Biol. 6, a009191.

Cani, P.D., Possemiers, S., Van de Wiele, T., Guiot, Y., Everard, A., Rottier, O., et al., 2009. Changes in gut microbiota control inflammation in obese mice through a mechanism involving GLP-2-driven improvement of gut permeability. Gut 58, 1091–1103.

Chavan, S.S., Pavlov, V.A., Tracey, K.J., 2017. Mechanisms and therapeutic relevance of neuro-immune communication. Immunity 46, 927–942.

Cobb, J., Gall, W., Adam, P.-K., Nakhle, P., et al., 2013. A novel fasting blood test for insulin resistance and prediabetes. J. Diabetes Sci. Technol. 7, 100–110.

Consolim-Colombo, F.M., Sangaleti, C.T., Costa, F.O., Morais, T.L., Lopes, H.F., Motta, J.M., et al., 2017. Galantamine alleviates inflammation and insulin resistance in patients with metabolic syndrome in a randomized trial. JCI Insight 2, 93340.

Dá Mesquita, S., Ferreira, A.C., Sousa, J.C., Correia-Neves, M., Sousa, N., Marques, F., 2016. Insights on the pathophysiology of Alzheimer's disease: The crosstalk between amyloid pathology, neuroinflammation and the peripheral immune system. Neurosci. Biobehav. Rev. 68, 547–562.

Dantzer, R., 2018. Neuroimmune interactions: from the brain to the immune system and vice versa. Physiol. Rev. 98, 477–504.

Defronzo, R.A., 2009. Banting lecture. From the triumvirate to the ominous octet: a new paradigm for the treatment of type 2 diabetes mellitus. Diabetes 58, 773–795.

DeFronzo, R.A., Tripathy, D., 2009. Skeletal muscle insulin resistance is the primary defect in type 2 diabetes. Diabetes Care 32, 157–163.

de Lartigue, G., Barbier, D.L.S., Espero, E., Lee, J., Raybould, H.E., 2011. Diet-induced obesity leads to the development of leptin resistance in vagal afferent neurons. Am. J. Physiol. Endocrinol. Metab. 301, E187–E195.

Del Coco, L., Vergara, D., De Matteis, S., Mensà, E., Sabbatinelli, J., Prattichizzo, F., et al., 2019. NMR-based metabolomic approach tracks potential serum biomarkers of disease progression in patients with type 2 diabetes mellitus. J. Clin. Med. 8.

Denver, P., English, A., McClean, P.L., 2018. Inflammation, insulin signaling and cognitive function in aged APP/PS1 mice. Brain Behav. Immun. 70, 423–434.

Duncan, G.E., Hutson, A.D., Stacpoole, P.W., 2001. QUICKI does not accurately reflect changes in insulin sensitivity with exercise training. J. Clin. Endocrinol. Metab. 86, 4115–4119.

Espinola-Klein, C., et al., 2011. Inflammatory markers and cardiovascular risk in the metabolic syndrome. Front. Biosci. 16, 1663–1674.

Everard, A., Belzer, C., Geurts, L., Ouwerkerk, J.P., Druart, C., Bindels, L.B., et al., 2013. Cross-talk between Akkermansia muciniphila and intestinal epithelium controls diet-induced obesity. Proc. Natl. Acad. Sci. U.S.A. 110, 9066–9071.

Farooqui, A.A., 2013. Metabolic Syndrome: An Important Risk Factor for Stroke, Alzheimer Disease, and Depression. Springer International Publishing, Switzerland.

Farooqui, A.A., 2017. Neurochemical Aspects of Alzheimer's Disease. Academic Press, An Imprint of Elsevier, San Diego, CA.

Floyd, R.A., Hensley, K., 2002. Oxidative stress in brain aging. Neurobiol. Aging. 23, 795–807.

Gall, W.E., Beebe, K., Lawton, K.A., Adam, K.-P., 2010. Alpha-hydroxybutryrate is an early biomarker of insulin resistance and glucose intolerance in a nondiabetic population. PLoS One 5, e10883.

Goldin, A., Beckman, J.A., Schmidt, A.M., Creager, M.A., 2006. Advanced glycation end products: sparking the development of diabetic vascular injury. Circulation 114, 597–605.

Gómez-Ambrosi, J., Silva, C., Galofré, J.C., Escalada, J., Santos, S., Gil, M.J., et al., 2011. Body adiposity and type 2 diabetes: increased risk with a high body fat percentage even having a normal BMI. Obesity (Silver Spring) 19, 1439–1444.

Guillemot-Legris, O., Masquelier, J., Everard, A., Cani, P.D., Alhouayek, M., Muccioli, G.G., 2016. High-fat diet feeding differentially affects the development of inflammation in the central nervous system. J. Neuroinflamm. 13, 206.

Gutch, M., Kumar, S., Razi, S.M., Gupta, K.K., Gupta, A., 2015. Assessment of insulin sensitivity/resistance. Indian J. Endocrinol. Metab. 19, 160–164.

Hamilton, M.K., Raybould, H.E., 2016. Bugs, guts and brains, and the regulation of food intake and body weight. Int. J. Obes. Suppl. 6, S8–S14.

Hampel, H., Mesulam, M.M., Cuello, A.C., Farlow, M.R., Giacobini, E., Grossberg, G.T., et al., 2018. The cholinergic system in the pathophysiology and treatment of Alzheimer's disease. Brain 141, 1917–1933.

Hirst, D.G., Robson, T., 2011. Nitric oxide physiology and pathology. Methods Mol. Biol. (Clifton, NJ) 704, 1–13.

Hölscher, C., 2019. Insulin Signaling Impairment in the Brain as a Risk Factor in Alzheimer's Disease. Front Aging Neurosci. 11, 88.

Hone, P.D., Pacini, G., 2008. Hepatic dysfunction and insulin insensitivity in type 2 diabetes mellitus: a critical target for insulin-sensitizing agents. Diabetes Obes. Metab. 10, 699–718.

International Diabetes Federation, 2011. The Global Burden Prevalence and Projections 2011 and 2030. Available from: <http://www.idf.org/diabetesatlas/5e/the-global-burden>.

Kaelberer, M.M., Buchanan, K.L., Klein, M.E., Barth, B.B., Montoya, M.M., Shen, X., et al., 2018. A gut-brain neural circuit for nutrient sensory transduction. Science 361, eaat5236.

Kaptan, Z., Akgün-Dar, K., Kapucu, A., Dedeakayoğulları, H., Batu, Ş., et al., 2015. Long term consequences on spatial learning-memory of low-calorie diet during adolescence in female rats; hippocampal and prefrontal cortex BDNF level, expression of NeuN and cell proliferation in dentate gyrus. Brain Res. 1618, 194–204.

Katsuki, A., Sumida, Y., Gabazza, E.C., Murashima, S., Urakawa, H., Morioka, K., et al., 2002. QUICKI is useful for following improvements in insulin sensitivity after therapy in patients with type 2 diabetes mellitus. J. Clin. Endocrinol. Metab. 87, 2906–2908.

Kawashima, K., Fujii, T., Moriwaki, Y., Misawa, H., 2012. Critical roles of acetylcholine and the muscarinic and nicotinic acetylcholine receptors in the regulation of immune function. Life Sci. 91, 1027–1032.

Kawashima, K., Fujii, T., Moriwaki, Y., Misawa, H., Horiguchi, K., 2015. Non-neuronal cholinergic system in regulation of immune function with a focus on alpha7 nAChRs. Int. Immunopharmacol. 29, 127–134.

Khan, S.H., Khan, A.N., Chaudhry, N., Anwar, R., Fazal, N., Tariq, M., 2019. Comparison of various steady state surrogate insulin resistance indices in diagnosing metabolic syndrome. Diabetol. Metab. Syndr. 11, 44.

Kim, J.Y., Hickner, R.C., Cortright, R.L., Dohm, G.L., Houmard, J.A., 2000. Lipid oxidation is reduced in obese human skeletal muscle. Am. J. Physiol. Endocrinol. Metab. 279, E1039–E1044.

Kim, M.-S., Hwang, S.-S., Park, E.-J., Bae, J.-W., 2013. Strict vegetarian diet improves the risk factors associated with metabolic diseases by modulating gut microbiota and reducing intestinal inflammation. Environ. Microb. Rep 5, 25–38.

Koopman, F.A., Chavan, S.S., Miljko, S., Grazio, S., Sokolovic, S., Schuurman, P.R., et al., 2016. Vagus nerve stimulation inhibits cytokine production and attenuates disease severity in rheumatoid arthritis. Proc. Natl. Acad. Sci. U.S.A. 113, 8284–8289.

Lark, D.S., Fisher-Wellman, K.H., Neufer, P.D., 2012. High-fat load: mechanism(s) of insulin resistance in skeletal muscle. Int. J. Obes. 2, 31–36.

Lebovitz, H.E., 2006. Insulin resistance—a common link between type 2 diabetes and cardiovascular disease. Diabetes Obes. Metab. 8, 237–249.

Lepretti, M., Martucciello, S., Burgos Aceves, M.A., Putti, R., Lionetti, L., 2018. Omega-3 Fatty Acids and Insulin Resistance: Focus on the Regulation of Mitochondria and Endoplasmic Reticulum Stress. Nutrients 10, E350.

Leung, P.S., 2016. The potential protective action of vitamin D in hepatic insulin resistance and pancreatic islet dysfunction in type 2 diabetes mellitus. Nutrients 8, pii: E147.

Logsdon, A.F., Erickson, M.A., Rhea, E.M., Salameh, T.S., Banks, W.A., 2018. Gut reactions: how the blood-brain barrier connects the microbiome and the brain. Exp. Biol. Med. (Maywood) 243, 159–165.
Lusis, A.J., 2000. Atherosclerosis. Nature 40, 233–241.
Maciejczyk, M., Żebrowska, E., Chabowski, A., 2019. Insulin resistance and oxidative stress in the brain: what's new? Int. J. Mol. Sci. 20. pii: E874.
Maddux, B.A., Chan, A., Mandarino, L.J., Goldfine, I.D., 2006. IGF-binding protein-1 levels are related to insulin-mediated glucose disposal and are a potential serum marker of insulin resistance. Diabetes Care 29, 1535–1537.
Malin, S.K., 2019. Insulin at the crossroads of metabolic, neurologic, and vasculature disease. Exerc. Sport Sci. Rev. 47, 64–65.
Marino, F., Cosentino, M., 2013. Adrenergic modulation of immune cells: an update. Amino Acids 45, 55–71.
Martin, S.D., McGee, S.L., 2014. The role of mitochondria in the aetiology of insulin resistance and type 2 diabetes. Biochim. Biophys. Acta 1840, 1303–1312.
Martínez, I., Stegen, J.C., Maldonado-Gómez, M.X., Eren, A.M., Siba, P.M., Greenhill, A.R., et al., 2015. The gut microbiota of rural Papua New Guineans: Composition, diversity patterns, and ecological processes. Cell Rep. 11, 527–538.
Martín-Timón, I., Sevillano-Collantes, C., Segura-Galindo, A., Del Cañizo-Gómez, F.J., 2014. Type 2 diabetes and cardiovascular disease: have all risk factors the same strength? World J. Diabetes 5, 444–470.
Matsuzawa-Nagata, N., Takamura, T., Ando, H., Nakamura, S., Kurita, S., et al., 2008. Increased oxidative stress precedes the onset of high-fat diet–induced insulin resistance and obesity. Metabolism 57, 1071–1077.
Menni, C., Fauman, E., Irte, E., Perry, J.R., et al., 2013. Biomarkers for diabetes and impaired fasting glucose using a non-targeted metabolomics approach. Diabetes 62, 4270–4276.
Mittal, K., Katare, D.P., 2016. Shared links between type 2 diabetes mellitus and Alzheimer's disease: a review. Diabetes Metab. Synd. 10 (2 Suppl. 1), S144–S149.
Muegge, B.D., Kuczynski, J., Knights, D., Clemente, J.C., Gonzalez, A., Fontana, L., et al., 2011. Diet drives convergence in gut microbiome functions across mammalian phylogeny and within humans. Science 332, 970–974.
Muniyappa, R., Lee, S., Chen, H., Quon, M.J., 2008. Current approaches for assessing insulin sensitivity and resistance in vivo: advantages, limitations, and appropriate usage. Am. J. Physiol. Endocrinol. Metab. 294, E15–E26.
Muntoni, S., Muntoni, S., 2011. Insulin resistance: pathophysiology and rationale for treatment. Ann. Nutr. Metab. 58, 25–36.
Newgard, C.B., An, J., Bain, J.R., Muehlbauer, M.J., et al., 2009. A branched chain amino acid related metabolic signature that differentiates obese and lean humans and contributes to insulin resistance. Cell. Metab. 9, 311–326.
Ormazabal, V., Nair, S., Elfeky, O., Aguayo, C., Salomon, C., Zuñiga, F.A., 2018. Association between insulin resistance and the development of cardiovascular disease. Cardiovasc. Diabetol. 17, 122.
Ouwens, D.M., Boer, C., Fodor, M., de Galan, P., Heine, R.J., Maassen, J.A., et al., 2005. Cardiac dysfunction induced by high-fat diet is associated with altered myocardial insulin signalling in rats. Diabetologia 48, 1229–1237.
Pacini, G., Mari, A., 2003. Methods for clinical assessment of insulin sensitivity and beta-cell function. Best. Pract. Res. Clin. Endocrinol. Metab. 17, 305–322.
Park, C.K., Xu, Z.Z., Berta, T., Han, Q., Chen, G., Liu, X.J., et al., 2014. Extracellular microRNAs activate nociceptor neurons to elicit pain via TLR7 and TRPA1. Neuron 82, 47–54.
Park, S.E., Park, C.Y., Sweeney, G., 2015. Biomarkers of insulin sensitivity and insulin resistance: past, present and future. Crit. Rev. Clin. Lab. Sci. 52, 180–190.

Pavlov, V.A., Tracey, K.J., 2012. The vagus nerve and the inflammatory reflex—linking immunity and metabolism. Nat. Rev. Endocrinol. 8, 743—754.
Pavlov, V.A., Tracey, K.J., 2017. Neural regulation of immunity: molecular mechanisms and clinical translation. Nat. Neurosci. 20, 156—166.
Pavlov, V.A., Wang, H., Czura, C.J., Friedman, S.G., Tracey, K.J., 2003. The cholinergic anti-inflammatory pathway: a missing link in neuroimmunomodulation. Mol. Med. 9, 125—134.
Pavlov, V.A., Chavan, S.S., Tracey, K.J., 2018. Molecular and functional neuroscience in immunity. Annu. Rev. Immunol. 36, 783—812.
Pérez-Carreras, M., Del Hoyo, P., Martín, M.A., Rubio, J.C., Martín, A., Castellano, G., et al., 2003. Defective hepatic mitochondrial respiratory chain in patients with nonalcoholic steatohepatitis. Hepatology 38, 999—1007.
Petersen, K.F., Dufour, S., Befroy, D., Garcia, R., Shulman, G.I., 2004. Impaired mitochondrial activity in the insulin-resistant offspring of patients with type 2 diabetes. N. Engl. J. Med. 350, 664—671.
Riccio, P., Rossano, R., 2015. Nutrition facts in multiple sclerosis. ASN Neuro 7, pii: 1759091414568185.
Ritov, V.B., Menshikova, E.V., He, J., Ferrell, R.E., Goodpaster, B.H., Kelley, D.E., 2005. Deficiency of subsarcolemmal mitochondria in obesity and type 2 diabetes. Diabetes 54, 8—14.
Roberts, L.D., Koulman, A., Griffin, J.L., 2014. Towards metabolic biomarkers of insulin resistance and type 2 diabetes: progress from the metabolome. Lancet Diabetes Endocrinol. 2, 65—75.
Rosas-Ballina, M., Ochani, M., Parrish, W.R., Ochani, K., Harris, Y.T., Huston, J.M., et al., 2008. Splenic nerve is required for cholinergic antiinflammatory pathway control of TNF in endotoxemia. Proc. Natl. Acad. Sci. U.S.A 105, 11008—11013.
Sanmiguel, C., Gupta, A., Mayer, E.A., 2015. Gut microbiome and obesity: a plausible explanation for obesity. Curr. Obes. Rep 4, 250—261.
Savolainen, O., Fagerberg, B., Vendelbo Lind, M., Sandberg, A.S., Ross, A.B., et al., 2017. Biomarkers for predicting type 2 diabetes development-Can metabolomics improve on existing biomarkers? PLoS One 12, e0177738.
Schwartz, M.W., Porte Jr., D., 2005. Diabetes, obesity, and the brain. Science 307, 375—379.
Singh, B., Saxena, A., 2010. Surrogate markers of insulin resistance: a review. World J. Diabetes 1, 36—47.
Srikanth, V., Maczurek, A., Phan, T., Steele, M., Westcott, B., Juskiw, D., et al., 2011. Advanced glycation endproducts and their receptor RAGE in Alzheimer's disease. Neurobiol. Aging 32, 763—777.
Steinberg, B.E., Silverman, H.A., Robbiati, S., Gunasekaran, M.K., Tsaava, T., Battinelli, E., et al., 2016. Cytokine-specific neurograms in the sensory vagus nerve. Bioelectron. Med. 3, 7—17.
Suganami, T., Tanaka, M., Ogawa, Y., 2012. Adipose tissue inflammation and ectopic lipid accumulation. Endocr. J. 59, 849—857.
Tateya, S., Kim, F., Tamori, Y., 2013. Recent advances in obesity-induced inflammation and insulin resistance. Front. Endocrinol. 4, 93.
Unoki, H., Yamagishi, S., 2008. Advanced glycation end products and insulin resistance. Curr. Pharm. Des. 14, 987—989.
Vandal, M., White, P.J., Tremblay, C., St-Amour, I., Chevrier, G., Emond, V., et al., 2014. Insulin reverses the high-fat diet-induced increase in brain Abeta and improves memory in an animal model of Alzheimer disease. Diabetes 63, 4291—4301.
Wallace, T.M., Levy, J.C., Matthews, D.R., 2004. Use and abuse of HOMA modeling. Diabetes Care 27, 1487—1495.

Wu, G.D., Chen, J., Hoffmann, C., Bittinger, K., Chen, Y.Y., Keilbaugh, S.A., et al., 2011. Linking long-term dietary patterns with gut microbial enterotypes. Science 334, 105–108.
Xiang, C., Wang, Y., Zhang, H., Han, F., 2017. The role of endoplasmic reticulum stress in neurodegenerative diseases. Apoptosis 22, 1–26.
Xu, Z.Z., Kim, Y.H., Bang, S., Zhang, Y., Berta, T., Wang, F., et al., 2015. Inhibition of mechanical allodynia in neuropathic pain by TLR5-mediated A-fiber blockade. Nat. Med. 21, 1326–1331.
Yamagishi, S., Maeda, S., Matsui, T., Ueda, S., Fukami, K., Okuda, S., 2012. Role of advanced glycation end products (AGEs) and oxidative stress in vascular complications in diabetes. Biochim. Biophys. Acta 1820, 663–671.
Zhang, Y., Zhang, H., 2013. Microbiota associated with type 2 diabetes and its related complications. Food Sci. Hum. Wellness 2, 167–172.
Zhang, Y., Proenca, R., Maffei, M., Barone, M., Leopold, L., Friedman, J.M., 1994. Positional cloning of the mouse obese gene and its human homologue. Nature 372, 425–432.

Index

Note: Page numbers followed by "*f*" and "*t*" refer to figures and tables, respectively.

A

Aβ-derived diffusible ligands (ADDLs), 256–257
Abnormal circadian rhythms, 157–158
Acetylcholinesterase inhibitors (AChE), 444–446
Activating transcription factor 6, 18–21
Adaptor proteins, tyrosine phosphorylation of, 3–5
Adepocytokines, 41–42
Adipocytes in metabolism, 210*f*
Adipokines, 92–93, 233–234, 234*f*
  abnormalities in, 119–121
  expression, 28–29
  and insulin resistance, 233–234
  in insulin resistance and, 275–277
Adiponectin-deficiency, 439
Adiponectin in obesity, 45–47
Adipose tissues, 41–42, 208–210, 439
  inflammation, 22–23
  oxidative stress, 40
Advanced atherosclerotic plaques, 133–135
Advanced glycation end products (AGEs), 18–21, 124–125, 349–352
  aging, formation and accumulation, 230–231
  hyperglycemia-mediated insulin resistance, 233
  and insulin resistance, 230–233
  neural and peripheral cells damage, 231–233
  and Parkinson's disease, 315–316
  receptors, 230–231
Aerobic exercise, 210–211
Age-related cognitive decline (ARCD), 100
AGEs. *See* Advanced glycation end products (AGEs)
AGEs cross-linking breaking compounds, 132–133
AGEs with their receptors (RAGE), 230–231
Aging, 349–352
  and cognitive function, 352–354
Agouti-related peptide (AgRP), 1–3
AgRP-expressing arcuate neurons, 1–3
Akt, 3–5
  activation, insulin signaling pathway, 251–254
  phosphorylation, 122–124
Alagebrium, 124–125, 132–133
Allicin, 419
Allicor, 419–421
Allium sativum,, 419
α-Synuclein, 293–294, 299–302
  aggregation, 299–302
  functions, 301*f*
  misfolded, 302–303
  mitochondria and, 306–308
  N-terminus of, 299–302
  oligomers and protofibrils, 299–302
Alzheimer's disease (AD), 349–352, 443–444
  adipokines in insulin resistance and, 275–277
  cellular level, 96
  ceramode-mediated insulin resistance, 273–277
  cerebral blood flow, 271–273
  exercise and insulin resistance, 262*f*
  food consumption, signal transduction pathways, 254*f*
  healthy and unhealthy diet, effects, 261*f*, 262*f*
  metabolic syndrome (MetS), risk factors for, 94–100
  neurochemical links among insulin resistance, 265–270
  and obstructive sleep apnea, 184–189
  pathogenesis, 252*f*, 255–264
  tryptophan metabolism, 263*f*

Alzheimer's type of dementia, 355–359
molecular mechanisms, 357*f*, 358*f*
Aminoguanidine, 124–125
Amyloid beta-derived diffusible ligands, 447–449
Amyloid deposition in brain, 256
Amyloid precursor protein (APP), 94–96, 187–188, 255–256
Amyotrophic lateral sclerosis (ALS), 365–366
Angina, 175–176
Angiopoietin-like (ANGPTL) proteins, 167–168
Angiotensin-converting enzyme inhibitor (ACEI), 124–125, 166
Angiotensin-II receptor blockers, 124–125
Angiotensin II type 1 (AT1) receptor, 119–121
Antidiabetic drugs, 269–270
Antiinflammatory molecules, 13–15
Antioxidant defense, 368
Antioxidant enzymes, 13–15
systems, 312–314
Antioxidant response elements (AREs), 320–321, 400–401
Anxiety and depressive-like behaviors, 368
Aortic pulse-wave velocity, 127–128
Aortic valve calcification, 128–132
Apnea – hypopnea index (AHI), 160–162
Apneas and hypopneas, 160–162
Apolipoprotein E (APOE), 255–256
Apoptosis-related proteins, 409–411
APP. *See* Amyloid precursor protein (APP)
Arachidonic acid (ARA), 220–224, 441–442
ARCD. *See* Age-related cognitive decline (ARCD)
ARE. *See* Antioxidant response elements (AREs)
Arrhythmias, 175–176
Arterial stiffness, 133–135
Atherosclerosis, 6–8, 79–80, 87, 115–119, 133–136, 168–170, 308–312, 442–443
Atherosclerotic plaques, 128–135
Atherothrombotic and embolic strokes, 179–182
Attention, 157–158
Autoantibodies, 71–72
Autoimmune type 1 diabetes, 71–72
Autophagosomes, 98–99
Autophagy, 3–5, 98–99, 259
Autophagy-lysosomal degradation, 299–302
Autophagy/lysosomal pathway marker p62, 304–306
Autophosphorylation, 116
Axodendritic pruning, 270

B

BACE1-mediated amyloid biogenesis, 72–73
Bacterial amyloids, 263–264
Bacteroides, 30–34
Basal lamina, 126–128
BBB. *See* Blood – brain barrier (BBB)
BCAAs. *See* Branched chain amino acids (BCAAs)
β cell depletion, 6
Bile-tolerant microorganisms, 30–34
Biomarkers
insulin resistance, 442*f*
for obstructive sleep apnea, 170–172
Parkinson's disease (PD), 319–320
for type 2 diabetes, 77–78
Biphasic dose, 128–132
Blood – brain barrier (BBB), 1, 30–34, 210–211, 386–389
permeability, 72–73, 93–94
Blood glucose levels, 77–78
Body mass index (BMI), 26–28, 79–80, 443–444
and weight, 26*f*
Brain
atrophy, 270
insulin, 1–3
insulin resistance, 270
pathological and neurochemical abnormalities, 269–270
plasticity, 157–158
Brain-derived neurotrophic factor (BDNF), 210–211, 369–370
in hippocampus, 450–453

Brain stroke, 218–230
insulin resistance and, metabolic links, 228–230
molecular mechanisms, 220–228
nucleic acids and, 227*f*
oxidative stress level on brain, 223*f*
stroke-mediated brain damage, 220–228
Branched chain amino acids (BCAAs), 17–18, 119, 454–455
Breathing abnormalities, 162
Brown adipose tissue (BAT), 9–12

C

Calcium signaling, 116–119
Camellia sinensis,, 404–405
CAMP response element binding protein (CREB), 389
Cardiac metabolic network, 136–137
Cardiometabolic risk factors, 359–361
Cardiomyocytes, 116
apoptosis, 116–119
Cardiovascular morbidity and mortality, 30
Cardiovascular ROS, 119–121
Carnitine, 12–13
Carnitine palmitoyltransferase 1 (CPT1), 12–13
Carotid artery compliance, 127–128
Caspases activation, 72–73
Catalase (CAT), 13–15
Cell-autonomous mechanisms, 299–302
Cell injury, 28–29
Cellular glucose uptake, 207–208
Cellular proliferation, 3–5
Central homeostasis, 253–254
Ceramides, 15–16, 273–275
Ceramide synthase (CERS), 15–16
Ceramode-mediated insulin resistance, 273–277
Cerebral amyloid angiopathy, 271–273
Cerebral autoregulation, 229–230
stroke patients, 173–174
Cerebral blood flow (CBF), 211–212
AD pathogenesis, 267*f*
Alzheimer's disease (AD), 271–273
Alzheimer's disease and type 2 diabetes, 271–273
endocannabinoid system and oxidative stress system, 269*f*
regulatory mechanisms, 229–230
Cerebral hemorrhage, 212–215
Cerebral hypoperfusion, 359–361
and hyperperfusions, 229–230
Cerebral microvessels expression, 1–3
Cerebrovascular autoregulation, 212–215
CERS. *See* Ceramide synthase (CERS)
Chaperone-mediated autophagy, 98–99
Chemokines, 73–74, 80–82, 94–96, 212–215, 234*f*
Cholesterol homeostasis, 133–135
Cholinergic compounds, 444–446
Chromatin remodeling, 157–158
Chronic hyperglycemia, 446–447
Chronic hypoperfusion, 359–361
Chronic inflammation, 28–29, 133–135
Chronic insulin resistance, 12–13, 443–453
Chronic intermittent hypoxia, 162, 173–174, 179–182
in OSA, 173–174
Chronic neuroinflammation, 298–299
Chronic peripheral hyperinsulinemia, 249
Chronic psychosocial stress, 366–367
Chronic stress, 366–367
Chronic sustained hypoxia, 173–174
Chronic systemic inflammation, 30–34
Cinnamaldehyde, 415–419
Cinnamomum aromaticum,, 415–419
Cinnamomum verum,, 415–419
Cinnamomum zeylanicum,, 415–419
Cinnamon, insulin resistance and insulin resistance – linked diseases, 415–419, 416*f*, 417*f*
Circadian misalignment, 184–186
Circadian rhythms, 157–158
Circulating hormones and adipokines, 254–255
Claudin-5, 1–3
Clinical overlapping, 296–298
CoA-dependent metabolic processes, 12–13
Coagulopathy and smoking, 79–80
Cognitive abilities and insulin receptors (IRs), 1–3

Cognitive decline, 250–251, 256–257, 271–273, 275–277, 324–326
Cognitive dysfunction, 352–353
in Parkinson's disease (PD), 324–326
Cognitive function, 75–76, 99
and aging, 352–354
Comma, 79–80
Compensatory hyperinsulinemia, 6, 249
Congenital leptin deficiency, 43–45
Continuous positive airway pressure (CPAP), 167–168, 174–175, 183–184
Core body temperature, 157–158
Coronary artery disease (CAD), 211–212
Coronary atherosclerosis and hypertension, 116–119
Coronary heart disease (CHD), 116–119
Coronary spasm, 175–176
C-reactive protein (CRP), 126–127
CRP. *See* C-reactive protein (CRP)
Curcumin
A53T α-syn-mediated cell death, 404
on insulin resistance and insulin resistance – linked diseases, 400–404, 402*f*
Cystine supplementation, 26–28
Cytokines, 72–74, 83–84, 94–96
in pathogenesis of stroke, 234*f*

D

Death, 28–29
Dementia, 48–49, 453–456
Alzheimer's type of, 356–359
classification, 355–356
diet and exercise effect on, 373–374
frontotemporal, neurochemical aspects, 365–366
Lewy body dementia, neurochemical aspects, 361–365
neurochemical aspects, 354–355
stages, 359–361
vascular, neurochemical aspects, 359–361
Dendritic spine density, 366–367
Dendritic sprouting, 1–3
De novo synthesis, 15–16, 16*f*
Depression, 366–373
diet and exercise effect on, 373–374
and obstructive sleep apnea, 189–191
Parkinson's disease (PD) and, 324–326
pathogenesis, 371*f*
Depression-mediated disruptions, 369–370
Diabetes, 71–72
type 1, 71–72
type 2. *See* Type 2 diabetes
Diabetes-related micro- and macrovascular complications, 6–8
Diabetic cardiomyopathy, 116–119
Diabetic retinopathy, 79–80
Diabetic stroke, 211–212
Diacylglycerol (DAG), 136–137
Diet
effects, healthy and unhealthy, 261*f*, 262*f*
and exercise effect on dementia, 373–374
on insulin resistance, 389–400
homeostasis, insulin and glucagon, 396–398
Mediterranean diet, 391–392, 391*f*
Okinawan diet, 394–396, 394*f*
PtdIns 3K dependent insulin signaling pathway, 396–398
vegetarian diet, 392–394, 393*f*
western diet on human health, 390–391, 390*f*, 397*f*
and microbiota, 30–34
Dietary fat, 137–138
Dietary supplements, 386–389
Dilator muscle activity, 160–162
Dimethylarginine, 122–124
Disease-causing mutations, 319
DJ1 gene, 298–299
Docosahexaenoic acid (DHA), 46–47, 137–138
Down syndrome cell adhesion molecule, 77–78
Dynamin-related protein 1 (Drp1), 320–321
Dysbiosis, 30–34, 135–136, 261*f*, 453
Dyslipidemia, 6–8, 21, 79–80, 439–440

E

Electroencephalogram, 157
Emotional regulation, 157–158

Endocannabinoid system and oxidative stress system, 269*f*
Endoplasmic reticulum (ER)
insulin resistance, 18
stress, 18, 73–74
stress-mediated insulin resistance, 449*f*
Endothelial dysfunction, 6–8, 76, 113–124, 126–127, 140, 175–176, 178–179, 208–210, 249–250, 308–312
and hypertension, 175–176
Endothelial-mediated relaxation, 115
Endothelial NOS (eNOS), 122–124
Endothelium-dependent vasomotor function, 229–230
Enhanced CB chemosensory discharge, 172–173
Enteric nervous system (ENS) synucleinopathy, 296
Enteroendocrine cells (EECs), 303–304
Episodic memory, 352–353
ER-associated degradation (ERAD), 18–21
Estrogen-responsive tissues, 444–446
Excitotoxicity, 299–302, 308–312
Exercise, 373–374
effect and cardiovascular function, 130*f*, 131*f*
and insulin resistance, 262*f*
Exercise-induced neurogenesis, 373–374
Exercise-related methylation, 373–374
Explicit memory, 352–353
Extracellular matrix remodeling, 6–8
Extracellular signal-regulated kinases (ERK), 3–5, 174

F

Familial AD, 184–186
Familial Parkinson's disease (PD), 297*f*, 298–299
Fasting plasma tCys, 26–28
Fatigue, 294–296
Fibrinolysis, 18–21
Fish, phytochemicals, 391–392
Flavanols, 386–389
Flavanones, 386–389
Flavonoids, phytochemicals, 386–389
Forkhead homebox type O (FOXO), 116, 312–314
FOXO1, 5–6
Free fatty acids (FFAs), 5–6, 72–73, 208–210
Free glucose, 116
Free radical attack
fatty acid chains, 224–225
proteins, 225–226
Free radicals production, 368, 409
Frontotemporal dementia/degeneration (FTD), 355–356, 365–366
Fructosamine, 77–78
Functional food, 386–389

G

Galactosyl-ceramidase (GALC), 316–317
Galactosylsphingosine, 316–317
Gamma-aminobutyric acid (GABA) receptors signaling, 182–183
Garlic
in insulin resistance and insulin resistance – linked diseases, 419–424, 420*f*
phytochemicals, 391–392
Gaucher's disease, 293–294, 318–319
Genotype and lifestyle interactions, 254–255
Germ-free animals, tight junction proteins, 217–218
GI tract symptoms, 296
Globoid cell leukodystrophy, 293–294
Glucocorticoid receptor immunoreactivity, 390–391
Gluconeogenesis, 46–47, 441–442
Glucose, 71
concentration, 87
homeostasis, 18–21, 211–212
infusion rate, 6–8
insulin signaling, 3–5
intolerance, 21, 361–363
oscillations, 79–80
Glucose-oxygen deprivation, 308–312
Glucose transporter type 1 (GLUT1), 71
Glucose transporter type 4 (GLUT4), 1–3, 116, 401–403
Glutathione peroxidase (GPx), 13–15

Glycated albumin, 77–78
Glycated hemoglobin (HbA1c), 77–78
Glycated proteins, 86–87
Glycation, 86–87, 267–269, 302–303
Glycogenesis, 158–160
Glycogenolysis, 308–312, 441–442
Glycogen synthase kinase-3β (GSK-3β), 356–359
Glycosylated hemoglobin (HbA1C), 6–8
Glycosylation, 302–303
GM_3 ganglioside, 16
G-protein coupled receptors, 35–37
  signaling, 119–121
Green tea, insulin resistance – linked diseases, 404–407, 406*f*
Ground substance, 127–128
Guanylate cyclase, 122–124
"Gut – brain axis" homeostasis, 30–34, 453
Gut dysbiosis, 363–364
Gut microbiota, 25–26, 30–34, 215–217
  alterations, 263–264
  gut barrier and immune activation, 263–264
  neuroinflammation, 263–264
  and obesity on brain, 38–42, 39*f*
  Parkinson's disease (PD), 303–308

H

Heart disease, 79–80
  Alzheimer's disease, 142–143
  atherosclerosis, molecular mechanism, 133–136
  carbohydrate metabolism and insulin resistance, 136–140
  fatty acids metabolism and insulin resistance, 137–140
  and insulin resistance, biochemical links, 140
  insulin signaling and nitric oxide production in, 116–121
  lipid mediators in, 140–143
  and sleep apnea, 172–175
  vasculature, insulin signaling in, 121–133
Heme oxygenase-1, 13–15
Hepatic gluconeogenesis, 177–178, 441
Hepatocyte gluconeogenesis, 308–312
Hesperetin, 386–389
High-calorie diet, 119–121
  consumption, 48*f*, 449–450
  and sedentary lifestyle, 115–116
High-fat diet (HFD)
  adipose tissue inflammation, 164
  feeding, 28
Hippocampal neurogenesis, 370–373
HOMA-IR, 453–454
Homeostasis model assessment (HOMA), 453–454
Hormone-sensitive lipase, 138–139
  activity, 439–440
Hospital-related hyperglycemia, 212–215
Huntington's disease, 349–352
Hybrid oligomers, 296–298
3-Hydroxykynurenine, 262–263
Hydroxyl radical attack, 226–228
Hypercoagulability, 18–21
Hyperglycemia, 6–8, 18–21, 86–87, 124–125, 132, 162–164, 212–215, 267–269, 361–363, 439–440
  neurovascular unit, effect on, 190*f*
Hyperglycemia-mediated brain damage, 212–215
Hyperglycemia-mediated metabolic derangements, 212–215
Hyperhomocysteinemia, 76, 122–124
Hyperinsulinaemic euglycemic clamp, 453–454
Hyperinsulinemia, 6–8, 18–21, 76, 211–212, 308–312
Hyperlipidemia, 6–8, 76
Hyperosmolar hyperglycemic state, 79–80
Hyperphosphorylated tau, 257–258, 277–278
Hypertension, 6–8, 21, 79–80, 115, 119–121, 168–170, 175–176
  and brain damage, molecular links, 234–236
  and obstructive sleep apnea, 175–176
  and pathogenesis of stroke, 212–215
Hypertension-mediated changes, brain damage, 235*f*

Hypertension-mediated white matter damage, 237*f*
Hypoglycemia, 71–72, 79–80
Hypo-ponectinemia, 6–8
Hypothalamic – pituitary – adrenal axis (HPA axis), 158–160
 activation, 215–217
Hypothalamus, 71
Hypothyroidism, 23–24
Hypovitaminosis D, 74–75
Hypoxia-induced vascular remodeling, 173–174
Hypoxia-inducible factor (HIF)-1, 164
Hypoxia – reoxygenation, 175–176
 produce oxidative stress, 175–176

I

IkB kinase catalytic subunit β (IKK-β), 5–6, 13–15, 449–450
Immune cell infiltrates, 71–72
Impaired endothelial cell – mediated vasodilation, 249–250
Impaired glucose tolerance, 6–8
Impaired insulin tolerance, 6–8
Inducible nitric oxide synthase (iNOS), 13–15, 177–178, 400–401
Infective dementia, 355–356
Inflammation, 12–13, 18–21, 210–212
 and insulin resistance, obstructive sleep apnea, 168–170
 in obesity and insulin resistance, 233–234
 oxidative stress and, 212–215
 postischemic, 218–220
Inflammatory bowel disease (IBD), 443–444
Inflammatory markers in plasma, 6–8
Inflammatory responses, 28, 133–135
Initiate chronic inflammation, 18–21
Inositol-requiring enzyme 1 (IRE1), 18–21
Insulin in brain, 2*f*
Insulin-like growth factor binding protein-1, 453–454
Insulin-like growth factor (IGF-I), 308–312
Insulin receptor – expressing neurons, 71
Insulin receptor signaling, 4*f*
Insulin receptor substrates (IRSs), 3–8, 75*f*, 116
 proteins, 249–250
Insulin resistance, 76, 136–137, 208–210, 366–373
 characterization, 207–208
 definition, 308–312
 factors associated with, 114*f*
 in heart disease, 442–443
 high calorie diet consumption and neurological disorders, 48*f*
 and insulin resistance – linked diseases
  cinnamon in, 415–419, 416*f*, 417*f*
  curcumin on, 400–404, 402*f*
  garlic in, 419–424, 420*f*
  green tea on, 404–407, 406*f*
  n-3 fatty acids in, 413–415
  quercetin in, 421–424
  resveratrol on, 407–413, 410*f*
 myocardium, 113–114
 and obesity, 349–352
 obesity and heart disease, 114*f*
 Parkinson's disease (PD), 303–316
 in skeletal muscle, 441–442
 and sleep apnea, 177*f*, 181*f*
 stroke in diabetic patients, 213*f*
 subcellular level, 440–441
 tissues respond to, 441
 type 2 diabetes, 439–440
 and visceral obesity, 113–114
Insulin resistance – linked visceral and neurological disorders, 398–400
Insulin resistance – mediated changes, 139–140
Insulin-responsive tissues, 12–13
Insulin sensitivity, 6–8
Insulin signaling, 1–3, 368–370, 448*f*
 in brain, 250–255
 cascade, 75–76
 free fatty acid and ceramide-mediated changes, 139*f*
 glucose transport and, in heart, 117*f*
 and nitric oxide production, 116–121
Insulin signal transduction, 356–359
Insulin-stimulated glucose transport activity, 21, 439–440

Intellectual functions, loss of, 99
Intercellular adhesion molecule-1 (ICAM-1), 125–126
Intermittent hypoxia, 160–164, 165*f*, 166, 170, 172–176
  effect, 165*f*
  episodes, 168–170
International Diabetes Federation, 73
Interstitial fluid (ISF) levels, 186, 215–217
Intestinal bacteria, 30–34
Intracerebral hemorrhage, 212–215
Intracerebroventricular treatment, 188–189
Intramyocardial triglyceride accumulation, 116–119
Irritable bowel syndrome (IBS), 443–444
IRS-1 tyrosine phosphorylation, 6–8
Ischemia, 115–116, 212–215
Ischemic brain injury, 212–215
  glial cells activation, 220–224
Ischemic stroke, 218–220
Ischemic/vascular lesions, 352–353
Isoleucine, BCAAs, 17–18

J

Jet lag, 184–186
JNK-1 pathway, 16–17
C-Jun amino-terminal kinases (JNKs), 13–15
Jun kinase (JNK), 449–450
Jun N-terminal kinase (JNK), 207–208

K

Ketoacidosis, 79–80
Krabbe's disease (KD), 293–294, 316–317
Kynurenine pathway, dysregulation, 262–263

L

Lactic acid – mediated brain acidosis, 212–215
Learning and memory, 157–158
Leptin, 1–3, 275–277
  obesity and, 43–45
  role in brain, 44*f*
Leptin receptors (LepRs), 22–23, 43–45
Leucine, BCAAs, 17–18
Levodopa (L-DOPA) therapy, 361–363
Lewy bodies (LBs), 293–294
Lewy body dementia (LBD), 293–294, 355–356
  neurochemical aspects, 361–365
Ligand-activated nuclear transcription factors, 269–270
Lipid mediators, 18–21, 133–135, 267–269, 439–440
  chemical structures, 10*f*
  in heart disease, 140–143
  insulin resistance and metabolic syndrome, 75*f*
  pathogenesis of insulin resistance, 11*t*
  TAG-mediated insulin resistance, 9–12
Lipids
  and lipid-derived lipid mediators, 133–135
  permeability, 122–124
  peroxidation, 267–269
Lipogenesis, 441
Lipoprotein-cholesterol (LDL-c), 113–114
Lipoprotein lipase (LPL), 167–168
Lipoprotein receptor-related protein-1 (LRP1), 271–273
Lipotoxicity-mediated damage, 21–22
Long-term depression (LTD), 1, 308–312
Long-term potentiation (LTP), 308–312
Low-density lipoprotein (LDL), 86–87
Low-grade inflammation, 115–119
Low-molecular-weight antioxidants, 312–314
Lymphocyte-dependent mechanism, 217–218
Lysophospholipids, 220–224
Lysosomal β-glucocerebrosidase, 319
Lysosomal dysfunction, 319

M

Macrophage chemoattractant proteins, 18–21
Magnesium deficiency, 76
Mammalian epigenome, 30–34
Mammalian target of rapamycin complex 1 (mTORC1) pathway, 115

Mammalian target of rapamycin (mTOR), 3–5, 98–99, 116
mTOR complex 1 &2 (mTORC1 & mTORC2), 253–254
MAPK activation, 132–133
Matrix metalloproteinase-9 (MMP-9), 166
Mediobasal hypothalamus inflammation, 28
Mediterranean diet, insulin resistance, 391–392, 391*f*
Megalin expression, 96–97
Melanoma cell adhesion molecule, 77–78
Membrane-initiated signaling pathways, 224–225
Memory, 157–158, 352–353, 389
Memory-related processes in hippocampus, 368
Metabolic diseases, 115–119
obstructive sleep apnea and, 182–184
Metabolic endotoxemia, 30–34
Metabolic syndrome (MetS), 6–8, 79–80
cardiovascular disease (CVD), 441–442
and obstructive sleep apnea, 176–179
pathogenesis, 89–92
risk factors, 88–89, 94–100
type 2 diabetes and, 92–94
Metabolomics, 454–455
MetS. *See* Metabolic syndrome (MetS)
MetS-mediated metabolic damage, 446–447
Microbiota
composition in intestine, 450*f*
and short chain fatty acids, 34–38
stroke, composition effects, 236–238
Microbiota-improving methods, 217–218
Microglia activation, 256
Microvascular disruption, 249–250
Middle cerebral artery (MCA) blood velocity, 271–273
Mild cognitive impairment, 354–355
miRNAs, type 2 diabetes, 78–79
Mitochondrial biogenesis and function, 3–5
Mitochondrial DNA (mtDNA), 226–228
Mitochondrial dysfunction, 116–119, 207–208, 299–302, 441–442
insulin resistance, 13–15
Mitochondrial permeability transition pore (MPTP)-induced model of PD, 404
Mitochondrial respiratory chain proteins, 132
Mitogen-activated protein kinase (MAPK) signaling, 71
Mitogen-activated proteins, 6–8
Monoamine oxidase (MAO)-mediated abnormal dopamine metabolism, 364–365
Monocyte chemoattractant protein-1 (MCP-1), 304–306
Monocyte/macrophage caveolin-1, 133–135
Monocytes and lipoproteins, 126–127
MyD88 adaptor-like (Mal), 304–306
Myocardial energy metabolism, 116–119
Myocardial matrix, 132
Myocardial remodeling, 115–116
Myocardial stiffness, 132
Myocardin (MYOCD) expressions, 271–273
Myocardium, insulin resistance, 446–447

N

Naringenin, 386–389
Nasal CPAP, 183–184
obstructive sleep apnea, 183–184
Nephropathy, 79–80
Netrin, 77–78
Neural membrane phospholipids, 224–225
Neuritic dystrophy, 257–258
Neuroactive metabolites, 25–26
Neurodegeneration, 218–220
Neurodegenerative diseases, 93–94
Neurofibrillary tangles (NFTs), 184, 255–256
Neurogenesis, 373–374
Neuroinflammation in Parkinson's disease (PD), 322–324
Neuroinflammatory marker-mediated processes, 294–296
Neuronal – glial – endothelial interactions, 215–217
Neuronal plasticity, 366–367

Neuronal-specific knockout of insulin receptors (NIRKO), 368
Neuronal stem cell activation, 1–3
Neuronal synchrony, 157
Neuropeptide Y (NYP), 1–3
Neuropsychiatric diseases, 349–352
Neurotransmitters, 25–26
Neurovascular coupling response, 229–230
Neurovascular unit (NVU), 30–34, 173–174, 190*f*
n-3 fatty acids, 413–415
Nicotinamide adenine dinucleotide phosphate (NADPH) oxidase, 119–121
Nitric oxide (NO)
  bioavailability, 76
  Parkinson's disease, 314–315
  production and insulin signaling, 116–121
Nitrosative stress-mediated neuroinflammation, 314–315
Nitrotyrosine formation, 368
Nonesterified fatty acids, 439–440
Nonmotor symptoms (NMS), 293–294
Non – rapid eye movement (non-REM) sleep, 157
Nonvascular complications, type 2 diabetes, 79–80
NO synthase (NOS), 119–121
Nuclear α-synuclein, 293–294
Nucleic acid – derived metabolites, chemical structures, 227*f*
Nucleic acids
  brain stroke and, 227*f*
  nonenzymic oxidation, 227*f*
Nutraceutical, 386–389

O

Obesity, 21
  adiponectin in, 45–47
  comorbidities and, 25–26
  diet effect on microbiota population, 30–34
  factors effecting, 29*f*
  factors modulating insulin resistance and, 8*f*
  gut microbiota effects, 38–42
  insulin resistance, 6–21, 42–43
    candidate genes contributing to the pathogenesis, 9*t*
    as protective mechanism, 21–30
  insulin signaling, 42–43
    in brain, 3–6
    and neurological disorders, 47–49
  leptin and, 43–45
  microbiota and short chain fatty acids, 34–38
  molecular aspects of, 22–30
Obesity-associated dysbiosis, 30–34
Obesity-induced ER stress, 18–21
Obesity-mediated inflammation, 12–13
Obstructive sleep apnea (OSA), 160–162
  and Alzheimer's disease, 184–189
  biochemical changes in, 162
  biomarkers for, 170–172
  and depression, 189–191
  and hypertension, 175–176
  inflammation and insulin resistance, 168–170
  lipid metabolism, changes in, 167–168
  metabolic disease and, 182–184
  and metabolic syndrome, 176–179
  nasal CPAP, 183–184
  oxidative stress and insulin resistance, 162–166
  and stroke, 179–182
Occludin, cerebral microvessels expression, 1–3
Okinawan diet, insulin resistance, 394–396, 394*f*
Oligomeric proteins, 304–306
Olive oil, phytochemicals, 391–392
OSA-accelerated atherosclerosis, 174
Overconsumption, 439
Overnutrition, 24–25, 115
Oxidative damage, mitochondrial DNA (mtDNA), 226–228
Oxidative stress, 28–29, 115, 122–124, 166, 267–269, 299–302, 308–312, 364–365, 368
  curcumin and, 403–404
  excitotoxicity and, 229–230
  induction, 23–24
  level on brain, 223*f*

levels on brain, 223*f*
neurodegeneration, 14*f*
neuroinflammation, 212–215
nucleic acids, nonenzymic oxidation, 227*f*
in Parkinson's disease (PD), 320–322
Oxidative stress-mediated damage protein, 271–273
Oxysterols, 224–225

P

Pakinson's-linked dementia, 362*f*
Palmitic acid, 12–13
Parkinson-plus syndromes, 293–294
Parkinson's disease dementia (PDD), 293–294
Parkinson's disease (PD), 349–352, 443–444
biomarkers, 319–320
chronic inflammation in, 314–315
clinical overlapping, 296–298
cognitive dysfunction in, 324–326
depression and, 324–326
familial, 298–299
Gaucher's disease and, 318–319
gut microbiota, 303–308
insulin resistance, 303–316
Krabbe's disease and, 316–317
molecular level, 296
neuroinflammation in, 309*f*, 322–324
nonmotor syndromes, 294–296
oxidative stress in, 320–322
pathogenesis, 306*f*
progression, 94–96
sporadic, 299–303, 313*f*
symptoms, 295*f*
type 2 diabetes, 308–316
Pathogen-associated molecular patterns, 303–304
PD associated with GBA1 mutations (PD-GBA1), 318–319
Peroxiredoxins, 13–15
Peroxisome proliferator activator receptors (PPARs), 269–270
Peroxynitrite, 40–41
Phagocytic caveolae, 133–135
Phosphatidylinositol 3-kinase (PtdIns 3K), 116
Phosphoinositide-dependent kinase-1 (Akt), 116
Phosphoinositide-dependent protein kinase-1 (PDK1), 3–5
Phosphoinositide 3- kinase (PtdIns 3K), 3–5, 71, 306*f*
Phospholipid-derived lipid mediators, 133–135
Phosphorylation and dephosphorylation, insulin resistance, 6–8
Phosphorylation GLUT4, 308–312
Physical inactivity genes, 370–373
Phytochemicals
analogues synthesis, 386–389
chemical structures, in plants, 399*f*
classes, 386–389, 387*f*
consumption, 386–389
effects, 389
flavonoids, 386–389
health benefits, 386–389
on insulin resistance – linked visceral and neurological disorders, 398–400
medicinal system, 386–389
neurogenesis in hippocampus, 389
roles in plants, 386*f*
PKR-like ER kinase, 18–21
Plant-based diet, 450–453
Plant-based food, 30–34
Plant polysaccharides, 30–34
Plaques and tangles, 255–256
Plasma lipids, 254–255
Plasma total cysteine (tCys) levels, 26–28
Polyol pathway, 84–87
Polysaccharide Utilization Loci (PUL), 30–34
Polysomnography, 157
Polyunsaturated fatty acids (PUFAs), 137–138
Postischemic inflammation, 218–220
Postprandial hyperglycemia, 6–8
Postural instability, 364–365
Predementia, 354–355
Pro- and antiinflammatory molecules, 121–122
Pro- and antioxidants, 121–122

Pro- and antithrombotic signals, 121–122
Proatherosclerotic molecular events, 122–124
Proinflammatory cytokines, 12–13, 72–73, 267–269, 353–354, 369–370
Proinflammatory enzymes, 79–80
Proinflammatory mediators, 177–178
Proinflammatory signaling pathways, 168–170
Proinflammatory transcription factor nuclear factor kappa B (NF-κB), 116–119
Proinsulin, 1
Pro-opiomelanocorticotropin (POMC), 1–3
Protein clearance pathways, impairment, 299–302
Protein kinase C (PKC), 116
  pathway, 16–17
Protein trafficking, 318–319
Protein tyrosine phosphatase 1B (PTP1B), 22–23
Psychomotor retardation, 366–367
Psychosine, 316–317
PtdIns 3K – Akt – eNOS pathway, 124–125
PtdIns 3K dependent insulin signaling pathway, 396–398

Q

Quantitative insulin sensitivity check index (QUICKI), 453–454
Quercetin
  in insulin resistance and insulin resistance – linked diseases, 421–424, 423*f*
Quinolinic acid, 262–263

R

RAGE gene promoter, 132–133
Rapid eye movement (REM) sleep, 157
Ras – mitogen-activated protein kinase (MAPK) pathway, 3–5
Reactive nitrogen species (RNSs), 115–116
Reactive oxygen species (ROS), 13–15, 72–73, 115–116, 349–352
Receptor-dependent agonists, 122–124
Receptor for advanced glycation end products (RAGE), 1–3, 220–224, 230–231, 249–250
Redox cycling compounds, 312–314
Redox homeostasis, 267–269
Redox imbalance, 115–116
Red wine in Mediterranean diet, 391–392
Regulatory T cells, 71–72
REM behavior disorder (RBD), 182–183
Renin – angiotensin system, 115
Resveratrol
  antiaging agent, 411–412
  on insulin resistance – linked diseases, 407–413, 410*f*
  SIRT1-dependent cellular processes, 412–413
ROS-activated proinflammatory transcription factors, 174
Rostral ventrolateral medulla (RVLM), 119–121

S

Sedentary lifestyle, 115–116
Serotonin levels, depression, 369–370
Serum response factor (SRF), 271–273
Sexual dysfunction, 79–80
Shear stress, 122–124
Short chain fatty acids (SCFAs), 25–26, 34–38, 443–444
  microbiota releasing, 34*t*
Sleep
  and brain glymphatic system, 186–187
  deprivation, 158–160
  disturbances, 366–367
  fragmentation, 162
  loss, 158–160
Sleep apnea
  biochemical changes associated with, 163*f*
  and heart disease, 172–175
  and insulin resistance, 181*f*
  metabolic consequences of, 160*f*
  obstructive sleep apnea, 160–162
    and Alzheimer's disease, 184–189

biochemical changes in, 162
biomarkers for, 170–172
and depression, 189–191
and hypertension, 175–176
inflammation and insulin resistance, 168–170
lipid metabolism, changes in, 167–168
metabolic disease and, 182–184
and metabolic syndrome, 176–179
nasal CPAP, 183–184
oxidative stress and insulin resistance, 162–166
and stroke, 179–182
potential biomarkers for, 171*t*
risk factors for, 161*f*
Sleep-deprived patients, 187–188
Sleep-disordered breathing, 177–178
Sleep – wake cycle, 188–189
Slow wave sleep (SWS), 157
Sporadic Parkinson's disease (PD), 299–303, 313*f*
S6 ribosomal protein, 3–5
Streptozotocin-treated mice, 1–3
Stress, 366–373
Stroke
adipocytes in metabolism, 210*f*
adipokines and insulin resistance, 233–234
advanced glycated end products and insulin resistance, 230–233
alterations in microbiota and risk of, 216*f*
biochemical changes in associated with, 209*f*
in brain, 218–230
insulin resistance and, metabolic links, 228–230
molecular mechanisms, 220–228
nucleic acids and, 227*f*
oxidative stress level on brain, 223*f*
stroke-mediated brain damage, 220–228
hypertension and brain damage, molecular links, 234–236
hypertension and pathogenesis of, 212–215
in insulin resistance diabetic patients, 213*f*
microbiota composition effects, 236–238
and obstructive sleep apnea, 179–182
pathogenesis, molecular mechanisms, 215–218, 232*f*
rehabilitation, 179–182
Stroke-mediated brain damage, 220–228, 221*f*
Stroke-mediated injury, 179–182, 218–220
Subfornical organ-paraventricular (SFO) nucleus, 234–236
Sulfur amino acid – mediated adiposity in human, 26–28
Sulfur-containing compounds, 419, 420*f*
Sulfur switches, 13–15
Sumoylation, 302–303
Superoxide
anions, 9–12
and hydroxyl radicals, 224–225
production, 249–250
Superoxide dismutase (SOD), 13–15, 220–224
Suppressor of cytokine signaling (SOCS) protein, 5–6
Sustained cerebral hypoperfusion, 271–273
Swedish Obesity Subjects (SOS) study, 30
Sympathetic nervous system (SNS), 119–121
regulation, 120*f*
Sympathoinhibition, 119–121
Synaptic plasticity, 370–373, 386–389
and LTP modulation, 2*f*
Synaptophysin (SYN), 210–211
Synucleinopathies, 303
Systemic inflammation, 168–170

T

TAC. *See* Total antioxidant capacity (TAC)
Tau, 257–258, 296–303
Thiazolidinediones, 8–9
Thioredoxin, 13–15
Thyroid-stimulating hormone (TSH) prolactin, 158–160

TIR domain-containing adaptor-inducing interferon-beta (TRIF), 304–306
Tissue-type plasminogen activator (tPA), 212–215
Toll-like receptors (TLRs), 73–74, 168–170, 303–304, 353–354
  signaling, 18–21
Total antioxidant capacity (TAC), 179–182
Trans fatty acids, 137–138
TREM2 expression, 260
Triacylglycerol (TAG), 136–137
Triglyceride-rich lipoprotein, 167–168
Trimethylamine N oxide (TMAO), 37–38
Trimethylamine (TMA), 37–38, 217–218
Tryptophan metabolism, 262–263, 263*f*
Type 2 diabetes, 21, 30, 72–73, 308–312
  biomarkers for, 77–78
  ceramode-mediated insulin resistance, 273–277
  cerebral blood flow, 271–273
  complications, 79–82, 80*f*
  diabetes, 78–79
  genes and, 76
  insulin resistance, 18–21
  metabolic syndrome and, 92–94
  miRNAs, 78–79
  molecular level, 73–74
  molecular mechanisms to complications, 83–88
  neurochemical links among insulin resistance, 265–270
  Parkinson's disease (PD), 308–316
  pathogenesis, 74–75, 74*f*
  risk factors for Alzheimer's disease, 94–100
  vascular complications, 78–79
Tyrosine kinase activity, 3–5

U

Ubiquitination, 302–303
Ubiquitin-proteasome system dysfunction, 296
Unfolded protein response (UPR), 18–21, 449–450
Upper airway inflammation, 168–170
UPR. *See* Unfolded protein response (UPR)
Uric acid – induced RAGE signaling, 18–21
Uric acid levels, neurodegenerative disorders, 9–12

V

Valine, BCAAs, 17–18
Vascular cell adhesion molecule-1 (VCAM-1), 125–126
Vascular complications, type 2 diabetes, 78–80
Vascular dementia, 355–356, 359–361
Vascular endothelium, 121–122
Vascular injury, 126–127
Vascular oxidative stress, 122–124
Vascular smooth muscle cells (VSMCs), 122–124
  relaxation, 128–132
Vasoconstriction, 249–250
Vasoconstrictors, 405–407
Vasodilation, endothelium-dependent impairment, 173–174
Vasodilators, 405–407
  and vasoconstrictors, 121–122
Vegetarian diet, 452*f*
  on insulin resistance, 392–394, 393*f*
Ventricular hypertrophy, 162
Very low-density lipoprotein (VLDL), 128–132
Visceral and neurological disorders, 443–453
Visceral fat, 442–443
Vitamin D deficiency, 74–75
Vitamins, 124–125

W

Weight loss, 30
Western diet, 30–34
  advanced glycated products, generation, 85*f*
  consumption, 444*f*
  consumption, insulin resistance, 13*f*
  on human health, 390–391, 390*f*, 397*f*
  insulin resistance and, 96–97
  and vegetarian diet, 452*f*

White adipose tissue (WAT), 9–12
  browning, 407–408

X

Xanthine degradation, 9–12

Z

ZO-1, cerebral microvessels expression, 1–3

FSC
www.fsc.org
MIX
Papier | Fördert
gute Waldnutzung
FSC® C083411

Kolibri 360 GmbH
Ferdinand-Jühlke-Straße 7
99095 Erfurt, Deutschland
produktsicherheit@kolibri360.de